America's Ancient Forests

From the Ice Age to the Age of Discovery

THOMAS M. BONNICKSEN PH.D.

Department of Forest Science
Texas A&M University

John Wiley & Sons, Inc.

New York • Chichester • Brisbane • Toronto • Singapore • Weinheim

To My Wife Dolly

Library of Congress Cataloging-in-Publication Data:

Bonnicksen, Thomas M.
   America's ancient forests : from the ice age to the age of
discovery / by Thomas M. Bonnicksen.
     p.   cm.
   Includes bibliographical references and index.
   ISBN 0-471-13622-0 (cloth : alk. paper)
   1. Forest ecology—United States. 2. Old growth forests—United
States—History. I. Title.
QH104.B635 2000
577.3′0973—dc21   98-24396

Contents

On the third day we ... came upon a scene of beauty and magnificence combined, unequaled by any other view of nature that I ever beheld. It really realized all my conception of the Garden of Eden. In the west the peaks and pinnacles of the Rocky Mountains shone resplendent in the sun. The snow on their tops sent back a beautiful reflection of the rays of the morning sun. From the sides of the dividing ridge between the waters of the Missouri and Columbia, there sloped gradually down to the bank of the river we were on, a plain, then covered with every variety of wild animals peculiar to this region, while on the east another plain arose by a very gradual ascent, and extended as far as the eye could reach. These and the mountain sides were dark with Buffalo, Elk, Deer, Moose, wild Goats and wild Sheep; some grazing, some lying down under the trees and all enjoying a perfect millennium of peace and quiet. On the margin the swan, geese, and pelicans, cropped the grass or floated on the surface of the water. The cotton wood trees seemed to have been planted by the hand of man on the bank of the river to shade our way, and the pines and cedars waved their tall, majestic heads along the base and on the sides of the mountains. The whole landscape was that of the most splendid English park. The stillness, beauty and loveliness of this scene, struck us all with indescribable emotions. We rested on the oars and enjoyed the whole view in silent astonishment and admiration. Nature seemed to have rested here, after creating the wild mountains and chasms among which we had voyaged for two days.

Trapper Thomas James, with three men, paddling in two dugout canoes down the Missouri River between the Gates of the Mountains and the Great Falls, Montana
April, 1810.

Preface

This book ends where most books on forests begin. It sweeps across vast reaches of time and space to tell the story of North America's forests from the Ice Age to the age of European discovery. It tells how the movement of planets and fluctuations in the sun's intensity affected the earth's climate and, in turn, the disassembly and assembly of forests. But this saga is not just about climate and trees. Native Americans were an integral part of America's forests. The forests and the people who lived there formed an inseparable whole that developed together over millennia. The book describes this relationship and shows how Native Americans helped to create and sustain the ancient forests that Europeans found beautiful enough to set aside in national parks.

The story of America's forests is told in two parts. The origins of the forests and the native peoples who influenced their development are the topics covered in Part I. The appearance of the forests and how they functioned when they were first seen by Europeans are the topics covered in Part II. The book weaves together eyewitness accounts and scientific knowledge to reconstruct America's forests and the events that shaped them.

Part I begins with the physical forces that regulate climate, ice ages, and forests. Next it describes the forests as they were when glaciers covered one-half of the continent. The massive ice sheets had pushed the trees southward and squeezed them against the Gulf of Mexico. There they assembled into forests unlike anything that exists today. These were America's primeval forests. Then come the first Americans. The climate was warming, the glaciers were retreating, and the Ice Age forests were already breaking apart when Paleoindians arrived. They walked into the forests as the trees were moving northward and added their influences to the climatic forces that were guiding them over the landscape. Then the trees reached their destinations and reassembled into modern forests. The remainder of Part I examines the ways native peoples hunted animals, harvested plant foods, cut trees, set fires, planted crops, and built civilizations from one end of the continent to the other. In so doing, it shows how they helped to change the future of America's ancient forests.

Part II begins with a few timeless qualities that all forests share. These qualities help in understanding how forests are constructed and the processes that sustain them. They also provide a foundation for comparing different forests. The last four chapters look at ancient forests through the eyes of people who were there. Unlike the first Americans, Europeans found a settled land where forests covered nearly one-half of the continent. Their explorations start with the Spanish conquistadors because they came first. They stayed in the south, traveling through forests from Florida to California. Next come the English and Dutch colonists who ventured into the forests along the eastern seaboard and the Appalachian Mountains. French explorers, soldiers, traders, and priests followed. They saw the forests that stretched across Canada, the upper Great Lakes, and along the edge of the Mississippi River. Finally, American trappers and fur traders went beyond the Mississippi to explore the forests of the western mountains and the Pacific Northwest. Wherever they went, no matter what they suffered along the way, these European explorers marveled at the sight of America's forests, and for good reason. America's ancient forests were among the most spectacular and diverse in the world.

This book was written for anyone who loves forests. My love of forests began when I first dipped a canoe paddle into a northern Wisconsin river as a boy. A time that now seems very long ago. I knew little about the forests I saw back then. I owe part of my enthusiasm to my father who seemed to love being outdoors more than anything. Native Americans were part of my mother's ancestry. So a little Native American blood flows through my veins. I like to think it is enough to make me keenly aware of my natural surroundings and deeply appreciative of this land. Therefore, I went to the University of California–Berkeley to study forests and became a professor of forest science at Texas A&M University. This book represents a lifelong enthusiasm for ancient forests and over 30 years of study. I hope the reader feels the same excitement and sense of discovery that I felt when I read the accounts of early explorers and wrote this book.

Many people contributed indirectly to this work. First among these is Dr. Edward C. Stone, my mentor at U. C. Berkeley. Other professors also shared their insights and inspired my enthusiasm for the history of America's forests, especially the late Dr. Harold H. Biswell and the late Dr. A. Starker Leopold. However, a book like this builds on the work of countless scholars. Each bit of knowledge woven into the story told here represents hundreds of hours of sweat, toil, and discomfort by someone willing to spend the effort to answer a critical question. It took many hundreds of such dedicated people to make it possible for us to have this modest glimpse into the past. We also should never forget our debt to those who risked their lives exploring America's ancient forests, and then took the time to record what they saw. The eyes of the explorer helped me to see, and the eyes of science helped me to understand. I needed both to write this book.

Writing a book takes time, a lot of time. I must thank my wife Dolly for her understanding and patience. But Dolly did more than tolerate my long hours, she read everything, and reread it. Her insights were invaluable. I am also deeply indebted to Texas A&M University for allowing me to spend so much time on a single work. The people at a great university know that it takes time to advance our understanding of the world around us. They also provided the equipment and support needed to make this book possible. I also thank those who allowed me to publish their photographs and other illustrations. Some of them took the time to make extensive searches of their collections and even to make photographs just for this book. A special thanks to Ron Larson and Mike McMurray for their efforts.

Finally, what is written here is my responsibility. Not everyone agrees on all points of science or history. There is much controversy even among equally knowledgeable scholars. So I read the various arguments in the literature and came to my own conclusion about what represented truth. Therefore, I am solely accountable for the interpretation and synthesis of the information that went into *America's Ancient Forests*.

PART ONE

The Making of America's Ancient Forests

Whoso desireth to know what will be hereafter, let him think of what is past, for the world hath ever been in a circular revolution; whatsoever is now, was heretofore; and things past or present, are no other than such as shall be again: Redit orbis in orbem.
Sir Walter Ralegh (1552–1618)[1]

ONE

The Great Cold

Snow begins to fall, warning us. . . . There is darkness upon the earth.
Pierre Esprit Radisson, soldier and explorer
Winter of 1658–1659[1]

GLACIAL AGES

We live in the age of glaciers. It is warm now, but the glaciers will return. The movement of planets, fluctuations in the sun's intensity, and drifting continents will always influence cycles in the earth's climate, ice ages, and landscapes. Glaciers sprawled across the top of North America as the world's temperature dropped. Then they melted away when the climate warmed again. Towering sheets of ice slid southward about 17 times during the Pleistocene, our most recent glacial age, which lasted 1.65 million years.[2]

The expanding glaciers scraped away forests and rocks. Even the land sank beneath their crushing weight. However, some kinds of trees escaped the wall of advancing ice and took refuge in the south. Then the temperature rose abruptly, and the glaciers retreated like a defeated army, leaving behind a barren landscape. Now the trees could advance northward from their refuges, occupy the land, and gradually assemble into new forests to replace those lost beneath the ice. These victories were brief, lasting only 10,000–20,000 years before the ice pushed the forests back again. Today's forests are only the latest victors in this ongoing battle between ice and trees.

The last glacial age ended about 250 million years ago. Massive ice sheets sprawled across the supercontinent of Gondwana as it rested over the South Pole. Gondwana consisted of India, Australia, South Africa, and South America before they divided into separate continents. North

America escaped this glacial age because it was moving across the equator toward its present location in the north.[3]

During most of the earth's history the continents were arranged in a way that allowed warm ocean currents to reach the polar regions. During a glacial age, however, continents drift toward the poles and block the flow of tropical waters, temperatures drop, and glaciers begin to form. That is the case today. Antarctica sits on the South Pole and North America and Eurasia surround the North Pole. Thus the continents can stay cold during the summer, which provides the conditions needed to produce the Pleistocene glacial age.

The Wisconsin, the last period of major glacial advances during the Pleistocene, began 100,000 years ago. The ice sheets reached their maximum southern extent south of the Great Lakes about 18,000 years ago. Smaller advances of the ice fronts occurred 15,000 years ago in eastern North America, 14,000 years ago in the Pacific Northwest and Iowa, and continued in the arctic until 10,000 years ago when the Pleistocene ended. During that time two ice sheets formed. The Cordilleran Ice Sheet covered all but the tallest peaks of the Canadian Rockies and Coastal Mountains in the Northwest. Tongues of ice protruded between the peaks and pushed down valleys onto the lowlands and into the Pacific Ocean. The larger Laurentide Ice Sheet sprawled across eastern Canada and the northeastern United States, and westward to the foot of the Rocky Mountains. It was nearly 2 miles thick over Hudson Bay and reached as far south as central Illinois, gouging out the Great Lakes on its way.[4]

Together the two massive glaciers covered 6 million square miles of North America and extended into the sea as shelves of solid ice, while pack ice covered the northern oceans. The enormous weight of the glaciers pushed the earth's crust down nearly 1000 feet. Small ice caps also dotted islands in the Arctic, and glaciers streamed down valleys on all sides of the Sierra Nevada, Cascade, Olympic, and Rocky Mountains. A heavy ice cap also blanketed the Yellowstone Plateau, forcing glaciers through narrow gaps between peaks and into the surrounding lowlands.[5]

CLIMATE AND ICE

Ice sheets advance whenever summers are too cool to melt the past winter's snow. When the summers grow warm, they retreat. Average temperatures were 9–13°F colder in North America during the Ice Age than today, but they dropped 18°F or more near the ice sheets.[6] During a cold period, snow builds up layer by layer over the years to form a dome. The white surface of the snow reflects the sun's energy, further reducing the temperature and accelerating the rate of accumulation. Snow underneath the dome gradually compresses and turns to ice. Eventually, the weight of the thickening dome presses down on the underlying ice and squeezes it out in an ice flow. The ice flow thickens and spreads over the landscape. The edge of the flowing ice sheet halts when it reaches the sea or meets with temperatures that are warm enough to melt the ice as quickly as it arrives.

Climate drives the ebb and flow of glaciers and forests, and the sun drives the climate. If the sun's output of energy did not vary, and it does, the amount of sunlight reaching different areas still changes because of the way the earth moves around the sun. This causes ice sheets to spread and shrink within a 100,000-year cycle. Glaciers dominate the land for 60,000–90,000 years during the cold phase of the cycle, and then they all but disappear for 10,000–40,000 years during the warm phase.[7]

Yugoslavian astrophysicist Milutin Milankovitch discovered the connection between Earth's cycles of rotation and climate. He demonstrated that these cycles control the expansion and contraction of glaciers by regulating differences between summer and winter temperatures.[8] A decrease in summer temperatures reduces glacial melting, increases annual snowfall, and causes glaciers to expand. An increase in summer temperatures has the opposite effect.

The Milankovitch Cycle that drives the climate results from an interaction among changes in the orbit, tilt, and wobble of the earth (Fig. 1.1). The earth's orbit does not form a circle as it moves around the sun; it forms an ellipse, passing farther away from the sun at one end of its orbit than it does at the other end. During a 100,000-year cycle the tug of other planets on the earth causes its orbit to change shape. It shifts from a short broad ellipse that keeps the earth closer to the sun to a long flat ellipse that allows it to move farther from the sun, and back again. At the same time the earth is orbiting, it also spins around an axis that tilts lower and then higher during a 41,000-year cycle. Close to the poles the contrast between winter and summer is greatest when the tilt is large.

The earth wobbles because it is spinning around an axis that tilts back and forth. Thus temperatures drop in the Northern Hemisphere when it tilts away from the sun. Then the same thing happens in the Southern

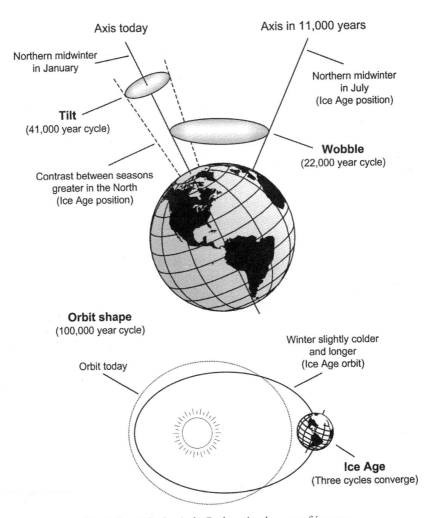

Axis today

Axis in 11,000 years

Northern midwinter
in January

Northern midwinter
in July
(Ice Age position)

Tilt
(41,000 year cycle)

Wobble
(22,000 year cycle)

Contrast between seasons
greater in the North
(Ice Age position)

Orbit shape
(100,000 year cycle)

Winter slightly colder
and longer
(Ice Age orbit)

Orbit today

Ice Age
(Three cycles converge)

Fig. 1.1   Milankovitch Cycle—timekeeper of ice ages.

Hemisphere and again in the north in a 22,000-year cycle. This means that about 11,000 years from now northern midwinter will fall in July instead of January, and the glaciers may return. It also means that summer temperatures peak in the tropics twice as often as the concentrated heat of the sun passes back and forth across the equator. This higher frequency of warming can disrupt weather and modify the behavior of glaciers. Some scientists believe that it also is responsible for the 10,000-year cycle of Heinrich events; the partial collapse of ice sheets that release fleets of icebergs into the oceans.[9]

Short warm gaps or interglacials break up glacial ages. Glaciers retreat when the Milankovitch Cycle causes summer temperatures to rise in the north. The last interglacial ended about 122,000 years ago.[10] The interglacial in which we live, the Holocene (*holos* is the Greek word for "recent," so it means postglacial times), began 10,000 years ago, and it is half over. The warming trend that melted the glaciers started much earlier, but it peaked about 8000 years ago when the earth was closest to the sun.[11] Then the climate began to cool.[12]

Many forces influence climate; thus temperatures do not rise and fall uniformly. Small cycles occur within larger cycles producing spikes of warming and cold snaps. Cycles in the energy output of the sun, changes in carbon dioxide concentration in the atmosphere, and shifting ocean currents contribute to these swings in the earth's temperature.[13] Ratios of heavy and light oxygen in air bubbles in ice cores taken from deep within Greenland's glaciers show that temperatures can rise and fall by as much as 14°F within a few decades.[14] Thus short periods of warmth can occur during full glaciation, causing ice sheets to retreat, and near ice age conditions can temporarily develop within a warm interglacial period.[15]

Meltwater and icebergs from thawing glaciers probably caused the temperature to drop abruptly between 12,900 and 11,600 years ago; so the glaciers started to return.[16] Melting icebergs cooled the ocean as the Laurentide Ice Sheet retreated from the sea. These icebergs could be huge. Some were probably as large as the one that broke off the Ross Ice Shelf in Antarctica in 1987. It was 100 miles long, 25 miles wide, and 750 feet thick.[16] When the Ice Age glaciers thawed enough to fall behind the Great Lakes, the meltwater stopped draining down the Mississippi River and shifted into the St. Lawrence River. The icy water rushed into the North Atlantic and blocked northward flowing warm ocean currents.[18] Thus, icebergs and ice water from the glaciers helped to start a long temporary cold period known as the Younger Dryas (named after a flower that lives in the Arctic). During that time a full glacial climate returned to North America for over a thousand years. Then the temperature rose as abruptly as it had fallen.[19]

The most recent cold snap, known as the Little Ice Age, began in 1450 AD and lasted four centuries (Fig. 1.2). Summers in the north grew

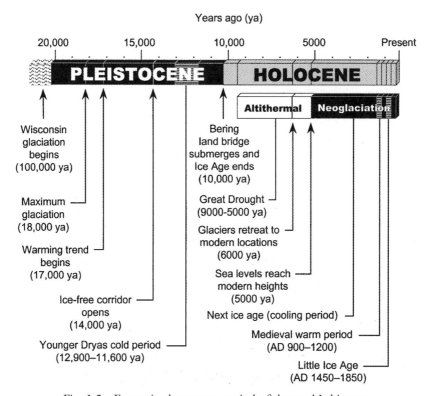

Fig. 1.2   Events in the present period of the earth's history.

cooler and wetter. As a result, small glaciers moved down mountain valleys in the Rocky Mountains, a small ice cap grew on Iceland, northern sea ice expanded southward, and the North Atlantic became stormier. The Little Ice Age ended between 1850 and 1880.[20] Retreating mountain glaciers left behind sheared trees and broken stumps, and shrinking ice caps on arctic islands uncovered crushed plants.[21] The warming trend continued until 1940 and then the temperature began to drop. This cooling trend reversed again in 1970 and the climate is still growing warmer.[22]

LAND OF THE GREAT COLD

The Ice Age landscape differed dramatically from today. North America grew larger as vast quantities of water were drawn from the oceans into clouds, dropped on the land as snow, and stored in glaciers. Sea level fell between 328 and 492 feet as the water evaporated. As a result, the coastline on the west side of North America reached out 31 miles farther than it does today. The eastern coastline extended as much as 62 miles farther out in places, creating a vast coastal plain. Rivers cut deep channels into the newly exposed land as they flowed downward to reach the lower level of the ocean.

Over the past million years or more, ice piled up on North America many times, draining away the shallow waters of the Bering Strait and uniting Eurasia with North America to create one huge continent. Once exposed, the flat bottom of the strait became a vast treeless plain nearly 1000 miles wide and only about 134 feet above sea level that temporarily connected Siberia with Alaska. When the ice melted, the Pacific and Arctic Oceans reunited to reclaim the land, forcing the coastline back and severing the land bridge between the old world and the new.[23]

Farther inland, desert basins in the American West contained more than 100 freshwater lakes, some immense (see Fig. 1.3). Heavy rainfall and meltwater from distant mountain glaciers poured into the basins to form inland seas, and the lack of outlets and low evaporation rates prevented them from shrinking. The lakes also helped to maintain themselves because up to 60% of the rainfall came from moisture that evaporated from the lakes and remained trapped in the atmosphere of the basins. Lake Bonneville in Utah was the largest lake, covering nearly 19,000 square miles with depths reaching more than 1000 feet. Today the Great Salt Lake is simply a tiny remnant of this once vast sea. Lake Lahontan in Nevada was smaller, but it still covered nearly 6000 square miles. Even Death Valley in California, which is now one of the hottest and driest places on the earth, filled with water during the Ice Age. All that remains of most of these huge Ice Age lakes are dry salt flats, and old wave-beaten shorelines perched high up on the sides of surrounding mountains.[24]

The massive ice sheets of the Wisconsin period changed the weather around them, which influenced everything else. They generated squalls that blasted the land with sand and silt. Cold air is denser than warm air, so winds flow off the summits of glaciers and down their sides. The high-level jet stream also swirled around the glaciers, further strengthening surface winds, drying the air, and pushing more rain into the Southwest. These bitterly cold winds probably reached 100 mph at the edge of these mountains of ice.[25]

In the winter when conditions were dry, these powerful winds picked

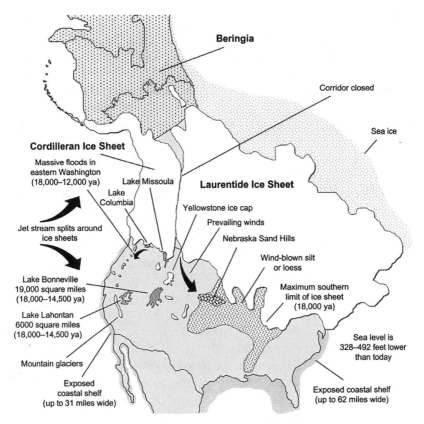

Beringia

Corridor closed

Sea ice

Cordilleran Ice Sheet

Massive floods in
eastern Washington
(18,000–12,000 ya)

Lake Missoula

Lake
Columbia

Laurentide Ice Sheet

Yellowstone ice cap

Prevailing winds

Jet stream splits around
ice sheets

Nebraska Sand Hills

Wind-blown silt
or loess

Lake Bonneville
19,000 square miles
(18,000–14,500 ya)

Maximum southern
limit of ice sheet
(18,000 ya)

Lake Lahontan
6000 square miles
(18,000–14,500 ya)

Sea level is
328–492 feet lower
than today

Mountain glaciers

Exposed
coastal shelf
(up to 31 miles wide)

Exposed coastal shelf
(up to 62 miles wide)

Fig. 1.3   Ice Age North America about 18,000 years ago.

up finely ground rock washed from beneath the shifting glaciers during the previous summer. This produced monstrous sand and dust storms that blew across the Great Plains and upper Midwest, and down the Mississippi River Valley. Gradually vast sand dunes built up that covered nearly 35,000 square miles of north-central Nebraska and South Dakota, and lesser dunes formed elsewhere in the Great Plains. Wind-blown silt also piled up at the base of trees near the glaciers and blanketed the land in layers as much as 32 feet thick. Today Ice Age silt covers 30% of the United States, but it lies hidden beneath forests and grasslands. It extends in a broad belt from the Rocky Mountains to Pennsylvania, with fingers jutting into west Texas and down the Mississippi River Valley to the Gulf of

Mexico. The Nebraska Sand Hills still serve as a visible reminder of these awesome dust storms.[26]

END OF THE ICE AGE

The end of the Great Cold began about 17,000 years ago. By 14,000 years ago the edges of the ice sheets no longer extended into the sea where they could release icebergs to cool the water. So oceans warmed and the ice melted more rapidly. However, the enormous mass of the glaciers still caused them to thaw slowly. The northward retreat of the glaciers was also erratic, with several halts and readvances. Thus it took 4500 years for the ice margin to back into Canada. The retreating ice sheets did not disappear until about 6000 years ago. Even today remnants of Ice Age glaciers remain at the heads of some valleys in the northern Rocky Mountains. Small ice caps also cover Baffin and Devon islands in the Arctic, and a large ice cap rests on Greenland. Some of this ice is 100,000 years old.[27]

Glaciers left behind strange landscapes after they wasted away. A jumble of rock fragments, and sand and silt that collected under the ice, flowed out and covered vast areas in sheets, piles, and ridges. Glaciers carried boulders the size of small houses hundreds of miles from their sources. These boulders still lie scattered over the ground where they fell when the ice melted out from under them. Ice pushed up low volcano-shaped hills called pingos in small lake basins and a profusion of tiny hummocks that looked like fields of soft cobblestones. Freezing and thawing soil also pushed stones up to the surface and gradually shoved them out to form rings. The outer edge of the rings produced a pattern on the landscape that looked like the honeycomb of a beehive.[28] Remnants of these patterns still exist in many places.

A 60-mile-wide strip of stagnant ice nearly 300 feet thick rested on North and South Dakota when the Laurentide Ice Sheet backed away. Then the glacier returned briefly, dragging debris over the ice and retreating again about 12,000 years ago. It left the stagnant ice covered with an insulating layer of rocks, sand, and silt that kept it from melting for another 3000 years. Chunks of ice also broke off the glaciers and sank into the rocky debris that poured around them as the glaciers backed away, which insulated the ice from the sun. The ice blocks finally melted and created kettle-shaped ponds. Other ponds appeared when water trickled into cracks in the ground and froze, becoming ice wedges. Holes formed and filled with water when the ice wedges thawed. These Ice Age ponds still dot the Midwestern plains providing wetland habitats for millions of migrating waterfowl.[29]

Continental ice sheets scraped away old landscapes and created new ones when they plowed forward and slithered backward toward their origin. The Great Lakes, one of the most spectacular and important features in North America, did not exist before the end of the Pleistocene. It was a land of broad stream valleys cutting through an otherwise flat terrain before the glaciers came. The glaciers moved sluggishly over the land but sped up as they slipped into the valleys. Thus fingers of rapidly advancing ice spread out from the glaciers and gouged deep trenches in the earth, carrying boulders and soil with them as they went. When the glaciers stopped, and the ice front began melting, huge loads of debris were left behind that formed dams that blocked the ends of the newly enlarged valleys. The meltwater from the retreating glaciers then filled the valleys and created the Great Lakes. Eastern glaciers also dropped their massive loads of rock and soil at the ice front as they backed away from the sea, constructing ridges or terminal moraines. These piles of glacial debris now form Long Island in New York and Cape Cod in Massachusetts.[30]

Melting ice created many giant freshwater lakes at the foot of the glaciers. Sometimes these lakes emptied with titanic force, especially when they formed behind ice dams. Eventually ice dams break and produce powerful floods that rip and tear the land. Such floods occurred often, leaving behind deep channels or spillways that still mark their passing. Ice Age potholes 15 feet wide still exist above the Mohawk River at Little Falls, New York. Ice dams in the northern part of Yellowstone National Park broke at least twice releasing floods up to 148 feet deep. Lake Tahoe in California's Sierra Nevada grew in size behind an ice dam that eventually broke. Flood waters up to 82 feet deep poured out of the lake and down the Truckee River. This massive wall of water hurled 32-foot-diameter boulders like bowling balls through the canyons below.[31]

The most spectacular floods washed over eastern Washington 18,000–12,000 years ago. A lobe of the Cordilleran Ice Sheet plugged the Clark Fork Valley in Idaho. Meltwater filled the valley as other parts of the ice sheet retreated, forming glacial Lake Missoula in western Montana. The lake grew until it covered more than 3000 square miles. Rising water floated the ice dam. It failed and released a raging torrent that drained Lake Missoula. Water and chunks of ice poured across Idaho and eastern Washington and down the Columbia River to the Pacific Ocean. The flood lasted a month, and at times flowed 55 times the rate of the Amazon River. A wall of churning water peeled away layers of volcanic bedrock, scoured out enormous channels, and piled up gigantic 16-foot-high versions of the gravel ripples found on the bottom of small streams. This process was repeated at least 4 and maybe 40 times, creating a desolate landscape known as the Channeled Scablands.[32]

The largest glacial lakes were Lake McConnell and Lake Agassiz in Canada, which formed about 12,000 years ago. Lake McConnell was 683 miles long, the longest lake known. However, Lake Agassiz, named after Swiss naturalist Louis Agassiz who formulated the theory of the great Ice Age, was the largest, covering more than 135,000 square miles. This is four times the size of Lake Superior, now the biggest freshwater lake in the world.[33]

Great rivers connected these glacial lakes to the Arctic and the Gulf of Mexico. Fish such as northern pike migrated into Lake Agassiz from both directions, although most species came from the south. Some entered through the Mississippi River, which flowed out of the southern end of the lake. Others came up tributaries that flowed out of Lake Agassiz and into the upper Missouri River or Lake Superior. Still others swam up the MacKenzie River, passed through Lake McConnell, and then entered Lake Agassiz from the northern end. Today, 106 species of fish live in the basin that used to hold Lake Agassiz. Most of them probably had ancestors that lived in the Ice Age lake 10,000 years ago.[34]

The ice front loomed over the north shore of these huge lakes where waves pounded them, chipping off chunks that fell into the water as icebergs. These icebergs floated south, melting as the water warmed, which quickened the withdrawal of the ice sheets and increased the size of the lakes. By 9500 years ago the ice retreated enough to allow Lake Agassiz to drain northeast into Hudson Bay. The lake poured out in one gigantic flood, leaving behind thousands of square miles of mud. Today Lakes Winnipeg, Winnipegosis, and Manitoba rest at the center of what used to be the bottom of Lake Agassiz. Boulders also lay strewn over the landscape to the northeast, reminders of the terrible power of the great flood. Sand and gravel beaches that piled up along the southern shore of this immense Ice Age lake also remain visible as far away as northwestern Minnesota.[35]

Water from the thawing ice sheets found its way back into the oceans from which it came. It rushed down the Connecticut River and poured into Long Island Sound. Meltwater also swelled the Hudson, Mississippi, Columbia, and other rivers to produce raging torrents that swept everything in their paths into the sea. The deluge in the Mississippi River reached four times the size of any flood in recent history. As a result, sea level inched upward engulfing beaches, killing coral reefs, and battering the retreating shore. Augusta and Bangor, Maine, sit on land that slipped beneath the rising water because the great weight of the glaciers pushed it down below sea level. Removing the weight of the ice allowed the land to rebound, like removing a finger after pressing it into a rubber ball. The sea flowed off the uplifted area about 12,000 years ago. The same thing is still happening at Hudson Bay, where the ground moves higher at the rate of about one-half inch a year. Eventually the water will drain away as the bottom of the bay rises above sea level.[36]

At the end of the Ice Age, when the sea rose again, it pushed the rivers back and submerged their lower channels. The flooded ends of the channels formed spectacular submarine canyons on the steeply rising West Coast, and broad harbors such as Chesapeake Bay on the gently sloping East Coast. North of Cape Cod a series of large hills popped above the surface of the sea during the Ice Age and drowned again when the glaciers melted. Today these submerged hills surrounded by sandy outwash plains give us great quantities of seafood, such as the Georges Banks off the coast of Massachusetts where trawlers often drag up mastodon and mammoth teeth with their fish.[37]

Glaciers paint with bold strokes instead of subtle brushwork. They crush forests, remove soil, redirect rivers, create new lakes, and destroy others. They build hills, transport boulders, grind rock into sand and silt, widen valleys, slice away slabs of mountains, and carve canyons. The gravel they drag over the land cuts shallow grooves in some rocks and polishes others to shiny smoothness. They modify the weather and push plants and animals around, mixing them up into new communities that decorate the landscape with color and life. These strange and beautiful Ice Age communities may last for tens of thousands of years, but like the ice itself they melt away when the climate warms again.

Ice Age Forests

> *This is the forest primeval.*
> "Evangeline"
> Henry Wadsworth Longfellow (1807–1882)

Forests only exist in human minds. Groups of animals and plants that we call forests come together for a short time; then each species goes its separate way when conditions change. Constant warming and cooling of the climate, and the ebb and flow of glaciers, caused the disassembly of old forests and the reassembly of new forests. Some species thrive in glacial cold while others do best in the warm periods between glaciations. So different forests dominated the land under different climates.

Glacial or cold-weather species such as spruce, fir, and bristlecone pine expanded their territories during cold periods and drew back when it became warmer. Interglacial or warm-weather species such as oak, hickory, and ponderosa pine spread northward when the glaciers withdrew and retreated toward the south when the glaciers returned.[1] Since each species advanced and retreated at its own pace, the trees shifted and sorted themselves into unique forests while moving around the landscape. The most recent sorting process began when the climate warmed shortly after the Wisconsin glaciers reached their maximum size about 18,000 years ago. Until then, North America's Ice Age forests (Fig. 2.1) dominated the landscape, reaching an uneasy stability that endured for thousands of years.

LIFE NEAR THE ICE

Life perished beneath the glaciers, but it flourished along the edge of the ice cliffs at the front of the glaciers and throughout the rest of North America during the Ice Age. The landscape beyond the glaciers would

15

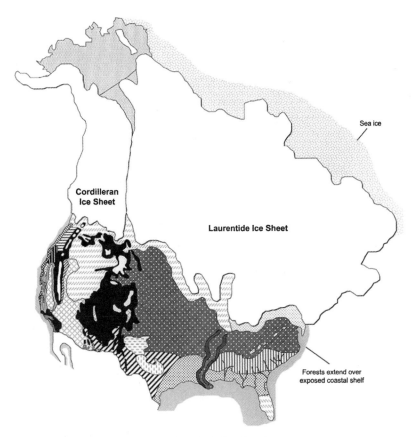

Fig. 2.1 America's Ice Age forests.

not be familiar to modern eyes. It was an alien world of modern and extinct animals living among well-known plants mixed in unusual ways. They could coexist back then because summers were cooler; so differences between seasons were less extreme than today.[2] These exotic mixtures of plants and animals formed a complex patchwork of communities. As a result, Pleistocene communities had a greater diversity of species, higher numbers of animals, more large animals, and larger animals than any that existed from then until now.[3] Many of these primeval communities fell apart after the ice sheets melted, although most of the species survived and reassembled to form the less diverse landscapes that we recognize.

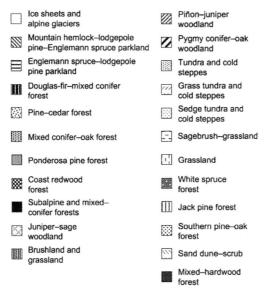

Fig. 2.1 (*Continued.*)

Half the continent escaped the massive ice sheets, even in the far north. Alaska's coastal and interior lowlands extending up the Yukon River into Canada were ice free and full of life during the Pleistocene. Plants and animals took advantage of lower sea levels by crossing land bridges to invade what are now islands. Many remained trapped there when the sea rose again. Some animals on these islands, such as mammoths, became smaller from generation to generation because food was scarce, and then they gradually disappeared. For instance, dwarf Columbian mammoths that stood only $5\frac{1}{2}$ feet tall survived on Santa Rosa Island, off the coast of southern California, until 10,000 years ago. Even more startling, dwarf woolly mammoths survived on Wrangle Island off the north coast of Siberia until as recently as 3700 years ago.[4] Plants and animals also migrated back and forth across the newly exposed land that joined Alaska to Siberia, and they invaded the exposed coastal shelves along the edge of the continent. They also clung to mountain summits that were tall enough to protrude through the ice sheets, called nunataks, and on windswept ridges separating glaciers that flowed into the Pacific Ocean.

Northern nunataks were an unlikely place for life to survive, but it did. They were cold dry islands of rocky ground surrounded by a sea of ice that stretched to the horizon. Trees could not live in such a harsh environment. However, scattered over the stony surface grew louseworts, small plants with a featherlike leaf, while other plants, such as sandworts, hugged the ground to escape the bitter wind. A nunatak poking through an ice sheet near the Pacific Ocean was still cold and bleak, but warmer and wetter than those farther inland. Here on the fog-shrouded coast the same species of ferns thrived that now cover extensive areas high up in western mountains.[5]

Subzero temperatures that averaged about 21°F near the advancing ice sheets kept all but the surface soil permanently frozen as permafrost. Trees became stunted because they could not push their roots deeply into the frozen ground. Constant freezing and thawing of the thin soil above the permafrost tilted the stunted trees in all directions. In a few protected areas where soils avoided freezing, pockets of spruce managed to prosper near the base of the massive glaciers, such as in parts of Minnesota. In most areas near the ice, however, young trees failed to replace those that died. So forests gradually gave way to grasses, sedges, and shrubs.[6]

A narrow band of treeless land sparsely carpeted with low-growing plants lapped against the southern edge of the ice sheets and covered ice-free areas of Alaska as well. Tundra covered much of this land. It consisted of grasslike sedges, with their triangular shaped coarse leaves, lesser amounts of grass, herbs, lichens, and mosses, and widely scattered small shrubs such as sage. Other areas were covered by cold steppes that contained mostly sagebrush and grass, and very little sedge. Many bare areas also dotted the cold steppes because the dry soils could not support a thick cover of plants. Stringers of willow also wound their way along streams and rivers through the cold steppes and tundra.[7]

Tundra and cold steppes intermixed to form a mosaic of patches, but tundra dominated the soggy soils below the southeastern edge of the glaciers. Tundra also extended down the crest of the Appalachian Mountains to Georgia, although white spruce forest interrupted it here and there.[8] However, cold steppes became more important than tundra west of the Great Lakes where the climate was drier than in the East, and in places that escaped moving ice.[9]

In the West, tundra and cold steppes dipped south into the valleys between the glacier-draped mountains of the Rockies.[10] They also pushed down into the Columbia Basin between the Rocky Mountains and the

Cascade Range, and into the Great Basin and the Bonneville Basin of northern Nevada and Utah. Tundra and cold steppes also descended the Puget Trough, which separates the Cascade and Coast Ranges in Washington and Oregon. However, scattered patches of Englemann spruce and lodgepole pine grew within the tundra and steppes, which formed open forests or parklands in some places.[11]

Cold steppes originated in central Asia, but they expanded and nearly encircled the globe during the Ice Age.[12] Even today, more than 90% of the species of flowering plants in cold steppe and tundra communities occur in both Asia and North America.[13] During the Ice Age they extended from Spain and England, across Europe, Russia, and Siberia. They joined similar communities in Alaska when the shallow waters of the Bering Strait drained away creating an enormous flat land bridge between the continents called Beringia.

Rivers and streams colored milky-gray with powdered rock flowed from the glaciers and spread across Beringia and the lowlands of Alaska, creating oases of marshland within a bone dry landscape. They left the ground covered with sand and silt as flowing water curved back and forth across the land. Frigid winds picked up the debris and filled the air with a choking pall of dust, piling up sand dunes as they went.[14] This vast, dry plain, with only an occasional clump of poplar trees to provide protection from frigid winds, extended to the northern edge of the Cordilleran and Laurentide Ice Sheets that covered most of North America.[15]

These two ice sheets backed away from each other at least three times during the late Pleistocene (Wisconsin). This opened a narrow ice-free corridor from Alaska into the center of the continent. This corridor first opened at a time that we cannot date, but it opened again between 28,000 and 23,000 years ago. It opened for the last time between 14,000 and 12,000 years ago after the glaciers began their final retreat.[16] Each time the corridor opened, the Bering land bridge also lay exposed, which created an unobstructed passage from Siberia into the lower United States. Cold steppe and tundra communities worked their way through this passage and down into the middle of the continent. This doorway to the south slammed closed when the ice sheets again expanded, pinching off pieces of these communities below the glaciers.[17] These trapped communities of tundra and cold steppes gradually spread along the southern edge of the glaciers.

As trees shifted toward the south to escape the glaciers, they sorted themselves into wide bands of forest below the tundra and cold steppes. These forest bands stretched in an east-west direction across the continent.[18] Boreal or northern trees that could tolerate bitter winters and short cold summers such as spruce, tamarack, and fir stayed in the North near the tundra and cold steppes. The cold glacial air pushed other trees such as jack pine and red pine out of the northern states, and oak and hickory out of the Midwest, into warmer weather farther south. These trees then pushed southern pines and magnolia to the shore of the Gulf of Mexico where they huddled by the warmth of coastal waters. The trees intermingled as they moved, which also created odd mixtures of species that have no modern counterpart.

Frigid air flowing off the ice sheets and down the Mississippi River Valley cut the bands of Ice Age forests in half.[19] The forest bands retained much of their identity in the East and Midwest, but rows of mountains chopped them into a complex mosaic in the West. Thus, rumpled mountains, jagged peaks, and tundra and cold steppe-covered basins, in the Rocky Mountains and Cascade and Coast Ranges, broke the forests into a patchwork of fragments containing different mixtures of species. Each species grew where the soil, temperature, and moisture conditions were favorable. Windstorms, floods, and fires also broke up the forests into even smaller patches containing various sizes and kinds of trees intermixed with grasses, sedges and other herbs, and shrubs.

THE SPRUCE FOREST

One tree more than any other defined forests near the ice sheets when the glaciers were at their maximum about 18,000 years ago—white spruce. It is a short tree with a thin trunk and low-hanging branches that form a narrow cone of pointed blue-green needles. It grows on relatively dry soils, while black spruce, an even shorter tree, grows in swamps and bogs within the same forest. Spruce forests grew in a band hundreds of miles wide from the Rocky Mountains to the East Coast. They hugged the southern edge of the tundra and cold steppes at the foot of the glaciers. Occasionally fingers and patches of spruce also protruded into the tundra and worked their way up to the edge of the ice.

Finding out where forests grew and how they looked thousands of years ago requires detective work. Most evidence comes from cores taken from deep within sediments that build up at the bottom of very old ponds.

Sand and silt settles in layers as rainwater washes soil into the ponds each year. Pollen from trees and other plants blows into the ponds and mixes with the sediments as they fall to the bottom. Sometimes cones and needles, charcoal, and even insect parts stay preserved in these sediments. These pieces of plants, grains of pollen, and insect parts gradually become fossilized. This creates a nearly permanent record of the surrounding vegetation.

Fossil pollen is especially useful in determining the number of different kinds of trees, shrubs, and other plants that grew around a pond when each layer formed. The core samples that scientists take from the bottom of these ponds capture this record in long cylinders of mud. Then scientists date each layer using radiocarbon methods and identify and count the fossilized pollen grains. Thus the ooze at the bottom of a pond can provide the key to unlocking the ancient history of a forest.

Scientists reconstructed the vegetation that grew on the Great Plains and elsewhere during the Ice Age using fossil pollen and macrofossils, or fossilized plant parts such as seeds, needles, cones, wood, and twigs. Much of this evidence comes from cores, but plant parts also accumulate in pack rat dens, glued together with their droppings into a mass called a midden. These middens provide valuable evidence of past vegetation because they last for thousands of years. Additional evidence comes from logs and land snails that became buried beneath the thick layers of soil that glacial winds spread across the land. Scientists know from this kind of evidence that the enormous expanse of grasslands so familiar to us did not exist on the Great Plains during the Ice Age. White spruce forests covered the plains, although grasses still grew in openings among the trees.[20]

A white spruce forest blanketed the Great Plains 18,000 years ago. It extended from the base of the Rocky Mountains through eastern Kansas and up on to the Ozark Plateaus. Colorado blue spruce and stubby limber pine seeking refuge in the lowlands from Rocky Mountain glaciers also spread over the plains to add diversity to parts of the spruce forest.[21] The white spruce forest also extended east of the Mississippi River and into the Southwest, mixing with an unknown species of pine in the Texas panhandle and eastern New Mexico.[22] Scientists found enough herb pollen to show that the forest was relatively open because these plants cannot grow well in shade. Thus the Ice Age white spruce forest might have looked similar to the open spruce forests or parklands that occupy south-central Alaska today.[23]

The Ice Age white spruce forest differed somewhat in species from east to west and from north to south, but overall it consisted of groups of short spindly trees scattered over prairies and swamps. The red-brown bark of juniper added streaks of color here and there among the dark green trees. Interspersed with the groups of spruce were groves of aspen and birch that brightened the forest with their white trunks and light green leaves, which turned vivid yellow in the fall. Clumps of eastern larch or tamarack also flashed golden yellow in the fall before their needles dropped, like bursts of light in the shadowy swamps and bogs. Sagebrush with its silvery leaves and bright pink patches of fireweed dotted the clearings between trees providing accents and highlights to an already picturesque landscape.[24]

The immense tracts of open spruce forest were not just lifeless art hanging in our imagination. They were as real as today's forests, full of smells, sounds, and motion. Bitter winds poured from the glaciers, roared through openings between the clumps of little spruce trees, and whipped dry grasses and sedges back and forth into swirls. Trees bent to absorb the impact as cold winds shrieked through their branches carrying along bits of bark and needles, and dusting the ground with cottony willow seeds in late spring. Shaking limbs filled the air with clicking and scratching as they dragged over one another. Trees fell here and there, hitting the dry musty smelling ground with muffled crashes because their shallow roots could no longer resist the onslaught. During lulls, other sounds would fill the air, such as the snapping of twigs, chomping and grinding, swishing branches, and the dull rhythmic thud of mammoth and mastodon feet (Fig. 2.2).

These primeval forests were rich in species beyond anything known today. The open patchy character of spruce forests and the mosaic of different species and sizes of trees and shrubs, intermixed with prairies and swamps, provided an incredible diversity of habitats. Mammoths coexisted with horses, camels, and mastodons, which looked similar but they lacked the mammoth's enormous curved tusks and dome-shaped head. Their ranges overlapped, but mammoths lived mainly in the West where grass was more plentiful than in the East.[25] They could live together because mastodons browsed on shrubs and tree needles and mammoths ate mostly grass that grew in the clearings.[26] Neither species could live in a closed forest. Grass cannot grow in the shade of trees, and dense shade causes lower limbs to die, leaving green foliage too high on the trees for

Fig. 2.2 Mastodons in the midwestern Ice Age spruce forest. (Courtesy of the Illinois State Museum. Photograph by Marlin Roos)

the mastodons to reach. Smoky shrews, thirteen-lined ground squirrels, heather voles, and Ungava lemmings scurried beneath the feet of these massive beasts, only to go their separate ways when the ice melted and the forest changed.[27] Thus the diversity of the Ice Age spruce forest made it possible for mammoth and mastodon to live with each other and with many other kinds of extinct and living animals that no longer coexist.

In the East, the band of spruce forest extended down to northern North Carolina. It also spread onto the broad continental shelf beyond where the waves of the Atlantic Ocean now pound the shore.[28] It pushed eastern white pine and hemlock trees even farther out on the exposed continental shelf, and down into some unknown location in the Appalachian foothills.[29] West of the Appalachian Mountains spruce grew as far south as central Missouri, Arkansas, and southeastern Oklahoma. Black ash, a small hardwood with a crooked trunk and an open crown, probably grew among spruce as scattered trees in wet areas just as it does today. Other hardwoods, such as hornbeam or ironwood and northern red oak, probably lived with spruce in some areas as well.[30] A narrow stringer of white spruce forest also followed the cold air draining from the glaciers down the Mississippi Valley, reaching into Louisiana. A thin ring of mixed-hard-wood forest, including oak, hickory, beech, and walnut, bordered the spruce forest on the warmer south-facing bluffs above the lower reaches of the Mississippi River. Hardwoods also grew along the bluffs of other rivers in the South.[31]

WESTERN FORESTS

Valley glaciers pouring from mountain summits in the West pushed high-elevation or subalpine trees such as Englemann spruce, subalpine fir, and lodgepole pine to the lower slopes surrounding the valleys. These sub-alpine species probably mingled with white spruce in the northern Rocky Mountains of Montana and Wyoming where some white spruce trees still grow. Treeline, or the highest place trees can survive, also moved more than 2000 feet lower because of the terrible cold.[32] This compressed subalpine forests into narrow belts. Englemann spruce and subalpine fir moved down the mountain sides and squeezed lodgepole pine forests against the edge of the dry windswept valleys below. Warmer weather forests of ponderosa pine that originally grew around the valleys had to migrate out of the northern Rocky Mountains entirely.[33] Similarly, west-ern larch probably found refuge somewhere in the Pacific Northwest, but no one knows for certain.[34] Thus an unusual mosaic of open subalpine forests intermixed with tundra and cold steppes wound through the west-ern mountains and valleys below the glaciers.[35]

In the Pacific Northwest, the mild and moist climate that seems nor-mal today was far different when glaciers were at their maximum during the late Ice Age. The shoreline moved westward as ocean levels fell and Puget Sound became a dry basin filled with glacial ice at its upper end. Thus the climate in the Puget Trough became continental rather than maritime. Ocean temperatures also dropped, and the tall glaciers split the jet stream winds causing one branch to follow the distant coastline toward the south. A southern shift in the jet stream also robbed the region of its moisture. As a result, average temperatures fell 13°F in the Pacific North-west and rainfall only reached 50% of current levels.[36]

The dry cold air pushed temperate or mild climate trees such as Douglas-fir and red alder out of the Coast Range and into small pro-tected areas at the southern end of the Puget Trough.[37] The area of Douglas-fir forests even shrank in northern California.[38] However,

Douglas-fir also probably took refuge on the exposed coastal shelf to as far south as southern California. Even today, 17,000-year-old Douglas-fir logs are still eroding out of gullies on the Channel Islands.[39] Thus the vast forests of huge Douglas-fir trees seen by early European explorers did not exist in the Pacific Northwest during the Ice Age.

The cool moist climate along the Oregon coast sheltered Sitka spruce and western hemlock trees in some of the same places they still grow. However, these temperate trees only survived in protected areas and possibly on the now submerged coastal shelf near the sea. Subalpine forests that normally live near the upper limit of tree growth on mountains pushed them aside as they occupied the lowlands. As a result, mountain hemlock, lodgepole pine, and Englemann spruce formed a parkland that covered most of the Washington and Oregon coast. However, mountain hemlock only remained within the forest during moderately cold periods when conditions were moist and foggy. It moved south during the dry cold periods that only Englemann spruce and lodgepole pine could tolerate. Tundra and steppe vegetation temporarily pushed even these cold-weather trees farther south during the coldest periods.[40]

Jet stream winds split by the continental ice sheets carried moisture from the Pacific Ocean into the Southwest. As a result, winter rainfall increased, allowing pygmy conifer-oak woodlands to invade the deserts of West Texas and parts of the Southwest.[41] They even moved into the Chihuahuan Desert and the Sonoran Desert where cactuses grow today.[42] The giant saguaro cactus and the palo verde, in turn, retreated far into Mexico.[43] Woodlands also expanded into the southern Mojave Desert in California and Nevada and northward into the lower reaches of the Great Basin. However, piñon pine and evergreen oaks could not grow with juniper on the dry valley floor of the Mojave Desert or around the edge of the Grand Canyon in Arizona. Mainly sagebrush grew among junipers in these areas, which formed a juniper-sage woodland.[44] This woodland may have extended along the eastern edge of California's Sierra Nevada to as far north as the shores of Ice Age Lake Lahontan.[45] On the other hand, piñon pine did grow among junipers on the lower slopes of the mountains near Flagstaff, Arizona. They also grew together in the more humid climate around the edge of Owens Lake on the east side of the Sierra Nevada.[46]

Pygmy pine woodlands grew on the mountainsides directly above the juniper-sage woodlands that sat on the valley floors of the Mojave and southern Great Basin during the late Wisconsin. They consisted of limber pine and bristlecone pine, the world's longest living tree. Both conifers have thick drooping branches, with little plumes of needles, connected to a short, thick trunk that tapers toward the top. They prefer to grow on dry, rocky slopes and ridges. Englemann spruce grew on the moister slopes above them until it reached treeline, which was much lower than today. A closed Englemann spruce forest also probably grew on the Mogollon Rim in northern Arizona where ponderosa pine forests now grow.[47] Fir, spruce, sagebrush, and limber pine coexisted above the Grand Canyon and on parts of the Colorado Plateau in Arizona, New Mexico, and Utah.[48] Mixed-conifer forests also grew within canyon bottoms such as Chaco Canyon in northwestern New Mexico, but cold steppes dominated the surrounding mesa tops.[49]

Today ponderosa pine forests dominate many parts of the Colorado Plateau where fir, spruce, sagebrush, and limber pine grew during the Ice Age. Ponderosa pine is a majestic tree with a tall barrel-shaped trunk covered by plates of yellow-brown bark. It now grows on lower mountain slopes in all states west of the Great Plains. During maximum glaciation, however, the cold climate forced ponderosa pine out of the Rocky Mountains. Some may have survived in a few favorable and yet unknown areas in the southern part of the mountains. However, most of the ponderosa pine trees retreated south to the Santa Catalina Mountains in southern Arizona, the San Andres Mountains of southern New Mexico, and places farther south.

California's weather became cooler, drier, and more continental during the late Wisconsin. Santa Ana winds also increased, blowing great clouds of sand from the Mojave Desert westward over southern California and piling it up in dunes near the coast. Even so, California's Mediterranean climate buffered forests from the extreme cold that lay over most of North America during the Ice Age. Thus, forests of giant redwoods probably grew along the northern California coast at this time just as they had done for at least the past 5 million years.[50] However, the coast was farther west during the Ice Age than it is today. In addition, Douglas fir and other trees that now grow among the redwoods most likely grew with them back then as well. More frequent coastal fogs may also have increased moisture in southern California.[51] Here forests of pine and Douglas-fir extended onto the exposed coastal shelf and over southern California's Channel Islands. Today only a few small relics of the Ice Age

Douglas-fir forests still exist in southern California's coastal mountains.[52] Likewise, ponderosa pine retreated into a few isolated areas high in these mountains.[53]

The mild, but still cold and dry, Ice Age climate in California filled the upper valleys of the Sierra Nevada with glaciers. So forests shifted as much as 2000 feet down the mountains to escape the cold.[54] As a result, trees such as ponderosa pine, juniper, sugar pine, incense-cedar, and fir moved to the lower slopes to form diverse forests that also included subalpine species. These forests also filled canyon bottoms in the Sierra Nevada, and giant sequoia found refuge in shady ravines around the canyons. Sagebrush-grass communities mixed with pine and juniper occupied mid-elevation slopes on the west side of the southern Sierras.[55] A mixed conifer-oak forest also fanned out on the foothills farther down the Sierras and over much of the California Coast Range.[56] Fingers of juniper-sage woodland from the Great Basin extended beyond the foothills along streams and rivers and into the San Joaquin valley.[57] These migrating forests and woodlands probably pushed grasslands and brushlands out of the foothills and valleys and onto the exposed coastal shelf.[58] A pine-cedar forest that consisted of a mosaic of open and closed patches of pine, cedar, juniper, fir, and sagebrush also spread over inland areas in northwest California during the coldest periods.[59]

SOUTHERN FORESTS

Jack and red pine grew at the southern edge of the white spruce forest. Jack pine is a small-diameter tree that looks scraggly because of a shaggy crown and a trunk that usually bristles with dead branches. Red pine is tall and thick with a clear trunk covered by flaky orange-red bark topped by an oval crown of dark-green needles. Red pine looks majestic by comparison to jack pine, but jack pine dominated the forest. Fingers of spruce also penetrated the jack pine forest in a few places, reaching down to the Gulf of Mexico.[60]

Only small numbers of temperate hardwoods such as ash and maple grew in favorable spots throughout the jack pine forest. They probably settled on warmer south-facing slopes because the climate then was like Maine or Minnesota is today. The open and patchy character of the forest allowed grasses, sedges, sagebrush, daisies, and many other plants to grow among the pines. Scattered prairies also broke up the forest and increased its diversity.[61] This Ice Age jack pine forest formed a short third band of vegetation below the glaciers.[62] It extended eastward onto the now submerged continental shelf of the Atlantic Ocean. It reached its southern limit in central Georgia, but it only grew westward into Tennessee, where it mixed with spruce.[63] The jack pine forest stopped at the edge of a mixed-hardwood forest toward the west, near Memphis. Here a pocket of oak, hornbeam or ironwood, ash, hickory, maple, beech, and chestnut found refuge in the warmth of the lower Mississippi Valley.[64]

An open southern pine-oak forest grew in a band on the lower edge of the jack pine forest. The widely spaced and tall pine trees stood like columns supporting a thin green canopy over a carpet of grass. A scattering of oak and hickory also grew among the pines. This final strip of forest below the glaciers extended downward into north-central Florida.[65] The pine forest in Florida probably consisted of sand pine, which is a short two-needle species of southern pine that looks like jack pine.[66] The southern pine-oak forest did not penetrate farther south in Florida because conditions were too dry to support trees. However, thin lines of small trees and shrubs, or scrub, probably snaked through the parched landscape along streams. Even so, active sand dunes and prairies covered most of southern Florida during the coldest part of the Ice Age.[67]

The southern pine-oak forest continued toward the west into Texas. In Texas, however, a semidesert pygmy conifer-oak woodland composed of piñon pine, juniper, and shrubby evergreen oaks replaced the southern pine-oak forest.[68] This pygmy conifer-oak woodland continued through Texas and into parts of New Mexico and Arizona. In Texas, however, it grew around the edges of two small areas of grassland that later expanded to cover the Great Plains.

FINDING THE LOST PRAIRIES

Until recently, no one could find evidence that prairies existed in the center of North America during the Ice Age. Many scientists thought that the Great Plains shrank and eventually disappeared in the glacial cold.[69] Then the Ice Age prairies reappeared as if rising from the dead. Scientists found fossil pollen that had blown into a cave, and settled to the bottom of an old lake, from Ice Age prairies in Texas. These pollen grains showed that one small piece of the Great Plains grassland took refuge from the cold in the Llano Estacado, or High Plains of the Texas panhandle and eastern

New Mexico. Another piece of the Great Plains grassland found safety just below the High Plains on the Edwards Plateau in central Texas. Even so, these Ice Age prairies did not look like the modern shortgrass and tallgrass prairies that we know. Both of them contained more herbs and fewer trees than today. In addition, sagebrush grew among the grasses and herbs in the Ice Age High Plains prairie, but they did not grow in the Edwards Plateau grassland.[70]

These Ice Age prairies sat within the bands of forest like lakes of grass. The High Plains sagebrush-grassland rested within the southwestern part of the white spruce forest that covered the Great Plains, but the Edwards Plateau grassland cut through the band of pygmy conifer-oak woodland farther south. This woodland wound around the edge of the plateau within deeply eroded stream channels. A few hardwood trees such as hickory, ash, elm, walnut, alder, and sweetgum also survived the cold in these moist valleys within the pygmy conifer-oak woodland.[71] Other areas of prairie may also have existed during the Ice Age, perhaps in the northwestern Great Plains and even in the central plains.[72]

CREATURES OF THE ICE AGE

Small mammals, such as the arctic ground squirrel, arctic shrew, yellow-cheeked vole, lemmings, least weasel, and ermine, followed the tundra and cold steppes southward through the ice-free corridor between the glaciers. There they temporarily mixed with nonarctic species such as pocket gophers, eastern chipmunks, and sage voles that also took advantage of the new habitat.[73] Joining them were musk ox, reindeer, and caribou. The caribou even found its way from the far north down to northern Alabama and Georgia.[74] Only a few mammals migrated northward through the cold treeless plain within the corridor, including the giant short-faced bear, camel, badger, and black-footed ferret.[75]

The long-horned bison that came across the Bering land bridge into Alaska from Asia also pushed its way down the passage between the ice sheets into the United States. This southern population evolved into the giant bison after the glaciers closed the passage and separated it from the ancestral long-horned bison to the north. The range of the bison became smaller after the glaciers finally melted, leaving only three isolated populations. Horns grew shorter and bodies smaller as the bison's habitat shrank. Ultimately, these isolated populations evolved into the short-horned bison of Europe, the wood bison of eastern Siberia, Alaska, and Canada, and the plains bison of the United States. Similarly, the American bighorn sheep evolved from an Asian ancestor that traveled through the passage and then found itself trapped below the glaciers.[76]

No single plant or animal symbolizes the Ice Age better than the woolly mammoth. It evolved in Eurasia and migrated across the Bering land bridge into North America about one-half million years ago. The smaller and more primitive mastodon, a forest dweller that lived on shrubs and trees, joins the mammoth as a symbol of the Ice Age. Coarse golden-brown hair covered the animal, but it was not as shaggy as the hair of the woolly mammoth.[77] It roamed North America and parts of South America for several million years before the woolly mammoth arrived. Woolly mammoths plodded through cold steppes and tundra in the far north and at the southern edge of the glaciers. They also flourished within the open spruce and pine forests, and the Columbian and Jefferson mammoths lived to as far south as northern Florida, southern California, and Central America.[78]

Mammoths looked something like a modern elephant. However, instead of rough gray skin, thick brown hairs covered their entire body, even the trunk. A 6-inch woolly undercoat provided additional protection from the cold. Their eyes were like protruding saucers, and their dull white tusks stuck out 12 feet and curled sharply upward. They swung their tusks back and forth to scrape away snow so that they could eat the underlying grass much like bison use their noses for the same purpose. Their ears were small and round, and long hair draped over their dome-shaped heads like a bad toupee. Wide padded feet designed for snow or marshy ground supported the 8-ton weight of the shaggy beasts as they lumbered along in search of food.[79]

An abundance of other extinct herbivores lived among these giants, such as shrub-ox, stag-moose with antlers that combined the shapes of modern moose and elk, 7-foot-long armadillos, and a beaver the size of a black bear.[80] Giant ground sloths weighing several tons ranged as far north as Alaska, and Harrington's mountain goat lived with them in the juniper woodlands of the Southwest.[81] These primeval communities also included the Western camel, horse, and long-legged and short-legged llamas. Enormous herds of saiga antelope and reindeer also immigrated across the Bering land bridge into western Alaska to take advantage of the expansion of their cold steppe habitat. However, they became locally extinct when they withdrew into Eurasia at the end of the Ice Age.[82] A giant condorlike vulture with a 15- to 17-foot wingspan called Merriam's Teratorn, and a tortoise, also became extinct. Many carnivores also disappeared, including the dire wolf, saber-toothed cats, the American lion, and the American cheetah.[83] The most terrifying extinct carnivore of all is the giant short-faced bear. It stood as high as a moose and it used its long legs to run swiftly after prey across the tundra and cold steppes.[84]

Modern animals also lived among these extinct beasts. Ice Age animals common in North America today include elk (wapiti), white-tailed deer, mule deer, musk ox, caribou, wolves, cougars, black bears, and such small animals as red squirrels, the western jumping mouse, and voles. Moose and grizzly bear lived in Alaska and the Yukon, but they did not move south until the glaciers retreated.[85]

North America's Ice Age forests started to disintegrate roughly 17,000 years ago when the climate warmed and the massive ice sheets began to melt.[86] Occasional sharp increases and decreases in temperature did not reverse the warming trend. The glaciers reluctantly backed away, lurching forward here and there, and then retreating like a dying beast. Each time the glaciers lurched forward, trees moved back as if cautiously avoiding the beast in its last moments of life. Then the trees advanced again, reassembling into the ancient forests that European explorers found as the glaciers continued their inevitable decline into oblivion.

THREE

The Birth of Modern Forests

A work of art is static; and its value and its weakness lie in being so:
but the tuft of grass and the clouds above it belong to our own travelling
brotherhood.
Freya Stark (1893–1993)
British travel writer[1]

Modern forests only exist today. They do not look like Ice Age forests nor do they look like forests of the future. Forests represent a loose collection of species that grow together for a time as they pass each other on their way somewhere else. Each species arrives and departs independently from other species. Plants move very slowly; animals move more quickly; but they all continue to move either to escape an inhospitable environment or to take advantage of a new one. If they cannot move, they adapt. If they cannot adapt, they become extinct. Thus, forests redefine themselves as plants and animals continue their relentless shuffling.

Several tree species still common in North America became extinct in Europe when they could not escape the advancing glaciers during the Pleistocene. They include sweetgum, tuliptree, magnolia, hemlock, white or false-cedar, and cedar. Obstacles blocked the trees southward migration, such as the Pyrenees, the Alps, and the Mediterranean Sea. So they disappeared.[2] North America's mountain ranges have a north-south orientation, which allowed trees to migrate south along the lowlands and survive. They followed the same route northward 17,000 years ago when expanding their territories to take advantage of the warming climate and the new land left behind by the retreating glaciers.

TREES BEGIN TO MOVE

The migration and expansion of trees began as soon as the glaciers started to disappear. However, glaciers give up the land slowly, even though they rapidly lose weight as massive volumes of meltwater rush down rivers and coulees. The ice sheet in New Hampshire, for example, lost 6500 feet of thickness in a few hundred years.[3] Nevertheless, as late as 13,000 years ago a thin layer of ice still blanketed most of the area originally occupied by the glaciers.[4] During this time rising seas covered much of the coastal shelf squeezing plants and animals into a smaller area between the glaciers and the oceans. Eventually the glaciers gave up the land. The remnants of once towering walls of ice finally retreated across the border into Canada 9500 years ago. The ice sheets took 3000 more years to disappear completely from the North American continent.[5]

The glaciers left behind eroding mountains and a desolate, cold landscape as they backed away. Strong winds picked up loose sand and silt and dropped it on the barren ground near the ice where freezing and thawing mixed it into the stony soil. These winds also sprinkled the seeds of sedges and grasses over the soil, which created a fringe of tundra that followed the ice front as it retreated toward the north. Shrubs came next. Trees trailed closely behind, occupying bare patches scattered across the tundra.[6] More trees followed in a series of broad ranks of different species marching out of the south.

Trees arrived at their modern locations at different times because species started from different places and traveled at varied rates. They depended on favorable winds and the availability of birds to carry their seeds, especially blue jays and the now extinct passenger pigeon.[7] In addition, seeds could not germinate and survive unless they reached the proper soil. Large lakes and cold mountain ridges slowed the progress of some trees, while others could avoid such obstacles and moved steadily forward. Thus, the composition of forests constantly changed at the end of the Ice Age as species converged and separated while shifting around the landscape. Their movements choreographed by wind, birds, soil, physical barriers, the presence of other trees, and a fickle climate. Even disease outbreaks interfered with the migration of some trees. Eventually the trees and other plants and animals assembled into what we consider modern forests.

In northern Wisconsin and the upper Peninsula of Michigan, for example, spruce arrived first. Tamarack, aspen, and birch arrived shortly afterward, followed by balsam fir. Then such hardwoods as American elm, oak, maple, black ash, and hornbeam or ironwood entered the region. Jack and red pine came next. It took eastern white pine several thousand years more to reach the area, and another thousand years for eastern hemlock to arrive. Several thousand more years passed before beech made its appearance.[8] New arrivals increased the number of species available to create modern forests, and a changing climate and other forces constantly stirred them into new mixtures. Gradually they sorted themselves into the forests that characterize the region today. These include maple-basswood forests intermixed with pockets of eastern hemlock, and white-red-jack pine forests. Who knows how long today's forests will persist, but history shows that further changes are likely.

PIONEER AND SETTLER TREES

Trees use two main strategies to occupy land. One group uses a pioneer strategy, and the other uses an infiltration or settler strategy. Pioneers such as aspen, birch, tamarack, Douglas-fir, and pine usually grow in clearings because they need bare soil, shallow duff, light litter, or rotten wood for a seedbed, and abundant sunlight for rapid early growth. Thus pioneer species are the first to occupy land left exposed by a disturbance, such as the recent withdrawal of a glacier, volcanic eruptions, landslides, fires, floods, or windstorms that knock down groups of trees. However, some pioneers such as white spruce, eastern white pine, coast redwood, and Douglas-fir can grow in partial shade, but growth slows as shade increases. Other pioneers such as jack pine need full sunlight, and they grow well on sandy or dry soils poor in nutrients. Still others, such as quaking aspen, thrive within vast areas of North America because they can take advantage of a wide variety of soils. Pioneers usually occupy a site for one generation because they cannot reproduce in the deep litter and dense shade that they create on the forest floor. Thus pioneers lead the way and settler species gradually replace them.

Settler species such as beech, maple, hemlock, and fir infiltrate stands or groups of pioneer trees and wait in the shady understory for the overstory trees to die. Then they take over the stand. These species are self-replacing because their seedlings grow well on the thick moist litter that accumulates under older trees, and they do not need as much sunlight as pioneers. Nevertheless, some settler species also grow well on bare soil and join the pioneers when they occupy freshly cleared land. When the pioneers die, and the settler species dominate the stand, they continue to replace themselves until another disturbance clears the forest. Then the pioneers return and the process begins again.

The retreat of the glaciers at the end of the Ice Age uncovered new land that pioneer species could occupy. These new pioneer forests then created conditions needed by settler species, so they infiltrated the forest, replaced the pioneers, and settled the land. In other areas the Ice Age forest broke apart and left openings for pioneers, so they moved in and took their place. Then the settler species followed.

Fires and windstorms also sweep over most forests every few centuries or decades. This provides a constant supply of openings somewhere within a forest. As a result, some parts of a forest may contain pioneer seedlings while others may contain old pioneers with settler trees invading the understory. Settler trees that have already infiltrated and replaced the pioneers may occupy additional parts of a forest. Settler trees usually cover more of a forest when disturbances are infrequent, but the pioneers seldom disappear unless a change in the climate forces them to move elsewhere. Even then, new pioneer species usually arrive to take their place.

The warming climate at the end of the Ice Age favored the spread of interglacial, or warm-weather, species and restricted glacial, or cold-weather, species. Thus, some warm-weather trees such as beech remained in the South and just expanded their territory northward. This settler species infiltrated existing forests, forming waves that spread out from their refuges in the South like the ripples produced by throwing a stone into a lake. However, glacial, or cold-weather, species such as spruce had to leave the South and migrate northward. Even so, these trees also moved in waves, only instead of ripples they looked like the rows of dust that pile up in front of a broom as it pushes them forward. Thus spruce followed the retreating glaciers northward or up cool mountain slopes while their populations dwindled away in the heat at the rear.[9]

SPRUCE MIGRATION

East of the Rocky Mountains pioneer forests of white spruce usually invaded the tundra first because they grew close to the glaciers. Black spruce followed, and in New England red spruce left its now submerged refuge on the coastal plain and joined the invasion. However, red spruce probably stayed on steep rocky slopes and damp lowlands near the coast. Red spruce behaved as a settler species because of its tolerance to shade; so it replaced the less tolerant white spruce in some places.[10]

White spruce kept pace with the retreating ice sheets. Each time the glaciers stepped back a strip of new tundra formed, and white spruce invaded the strip of old tundra left behind. They moved out onto the tundra like small groups of soldiers that got ahead of the main force. Spruce seedlings occupied bare spots here and there to create an open, patchy community that still looked more like tundra than forest. Gradually the trees spread from their footholds, closed the gaps, and overwhelmed the tundra. Today alder leads the charge onto newly deglaciated land. This small hardwood takes nitrogen from the atmosphere and fertilizes the rocky soil left behind by the glacier, which helps other trees become established. This did not happen south of the ice sheets at the end of the Ice Age because tundra plants had already persisted long enough to enrich the soil for the pioneers (Fig. 3.1). So spruce led the invasion and alder arrived after the battle ended.[11] Much later, spruce also arrived before alder in the interior of Alaska, but alder came first in northwestern and southwestern Alaska.[12]

White spruce advanced northward at an incredible pace as it invaded tundra above the Great Plains. About 14,000 years ago the outer edges of the Cordilleran and Laurentide Ice Sheets parted as they melted, creating a long narrow ice-free gap.[13] This gap between the walls of ice wound its way north along the eastern flank of the Canadian Rockies and connected the upper Great Plains to Alaska. The wind increased in velocity as it squeezed through the gap, flowing unobstructed over the flat open tundra. These strong winds carried spruce seeds northward and sprayed them over the ground. So white spruce spread into the tundra within the gap, traveling from the upper Great Plains to the MacKenzie River Delta 1250 miles north. Thus a finger of white spruce poked up into Canada about 9000 years ago, well ahead of the leading edge of the rest of the migrating spruce forest.[14]

In central Minnesota, shrubby vegetation that included dwarf birch, willow, poplar, juniper, silverberry, and soapberry prepared the way for spruce. They began invading tundra at the edge of the ice sheet about 14,500 years ago. An open white spruce forest intermixed with patches

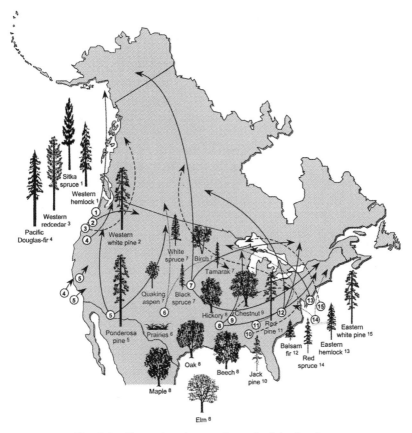

Fig. 3.1   Tree migrations at the end of the Ice Age.

of paper birch and, with black spruce, tamarack, and black ash growing in wet areas, took over the tundra more than a thousand years later.[15] Ice still filled deep depressions in the ground at this time even though the glacier was gone. Often a thin layer of wind-blown soil covered the ice so that spruce could grow there. However, these were only temporary forests. Eventually the ice melted and the spruce forest sank into the water and disappeared below the surface of a newly formed lake.[16] The white spruce forest was moving slowly at this time. It took 1500 years to reach northeastern Minnesota. Then it hesitated during a temporary cold period known as the Younger Dryas.[17] Finally, white spruce began moving in earnest again. It swept over the shrubs and tundra so quickly that it reached the foot of the glacier on the border of Canada 10,000 years ago.[18]

White spruce forests moved out of central Illinois about 14,000 years ago as they shifted northward to escape the warming climate. Then the Lake Michigan ice lobe pushed forward, crushing the front line of the invaders and forcing them to retreat southward, so they reappeared briefly 11,600 years ago in central Illinois before advancing north again. Thus spruce reached its present position in the upper Great Lakes region about 10,000 years ago. The same changes occurred farther east in Indiana and Ohio.[19] Spruce also invaded tundra in a line that stretched from Pennsylvania to Massachusetts at about the same time.[20]

Spruce trees quickly marched up the mountains when they arrived in New Hampshire 12,000 years ago. They anchored their roots in the barren soil scraped clean by the glaciers, stopped the slumping and erosion, and remained there until today.[21] The ice sheet shrank away from the southern coast of New Brunswick about the same time. So spruce also migrated northward along the coast of Maine and New Brunswick and into Nova Scotia.[22] It paused, and even retreated somewhat, during the Younger Dryas cooling event. Spruce resumed its northward advance when the climate warmed again.[23] However, a lack of soil rather than climate slowed the migration of pioneer spruce forests in central Maine. Even though the ice sheets had withdrawn, tundra stayed in the area. It took 2000 years for enough soil to accumulate to allow spruce, poplar, and birch to invade the tundra.[24] It took a 1000 years to complete the invasion, which ended when tundra finally disappeared from Maine 10,000 years ago.[25]

Migrating white spruce became trapped between the ice sheets to the north and the other invading trees from the south. The massive fields of ice could not melt as fast as the climate warmed and trees moved. So spruce pressed tighter and tighter against the edge of the glaciers. Finally, only a narrow strip of the once great white spruce forest remained. It hugged the retreating wall of ice for 4000 years. Then the ice sheets melted away, leaving behind new land for spruce to colonize. The climate began to cool as well, and spruce could expand its range once more, moving toward the north and west into Canada and Alaska. By 4000 years ago spruce forests had spread over much of Canada and regained their former glory.[26]

TREES ABANDON THE GREAT PLAINS

As the climate warmed, and summer droughts became more common, spruce began to die over vast areas of the Great Plains. Even so, the wave of advancing spruce trees continued northward, leaving their wounded companions behind. Prairie grasses also crept northward and eastward from their refuge in Texas and occupied the ground beneath the dying trees. The prairie reached southeastern Oklahoma about 12,000 years ago.[27] No doubt the hotter weather and expanding grasslands helped to increase the frequency of massive fires, which hastened the declined of spruce and accelerated the spread of the prairies. By 11,300 years ago mostly fire blackened and bleached skeletons of spruce trees stood in the grasslands of Kansas.[28] The massive dieoff of trees on the southern edge of the white spruce forest continued to spread north through Nebraska and South Dakota during the next 2000 years.[29] Small numbers of white spruce took refuge on cool mountain slopes. The other trees perished on the surrounding lowlands as the rest of the forest migrated to safety in Canada.

Remnants of the Ice Age white spruce forest still survive in the mountains of northwest Montana and in the Black Hills of South Dakota.[30] Today the Black Hills rise 4000 feet above the northern Great Plains to form a rocky island in a sea of grass. Dry grasslands encircle this island like shark-infested waters, holding spruce trees, thimbleberry, and other cold-weather plants and animals captive. Even yellow-bellied marmots and long-tailed voles cannot cross the prairies and return to the Rocky Mountains. These relicts of the Ice Age now live far away from the rest of their kind that escaped to other places when the climate warmed.[31]

Ponderosa pine, a warm-weather tree, now dominates the Black Hills, but relicts of the Ice Age spruce forest still persist in canyons and on steep north-facing slopes. Eastern hardwoods that migrated into the Great Plains behind the departing white spruce forests, such as hackberry, green ash, American elm, and hophornbeam, found refuge around streams. Willows from the northeast and East still crowd along streams in the Black Hills. A few small patches of lodgepole pine and limber pine that lived on the Great Plains when the glaciers were at their maximum cling to the hills as if they were life rafts. Another tiny relict forest of limber pine survives in the Badlands of North Dakota. White spruce and paper birch also remain trapped in parts of the Black Hills more than 600 miles away from the main forest that migrated northward with the retreating glaciers. However, pine and spruce no longer grow together on this mountain island as they did long ago. The remaining white spruce forests also lack the openness of the primeval forest, and the sounds of mammoths and camels grazing in the clearings. Nevertheless, "old-man's-beard" lichen still drapes from branches, and northern flying squirrels continue to glide among the spruce trees, just as they did during the Ice Age.[32]

TREES ADVANCE IN THE MIDWEST AND EAST

Eastern larch, or tamarack, did not find a refuge on the cooler Black Hills when the climate became warmer and drier on the Great Plains. So it followed white spruce northward, migrating deep into Canada. It arrived early with spruce at many places in the northern states and only lagged behind for a short time at other places. Tamarack bumped into the west side of the Appalachian Mountains as it migrated toward the Northeast, but it found a passage around the north end of the mountains into New York and beyond. It traveled quickly, but it increased in density slowly because of the lack of swamps and bogs. It picked up speed 8000 years ago as more wet areas became available.[33] Tamarack now lives with black spruce primarily in cool bogs and swamps, as it did during the Ice Age, but farther north within a broad belt that cuts across Canada and Alaska. It also grows in New England and the Great Lakes region.

Balsam fir, a settler species, migrated out of its refuge in the Southeast and followed spruce and tamarack northward. Fraser fir, a close relative, remained behind and moved off the lowlands and into the southern Appalachian Mountains where it lives in a few small areas today, primarily in North Carolina. Balsam fir quickly traveled north along the east side of the Appalachian Mountains and possibly along the west side as well. It arrived in the northern United States along a front that extended from southern New England to the Great Lakes region 12,000–11,000 years ago. Then it invaded the white spruce forest.[34] Balsam fir's tolerance to shade also allowed it to invade groves of aspen and birch that had occupied openings in the white spruce forest.[35]

Aspen, birch, and balsam fir also followed spruce into eastern and central Canada. However, the Rocky Mountains blocked the spread of balsam fir, so it could not move into western Canada and Alaska with the other trees.[36] Now balsam fir grows within the modern white spruce-balsam fir forest in southern Canada east of the Rocky Mountains. It also extends down into northern Minnesota, Wisconsin, and upper Michigan to as far east as New York and Maine. Balsam fir also survives farther south in small isolated areas within the mountains of Virginia and West Virginia.[37]

Hardwoods also spread out in the Southeast when the climate started to warm. They appeared within white spruce forests in the Ozark Mountains of Missouri as early as 16,500 years ago.[38] By 12,700 years ago hardwoods made up a prominent part of what remained of the primeval jack pine and white spruce forests in the Southeast.[39] Hardwoods also spread rapidly up the Mississippi River Valley and across the Great Plains from their refuge in Tennessee. They bumped against the declining southern edge of the migrating spruce forest, finally arriving in the Great Lakes region about 11,000 years ago. Thus, a band of hardwoods, including American elm, oak, maple, and black ash grew south of the white spruce forest and migrated with it toward the north. The shrubby hornbeam, also known as ironwood because of its hard tough wood, thrived in the understory along streams and in ravines, and hophornbeam took its place on the uplands.[40]

Black ash, which grows in swamps, and northern red oak, which grows on uplands, arrived in the North earlier than most hardwoods. This leads some scientists to believe that these pioneer species may have gotten a head start by surviving in scattered spots within the Ice Age white spruce forest.[41] Hickory, also a pioneer, arrived a few centuries later. Hardwoods also moved northeast in a broad front through passes in the Appalachian Mountains and up into New England. Oak, a pioneer species, and elm and maple, settler species, arrived first, about 10,000 years ago. However, hickory did not reach New England until 5000 years ago even though it is a pioneer species.[42]

Oaks quickly spread northward over vast areas as the climate warmed. Blue jays played a major role in rapidly expanding the range of oak. These amazing birds can hold three to five acorns when flying and carry them nearly $2\frac{1}{2}$ miles. Equally important, blue jays store 67% of the acorns in open forests and 12% in nearby grasslands.[43] Oaks do well in such environments. Thus blue jays not only carry the acorns long distances, but they also bury them in just the right places to favor the establishment of oaks.

As they expanded their ranges, oaks sorted themselves into regions where conditions were favorable for the reproduction of each species.

Thus, pin oak formed a broad band on wet bottomlands and upland flats in the north central and northeastern United States. White oak spread throughout the midwestern and eastern states, where it grew particularly well on deep moist soils. Bur oak only spread west of the Appalachian Mountains, but it also reached as far north as southeastern Canada. Bur oak also penetrated the Great Plains because of its ability to resist drought. Likewise, post oak and blackjack oak are drought hardy, so they moved into the Great Plains as well, but they only spread northward to southern Illinois. However, they covered a large area that extended down to the Gulf of Mexico and from the Great Plains to the East Coast. Southern live oak stayed in the South. It formed a narrow east–west belt on dry, sandy plains near the Gulf Coast and up the Atlantic coast to Virginia, with a bulge into northern Florida.[44]

Jack pine and possibly red pine moved more quickly than any other species. These pioneers invade fresh burns, so wildfires must have cleared the landscape for their migration. They began leaving the Southeast 15,000 years ago. Jack pine finally abandoned eastern Tennessee west of the Appalachian Ridge 12,750 years ago, though spruce and fir still grew there.[45] It only took jack pine a few thousand years to sweep north over and around the Appalachian Mountains into southern New England. Jack pine entered the white spruce forest in southern Connecticut 12,000 years ago, about the same time that it disappeared from the South. Eventually it penetrated southeastern Canada, but it is not plentiful there or in New England today.[46]

Jack pine also turned westward, taking only a thousand years to reach Wisconsin. It slowed its westward assault as the glaciers moved back and forth, entering the declining southern edge of the white spruce forest in Minnesota about 10,700 years ago. Paper birch and alder followed jack pine about 1000 years later. Then the spruce forest collapsed, and jack pine replaced it within only a few hundred years.[47] This was a relatively open forest in which a thick carpet of bracken fern covered the ground underneath the trees.[48] Jack pine could not move farther west due to the warm, dry conditions on the Great Plains. However, it did not hesitate in its aggressive migration toward the north into Canada.[49]

Jack pine inserted itself like a wedge between the hardwood forest and the spruce forest when it arrived in the Great Lakes region. It invaded both forests, replacing spruce as they declined and mixing with hardwoods. This created a distinctive band of mixed pine-hardwood forest below the spruce forest about 10,000 years ago. Going westward, these three forest bands zigzagged across northern Ohio, southern Michigan, and northern Indiana and Illinois. Then they curved upward in a northwest direction, cutting diagonally across Wisconsin and Minnesota. Finally, the bands of forest bent sharply north at their western end where they hit an eastward bulge in the Great Plains grasslands. As the climate warmed, this bulge in the Great Plains expanded farther east, and the Canadian prairie in Alberta spread north about 50 miles. By 8000 years ago prairies grew in most of Minnesota and Iowa, and reached as far east as Illinois. The swelling prairie compressed the bands of forest as it shoved them northward, and it pushed the western end upward even more sharply.[50] Remnants of these bands of primeval forest remain faintly visible today.

It took eastern white pine several thousand years to join jack and red pine in Minnesota. Like most other pines, white pine is a pioneer that needs fire, and there were plenty of fires clearing the way in the warm climate. However, white pine probably arrived in Minnesota later than jack and red pine because it spread from the East. So it had to cross the Appalachian Mountains and curve around the Great Lakes on its way west.

White pine moved from its refuge on the eastern continental shelf to escape the rising sea and made its first appearance 12,700 years ago in the Shenandoah Valley of Virginia.[51] It expanded in Virginia and Maryland, and then dispersed north into the White Mountains of New Hampshire 9000 years ago. It also spread west through southern Canada and the northern Great Lakes region, reaching its modern location in northeastern Minnesota about 7200 years ago. As the climate grew warmer, the density of eastern white pine declined in New England while it was still making its westward advance.[52] It was still slowly advancing westward in Minnesota at the time of European settlement.[53]

Oak, ash, maple, and ironwood moved in and gradually replaced the jack pine forest as it departed from the South. Hardwoods also succeeded jack pine along the southern coastal plain about the same time.[54] By 10,000 years ago what remained of the white spruce forest finally disappeared from eastern Tennessee, and ironwood decreased in abundance. However, remnants of the white spruce forest probably survived high in the southern Appalachian Mountains.[55] Ash and maple remained in the area; oak continued to expand; and sweetgum, hickory, and chestnut joined them as they spread north. A similar sequence of events took place in northwestern Georgia.[56]

While these changes occurred, sand pine moved farther south in Florida and such hardwoods as oak, hickory, hackberry, and juniper or cedar took its place in northern Florida. Open grassy oak woodlands and pine forests dominated the Southeast at this time. Then the hardwoods declined, and pine increased about 8400 years ago in South Carolina and about 6700 years ago in southern Georgia. Okefenokee, Everglades, and other great swamps also emerged as the oceans continued rising. Thus, in just a few thousand years southern forests began to take shape and nothing remained of the Ice Age forest. The southern pine and oak-gum-cypress forests that characterize the Southeast today finally developed their modern appearance about 5000 years ago.[57]

Eastern hemlock spread toward the north from the Appalachian foothills, the Coastal Plain, or the continental shelf. No one knows the exact location of its refuge. It moved slowly, though it had light-winged seeds easily carried by wind. It climbed mountains in New England 8000 years ago. Then it turned westward, arriving in small scattered groups in southern Michigan about 7000 years ago. Hemlock moved north, blew across the narrow, northern end of Lake Michigan and entered the Upper Peninsula of Michigan. Then its westward advance slowed dramatically. Possibly because the hemlock population crashed about 4800 years ago due to what may have been the outbreak of a fungal disease. Up to 80% of the trees died within 30 years throughout its range. Hemlock did not recover until 2000 years later.[58] Thus, it took hemlock 5000 years from the time it crossed Lake Michigan until it reached its present location within northern Wisconsin.[59]

Beech arrived in central Michigan about the same time as hemlock, though it had to travel a much greater distance because it spread around the eastern side of the Appalachian Mountains. It also had to rely on blue jays and passenger pigeons to carry the heavy beech nuts while wind dispersed the lighter hemlock seeds.[60] Beech is a settler species that reproduces best in the shade of other trees. It headed west through Ontario after it reached the northern edge of the Appalachian Mountains. Beech arrived in southern Michigan 8000 years ago to form the modern beech-maple forest in that area. Then beech moved slowly southward to complete the beech-maple forest in Indiana and Ohio.[61] It also moved northward through Michigan, but low water levels in Lake Michigan at this time made it easy for beech to cross the lake into Wisconsin. Lake Michigan became narrower and many islands popped up in the center of the lake, which allowed birds to fly from island to island, planting groves of trees as they went. Eventually beech arrived on the Wisconsin shore, where it became established about 6000 years ago.[62] Birds may also have carried beech nuts across the Illinois prairies into Wisconsin.[63] However, once beech arrived in Wisconsin it stayed on the eastern side of the state.

It took beech much longer to jump over the northern end of Lake Michigan. Water levels in the lake rose 23 feet higher than today by the time beech arrived in northern Michigan. Many islands flooded, so birds had few places to land. Beech halted its advance until the water levels dropped again about 3300 years ago; then it crossed into the Upper Peninsula of Michigan, where it remains today.[64] Now beech grows on moist soils over a wide area east of the Mississippi River. In the north its principal associates within the beech-maple forest are sugar maple and yellow birch. Beech grows with southern magnolia, basswood, and sweetgum in the South.

Chestnut expanded northward at the slowest rate of all the trees, averaging only 300 feet per year. It left its refuge in the lower Mississippi River Valley and followed the Appalachian Mountains toward the northeast. It spread over a swath of land reaching as far west as southern Missouri and extending in length from Mississippi to southern Maine. Chestnut moved so slowly that it did not reach the central Appalachians until about 5000 years ago. Chestnut did not become abundant in southern New England until about 2000 years ago. Then it increased abruptly, probably because of intensified burning and land clearing by Indians.[65] Once chestnut arrived, it intermixed with oak on dry ridge tops and ultimately became a dominant tree in the oak-chestnut forest.[66]

TREES ADVANCE IN THE WEST

Lodgepole pine, a species closely related to jack pine, played the role of pioneer in the lowlands of the West. It spread rapidly in tundra and on the gravelly soils or outwash left behind by melting glaciers. Lodgepole pine spread more slowly as it approached its northern limit, reaching the central Yukon Territory less than a century ago. Englemann spruce, an emigrant from the mountains closely related to white spruce, joined lodgepole pine to invade tundra along the Pacific Coast. Sitka spruce, the largest of the spruce trees, also invaded tundra.[67]

Red alder, a pioneer species, and bracken fern usually followed spruce and lodgepole pine into tundra on the Pacific slope of the Olympic Mountains. They invaded the tundra as early as 16,800 years ago because of the warm moist weather by the sea. Starting about 15,400 years ago, western hemlock and mountain hemlock, both settler species, spread within the pioneer forest and became an important part of it. Together these trees and ferns formed a patchy, open parkland of forest and tundra.[68]

Rainfall increased dramatically and the temperature rose even more about 14,500 years ago because the ice sheets could no longer split the jet stream.[69] So the jet stream moved back into the Pacific Northwest, gradually robbing the Great Basin of much of its moisture and drying up Lake Bonneville and other basin lakes.[70] The wetter climate on the Pacific slope of the Olympic Mountains allowed trees to spread out from various places until they merged and engulfed the tundra. The higher temperatures also forced mountain hemlock to abandon the lowlands and migrate up into the Olympic Mountains.[71]

By 10,500 years ago a pioneer forest of Sitka spruce mixed with western hemlock began developing along the Pacific Northwest coast.[72] It became an almost modern coastal forest about 6000 years ago. However, western redcedar, another settler species and one of the giants of this forest, moved more slowly because it requires very moist soils. It grows especially well in stream bottoms and gulches, on moist flats, and on cool north-facing slopes.[73] So it was not common in the northern Puget Trough until about 5000 years ago, and it took another 2000 years to become a significant part of the coastal forest.[74] Because western redcedar can tolerate dense shade, it steadily invaded the understory of Sitka spruce and Pacific Douglas-fir forests until it equaled up to 50% of the trees in some places.[75] Therefore, the Sitka spruce-hemlock forest, and the Pacific Douglas-fir forest, started to look like the ancient forests explorers saw about 5000–3000 years before they arrived.[76]

It took a long time for pioneer trees to invade tundra in the Puget Trough on the east side of the Olympic Mountains, away from the coast. The temperature stayed cold because the lowered sea level kept the Pacific Ocean too far away to have much influence. A tongue of ice also protruded into the lowlands until 13,100 years ago.[77] An odd mixture of mountain and lowland species invaded the tundra as the ice withdrew. Lodgepole pine, Sitka spruce, and Englemann spruce started the invasion. Sitka spruce reached the foothills of the North Cascade Range in northwestern Washington before 12,000 years ago.[78] Mountain hemlock, western hemlock, and subalpine fir, all settler species, followed close behind these pioneers.[79] However, as the climate warmed, Englemann spruce, mountain hemlock, and subalpine fir moved off the lowlands and retreated back into the surrounding mountains. Englemann spruce typically led the way into the mountains, often advancing ahead of lodgepole pine. Sitka spruce and western hemlock stayed on the lowlands, but Sitka spruce eventually concentrated in the higher rainfall areas near the coast. On the other hand, western hemlock slowly spread east as well. It reached northeastern Washington about 2500 years ago and northern Idaho about 1500 years ago.[80]

Pioneer forests of lodgepole pine also took a long time to invade tundra in the inland West, east of the Cascade Range. The invasion slowed because it stayed cold and dry here until the beginning of the Holocene. The Cascade Range created a rain shadow by blocking the eastward flow of warm, moist air into the upper Great Basin. Ocean air moved up the western side of the mountains, cooling as it rose so that the moisture condensed and fell as rain. Now robbed of its moisture, the dry ocean air flowed down the eastern side of the mountains and into the basin below. Cold, dry air also poured off the retreating glaciers and concentrated in the upper Great Basin because the encircling mountains acted like a funnel.[81] Thus, remnants of tundra and cold steppe, perhaps with a scattering of small whitebark pine trees, held out for several thousand years after the glaciers began withdrawing. Then about 10,000 years ago the glaciers backed away far enough to allow a warm steppe community of grass and sagebrush to replace the cold steppes of the Ice Age. Lodgepole pine invaded the moister areas within the Columbia Basin about the same time. Warm steppes had begun to displace Ice Age forests farther south within the Bonneville Basin about 800 years earlier.[82]

Jagged mountains and dry basins in the West made it impossible for trees to flow smoothly over the land as they could in the Midwest and East. Many trees simply abandoned the lowlands and moved into the mountains as the climate warmed. They shifted up the mountain sides in belts like steps in a moving staircase. Each lower belt of trees took over the space left behind by the next higher belt when it moved up the staircase. Englemann spruce sat at the top of the staircase in the northern Rocky Mountains, invading tundra at the foot of the retreating glaciers as early as 12,000 years ago. Whitebark pine and subalpine fir followed closely behind. Fir needs less light than spruce to grow, so it kept creeping into the understory of the spruce forest as well. By 10,500 years ago lodgepole pine expanded its range on the slopes beneath the spruce-fir forests, and sagebrush filled in the basins below. However, the Yellowstone Plateau differed from the overall migration pattern. The ice cap did not leave the plateau until 14,000 years ago, and then only lodgepole pine could invade because the volcanic soils contained too few nutrients to support spruce and fir.[83]

Similarly, in New Mexico, Arizona, and southern Nevada, piñon pine moved off the lowlands between 11,000 and 8000 years ago, followed by juniper and oak. Desert scrub moved in to replace them. Papershell piñon declined in abundance at the same time, while single-needle piñon and Colorado piñon expanded because they grow better in a warmer climate.[84] Thus piñon-juniper woodlands changed as they moved up into places such as Chaco Canyon in northwestern New Mexico where mixed-conifer forests used to grow.[85] The mixed-conifer and ponderosa pine forests also moved higher and replaced limber pine, and limber pine migrated higher still.[86]

One tree spread more widely than any other in the West at the end of the Ice Age—ponderosa pine. This stately and colorful warm-weather tree seems well adapted to its pioneer role. It grows on most soils, and seedlings establish readily on bare areas in full sunlight. Seedlings also resist drought by sending down a fast-growing taproot that takes advantage of moisture deep within the soil. Once ponderosa pine reaches maturity, it can resist fires and dominate a site by living 700 years or more.[87] No wonder ponderosa pine became the most important tree species in ancient forests of the West.

Ponderosa pine left its southwestern refuge at the end of the Ice Age and moved north. The trees wound along the lower slopes of mountains like the probing tentacles of a huge creature. They reached the San Andres Mountains of southern New Mexico 14,920 years ago, and the Santa Catalina Mountains near Tucson, Arizona, just a few hundred years later. Then they fanned out, sweeping north around the Grand Canyon about 10,000 years ago and into southern Nevada and Utah. They arrived in eastern Nevada 6100 years ago and took a few thousand years more to work their way into the Cascade Range in Washington.[88] Ponderosa pine also spread northeast through Chaco Canyon, New Mexico, about 9500 years ago and into Colorado near Fort Collins 5090 years ago. They reached southeastern Wyoming a thousand years later and continued northward around the upper Great Basin.[89] Ponderosa pine also spread east from the Rocky Mountains onto the Great Plains as far as eastern New Mexico and Colorado, central Nebraska, the Black Hills of South Dakota, and eastern Montana.[90]

In southern California, however, ponderosa pine and other conifers simply moved off the coastal lowlands and back up mountain slopes as the climate warmed. They reached their present locations 12,000–11,000 years ago. Ponderosa pine also climbed at least 3000 feet higher in the Sierra Nevada as the valley glaciers retreated, reaching their modern elevation about the same time.[91] As ponderosa pine moved about half way up the western flank of the Sierras it temporarily intermixed with other tree species that were also shifting over the landscape in route to their modern locations. Together these different species formed a very unusual and thick forest that does not exist today. This forest consisted of lodgepole pine, western white pine, ponderosa pine, white and red fir, incense-cedar, mountain hemlock, and western juniper. Then the trees separated about 3000 years later and western white pine, mountain hemlock, western juniper, and lodgepole pine moved to higher elevations. Thus the modern lodgepole pine forest of the Sierra Nevada formed about 8500 years ago. Red fir moved upslope as well, reaching its present location about 6300 years ago. But ponderosa pine, white fir, and incense-cedar stayed behind to form the modern ponderosa pine and mixed-conifer forests of the Sierra Nevada.[92]

Today ponderosa pine forests blanket the lower flanks of many western mountains. Small forests of ponderosa pine and Rocky Mountain juniper also cling to the tops of steep ridges surrounded by prairie in the western Great Plains. These relics of the late Ice Age survived because the steep, moist ridges protected them from wildfires that swept over the prairies.[93] Ponderosa pine lives nearly alone in some places, especially the Southwest and on the Black Hills of South Dakota, where the forests are open and grass grows beneath the trees. Elsewhere it lives in forests mixed with such species as Douglas-fir, white fir, sugar pine, Jeffrey pine, and incense-cedar.

THE GREAT DROUGHT

The climate warmed most rapidly and trees moved quickest from 12,000 to 10,000 years ago, which marks the beginning of our current interglacial period—the Holocene.[94] However, by 9000 years ago the climate not only continued to grow warmer it also became drier. Rainfall gradually decreased, dropping as much as 25% in the upper Midwest. Temperatures peaked at nearly 4°F warmer than today in some places. This warm dry period, known as the Hypsithermal in the East and the Altithermal in the West, lasted 4000 years.[95] The drought forced some trees to retreat from land already taken, while others had an opportunity to expand their range.

A lack of moisture and higher temperatures caused sand dunes to begin moving over the Great Plains.[96] It also pushed forests off the edges of the plains and forced some trees to retract into small areas or move higher into the mountains. Trees also abandoned some of the lower passes in the Rocky Mountains, and grass advanced upward and replaced them.[97] Open oak-hickory forests bordering the east side of the Great Plains also withdrew as the prairie expanded eastward into central Illinois and the west side of the Ozarks in Missouri. These woodlands did not move westward again until the climate began to cool several thousand years later.[98] However, when they did, they pinched off a piece of the prairie, which still remains in central Illiniois as a vivid reminder of the Great Drought.[99] Ponderosa pine also backed off the western Great Plains and never returned. Similarly, western larch vanished in the northern Rocky Mountains in the United States, where it had arrived from the Pacific Northwest perhaps only a few thousand years earlier. It probably contracted into the Canadian Rockies and coastal areas of British Columbia and Washington.[100]

Lake levels also dropped in California's Sierra Nevada, and many meadows went dry. The drought also killed many trees in the Sierras, which created openings that allowed more shrubs to grow within the forest than today. An increase in wildfires further opened the forest.[101] Pine and fir moved slightly higher on the mountain slopes, and trees that require more moisture, such as giant sequoia, shrank into smaller areas and clustered around streams and wet meadows.[102] Similarly, other trees, including spruce, balsam fir, and white pine in New England, and Englemann spruce and subalpine fir in the West, had to move higher up the mountains to escape the heat.[103]

Not all trees suffered from the heat. The seas rose to their modern levels about 5000 years ago, which expanded mangrove swamps along the shoreline of southern Florida. The short mangrove tree has roots that spread out above the water to prop it up in saturated soil. Pores also dot the surface of these prop roots so that the tree can breathe. Higher sea levels also raised the underground water tables along the coast. This allowed fresh water to bubble up into a 700-square-mile depression in southern Georgia and create a vast forest of bald cypress in Okefenokee Swamp. The cypress swamps and the Everglades of southern Florida also developed at about this time.[104]

Other trees took advantage of the warmer weather and occupied land vacated by trees that perished in the heat. For instance, juniper, incense-cedar, sage, and especially oak, expanded along the middle slopes of the Sierra Nevada to fill in areas vacated by pine and fir.[105] In southern California, oak grew among conifers on the lowlands during the Ice Age, but they expanded dramatically about 7800 years ago as the conifers withered in the warming climate.[106] Oak also expanded on the lowlands in northern California as the climate became warmer and drier. This forced many conifers to move up into the mountains.[107]

In the Northeast, white pine continued its westward advance during the Hypsithermal even though it contracted in the northern part of its range. Paper birch became more abundant in the Great Lakes forests at this time. Similarly, oak woodlands spread over a large area within the Puget Trough of Washington and Oregon between 10,000 and 7000 years ago. Douglas-fir, western hemlock, and alder also spread over vast areas in the Pacific Northwest. They displaced Englemann spruce, mountain hemlock, and lodgepole pine in the central Coast Range of Oregon, the Puget Trough, and on the west slope of the Cascade Range.[108]

THE NEXT ICE AGE

About 5000 years ago the warming trend reversed and the climate became cooler and wetter. Mountain glaciers began advancing, flooding increased, especially in the Southwest, and arctic pack ice expanded.[109] This marked the beginning of the next ice age—the Neoglaciation—and the end of the Great Drought. As expected, some trees benefited from the shift to a cooler climate and expanded their range, while others declined.

The Neoglaciation made the most recent adjustments to the locations of our modern forests. Interglacial, or warm-weather, trees, such as eastern white pine, had to halt their northward advance and even give up ground because of the cooler climate.[110] Treeline also moved lower in the mountains.[111] Glacial species, such as white spruce, gained ground in the northern United States while shrinking their range in northern Canada. Beech and hemlock declined in abundance as white spruce moved south. Red spruce left the rocky areas along the northeast coast about 2000 years ago and radiated outward toward the north, and up mountain slopes in Maine and elsewhere in New England. It displaced the less shade-tolerant white spruce as it spread.[112] Likewise, red spruce turned southward and expanded down the crest of the Appalachian Mountains, where it also displaced the last remnants of the Ice Age white spruce forest.[113]

Englemann spruce and subalpine fir also moved back down to middle elevations in western mountains and mixed with lodgepole pine, which had taken over these sites during the Great Drought.[114] Thus, spruce-fir and lodgepole pine forests reached their modern locations in the northern Rocky Mountains about 4000 years ago. Even so, they looked like modern forests more than 9500 years ago. They just moved up in elevation during the Great Drought, and then they moved down to their modern locations at the beginning of the Neoglaciation.[115]

Western juniper also moved lower and expanded its range within the northern Great Basin in eastern Oregon at this time. Grass between the junipers became thicker because of increased rainfall.[116] The cooler, wetter climate also made it possible for western larch to move out of coastal areas in the Pacific Northwest and return to its present location in the North Cascade Range. It arrived in the northern Rocky Mountains about 3500 years ago.[117] The modern larch-fir and white pine forests formed when western white pine entered the larch forest about 1500 years ago.[118]

In southern California, oaks declined on the lowlands and shrubs began expanding as the climate started to cool. By 2300 years ago modern coastal sage scrub, chaparral, and grassland communities covered the lowlands where mixed-conifer–oak forests grew during the Ice Age.[119] Northern California also became wetter about 3800 years ago, so oaks started to decline there as well. As a result, Douglas-fir spread in the mountains near the coast and ponderosa pine regained some land it had lost farther inland. Thus the modern oak woodlands and Pacific Douglas-fir forests formed in northern California at about this time.[120] Oak and alder also began declining in the Sierra Nevada.[121] However, about 4500 years ago increased rainfall allowed conifer forests to thicken along the west slope, and giant sequoia quickly spread into a string of groves that became part of the California mixed-conifer forest.[122]

The transition to the next ice age has not been smooth during the last several thousand years. Small shifts in temperature gave interglacial and then glacial tree species a temporary advantage. For instance, the climate became warmer during the Medieval Warm Period that extended from AD 900 to 1300 because the sun became more active.[123] Then temperatures plummeted in AD 1450, taking the world into the Little Ice Age, which did not end until 1850.[124] This unusual cold period forced eastern hemlock to move 1300 feet lower in elevation on the White Mountains of New England. White pine made an even more dramatic drop in elevation. It migrated 3000 feet lower on the White Mountains, where it now grows.[125] However, beech took advantage of the cool moist climate and expanded westward about 43 miles on the Upper Peninsula of Michigan in a few hundred years.[126] Because of these recent movements, parts of today's forests contain trees that have only lived together for a short time.[127]

The forests we see today have an aura of permanence. They seem like finished works of art. However, history shows that they assembled from species that moved from place to place in response to climatic changes. The climate may again be shifting to a temporary warming trend. We do not know. We do know that a new glacial age has begun and that our forests will change as the climate changes. Each species will move to a favorable environment independently of other species. When they meet one another, a new forest will form. Then today's forests will look primeval and the new forests will be modern to those who see them.

FOUR

Ancient People in a New World

I have been a stranger in a strange land.
Hebrew Bible
Moses, in Exodus 2:22.1

The glaciers left behind a raw canvas. A tortured and lifeless landscape littered with debris. However, life returns. It covers the wreckage, softens sharp edges, brightens colors, and adorns the land. Trees, shrubs, grasses, and other plants provided the palette of shapes and colors. They flowed over the land like spilled paint, mixing as they went. Climate and ice stirred the mixture many times, but everything changed at the end of the last ice age. Humans arrived in North America.

FIRST FOOTPRINT

Ancient people, or Paleoindians, left the first footprint on North America about 15,000 years ago. Some scientists think it might have been much earlier, yet no convincing evidence exists to support the claim.[1] This Paleoindian footprint lacked the drama of the first person to step on the moon. However, it altered our destiny in ways that are no less profound.

Unlike astronauts, Paleoindians did not know this was a new land. Wherever they looked, the cold, dry, and nearly treeless landscape appeared the same. All they saw were patches of grass, shrubs, and bare ground extending in all directions. Small bands, each consisting of about 50–100 people, wandered eastward from Siberia in search of game and perhaps adventure until the massive ice sheets temporarily blocked their way.[2] Then the ice sheets parted and they walked into the heart of North

America. Some may even have paddled small boats around the great glaciers that protruded into the ocean and entered the interior of the continent from the western coastline.[3] Regardless, the first people to arrive found a wild and untouched landscape that they unintentionally and deliberately changed forever by simply doing what was necessary to survive. These were probably Clovis hunters—the first Americans.

The ground upon which these immigrants from northern and perhaps southern Asia walked to reach North America now rests about 120 feet below the surface of the choppy waters of the Bering Strait.[4] Back then it seemed like part of Siberia rather than a bridge between continents. This lost land that we call Beringia covered a vast area. It spanned the 56 miles between Siberia and Alaska with a 1000-mile-wide rolling plain that extended from Wrangle Island in the north down to the Alaska Peninsula.[5] Paleoindians living on this plain could not know it used to lie beneath the sea or that it would disappear again when the glaciers melted.

This bleak landscape of shrub tundra and cold steppes, shifting sand dunes, icy streams, and swamps, raked by frigid winds and blowing dust, and swarming with mosquitoes would seem almost uninhabitable to modern people.[6] The ever-present danger of attack by American lions, scimitar cats, wolves, and grizzly bears that would eat either a human or a bison made life even more treacherous. The giant short-faced bear was the most fearsome predator of all. It stood 5 feet tall at the shoulder and over 11 feet tall when upright, and it could run fast enough to prey on caribou, bison, and horses.[7] However, to Paleoindians, ancestors of American Indians, this land represented home—a familiar, if harsh, place that provided for all of their needs.

Paleoindians were big-game hunters. So they moved through the interior of Beringia where occasional patches of grass and shrubs within the tundra supported a wide variety of game.[8] They probably hunted steppe antelope or Saiga and caribou, as well as horses, giant 2-ton bison, and woolly mammoths.[9] Ancestral Aleuts and Eskimos followed the Paleoindians thousands of years later, arriving in Alaska about 8400 years ago. The land bridge had long since disappeared beneath the rising waters. However, the channel between Siberia and Alaska remained narrow enough to see across because sea level was still nearly 50 feet lower than today.[10] These sea-faring people probably found it easy to paddle to North America by boat.[11] Here they remained near the coast so that they could continue to hunt whales, seals, sea otters, fish, birds, and urchins.[12]

Paleoindians survived in the frozen north because of their intelligence, as Roger Williams learned when observing New England's Indians. Speaking in the seventeenth century, he said about the Indians that, "for the temper of the braine in quick apprehension and accurate judgments to say no more, the most high and Sovereign Creator hath not made them inferior to Europeans."[13] The intelligence of Paleoindians and their descendants is not surprising because they came to the New World prepared with the knowledge, skills, and creativity of thousands of generations.

During the cold, dark arctic winter Paleoindians lived within hide-covered shelters partially dug into the ground for insulation and supported by mammoth bones and antlers.[14] They cooked food and heated their smoke-filled huts with fires fueled with mammoth bones, dung, fat, oil, or twigs from dwarf birch, and an occasional stick of wood from a willow or a poplar tree.[15] They most likely slept on hides supported by a springy mat of dwarf birch twigs.[16] They also kept warm with skin clothing that consisted of hooded parkas, caps, shirts, jackets, trousers, and moccasins tailored with bone needles.[17] They probably decorated their clothes with mammoth ivory beads, which their ancestors had done at least 7000 years earlier. The women most likely wore ivory bracelets and other jewelry. They even made room in their heavy packs for artistic images of animals and people carved or etched in bone and ivory.[18]

Whenever possible, Paleoindian hunters watched for game from hilltops, waiting in the warmth of campfires. They used this time to make new stone spear points and to resharpen old points damaged during the last hunt. Some of their tools and campfire pits still sit on the hills where they left them thousands of years ago.[19] These ancient remains show that Paleoindians carried a sophisticated stone toolkit similar to those used throughout northeast Asia, although it varied from region to region so that it fit local conditions.[20]

One weapon characterized the Paleoindian toolkit more than any other in the Arctic—microblades. These razor-sharp slivers of stone sat in narrow grooves on two sides of spearheads made from mammoth ivory or antlers.[21] Their toolkit also included advanced bifacial (flaked on both sides) stone projectile points and knives, stone choppers and scrapers, and many other weapons and tools. Here again Paleoindians benefited from 2 million years of tool making.[22] Even Neanderthals used 60 types of stone tools as long as 120,000 years ago.[23] So the Paleoindian's tools and weapons represented many millennia of additional innovation and perfection.

One innovation revolutionized the hunter's effectiveness—the *atlatl*, or spearthrower. It originated in Europe perhaps 17,000 or more years ago. Knowledge of such a powerful new weapon could not help passing quickly from band to band. Therefore it must have reached the Paleoindians in Asia before they crossed into North America. They undoubtedly carried it with them because they used it in the lower Snake River in southeastern Washington about 10,500 years ago and possibly a little earlier near the Yellowstone River in Montana.[24] Paleoindians still used thrusting spears to hunt and to protect themselves from charging animals, but the *atlatl* improved their ability to kill game from a distance. It increased leverage by extending the length of the throwing arm. Thus the spear not only flew farther but also faster and more accurately than a hand-thrown spear.[25] This sophisticated weapon could easily penetrate the tough hide and even the bone of a mammoth or mastodon.[26]

The *atlatl* consisted of a stick about 2 feet long with straps made of hide to keep it from slipping out of the hunter's grip. Stone weights attached to the stick provided balance and a hook carved at the end held the spear. The spear consisted of one or two parts.[27] A two-part spear included a wood mainshaft and a bone or wood foreshaft, or dart. The foreshaft held a stone point lashed to one end with animal tendon, or sinew, and sealed in pitch. A socket or split in the mainshaft held the detachable foreshaft. The foreshaft broke free of the mainshaft when it penetrated a game animal, which allowed the hunter to retrieve the spear, quickly attach a new foreshaft, and throw again. Thus a hunter could carry one spear with several light foreshafts instead of several heavy spears.[28]

The Paleoindian population grew and forced some bands to seek new territory. Eventually they reached an impenetrable wall of ice that formed when the Laurentide and Cordilleran glaciers pressed against each other.[29] These massive ice sheets blocked their way south into the heart of the continent and prevented them from expanding their hunting territories. So competition for game had to grow more severe. As the climate changed, summers became warmer and wetter, winter snows deepened, and dwarf birch, shrubs, and moss tundra gradually replaced the grasslands of the cold steppes.[30] As a result, many game animals found it difficult to find grass to graze at the same time that Paleoindians needed to support a larger population. As a result, hunting pressure continued to increase and game grew scarce.

Trampling by animals and grazing reduces mosses and increases grass, but the reverse also is true. Fewer grazing animals also make it possible for mosses and shrubs to increase. Hence a vicious cycle set in where a warmer climate increased mosses and shrubs, which reduced the number of grazing animals. Hunting pressure further reduced grazing animals, which accelerated the rate at which shrubs and mosses replaced the grasses. This also sped up the decline of game animals.[31] Thus both game animals and their grassland habitat spiraled downward toward extinction.

The rising seas crept over the land bridge, further reducing habitat for game animals and severing the tie with Asia.[32] The land bridge briefly opened and closed twice before permanently disappearing about 10,500 years ago.[33] Consequently, the Paleoindians remained trapped in Alaska for at least a thousand years after they arrived, and life became increasingly difficult. Life became even more difficult during the Younger Dryas cold period. Full glacial conditions returned about 12,900 years ago and lasted 1300 years.[34] Then the warming trend abruptly resumed.

PASSAGE SOUTH

An escape route out of Alaska appeared about 14,000 years ago when the Cordilleran and Laurentide Ice Sheets finally separated from each other. This opened a long, narrow ice-free passage along the Mackenzie River Valley, down through Alberta, Canada, and into the upper Great Plains.[35] The Paleoindians leaned into a bone chilling wind as they walked through this crack between the glaciers (Fig. 4.1). This dreary landscape probably seemed worse than the land left behind, but some of them had to seek new territory to survive. Grizzly bears followed them southward and doubtless made their lives even more precarious.[36] Nevertheless, patches of sage, dwarf birch, and grass dotted the ground, and sedges and willows hugged the edges of streams and swamps. So camels and other game animals also found their way into the gap, not many, but enough to support adventurous bands of Paleoindian hunters.[37]

Paleoindians just followed the game, each generation moving deeper into the passageway. It took them only about a century to emerge from the other end, 1250 miles farther south.[38] This part of the passageway widened like the mouth of a river, but giant Lake McConnell and Lake Agassiz blocked the east side because they were forming in southern Canada at the time. Meltwater from the retreating glaciers poured icy water into the lakes, expanding them until they became several times the

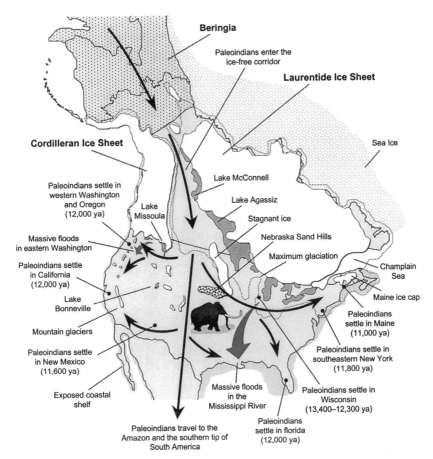

Fig. 4.1 Paleoindians settle North and South America.

size of Lake Superior.[39] Anyone gazing from the lakeshore and only see-ing windswept water dotted with glassy blue icebergs could not help think that this was another ocean. They probably walked around the southwest-ern edge of these huge lakes because cliffs of ice jutted out of the water on the northern and eastern sides.[40]

Those people who made it through the passageway and around the immense lakes could then cross the open tundra and cold steppes at the foot of the retreating glaciers. They did not walk onto a lush grassland as many people suppose.[41] Grass only grew in patches within a mosaic of bare ground, sagebrush, and dwarf birch. Stagnant ice sat just below the surface of the ground where it melted here and there to form pools of frigid water waiting to trap a careless traveler.[42] Rushing rivers of glacial meltwater blocked their path, and monstrous sand and dust storms obscured the way in winter and added to their difficulties.[43] Even so, Paleoindians only had to travel southward a little more than 50 miles before they saw a few scattered white spruce trees, pioneers at the leading edge of a forest advancing northward.[44] These stunted trees tilted in all directions as the freezing and thawing ice undermined their shallow roots.[45] Eventually the Paleoindian explorers entered the vast open forests of spruce, tamarack, black ash, and paper birch that covered most of the northern Great Plains.[46]

SETTLING THE WEST

In spite of the hardships they experienced during their trek between the ice sheets, what the Paleoindians found at the other end could not help delighting them—big game abounded. They saw woolly mammoths, mastodons, horses, bison, camels, musk ox, caribou, stag-moose, elk, deer, and many other game animals wandering through the tundra, steppes, and open spruce forest.[47] They also found smaller game to hunt, such as rabbits, gophers, turtles, and birds.[48] All of them easy prey because they had no experience avoiding the hunter's spear.

The abundance of game allowed Paleoindian populations to expand rapidly.[49] They also could roam through the center of the continent without fear of encroaching on another band's hunting territory.[50] Disease, accidents, voracious predators, and starvation undoubtedly reduced their numbers periodically, but the trend was unmistakably upward.[51] Some scientists estimate that their population in the Americas grew into the millions in a little more than 1000 years.[52]

Paleoindians did not know where they were going. They simply stayed in one area until game became scarce and then they moved.[53] During winter, they froze excess meat so they could stay longer.[54] During summer they sun-dried the meat.[55] Sometimes they stored meat by anchoring it at the bottom of a pond with intestines filled with sand and gravel. Apparently a bacterium now used to make cheese and yogurt kept the meat edible, even in summer.[56] Eventually their supplies became exhausted and they had to leave. So they took what food they could carry and renewed the search for more game.

Each new search led them farther into the continent. Some bands stayed behind but others headed east and west, and still others kept going south. They entered a dangerous and rapidly changing landscape covered with forests that also were on the move. Their trek southward between the ice sheets and across two continents represents one of the great feats of exploration. The hardships Paleoindians must have endured exceed those met during the settlement of the West by Europeans more than 12,000 years later. Guides led the way for European settlers. They also used oxen to pull wagons filled with food, metal knives and tools, ropes, furniture, books, and clothing. Unlike Paleoindians, some settlers also rode horses. They also carried guns, wore heavy boots, and often herded domesticated cattle. Glaciers did not block the way, open forests and rich grasslands covered the land, and the climate had warmed. Likewise, the giant short-faced bear, American lion, and dire wolf of the Ice Age were no longer a threat. Certainly European settlers suffered greatly during their journey, but they had domesticated animals, powerful weapons, luxuries, and periods of warm weather that were beyond the dreams of the first Paleoindian explorers.

Those Paleoindians who went west had to cross the Rocky Mountains. Then trees were already shifting up the mountain sides to avoid the warming climate.[57] Valley glaciers blocked their way in many places, including Yellowstone.[58] One day the ice cap over Yellowstone would disappear and reveal a long cliff of obsidian, or volcanic glass, a treasure house of material for making spear and arrow points. This obsidian cliff became famous throughout North America. At this time, however, Paleoindians had to find trails around the ice. Surely they moved up some canyons without knowing that reservoirs of meltwater held back by ice dams blocked their way. The dams broke occasionally and hurled walls of water and boulders down the canyons.[59] Paleoindians may have walked into the path of these floods as they sought a passage over the mountains. Eventually some of these ancient explorers found their way around the glaciers and across the jagged peaks of the Rocky Mountains. Those who survived the ordeal likely felt a sense of relief when they dropped down the other side and into the upper Great Basin.

Cold, dry air poured into the upper Great Basin from the surrounding mountains and off the glaciers to the north. It concentrated at the bottom of the basin, creating near arctic conditions that lasted for several thousand years after the ice sheets began melting. Patches of tundra, cold steppes, and bare volcanic rock covered this cheerless landscape.[60] The basin certainly appeared harsh, desolate, and forbidding to Paleoindians, but real terror lurked on the other side of the hills toward the north.

We can never be certain, but it is likely that Paleoindians stood in that part of the basin now known as the Channeled Scablands and heard the sound of distant thunder. Instead of subsiding, however, it grew louder and louder until the roar became deafening. They could not know that an ice dam had failed in Montana and emptied giant Lake Missoula. The churning wall of muddy water rushing toward them had traveled hundreds of miles before it appeared on the horizon. This colossal torrent contained nearly half the volume of water in present-day Lake Michigan, and it took a month to pass.[61] So the 30-foot boulders and chunks of ice carried along in the flood probably made little difference to the poor souls crossing the basin. Nothing could survive.

These titanic floods swept across the upper Great Basin many times while Paleoindians were on their way west.[62] The eruptions of Mt. St. Helens and Glacier Peak, volcanoes in the Cascades, added to their woes by pouring molten lava over the countryside and filling the air with hot ash.[63] However, in spite of the danger and hardships some Paleoindians eventually made it to the West Coast. They arrived in western Washington and Oregon about 12,000 years ago, perhaps earlier. We know because they killed a mastodon near Sequin, Washington, and they left projectile points in Fort Rock Cave in south-central Oregon at about that time.[64]

The temperature in Washington and Oregon was still somewhat cold when the Paleoindians came, but the jet stream returned and the climate became wetter. An unusual mixture of subalpine fir, western hemlock, mountain hemlock, and alder grew in the Puget Trough. These trees gradually replaced Sitka spruce, which was moving to the coast, and Englemann spruce, which was moving into the surrounding mountains. Lodgepole pine also headed north onto the stony ground left behind by the retreating glaciers, and grass and sagebrush declined. Douglas-fir was just beginning to spread out of its refuge farther south.[65] The cold, wet land the Paleoindians entered still looked much as it did during the coldest part of the Ice Age, but it was rapidly changing.

The route to the warmer climate south of Oregon presented few challenges compared with those already faced.[66] As a result, Paleoindians traveled so fast that they were already paddling boats to southern California's Channel Islands to hunt sea mammals, collect shellfish, and fish at least

11,500 years ago.[67] They were hunting Ice Age animals in the southern end of the San Joaquin Valley at about the same time.[68]

THE WAY EAST

Some Paleoindians ventured toward the East along the southern edge of the Laurentide Ice Sheet after emerging from the passageway between the glaciers. This route lacked the diversity of forests that other Paleoindians found in the Rocky Mountains on their way west. Those who went east probably spent most of their time in tundra and open spruce forests.[69] The way east also lacked the drama of crossing rugged mountains and canyons, but it was still dangerous. These people faced the usual array of challenges: swollen rivers and floods, frigid swamps, piercing winds and dust storms, and swarms of mosquitoes and black flies. They also had to defend themselves against giant beasts that regarded them as food, such as short-faced bears and dire wolves.[70]

Paleoindians reached southeastern Wisconsin perhaps as early as 13,400 years ago. They were definitely hunting mammoths there 12,300 years ago.[71] The Lake Michigan ice lobe started to readvance about this time,[72] so the Paleoindians had to curve southward around the huge lake at the foot of the glacier. They slogged their way through swamps and a myriad of icy streams and rivers that poured from the lower end of the lake as they crossed Illinois and Indiana. However, their endurance rewarded them with rich new concentrations of game, especially mastodons, caribou, elk, white-tailed deer, and stag moose, as well as mammoths, horses, and bison.[73] They also supplemented their diet with berries and fish whenever possible.[74] Some bands stayed behind to settle the land while others continued eastward. They lived near the ice and moved back and forth with the glaciers.[75] These were the people who settled southern Ontario when the ice sheets retreated.[76] Game remained plentiful for those who plodded onward to Ohio, where some of them settled in Sheriden Cave about 12,600 years ago.[77] Others kept going through Pennsylvania and on to New York and Maine.

Travel again became extremely difficult when Paleoindians tried to continue into New England. Meltwater from the glaciers filled the St. Lawrence Lowland to form the Champlain Sea, which blocked the direct route into New England. Again, they had to move south to curve around this barrier. However, torrents of meltwater rushing down the Hudson and Connecticut Rivers also blocked their way.[78] Nevertheless, those who went east still managed to move at nearly the same pace as those who went west. Paleoindians arrived in southeastern New York about 11,800 years ago where they left caribou bones in Duchess Quarry Cave. They killed caribou farther north in New Hampshire and Massachusetts as well, and here they left spear points behind for us to find with the bones.[79] They slowly worked their way north between the Champlain Sea and the Atlantic Ocean and settled in Maine about 11,000 years ago.[80]

The first Paleoindians to arrive in Maine found caribou to hunt and fine-grained chert rocks for making spear points and tools. Nevertheless, they had to be resourceful because everything was changing. This rough, barren, and cold land had barely begun to recover from the crushing weight of glaciers. The place where the cities of Augusta and Bangor now sit had just emerged from the sea because the great weight of the glaciers no longer pushed the land down beneath the water.[81] Huge pieces of the Laurentide Ice Sheet rested around Mount Katahdin in the center of Maine, separated from the main body of the glacier by the Champlain Sea. Tundra blanketed the northern part of Maine, but patches of poplar trees were invading the tundra near the melting ice masses. An advancing front of open spruce and hardwood forest was engulfing the south side of the tundra throughout the rest of Maine.[82] This restless landscape continued to change after the Paleoindians arrived. Just a thousand years earlier nearly all of Maine was ice and tundra, and a thousand years later it became forested.

TRAIL TO FLORIDA

Surely Paleoindians that traveled in the steppes and tundra near the ice sheets found the environment cold and familiar. However, temperatures became unusually warm for those people who traveled south after coming out of the passageway between the glaciers. Soon they entered forests and grasslands unlike anything they had ever experienced. In spite of that, their weapons and hunting techniques still worked well because many of the same large game animals lived in these areas.[83] Woolly mammoth, stag moose, musk ox, and elk stayed farther north, but they could hunt horses, bison, Columbian mammoths, Jefferson mammoths, and mastodons in more southerly parts of the continent at this time.[84] Camels also abounded in the Great Plains and the West but not in the East. They also could hunt white-tailed deer in the East and mule deer in the West.[85] Therefore, since Paleoindians relied primarily on large game for food, nothing prevented them from quickly spreading throughout the continent.

Those who headed toward the southeast reached Florida more quickly than the people who walked into Maine. Paleoindians found a land to explore nearly twice the size of today's Florida because the glaciers still stored much of the ocean's water. So sea level remained low and Florida stretched out east and west on to the coastal shelf.[86] Evidence of their presence around Little Salt Spring provides a poignant reminder of the hazards faced by these ancient settlers. About 12,000 years ago, a starving man trapped on a ledge after falling to the bottom of a sinkhole drove a stake into a land tortoise that also fell into the hole. Then he turned the tortoise on it's back, used the shell as a cooking pot and ate it for his last meal.[87] Other more fortunate Paleoindians killed a bison in northern Florida about 11,000 years ago. The broken tip of a spear point still protruded from the animal's skull when it came up from the bottom of the Wacissa River.[88] They also hunted mammoths in Florida until about 10,000 years ago.[89]

Paleoindians most likely worked their way to Florida by first traveling south and then east, although some surely turned southward after traveling east along the edge of the glaciers. Those who traveled south probably followed the level lands that are now the Great Plains before turning east. Others must have gone down the Mississippi River Valley because Clovis people were hunting mastodons 20 miles south of St. Louis, Missouri, about 12,000 years ago.[90] Regardless, they eventually turned eastward and went around the southern end of the Appalachian Mountains to Georgia and Florida. On their way to Florida they had to travel through several bands of forest, unaware that the trees also were moving, but in the opposite direction.

Once the Paleoindians found their way through the white spruce forest below the glaciers, they probably entered a hardwood forest that had already migrated up from the South. If they cut across Missouri or Tennessee, they found a mixture of woodlands and prairies that closely resembled the ancient oak-hickory forest seen by European explorers. A scattering of spruce and jack pine also grew here, remnants of the Ice Age forest that used to cover the area.[91] Jack pine, like spruce, abandoned the rest of the South about this time.[92]

As the Paleoindians entered Georgia, the forest changed a little more.

It now included beech and a scattering of butternut or white walnut. These ancient explorers undoubtedly added beech nuts and the oily nuts of the butternut to their diet, just as their descendants did thousands of years later.[93] The Iroquois, for example, crushed the nut meats of beechnut, boiled them, and then drank the liquid.[94] Paleoindians may have done the same. Moving deeper into the South, the climate and the forests eventually became more like today. Finally they walked into an open grassy forest of southern pines and oaks that reached down into north-central Florida. The climate became even hotter and drier in south Florida. As a result, broad-leafed trees disappeared, leaving only scattered pines to grow within the grasslands.[95] By now the great glaciers and heavy skin clothing they left behind surely represented nothing more than a faint memory: the subject of myths and legends.

JOURNEY TO ANOTHER CONTINENT

Many Paleoindians stayed in the North to hunt mammoths after emerging from the gap between the glaciers, tramping east and west near the ice.[96] Others ambled toward Florida, but many more just kept walking south. On their way south they had to cross vast sand dunes in western Nebraska, and pass through hundreds of miles of open spruce forest spread over the northern Great Plains.[97] These Paleoindians could not know that this forest was moving in the opposite direction to avoid the summer droughts of a warming climate. Likewise, they could not anticipate that the prairie grasses under their feat would soon replace the forest.[98]

The forest grew thinner and more open as the Paleoindians walked into eastern Colorado and Kansas. Here they emerged from the southern side of the forest where they most likely found a sea of grass filled with the bleached skeletons of dead spruce trees.[99] These early explorers left no description, but they undoubtedly saw what no human eyes had seen before, the beginnings of what later became the vast prairies of the Great Plains.

Paleoindians crossed the prairies, passed through the shrinking pygmy conifer-oak woodland farther south, and walked into Mexico. They continued south, funneling into the isthmus of Panama and then spreading out again when they reached South America.[100] They climbed the Andes Mountains and then trudged through tropical rainforests in the Amazon Basin, savannas in southern Brazil, and semideserts in Argentina.[101]

They still hunted mastodons, horses, and giant ground sloths.[102] In northern Venezuela Paleoindians left behind a partially butchered mastodon with a broken spear point in its pelvic area.[103] However, large game became less plentiful in some parts of South America, especially tropical rainforests. So they made short, triangular-shaped spear points to hunt smaller animals, such as deer, rabbits, and birds. They also relied more on fish and plants for food.[104] Wherever they went, they adapted to the new conditions and prospered.[105]

Bands of Paleoindians settled in territories here and there while others moved deeper into South America. They settled in Panama, Columbia, and Venezuela. They also settled the lower Amazon of Brazil at least 11,200 years ago, where they were living in a cave at Monte Alegre, and cutting and burning the surrounding forest. They also lived at other sites within the Amazon Basin at about the same time.[106] They moved into Ecuador, Peru, Chile, and Uruguay as well.[107]

Paleoindians spread so quickly that they reached Fells Cave on the windswept plateaus of Tierra del Fuego, near the southern end of South America, between 12,500 and 11,000 years ago.[108] If they emerged from the passage between the ice sheets before 13,400 years ago, when they most likely reached southeastern Wisconsin, then it took them as little as 900 years to travel the 8000 miles to Tierra del Fuego. Such a feat seems incredible. That is, until we realize that these nomadic and hardened people, whose populations were expanding, could have easily made the journey in 500 years by only moving their camps one day's walk southward once a year. Therefore, it probably took only one or two millennia for Paleoindians to settle two continents.[109]

Taming a Wilderness

We did not think of the great open plains, the beautiful rolling hills and the
winding streams with tangled growth as wild. Only to the white man was
nature a wilderness.
Chief Standing Bear, Ogalala Sioux[1]

MAMMOTH HUNTERS

Most people know the earliest Paleoindians that swept like a wave over the Americas as Clovis hunters because of their distinctive spear points. They designed these large points to kill their favorite prey, mammoths and mastodons. The effort required to make the points and pursue such huge beasts paid handsome returns. One animal could provide food for weeks, as well as hide for shelters, bones for tools and fuel, and fat for cooking and lamps.[2] The Paleoindians had lived on these beasts for thousands of years before they arrived in the Americas, and that is what they continued to do. So they pursued their quarry relentlessly, and established territories and modified the landscape along the way. These Clovis hunters and their descendants changed everything, and in doing so they tamed two continents.

The Clovis point (Fig. 5.1) originated in arctic Alaska, known there as the Mesa point.[3] Paleoindians carried it through the passageway between the ice sheets and onto the Great Plains. Then it spread throughout the United States and into Canada and Central America.[4] This stone point varies in size, the largest being about 6 inches long. It has gracefully curving edges that could be as sharp as a modern steel scalpel. A tapered flute, or rounded groove, extends up both sides of the point from the arc at the base so that it would fit in the end of a spear. A final grinding of sharp

68

Fig. 5.1 Clovis point. (Photograph courtesy of Ron Larson Indian Artifacts, Laurel, Maryland)

edges on the base prevented it from cutting through the sinew used to lash the point in place.[5]

Each Clovis point represents true art, patiently sculpted chip by chip from select materials using the knowledge learned over millennia. The expert artisans who created Clovis points chose only the best materials, usually fine-grained rock. They used dolomite from the Canadian River area in Texas, translucent brown chalcedony from the Knife River Valley of North Dakota, obsidian from as far away as southeast Idaho, and even quartz crystal.[6] They dug huge amounts of rock from quarries and pits with elk antlers, bison bones, and hammerstones in search of a few pieces that satisfied their exacting requirements.[7] Paleoindians assigned such great value to these beautiful spear points that they fashioned near perfect specimens for burial offerings.[8]

The beauty of the Clovis point cannot mask its lethal intent. Clovis people hunted to survive. Mammoths, mastodons, bison, horses, camels, and many other animals, big and small, fell before them, and then they moved on to find more game.[9] The effectiveness of their hunting techniques matched the perfection of their weapons. They became the ultimate predators, learning the unique behavior of each species and tailoring their hunting methods accordingly.

Mammoths and mastodons probably behaved much as elephants do today. They formed small families consisting of the eldest female or matriarch and her offspring. Males only joined the family during mating. Family members usually stayed close to the matriarch for protection, but some individuals, especially juveniles, occasionally drifted off a short distance. They also lived within territories, and they followed a regular schedule while foraging. Thus Paleoindians knew where to find them and they knew their habits.

Stalking a stray individual worked best for hunting mammoths. Therefore, a small group of Clovis hunters usually followed the mammoth family. They stayed down wind to avoid being detected and waited for one of the younger animals to wander out of the matriarch's sight. They rarely killed a mammoth outright, so the hunters aimed their spears at its chest to pierce a lung. One hunter probably attracted the animal's attention so that the other hunters could approach the beast close enough to throw their spears. The wounded mammoth would wander away with the hunters close behind. Then it would either lie down to die after a few hours or days, or the hunters would herd it into an arroyo where it could not escape, and kill it with their spears. It took eight spears to kill one mammoth in Naco, Arizona. We know because the Clovis hunters left their spear points embedded in the fallen mammoth when they left. Clovis hunters used this particular strategy many times at Slick Creek in the Big Horn Basin of Wyoming about 11,200 years ago and probably at Lehner Ranch in Arizona as well.[10]

Whenever a small band of Clovis hunters killed a mammoth, they had more meat than they could use. So they had no reason to kill more than one animal at a time. The hunters had to work fast because the animal became more difficult to butcher as it cooled. Lions, bears, wolves, and other dangerous predators also challenged them as they processed the meat. Therefore, Clovis hunters probably ate some of the best parts immediately, dried enough meat to take with them, and, in the winter, left other parts frozen in caches for future use. They probably camped next to a fallen animal in winter, taking frozen meat from the cache as needed. Bison hunters on the Great Plains did the same thing thousands of years later.[11] When the rest of the herd moved on, however, the hunters moved with them.[12] So piles of mammoth bones at several Clovis kill sites show that the meat remained unused.[13] Just the same, they still could not use some of the meat anyway, especially if the mammoth fell on its side. Mammoths weighed too much for a few hunters to turn one over to butcher the opposite side, so they had to leave nearly half of the meat behind. Predators and birds ate the remainder.[14]

Clovis hunters often camped near springs or small streams on the

Great Plains and in the Southwest and then ambushed their prey when the animals needed water. Mammoths and mastodons also concentrated in the lush vegetation that grew in these wet areas as summer droughts became more common. Thus an open forest of brush and trees, well on its way toward becoming the dry brushland that exists today, most likely covered the area around Slick Creek in Wyoming when Clovis hunters lived there.[15]

Little remained of the white spruce forest farther south in Kansas by this time, and prairies already dominated southeastern Oklahoma.[16] An open woodland interspersed with grassy areas also grew in place of spruce and tamarack around a Clovis mastodon kill site south of St. Louis.[17] Likewise, desert scrub began replacing juniper woodland in New Mexico, Arizona, and southern Nevada during this period.[18] So mammoth and mastodon numbers started to drop as their habitat withered in the sun. The same thing happened during previous ice ages, but now they also faced the clovis hunters. At the same time Clovis hunting increased because people needed more food to support a growing population.[19] As a result, the wildlife and the land changed forever.

ICE AGE EXTINCTIONS

Mammoth and mastodon habitat shrank throughout the continent during the Clovis period. Summers became warmer, but the growing season shortened, and the differences between seasons grew more extreme than earlier in the Ice Age.[20] This caused many trees and other plant species to move away from one another and into more favorable environments where they formed uniform vegetation types. On the Great Plains, for example, limber pine climbed back into the Rocky Mountains as the white spruce forest shifted north, and then a dry prairie moved in to replace them.[21] Lodgepole pine also engulfed the tundra and spruce parklands where mammoths lived in the Pacific Northwest, followed closely by thick Pacific Douglas-fir forests.[22]

In the Southwest, Harrington's mountain goat became extinct about 10,000 years ago. Its habitat of sagebrush-juniper woodlands, pygmy pine woodlands of limber pine and bristlecone pine, and mixed-conifer forests fell apart around the Grand Canyon.[23] Wildlife diversity depends on habitat diversity.[24] Hence when vegetation became denser, and less patchy and diverse, it favored only those species that could use the new habitat while others declined.[25] The hunting of Ice Age animals by Clovis people also reached its peak about this time, and then it declined as well.

The open white spruce and jack pine forests in which mammoths and mastodons thrived had already broken apart in the Southeast when Clovis hunting peaked 11,200 years ago.[26] In Florida, oaks and pines also spread over the grasslands, and oak savannas replaced scrubby vegetation on the lower peninsula.[27] Spruce also pushed caribou and musk ox out as it over-ran the tundra and pressed against the base of the melting glaciers near the Canadian border. In addition, the southern side of the white spruce forest continued to shrivel in the heat.[28] So migrating hardwoods quickly replaced the dying southern side of the spruce forest and shaded out the remaining grasses and shrubs upon which the mammoths and mastodons depended for food. Thus spruce not only displaced most of the tundra, but the spruce forest also shrank, squeezing the habitat of the mammoth and mastodon into a narrow strip that could support far fewer animals than before.

The loss of habitat in a warming climate and increased hunting rapidly depleted the mammoths and mastodons. Thus the Clovis people had to spread out to find what remained of these hairy beasts. Clovis hunters also placed greater pressure on other game to make up for the loss. They killed camels, horses, and the large Ice Age bison throughout the West, as well as turtles and turkeys in Texas.[29] They pursued bison high into the Rocky Mountains where they dropped a Clovis point along the Yellowstone River in Gardiner, Montana, just outside the northern entrance to Yellowstone National Park.[30] Clovis hunters even found it necessary to add jackrabbits, gophers, rats, and birds to their diet in the arid highlands of Mexico.[31]

Mammoths, mastodons, and other big game of the Ice Age, and the carnivores that preyed on them, finally became extinct in North America between 10,800 and 10,000 years ago.[32] Large mammals accounted for 91 percent of the extinctions.[33] They probably died out in the middle of the continent first where hunting was most intense, holding on a little longer in Alaska, Mexico, and Florida.[34] Horses, camels, and ground sloths survived longer than the other species before they finally vanished.[35]

Intense hunting pressure, and the disintegration of Ice Age forests and other habitats, caused the mass extinction of animals.[36] The evidence is written in mastodon tusks, which grow in layers like tree rings. The growth rings in the tusks become narrow when an animal is under stress. This usually signals sexual maturity in males because they undergo the stress of being driven from the family and forced to live alone. Scientists studied the annual growth rings in the mastodon tusks and found that from 12,000 to 10,000 years ago the time it took for males to become sexually mature decreased by 3 years. Males become mature at a younger age when the population adjusts to intense hunting pressure. The mastodon males would have matured at an older age if a changing climate and the loss of habitat were the only causes of extinction.[37] Thus the Paleoindian population on the Great Plains crumbled as they depleted what remained of the big game, and the Clovis culture gradually disappeared with them.[38]

THE HOLOCENE

The extinction of Ice Age animals, the passing of the Clovis culture, and a rapid rise in temperature all mark the end of the Pleistocene. Thus the modern epoch—the Holocene or recent interglacial warm period—began 10,000 years ago.[39]

The loss of so many animal species at the end of the Pleistocene severely and permanently reduced the diversity of North American wildlife. However, it also cleared the way for the survivors to expand their ranges and increase their numbers.[40] Thus, the extinction of the stag moose allowed the modern moose population to grow in the north. Mule deer numbers also grew as they replaced the extinct mountain deer in the West, and the pronghorn antelope population exploded on the Great Plains. Even the brown bear started moving south to replace the extinct giant short-faced bear. A tiny population of condors managed to hold on along the California coast by scavenging dead sea lions and other marine mammals even though they faded away in the Southwest. Over much of the Great Plains grasses became short and sparse as the climate dried, but bison survived because they adapted to the change by shrinking to their modern size. Consequently, the giant Ice Age bison became extinct only because it mixed with a smaller northern form and finally evolved into the American bison about 5000 years ago.[41]

BISON HUNTERS

Like the bison, Paleoindians did not become extinct, only the Clovis culture. Without the giant game of the Ice Age, the ways of the Clovis hunters could no longer dominate the North American landscape. Conditions changed, so Clovis hunters had no choice but to look elsewhere for food. Their populations sprang back as they splintered into many local cultures that developed better ways of exploiting the plants and animals that grew where they lived.[42] Still, the hunting tradition prevailed on the Great Plains, the Columbia Plateau, and in the Southwest for the next 10,000 years even though the game and the techniques changed over time. Thus the Goshen culture that either lived with Clovis people or followed them, and the Folsom, Midland, Cody, and other cultures that came later, relied on bison and pronghorn antelope for much of their food, clothes, and other necessities.[43]

The American bison congregated into vast herds unlike the small groups in which giant Ice Age bison lived.[44] Thousands of years later, in 1680, Father Louis Hennepin, a Franciscan priest with Robert Cavelier de La Salle's expedition, described the bison herds on the tallgrass prairies of Illinois:

> The buffalo herds change country according to the season. When they approach the northern lands and feel the beginning of winter, they pass to the south. They follow one another in a herd that sometimes extends for three miles. Their path is beaten like our great roads of Europe, and no grass grows on it.[45]

We can barely imagine not only the sight of such a herd but also the sounds these huge creatures must have made. Lewis and Clark camped near such an enormous herd of bison in 1806. The bison, they said, "kept up a dreadful bellowing during the night." They estimated that there were 10,000 bison, and commented with obvious annoyance that, "they are bellowing in every direction, so as to make an almost continued roar."[46]

These huge herds made it possible for several bands of Paleoindians to work together to kill large numbers of bison at one time. Thus communal hunts became more common.[47] Nevertheless, Paleoindians still killed many bison by hunting them individually or in small groups in a manner similar to what Clovis hunters must have done.[48]

Perhaps ancient bison hunters fasted to provoke dreams to ensure a successful hunt, just as their descendants did thousands of years later. We cannot know, but their rituals may have been similar to those described by Jesuit Father Claude Allouez in 1666–1667, when he lived among the Indians of the south coast of Lake Superior:

> They fast in honor of their . . . spirits, to learn some future outcome. I have seen men, planning for war or the hunt, pass a whole week eating nothing. They show such purpose that they will not stop their fast until they see in a dream that which they desire.[49]

The larger herds of American bison also led to more wasteful hunting practices because, according to Father Hennepin, "the Indians of this vast continent cannot exterminate the buffalo, for these beasts multiply in great numbers." So in describing how the Sioux hunted bison on foot in 1680, he said:

> Those Indians sometimes sent their swiftest warriors by land to chase the buffalo toward the water. As the animals crossed the river, the Indians sometimes killed 40 or 50, merely to take the tongue and most delicate morsels.[50]

Such waste probably occurred more often when war parties traveled and needed food because the women stayed in the village. According to Father Antoine Silvy, a French Jesuit priest who lived among the Illinois from 1710 to 1712: "The men then take only the tongues and flat ribs. . . . It is the women who must gather the rest of the meat" he said.[51] Father Sebastian Rasles also reported waste among the Illinois in 1692. He noted that when they "kill a buffalo that appears too lean, they take only the tongue and go in search of one that is fatter."[52] In contrast, in 1608 Captain John Smith observed that Virginia Indians killed and ate everything. He said, "hares, partridges, turkeys, or eggs, fat or lean, young or old, they devour all they can catch in their power."[53]

Presumably, ancient bison hunters showed respect for their prey much like their Indian descendants did, even though they wasted some of the animals. They probably worshipped many of the same spirits as well. These may have included the ones recorded by Father Allouez:

> The savages of these regions recognize no sovereign master of heaven and earth, but believe there are many spirits: some good, like the sun, moon, lake, river and woods; others evil, like the snake, dragon, cold and storms. Whatever seems helpful or hurtful, they call a *manitou*. . . . They appeal to these spirits whenever they go hunting, fishing, to war, or on a journey, by offering sacrifices to them.[54]

Indians usually sacrificed tobacco and other items that Father Pierre-Jean De Smet described as, "the most precious objects which they possess and which they esteem most."[55] However, they also sacrificed animals, as Father Allouez noted: "During storms and tempests, they sacrifice a dog by throwing it into the lake. 'This is to calm you,' they say to the lake. 'Keep quiet.'"[56] Occasionally, as Father De Smet observed in 1840, "they [Aricaras and Grosventres] even cut off one or two fingers, which they offer as a sacrifice to the Great Spirit, that he may grant them scalps in the warfare that they are about to undertake."[57]

Paleoindian bison hunters, like modern Indians, may also have believed that animals possessed spirits that deserved respect. In 1670, Father Clause Dablon, a Jesuit priest, lived among the people of the upper Fox River, in east-central Wisconsin. He wrote that, "each beast, fish and bird has a particular spirit who cares for it, who preserves it and defends it from the evil which anyone would do to it." Furthermore, he said, "to certain spirits they render far more respect than others, because the animals are more useful," especially the bear.[58] However, their respect for animal spirits did not apply to just the largest creatures, but also the smallest. He illustrated this point in the following story:

> These people have a peculiar consideration for animals, as for a mouse we had thrown out of doors. A girl seized it and wanted to eat it, but her father first took the mouse and gave it a thousand caresses. We asked him why. He said, "It is to appease the spirit who cares for the mice, so he will not distress my daughter."[59]

American Indians also held feasts as thanks for a successful hunt. Ancient bison hunters may have done the same. Father Allouez noted that during a feast, "a leading old man of the village acts as priest, haranguing the sun."

> In a loud voice he declares his thanks to the sun for lighting his way, to help him kill some animal. He exhorts the sun by this feast to continue its kind care of his family. During his plea, all the guests eat to the last morsel. Then a man breaks a cake of tobacco and throws it into the fire. Everyone cries aloud as the tobacco burns and smoke rises aloft. With these cries the sacrifice ends.[60]

These feasts were absolutely essential because, as Father Allouez said, "they believe that the most common cause of illness is failure to give a feast after success in fishing or hunting; that the sun, who takes pleasure in feasts, is angry with whoever fails in this duty, and so will make him ill."[61]

Post-Clovis hunters on the Great Plains invented many ingenious ways to hunt bison. Paleoindians understood their game well and could take advantage of the animals' behavior and use it against them. For instance, they knew that the split-hoof of the bison made it difficult for them to run in sand. So Paleoindians used the sand dunes that formed in the northwestern Great Plains during the Ice Age as animal traps. They herded small groups of bison into the soft sand between U-shaped dunes. Here they could kill the trapped animals more easily as they struggled to escape up the sandy slopes. In other areas they used shallow arroyos as bison traps. They herded the bison into the upper end of the arroyos where dirt cliffs blocked their escape as effectively as sand.[62]

The bison hunters also drove herds over cliffs and into sturdy corrals made of logs that they hauled into the grasslands. In the late prehistoric period, and probably much earlier, Plains Indians used stone tools to cut juniper and other trees to build their fences and corrals. They used antlers and bones to dig pairs of closely spaced holes in which to set posts. Then they stacked logs between the posts, just as some ranchers build corrals today. Having trapped the animals, the hunters could easily kill their game by hurling spears and darts into the swirling mass of frightened animals. The struggling bison usually tore up parts of the corral, which were repaired and used again.[63] Paleoindians also built fences and corrals to trap herds of pronghorns. One such trap near Medicine Hat in southern Alberta, Canada, dates back over 3000 years.[64]

Bison hunters concentrated on small herds of females and young animals, usually avoiding mature males that could more easily threaten them.[65] However, in the spring the hunters sometimes selected males because they often came through the winter in better condition than the females.[66] Occasionally they succeeded in killing 200 or more bison at one time. Such large kills precluded the option of selecting the sex and age of the animals.[67]

About 1854, Father De Smet witnessed a communal bison hunt as it must have been thousands of years earlier.[68] According to his description, the Flathead and Pend d'Oreille Indians used a trap to kill 600 animals on the northern plains. They established a camp of about 300 lodges containing perhaps 2000 people about 15 miles from the bison herd they were preparing to hunt. A band of warriors kept everyone in camp so that they would not alert the bison and, as Father DeSmet notes "The law against this is extremely severe."

Father De Smet says that the Indians then constructed a circular pen about an acre in size at which "everybody labors ... with cheerful ardor, because it is an affair of common interest." They cut many of the trees along nearby streams to build the pen. Then he says "stakes are firmly fixed in the ground, and the distance between them filled with logs, dry boughs, [and] masses of stone." They designed and located the pen in an arroyo just as their Paleoindian ancestors would have done. Wrote Father De Smet:

> The circular palisade has but one opening; before this opening is a slope embracing fifteen or twenty feet between the hills: this inclined plane grows wider as it diverges from the circle; at its two sides they continue the fence to a long distance on the plain.

Then, according to Father De Smet, the Indians elect "a distinguished personage ... belonging to the Wah-kon, or medicine band" who decides "the moment for driving the bisons into the inclosure." He "has four runners at his disposal, who go out daily and report to him ...; they tell at what distance from the camp the animals are, their probable number, and in what direction the herd is marching." "These runners frequently go forty or fifty miles in different directions," wrote Father De Smet. The hunters "often remain a month or more awaiting the most favorable moment." Finally, when the wind blows from behind the herd and the bison are in the right place, the Wah-kon gives the signal and the hunt begins. Then, Father De Smet wrote:

> Immediately all the horsemen mount their coursers; the foot soldiers ... forming two long ... diverging rows, from the extremity of the two barriers which spring from the entrance of the pen ... the horsemen continue the same lines ... so that the last hunter on horseback is found at about two or three miles distance from the pen ... one single Indian, unarmed, is sent upon the best courser ... in the direction of the buffalo.... He approaches against the wind and ... he envelops himself in a buffalo hide ... and also envelops his horse ... and then makes a plaintive cry in imitation ... of a bison calf. As if by enchantment, this cry attracts the attention of the whole herd ... [and] they push forward in full gallop.... When the buffalo arrive in the space between the extremities of the two lines, the scene changes.... At once the scent of the hunters is communicated among the frightened and routed animals, which attempt to escape in every direction. Then those on foot appear. The bisons surrounded ... except the single opening into the

circular pen before them, blow and bellow in the most frightful manner, and plunge into it with the speed of fear and desperation. Then the Indians commence firing their guns, drawing their arrows and flinging their lances. Many animals fall under the blows before gaining the pen: the greater number, however, enter ... amid the hurrahs and joyful shouts of the whole tribe, intermingled with the firing of guns.

"As soon as all are penned," Father De Smet recalled,

the buffalo are killed with arrows, lances and knives. Men, women and children, in an excitement of joy, take part in the general butchery and the flaying and cutting up of the animals. While men cut and slash the flesh, the women, and children in particular, devour the meat still warm with life—the livers, kidneys, brains, etc., seems irresistible attractions; they smear their faces, hair, arms and legs with the blood of the bisons; confused cries, clamorous shouts, and here and there quarrels, fill up the scene. After the butchery, the skins and the flesh are separated into piles, and these piles are divided among the families, in proportion to the number of which they are composed. The dogs also receive their portion of the feast, and devour the remains on the arena of the pen.

In 1797, Peter Fidler, a Hudson Bay Company trader who lived among the Piegan Indians on the Canadian Plains, described the smelly aftermath of such a bison drive. Shortly after he watched the hunters gather 250 bison into a pen and slaughter them, he said that "when the wind happened to blow from the pound in the direction of the tents, there was an intolerable stench of the large numbers of petrified carcasses, etc. on which account the reason of our leaving it."[69]

Large bison kills happened more often when the hunters drove big herds over bluffs or cliffs so that they would crush each other when they fell. Small herds of less than 100 bison could easily turn away from a cliff during a stampede. However, bigger herds formed more compact masses that could not turn, and the fleeing beasts just pushed each other over the cliffs. So this hunting method produced large kills mainly due to the number of animals needed to make it work. As a result, the hunters wasted much of the meat because the animals piled on top of each other and they could not reach the bison at the bottom of the heap.[70] Lewis and Clark floated by such a buffalo jump during their travels up the Missouri River in the late spring of 1805 and recorded such a scene:

On the north we passed a precipice about 120 feet high, under which lay scattered the fragments of at least 100 carcasses of buffaloes, although the water which had washed away the lower part of the hill must have carried off many of the dead. These buffaloes had been chased down the precipice in a way very common on the Missouri, by which vast herds are destroyed in a moment ... the herd is left on the brink of the precipice. It is then in vain for the foremost buffaloes to retreat or even to stop; they are pressed on by the hindmost rank, which, seeing no danger but from the hunters, goad on those before them till the whole are precipitated, and the shore is strewn with their dead bodies. The Indians then select as much meat as they wish; the rest is abandoned to the wolves, and creates a most dreadful stench.[71]

A band of Indians often had many jump sites located within their hunting territory in case a bison herd wandered nearby. Therefore, even though they might not use the same jump site every year, they still used it many times over hundreds or thousands of years. We know, for instance, that they repeatedly carried out mass kills at the Head-Smashed-In buffalo jump in the Canadian prairies, near the Montana border, for at least 7000 years.[72]

HUNTER-GATHERERS

Unlike Paleoindians living in the Great Plains, the Columbia Plateau, and the Southwest, those who lived elsewhere had few bison to hunt after the mass extinctions at the end of Ice Age. Fortunately, they could substitute a variety of smaller game, fish, and plant foods for the big game they used to hunt. So they began to diversify their food sources as the mammoths and mastodons dwindled away.

In Pennsylvania, Paleoindians caught fish, and they ate the seeds, nuts, and fruits of such plants as hackberry, wild plum, grape, blackberry, ground cherry, and goosefoot as early as 10,800 years ago.[73] They thrived in southern California and Florida by taking advantage of the abundance of fish, seals, and shellfish that lived along now-submerged coastal shorelines.[74] Paleoindians continued to prosper along the coast, even as the seas rose, by simply following the shoreline inland until it stabilized at the present level about 5000 years ago.[75] They also lived near the marshes along the edges of what remained of Ice Age lakes in the Great Basin. Here they found water birds, small rodents, rabbits, and hares to eat, and they gathered cattail and grass seeds, and the fleshy root stalk and seeds of the bulrush. They supplemented their diet by harvesting pickleweed seeds from the edge of salt pans and ground them into flour.[76] In Oregon, Paleoindians also fished for salmon at the Dalles of the Columbia River as early as 10,000 years ago.[77]

In the conifer forests of the Great Lakes and New England, Paleoindians continued to hunt big game, but they shifted to moose and woodland caribou instead of mammoths, mastodons, giant bison, and barrenground caribou. Farther south, white-tailed deer became the dominant game animal in the rapidly expanding broad-leaved forests.[78] The game they used to hunt roamed through tundra and open forests in herds. However, deer, moose, and woodland caribou behave in a more solitary manner that better fits the dense forests in which they live. In eastern forests today, and probably back then, deer only numbered about 10 per square mile.[79] So Paleoindians improved their spear points and revised their stalking tactics.

Calling, disguises, and decoying probably became more important during these early times since American Indians polished such methods to near perfection by the time Europeans arrived. Northeastern Indians,[80] such as the Ojibways of Ontario, Canada, made calls that mimicked a bull moose during the mating season, the bleat of a fawn, which attracted does, and the grunt of a caribou. Each call could bring animals close enough to a concealed hunter for a kill.[81] The Delaware (they called themselves "Real Men"[82]) also made a tiny 3-inch-long whistle that imitated the bleat of a fawn. It worked as well to attract deer for Indians on the Great Plains as it did for people living in New England's forests.[83] Southeastern Indians, such as the Caddoes (or "True Chiefs") of East Texas, like many other tribes, also disguised themselves with the antlers and hide of a deer so that they could approach their quarry close enough to shoot.[84]

In marshlands, American Indians used decoys as well designed as anything made today to attract waterfowl. For instance, scientists digging in Lovelock Cave in the Great Basin, near the Humboldt River in western Nevada, found lifelike duck decoys made of painted tule, and others covered with complete duck skins, including feathers. Some of these surviving decoys may be 4500 years old.[85] Even Plains Indians knew that feathers attached to the top of a spear stuck in the ground could easily bring a curious pronghorn antelope within shooting distance when they fluttered in the wind.[86] Apache scouts in Arizona taught soldiers how to use their handkerchiefs to decoy pronghorn in the same way.[87] California Indians used a similar technique.[88] Paleoindians must also have learned such calling and decoying skills and many others or they could not have survived.

DECLINE AND RETURN OF THE BISON

As the climate grew warmer and drier during the early Holocene, Paleo-indians had to continue to learn about their changing environment and develop new ways of getting food. By 9000 years ago summers were warmer and winters were colder than today.[89] Rainfall decreased dramatically, and the Great Plains became drier as the Great Drought spread from west to east across the prairies.[90] Many meadows went dry in the Sierra Nevada of California and lake levels dropped throughout the West.[91] The drought also robbed the Great Basin of much of its moisture and Ice Age Lake Bonneville and other basin lakes disappeared. Salts crystallized and settled as water evaporated from basin lakes until nothing but dry salt flats remained.[92] This drought lasted until about 5000 years ago.[93]

Drought meant disaster for the bison hunters, but people thrived east and west of the Great Plains. Oaks expanded over the middle slopes of the Sierra Nevada and throughout the lowlands of California, which provided more acorns for deer and other wildlife.[94] However, thick forests of Douglas-fir, western hemlock, and alder also spread over vast areas in the Pacific Northwest during this time, which greatly reduced game habitat. Even so, the more productive oak woodlands and prairies of the Willamette Valley spread northward as well and replaced some of the game lost to expanding forests.[95] In the East, expansion northward of nut-bearing trees such as oak, hickory, beech, and chestnut also provided a bountiful supply of food for people and wildlife. This further increased the availability of game. Oak and hickory also dominated forests in the Southeast, so food became more plentiful there as well.[96]

In addition, water levels increased in the Great Lakes and elsewhere, and water temperatures rose, which improved fisheries. The glaciers had already retreated into Canada by this time, so the rivers no longer carried massive amounts of meltwater. As a result, rich wetlands developed in the backwaters of rivers as they became more sluggish. Swamps and marshes also spread on the coastal plains of Georgia and Florida as the rising seas crept inland, which enlarged the habitat of water birds and shellfish.[97] These must have been times of plenty for many people living below the dwindling ice sheets, but others suffered.

The Great Plains became less hospitable to the bison hunters even though the grasslands expanded. The heat forced ponderosa pine off the western plains, which increased the habitat for grazing animals.[98] Prairies and oak savannas also replaced the pine forests of northwestern Minnesota during the drought.[99] Prairies also spread into central Illinois and western Missouri as the eastern woodlands withdrew; so bison herds and other Plains species moved eastward with the grasslands.[100] Nevertheless, the grasses became sparse in the parched landscape and bison populations decreased. Intense hunting pressure further reduced their numbers until the Paleoindian population itself started to fall.[101]

About 8000 years ago bison hunting declined dramatically on the Great Plains and in the Southwest, although hunting continued on the slopes of the Black Hills of South Dakota and Wyoming where lush grasslands remained.[102] Forest game, such as white-tailed deer, that lived on the eastern edge of the plains, and in the Ozarks of Missouri, also declined as the oak-hickory forest shrank eastward. Many people had to abandon the plains and move eastward as well, even though they returned long enough to hunt Plains animals that had also moved that way.[103] In the Southwest, Clovis hunters had lived as far west as southern California before the collapse of the Ice Age game populations. The later bison-hunting cultures continued to contract as the drying trend persisted, falling back into central Arizona, then western New Mexico, and then, finally, eastern New Mexico.[104] By then a nearly modern desert scrub vegetation replaced the juniper woodlands that used to cover much of the Southwest.[1105] Between 7500 and 4500 years ago bison abandoned the Texas High Plains completely.[106]

Some bison hunters moved off the scorched plains and into the cooler mountains nearby where game still thrived. They probed as high as 10,000 feet in the Rocky Mountains before the drought, but now their explorations for game took on a new urgency.[107] Here they found mule deer, mountain sheep, marmots, wood rats, rabbits, and an occasional bison to hunt. They also found plant foods such as roots, tubers, seeds, and berries that added even more diversity to their diet than was available on the Plains. They may also have eaten limber pine nuts since the grinding stones they left behind cluster near such trees.[108] Regardless, they continued to rely on their spears and darts to hunt deer, even though they found snares and clubs more useful for hunting sheep and smaller animals.

Paleoindians also used nets made of twisted juniper bark to trap small herds of mountain sheep as early as 8800 years ago. Since mountain sheep run uphill to escape predators, they stored the nets near sheep trails above the areas where the animals normally congregated. It only took a few people to hold the net while others simply drove the sheep uphill and into the trap. Then they killed the sheep with wooden and antler clubs once they became entangled in the net.[109]

They most likely built elaborate log fences and corrals to trap bigger herds of mountain sheep, just as their descendants did in later times. Paleoindians already knew how to cut the trees and build these structures because they did the same thing on the Plains for bison. It also took many live and fallen dead trees to build one trap, but they had plenty of trees available in the mountains. Even today a large number of these log structures still lay scattered over the Rocky Mountains, which shows that 8000 years or more of building traps must have kept some forests well thinned.

The sheep traps built by Indians in later times probably followed the same basic plan as those built by their Paleoindian ancestors. Often these sophisticated traps included a log ramp disguised with a covering of dirt and rock to make it look like the surrounding landscape. The vertical end of the ramp faced uphill and dropped into a corral made of logs. The Indians took advantage of the sheep's instinct to run uphill and guided them up the ramp so that they would drop into the corral hidden on the other side. The sides of the corral tipped inward to keep the sheep from climbing over the top, which they could do easily if the fence stood straight. Having trapped the animals, they killed the sheep with clubs in the same way they did when using nets.[110]

With plenty of food available, many people remained in the mountains until the climate began to cool again about 5000 years ago. Then rainfall increased and the Great Plains came back to life. As a result, the eastern edge of the prairie retreated westward a little as white pine and broad-leaved trees moved in from the East and shaded out the grass.[111] Nevertheless, the grass thickened over the rest of the Plains and too few hunters remained to keep the bison and pronghorn herds from growing to enormous size.[112] So the Indians came down from the mountains to begin the legendary period of the modern American bison and the colorful hunters of the Great Plains.

To be sure, the Plains Indians still used the same tactics to hunt bison that their Paleoindian ancestors devised many thousands of years before.

The spread of Spanish horses across the West between 1650 and 1730, and the arrival of the bow and arrow a thousand years earlier, did not change that.[113] However, it added the more picturesque way to hunt bison that most people associate with the Plains Indian, even though they were still using the old methods when Europeans arrived. Trapper Osborne Russell described such a classic hunt when traveling along the "great plain of the Snake River" in 1835:

> I lay all day and watched the Buffaloe which were feeding in immense bands all about me. I watched the motions of the dust for a few minutes when I saw a body of men on horse back pouring out of the defile among the Buffaloe. In a few minutes the dust raised to the heavens. The whole mass of Buffaloe became agitated producing a sound resembling distant thunder. At length an Indian pursued a Cow close to me running along side of her he let slip an arrow and she fell. I immediately recognized him to be a Bonnak with whom I was acquainted. He told me the village would come and encamp where I was. In the meantime he pulled off some of his Clothing and hung it on a Stick as a signal for the place where his squaw should set his lodge he then said he had killed three fat cows but would kill one more and stop. So saying he wheeled his foaming charger and the next moment disappeared in the cloud of dust. While the squaws were putting up and stretching their lodges I walked out with the Chief on to a small hillock to view the field of slaughter the cloud of dust had passed away and the prairie was covered with the slain. Upwards of a Thousand Cows were killed. . . .[114]

The cooler and wetter climate that followed the Great Drought had a less beneficial effect on the hunter-gatherer life in the East than on the Plains. Pine trees started to invade the hardwood forests, so oak and hickory became less abundant. This reduced the amount of food available to both game and people because nut production dropped.[115] Nevertheless, they continued to use acorns and nuts until the time when Europeans arrived. As the naturalist William Bartram, who traveled through the Southeast from 1773 to 1778, noted: "It [live oak] bears a prodigious quantity of fruit; the acorn is small, but sweet and agreeable to the taste when roasted, and is food for almost all animals. The Indians obtain from it a sweet oil, which they use in the cooking. . . ; and they also roast it in hot embers, eating it as we do chestnuts [sic]."[116] Indians throughout the East boiled white oak acorns as well, which also produced a clear, sweet oil that they used like butter.[117]

SETTLEMENT AND THE SEASONAL ROUND

The Great Drought forced most people to concentrate in river valleys, around lakes, and sink holes, and on the coastal plains during much of the year.[118] By this time bands of Paleoindians already occupied the best hunting grounds in the eastern woodlands and probably in the West as well, and their populations increased. So they could no longer wander unchallenged to search for game and plants. Since they could not move freely, the Paleoindians settled into territories.

On the south Atlantic slope in North and South Carolina and Georgia, each band occupied a major river drainage, and ridgetops served as the principal boundaries. Other landscape features undoubtedly served as boundaries as well, such as those described in the 1630s by Roger Williams in New England. Williams said that "the natives are very exact and punctual in the bounds of their lands, belonging to this or that prince or people, even to a river, brook, etc."[119] Their territories on the south Atlantic slope began along the coast and extended into the Appalachian Mountains, covering areas over 200 miles long and 60 miles wide. Thus each band could forage within a variety of habitats, including coastal shorelines and marshes, rivers and valleys, prairies, and the surrounding forested uplands.[120] Even so, they had to intensify their use of resources because of the limited size of their territories.[121] Little had changed by the time Captain John Smith wrote his journals centuries later, even after the introduction of agriculture:

> In their hunting and fishing they take extreme pains. . . . And by their continual ranging, and travel, they know all the advantages and places most frequented with deer, beasts, fish, fowl, roots, and berries.[122]

Thus, Paleoindians still lived a seminomadic life within their territories because foraging, like hunting, requires moving around as the seasons change.

Each band returned to a residential base camp or settlement during the winter as food declined elsewhere in their territory.[123] The rest of the year they moved from place to place. Thus Paleoindians followed a seasonal schedule dictated by ripening plants, the behavior of game, the need to quarry fine-grained rock and make new stone tools, and to exchange mates and trade with neighboring bands.[124] Windy Ridge, located at 9300 feet in the Colorado Rockies, served as one such seasonal camp that Paleoindians and their descendants used for at least 8000 years. Here they found a rich vein of high-quality quartzite and abundant game in the nearby prairie. So they stayed there during the summer to chip out the quartzite, make stone tools and weapons, and hunt. They did not abandon this quartzite mine until Europeans arrived with metal tools and points.[125]

In the winter, eastern Paleoindians who lived inland usually set up base camps in river bottoms. They trapped waterfowl migrating south, and they hunted deer that moved into the valleys during the breeding season. They also supplemented their diet with black bear which, like deer, are in prime condition during early winter. In the spring they might collect box turtles and move to a river or lake to spear, harpoon, or net fish. They even poisoned fish with powdered hickory nut husks. The Delaware Indians later used ground chestnuts to poison fish because they said it made them dizzy, while California Indians used crushed buckeye nuts for the same purpose. In the summer eastern Paleoindians collected freshwater mussels and snails, and harvested seeds and berries. In the fall they would move into the forest to harvest hickory nuts, beech nuts, and chestnuts. They returned to their settlements in the river bottoms when winter came, stored their nuts and other foods, and began the cycle again.[126]

Indians living in the Midwest, East, and Southwest continued the roving life of hunter-gatherers for thousands of years until they adopted agriculture. Even then the Missisquois Abnakis (or "Those Living at the Sunrise," easterners) who farmed around Lake Champlain still followed a hunter-gatherer tradition during part of the year by the time Europeans arrived. As the Swedish traveler Peter Kalm noted in 1749 during his trip to North America:

> At one time of the year they live on the small store of corn, beans, and melons, which they have planted; during another period, or about this time, their food is fish, without bread or any other meat; and another season they eat nothing but game, such as stags, roes, beavers, etc., which they shoot in the woods and rivers.[127]

Indians on the edge of the tallgrass prairie lived a similar life, mixing agriculture with foraging. Here they hunted in the winter so that they could plant their crops in the summer, as Father Antoine Silvy recorded in 1710:

> In winter, the Illinois leave their villages to hunt for buffalo, elk, deer, beaver and bear. Toward the end of April they return to their village to plant crops.... They remain there all summer, making only short hunting trips from time to time.[128]

People living in the Great Basin, the desert Southwest, south Texas, California, and the Pacific Northwest still depended on foraging when Spanish explorers landed in the South.[129] Foraging served people well in these environments; so agriculture probably seemed unnecessary.

The dry conditions in the desert Southwest, the Great Basin, and south Texas made agriculture difficult in most places, even in historic times. The Coahuiltecans, hunter-gatherer people who lived in the scrubby, semidesert mesquite brushlands of south Texas, did not practice agriculture. They did have a few gourd rattles that they used on important occasions. However, they found the gourds along the shore of the Pecos River where they washed up after floods. The Coahuiltecans thought they came from heaven. They had no idea that the gourds had floated down the river from Anasazi pueblos in New Mexico.[130]

Long droughts and the patchiness of the vegetation in south Texas made it hard to depend on any single source of food, especially agriculture. Survival required mobility, flexibility, skill, and an intimate knowledge of the seasonal abundance of game and plants.[131] The Coahuiltecans, a culture of many small bands, epitomize such survival skills. The Spanish explorer Alvar Nuñez Cabeza de Vaca lived among these people. Cabeza de Vaca was a member of the Pámfilo de Narvaez expedition. He was among 300 men that were left behind by the fleet near the entrance to Tampa Bay, on the west coast of Florida, in 1528. They explored the interior but failed to find the fleet, which searched for them for a year before returning to Mexico. It took Cabeza de Vaca 8 years of suffering and hunger to find his way from Florida to Mexico. He was one of only four members of the Narvaez expedition who survived.[132]

Cabeza de Vaca described what he had to eat when living among the Coahuiltecans:

> Occasionally they kill deer, and at times take fish; but the quantity is so small and the famine so great, that they eat spiders, and the eggs of ants, worms, lizards, salamanders, and vipers that kill whom they strike; and they eat earth and wood, and all that there is, the dung of deer and other things

that I omit to mention. They save the bones of the fishes they consume, of snakes and other animals, that they may afterwards beat them together and eat the powder.[133]

They also roasted fish without cleaning them, let them sit for 8 days, and then ate them rotten so that the fish contained added protein from fly larvae. However, such foods represented only part of a diet that also included an occasional bison that strayed too far south.[134] The Coahuiltecans relied primarily on the bulbs of the agave or century plant, which they roasted and ground into flour, mesquite beans, and the fruit or "tuna" of the prickly pear cactus that grew near San Antonio.[135] The prickly pear, Cabeza de Vaca said, "is the size of a hen's egg, vermilion and black in color, and of agreeable flavor." Like many other seasonal foods eaten by Indians, he noted that, "the natives live on it three months in the year, having nothing beside."[136]

Hunger plagued the Coahuiltecans constantly, as it did many hunter-gatherer people. Father De Smet, when traveling through the northern Rocky Mountains in 1842, wrote: "The Indian when he has nothing to eat does not complain, but in the midst of abundance he knows no moderation."[137] Likewise, Cabeza de Vaca wrote centuries earlier that, "it occurred to us many times while we were among this people, and there was no food, to be three or four days without eating, when they, to revive our spirits, would tell us not to be sad, that soon there would be prickly pears." However, their plight continued for "five or six months." Even so, he said: "They are a merry people considering the hunger they suffer."[138]

Cabeza de Vaca further praised the Coahuiltecans by saying "these people see and hear better, and have keener senses than any other in the world." "They are great in hunger, thirst, and cold," he said, "as if they were made for the endurance of these more than other men."[139] His amazement at their endurance shows clearly in his description of deer hunting:

These Indians are so accustomed to running, that without rest or fatigue they follow a deer from morning to night. In this way they kill many. They pursue them until tired down, and sometimes overtake them in the race.[140]

Nevertheless, the Coahuiltecan's culture, like that of the hunter-gatherers of the Great Basin, remained static for thousands of years because they were so busy supplying their minimum needs in a harsh land.

Prehistoric hunter-gatherers who lived farther west fared somewhat better than the Coahuiltecans because they had piñon-juniper woodlands, lakes, and other places to go for food. In northeastern Arizona they could use 80 percent of the plant species common to piñon-juniper woodlands for medicine, raw materials, and food, especially piñon nuts.[141] Piñon nuts helped to sustain people in the Southwest and Great Basin for thousands of years, and they still remain an important part of the diet of Southwestern Indians today. However, these people could not survive by relying on piñon nuts alone.

Piñon-pines mast, which means they produce large cone crops in the same year in widely separated areas. Then it takes several years to produce another large cone crop. Some scientists believe that masting keeps seed-eating predators, such as cone beetles and moths, from preventing trees from reproducing. Thus in poor seed years the predator population remains small, while in a heavy seed year there are too few predators to eat all the seed, so some escape and produce new trees.[142] Most likely this coordinated masting behavior also occurs because Southwestern Indians carried piñon nuts from masting trees to distant places through a trade network that criss-crossed the Southwest.[143] Indians even spread piñon pine far to the north. The Cheyenne and Arapaho may have dropped the nuts accidentally in Owl Canyon, north of Fort Collins, Colorado. An isolated grove of piñon pines grows here, and there is no other explanation for their presence since the closest grove is 155 miles farther south.[144] Therefore, the spread of piñon pines to places that nature could not reach, and their masting behavior, probably resulted from thousands of years of gathering and trade by Southwestern Indians.

Southwestern Indians welcomed the bumper crops of piñon nuts that came every few years, but the abundance of nuts brought a mixed blessing. The number of rodents such as pack rats increases dramatically during mast years, so people become more exposed to the plague and other life-threatening diseases. Their exposure increased even more when they removed piñon nuts from pack rat dens. Not surprisingly, even today the Zuni are aware that the death rate goes up during mast years.[145] We will never know, but this relationship between abundance and disease could go back many thousands of years.

Indians living in California and the Pacific Northwest had even a greater abundance of food; especially fish, shellfish, sea mammals, large game, and edible plants. Even so, people can starve in a land of plenty if conditions change more rapidly than knowledge. For example, coastal populations of California Indians dropped dramatically about 7000 years ago. During that time the rising sea drowned the forests and grasslands on the coastal plain where they lived, and the oceans became warmer. Kelp beds near the retreating shoreline faded away, probably shifting to the cooler waters around islands and in shallow areas farther out to sea. Fish and sea mammals that lived among the kelp most likely moved away from the mainland as well. Coastal Indians must have found it difficult to find enough fish and sea mammals to feed their people, so their population crashed.[146]

California Indians living along the coast had not yet learned how to process large numbers of acorns that were readily available in the oak woodlands around them. In all likelihood, they still ate the acorns one by one after burying them in mud for up to a year to leach out the tannic acid.[147] They also let the shelled acorns sit in a basket until moldy. Then they buried them in sand along a stream until the acorns turned black and were ready to eat whole. However, grinding stones appeared in California about 6000 years ago even though they first appeared at least 3000 years earlier in the Great Plains and the Great Basin. So the Indians most likely learned by then how to pound the acorns into meal and leach the tannic acid themselves. They heated the acorn meal in watertight baskets with hot rocks to make soup. This new source of food might have helped the California Indian population to rebound along the coast 6000 years ago. However, the ocean cooled and kelp returned with its abundance of fish and sea mammals around the same time.[148]

Many species of oak grow in California's oak woodlands and forests, but the Indians preferred the acorns from seven species.[149] They especially favored tanoak in the Coast Range, black oak in the Sierra Nevada, blue oak in the foothills, and valley oak in the lowlands. Often they traveled long distances for their favored species. Acorn crops cycle between abundance and scarcity, sometimes over large areas, so their favorite acorns were not always available. However, they could usually fall back on less desirable species during these lean years.[150] Hence, for thousands of years acorns remained a staple in the diet of most California Indians, and they continued to be important well after the arrival of Europeans.

HARVESTING THE FOREST
Nourishment and Healing

Trees not only furnished acorns and nuts during seasons of plenty, they also nourished the starving and healed the sick. Indians ate the woody parts of trees when they could not store enough food for the lean times and they had to live on what they could find. In the winter of 1660, Father Rene Menard, who lived among the Ottawa (which means "to trade") on the south coast of Lake Superior, said that they were forced to eat "the bark of oak, birch and basswood, well dried, pounded and mixed with fish oil." The aged and weak father survived the winter on such fare. However, he disappeared the following year near the headwaters of Wisconsin's Chippewa River while trying to reach a band of poor Hurons who were running from the Iroquois.[151] In 1805, Lewis and Clark also recorded that the pine trees around an abandoned Indian camp in Montana "had been stripped of their bark ... which our Indian woman says her countrymen do in order to obtain the sap and the soft parts of the wood and bark for food."[152]

Other tribes such as the Cree of east-central Saskatchewan also used the inner bark of trees for food, especially jack pine, and the Assiniboines and others made "bark cakes" from cedar and hemlock.[153] In July 1793, Alexander Mackenzie noted that "many hemlock trees" were "stripped of their bark" by Northwest coastal Indians to make their "cakes." They soaked the dried cakes in water and then doused them with "salmon oil" before eating them. "This dish is considered by these people as a great delicacy," wrote Mackenzie.[154]

Many drugs also came from trees. The Algonquin chewed the sap of white spruce as a laxative.[155] The Iroquois chewed pitch pine sap, and California Indians chewed sugar pine sap for the same purpose.[156] The Iroquois (or "We of the Extended Lodge") also made a laxative drink by boiling the bark of black ash.[157] Similarly, the Crow (or "Bird People") made a laxative tea from the crushed needles of subalpine fir, and the Algonquin did the same with balsam fir needles.[158] Boiled hackberry bark soothed sore throats for the Houmas of Louisiana.[159] Pacific Northwest coastal Indians chewed Douglas-fir sap for sore throats and the California Maidu chewed blue oak leaves.[160] The Flatheads in Montana also applied a poultice to burns made of heated lodgepole pine sap and bone marrow, and the Iroquois used boiled beech leaves to dress their burns and scalds.[161] Trees also provided drops for earaches and remedies for coughs, heartburn, toothaches, and headaches.[162] Then as now, however, trees served mainly as a construction material for dwellings and boats.

Temporary and Portable Shelter

Most American Indians framed their homes with logs or branches and then covered them with earth, grass, hides, matting, brush, or bark. They also used stone, adobe, daub and wattle, and even hand-hewn boards.[163] For example, the Mandans lived in circular, earth-covered lodges supported by large timbers. The floor sat 2 feet below the surface of the ground, which was hardened from use and swept clean of dirt.[164] Father De Smet described a similar Omaha earth lodge during his voyage up the Missouri River in 1839:

> The huts of the Omahas are built of earth, and are conical; their circumference at the base, 120 to 140 feet. To construct them, they plant in the ground long and thick poles, bend and join together all the ends, which are fastened to about twenty posts in the inside. These poles are afterwards covered with bark, over which they put earth about a foot in depth, and then cover the whole with turf. They look like small mounds. A large hole in the summit permits light to enter and smoke to escape. The fire-place is in the center and every hut holds from six to ten families.[165]

Other tribes simply piled up willow branches for shelter or built tepees entirely of poles.[166] On the other hand, the Coahuiltecans of south Texas had a small but sophisticated home that rivals modern backpack tents in its simplicity. "Their houses are of matting, placed upon four hoops," Cabeza de Vaca says, and they carry them "on the back, and remove every two or three days in search of food."[167] In contrast, Plains Indians used dogs and then Spanish horses to carry their large hide-covered tepees from place to place.

Pedro de Castaneda mentioned these dogs in his chronicle of the expedition of Coronado (1540–1542) while traveling on the Staked Plains of Texas. He said that the Plains Indians "travel like the Arabs, with their tents and troops of dogs loaded with poles and having Moorish pack-saddles with girths." The Indians trained the dogs, but the dogs also trained their masters. Castaneda noted that, "when the load gets disarranged, the dogs howl, calling some one to fix them right."[168] The Indians he described probably used juniper for their tepee poles, but most of the northern tribes such as the Cheyenne (a Sioux word for "red talkers," which means "people speaking language not understood"[169]), Dakota, Blackfoot, Crow, and the Paiute or Snake of the northern Great Basin preferred lodgepole pine.[170] On the other hand, the Cree, who lived farther north in southern Canada, used paper birch for their tepee poles instead of lodgepole pine, but they preferred black spruce.[171]

The Northern Plains Indians traveled high into the Rocky Mountains to cut close-growing and spindly, or small pole-size, lodgepole pine to frame their tepees and then hauled them down to the valleys and treeless plains.[172] They must have been cutting and hauling trees for at least 10,000 years because we know that some Paleoindian hunters also lived in tepees during Folsom times.[173] Certainly the number of trees cut each year for these dwellings was small by comparison to the size of the lodgepole pine forests. However, such logging still could have kept some forests thinned and patchy when practiced consistently in the same areas over many millennia.

Many other tribes carried the mat or skin coverings for their houses on their backs, but not the frames. They cut the trees needed to frame their tepees and wigwams in the forest where they stopped. For instance, in the Great Lakes region the Ojibwa and Potawatomi lived in oval-shaped wigwams built from hornbeam or birch poles that they cut locally and then overlaid with bark, mats made of sewn cattails or other plants or skins.[174] Father Hennepin described such houses in an Illinois village in 1680. He said: "It contained 460 lodges built like long arbors, covered with double mats of flat reeds, so well sewn that they kept out wind, rain, and snow." These lodges could be large, as Father Hennepin notes for the Illinois: "Each lodge had four or five fires, and each fire had one or two families."[175]

Still other tribes left their houses behind and built new ones from the materials available around their next camp. The Miwok, for example, used incense-cedar in the Sierra Nevada to erect tepee-shaped wooden huts that they could abandon and easily rebuild.[176] Similarly, the Pawnee (which means "Men of Men"), who lived along the Red River in west Texas, built a wigwam that George Catlin described as looking like a "straw beehive." He said they "build their wigwams by a sort of thatching of long prairie grass, over a frame of poles fastened in the ground and bent in at the tops." He added that "they are comfortable dwellings, easily constructed, and easily demolished when they are left, by putting a fire-brand to them."[177]

Men cut the trees to prepare the camp, but women carried the load when traveling. Soldier and explorer Pierre Esprit Radisson explained this in his account of a meeting with the Sioux, "whom we call the Nation of the Buffalo," in eastern Minnesota. An advance guard of 30 warriors carrying "nothing but bows and arrows" met him and his companion Groiseilliers in the spring of 1659. They built temporary "lodges" and then prepared the site for the arrival of the "elders."

> Snow is moved aside, the ground is covered with pine boughs, and tent poles are planted. The following day the Sioux arrive with incredible pomp. First come the young men with bows and arrows. Next come the elders, with great gravity and modesty, covered with buffalo robes that hang to the ground. Then come the women, loaded like so many mules. Their burdens make a greater show than they themselves.... The women are conducted to the appointed place where they unfold their bundles and fling the skins that make their tipis. They build houses in less than a half hour. The men rest, then come to the big meeting lodge.[178]

The Kalispels did the same thing in Montana. About 1844, Father De Smet said that "the place for wintering being determined" the "men cut down fir trees, the women brought bark and mats to cover them."[179]

Each time a band of Indians stopped they cleared a large patch of forest to set up camp. They also extended the patch farther by cutting trees for firewood and often for poles. They killed other trees by stripping the bark for food, to feed their horses, or to cover their houses and canoes. In 1846, Father De Smet again referred to these clearings in his journal when traveling through the territory of the Assiniboines in the northern Rocky Mountains. He said, "in various places on the river, we saw ravages of the beavers which I should have taken for recent encampments of savages, so great a quantity of felled trees was there."[180] Thus each time a tribe moved it created a new opening in the forest, some large and some small, but all scraped down to bare soil and fertilized with organic trash. This provided a rich sunny seedbed for young pioneer trees such as pine to occupy when the Indians left their camp. In other words, American Indians caused constant disturbances throughout America's forests for thousands of years just by setting up, and later abandoning, their camps. These disturbances helped to maintain a patchwork of different aged stands of trees in the forest, each stand as old as the number of years since the Indians left their camp.

Mostly, American Indians moved and cleared forests for new villages to find game, fish, and plant foods, or because of war. Even the more sedentary Pacific Northwest Indians relocated their villages from time to time.[181] Often they depleted nearby game by shooting whatever they found. Zenas Leonard, a trapper who wandered through the western mountains between 1831 and 1835, camped for a winter with the Crow. He was traveling with Captain Joseph Walker's party, but stayed behind with two companions "for the purpose of investigating them in the business of catching beaver and buffaloe." However, he also probably volunteered because, according to him, "I now found myself in a situation that had charms which I had many times longed for." As a result, Leonard had the opportunity "to minutely observe their internal mode of living." He wrote the following comments in his journal about Indians being forced to move camp and go to war because of overhunting:

> The Crows are a powerful nation, and inhabit a rich and extensive district of country. They raise no vegetation, but entirely depend on the chase for a living. This is the situation of nearly every tribe, and when game gets scarce in one part of the country claimed by a certain tribe, they remove to another part, until after a while their game becomes scarce, when they are induced to encroach upon the territory of a neighboring tribe, which will at once create a fearful strife, and not infrequently ends in the total destruction of some powerful nation.[182]

Father De Smet also notes the effects of such intense hunting by the Assiniboines (an Ojibwa word that means "ones who cook using stones") while traveling through the northern Rocky Mountains in 1846. He says that "the scarcity of game forced them to quit their land—since their departure the animals have increased in an astonishing manner."[183] Thus the Indians moved their villages to allow the game to recover, and then they returned. Later eastern Indians would nearly exterminate the deer for miles around remote European settlements in an attempt to starve the soldiers and settlers out of their forts.[184]

The constant warfare among tribes inadvertently compensated for some of the overhunting within Indian territories by providing game refuges between territories. For instance, Lewis and Clark noted that, "with regard to game in general, we observe that the greatest quantities of wild animals are usually found in the country lying between nations at war."[185] An anonymous author with La Salle's expedition submitted a report to the king of France in 1682 that showed that the same thing happened near the present site of Benton Harbor, Michigan.

> On the 28th they came to fine woods where they found plenty of food. Before that time they had lacked food more than once, and had often been forced to march until nightfall without eating. From this point on they suffered no lack, either of small game or of venison. They were in a country where the savages did not go to hunt, as it was the borderland between five or six nations who were at war and who did not enter it save by stealth, in order to surprise and kill some of their enemies. As a result, the shots fired by the Frenchmen and the carcasses they left behind soon put some of the barbarians upon their trail.[186]

Warfare may also have reduced hunting pressure in the Yellowstone River country during the early 1840s. Father De Smet made the following note in his journal about the abundance of game in the area after spending several days, and a sleepless night, avoiding Indian war parties:

> The Yellowstone country abounds in game; I do not believe that there is in all America a region better adapted to the chase. I was for seven days among innumerable herds of buffalo. Every moment I perceived bands of majestic elk leaping through this animated solitude, while clouds of antelopes took flight before us with the swiftness of arrows. The ashata or bighorn alone seemed not to be disturbed by our presence; these animals rested in flocks or frolicked upon the projecting crags, out of gunshot. Deer are abundant ... you will see him jump with all four feet at once, and his movements are so quick that he hardly seems to touch the ground. All the rivers and streams that we crossed in our course, gave evident signs that the industrious beaver, the otter and the muskrat were still in peaceable possession of their solitary waters. There was no lack of ducks, geese and swans.[187]

Along the Columbia River in the Pacific Northwest, where people relied primarily on seasonal salmon fishing, fleas often influenced their decision to build a new village. Lewis and Clark explain: "When they [fleas] have once obtained the mastery of any house it is impossible to expel them, and the Indians have frequently different houses, to which they resort occasionally when the fleas have rendered their permanent residence intolerable."[188] In any case, the Indians Lewis and Clark observed along the upper Columbia River still lived in "huts" that they could easily abandon. Three decades later European emigrants described these huts as "constructed of slabs split out of Cedar, hewn, and set upon end, around a frame of poles, and covered with bark."[189] However, that changed farther down the river.

Plank Houses and Canoes

Lewis and Clark came upon the Short Narrows on their way down the Columbia, where they heard "a great roaring." They said a "tremendous rock stretches across the river" thus "leaving a channel only 45 yards wide, through which the whole body of the Columbia must press its way. The water, thus forced into so narrow a channel, is thrown into whirls, and swells and boils in every part with the wildest agitation." The steepness of the surrounding cliffs made it impossible to carry the canoes around the rapid, so they went on, "and with great care were able to get through, to the astonishment of all the Indians." Equally astonishing to Lewis and Clark, the Indians watching them lived in "houses which are the first wooden buildings we have seen since leaving the Illinois country, are nearly equal in size, and exhibit a very singular appearance."[190] These are plank houses, one of the most sophisticated wooden structures ever built in prehistoric North America. Lewis and Clark could not know that the Northwest Indians started building these modern houses about the time that Khufu, or Cheops, the Egyptian king and founder of the IV dynasty, built the greatest pyramid at Gizeh.

Pacific Northwest Indians built their plank houses with western redcedar, which they logged in the surrounding Sitka spruce-hemlock-redcedar forest. In some cases they even had to "raft" logs down rivers to their villages if the trees they needed grew far away.[191] Northwestern California Indians such as the Yurok and Hupa also built plank houses, but they used coast redwood.[192] These houses consisted of split boards over a rectangular log frame. Plank houses in the Northwest could be as much as 200 feet long and 50 feet wide. The logs holding up a house reached $6\frac{1}{2}$ feet in diameter and the boards could be $5\frac{1}{2}$ feet wide and 2 inches thick.[193] John Meares saw them during a voyage to the Northwest coast in 1788–1789. He said: "The trees that supported the roof were of a size which would render the mast of a first-rate man of war diminutive." "Our astonishment was on the strength that must be necessary to raise these enormous beams to their present elevation," Meares added.[194] He did not know that they did it using the principle of the lever.[195] Totem poles were equally thick and up to 65 feet tall.[196]

During construction the Indians set two pairs of heavy log posts at opposite ends of the house and then fitted two long beams to the tops of these posts. Then they set split planks upright against the beams to make the walls of the house. According to Lewis and Clark the houses had gables with "a ridge-pole the whole length." In addition, "from this ridgepole to the eaves of the house are placed a number of small poles or rafters, secured at each end by fibers of the cedar" and "connected by small transverse bars of wood." Cedar planks also covered the roofs that had "a descent about equal to that common among us," said Lewis and Clark. They added that the houses were "sunk about four feet deep into the ground," so "the descent was through a small door down a ladder." They placed their house fires, "sometimes two or three," in the middle of the floor and let the smoke out by moving one of the roof boards with a pole. Each structure housed several families who slept on "small beds of mats placed on little scaffolds" that were "arranged near the walls." Large totem poles usually stood outside houses built near the coast. The totem pole often displayed the clan's family emblem such as a bear, wolf, whale, raven, owl, or beaver and the emblems of other clans that married into the family. Carvings also decorated the four large posts inside the house.[197]

Just before seeing the modern looking plank houses along the Columbia, Lewis and Clark said "we observed two canoes of a different shape and size from any which we had hitherto seen." Surprised by their unique design and "curious figures carved on the bow" they added that these "canoes are very beautifully made." What they saw was the smallest of several dugout canoes that "are cut out of a single trunk of a tree, which is generally white cedar." These canoes ranged in size from "about 15 feet long" to those "upward of 50 feet long" that could carry "8,000 to 10,000 pounds' weight, or from 20 to 30 persons." The expert Indian paddlers using such canoes "ride with perfect safety the highest waves, and venture without the least concern in seas where other boats or seamen could not live an instant" said Lewis and Clark.[198] Like the house poles and totem poles, the largest canoes reached a width of $6\frac{1}{2}$ feet.[199] Such enormous logs could only come from large old trees that they cut in the nearby forest.

James Swan, who lived among the Pacific Northwest Indians in

"Washington Territory" for 3 years, watched them build a dugout canoe in the early 1800s. He wrote: "A suitable tree is first selected, which in all cases is the cedar, and then cut down." "The tree was chipped around with stone chisels, after the fashion adopted by beavers, and looks as if gnawed off," he said.[200] This "slow and most tedious process" took "from two to three days," according to John Jewitt, who the Nootka held captive from 1803 to 1805.[201] California Indians used fire to speed up the process when they cut coast redwoods for their canoes and plank houses.[202]

In contrast to the crude means by which Indians cut trees, Swan noted their great skill in transforming them into canoes. "It is not every man among them that can make a canoe," commented Swan, "but some are, like our white mechanics, more expert than their neighbors." "When the tree is down," he wrote,

> . . . it is first stripped of its bark, then cut off into the desired length, and the upper part split off with little wedges, till it is reduced to about two thirds the original height of the log. The bow and stern are then chopped into a rough shape, and enough cut out of the inside to lighten it so that it can be easily turned. . . . the log is turned bottom up, and the Indian goes to work to fashion it out. This is done with no instrument of measurement but his eye. . . . the log is again turned, and the inside cut out with the axe. This operation was formerly done by fire. . . . Then the inside is finished, and the canoe now has to be stretched into shape. It is first nearly filled with water, into which hot stones are thrown, and fire at the same time of bark is built outside. This in a short time renders the wood so supple that the center can be spread open at the top from six inches to a foot. This is kept in place by sticks or stretchers. . . . the water is emptied out, and then the stem and head-pieces are put on. . . . the canoe is again turned, and the charred part, occasioned by the bark fire, is rubbed with stones to make the bottom as smooth as possible, when the whole outside is painted over with a black mixture made of burned rushes and whale oil. The inside is also painted red with a mixture of red ochre and oil. The edges all round are studded with little shells. . . . This was a medium sized canoe, and took three months to finish it.[203]

The Indians only had stone and antler tools, mussel shells, wooden wedges made from yew and crab apple trees, and fire to fall huge trees, cut them into the proper lengths, split them into boards, or hollow them out for canoes. Even more amazing, they probably started these lumbering operations about 5000 years ago because woodworking tools appeared at about that time in the Pacific Northwest.[204] The only thing that prevented them from building huge plank houses and canoes sooner than that was the lack of large redcedar trees.

Most trees in northwest forests decay rapidly in the humid climate. Western redcedar does not. Pacific Northwest Indians knew this, so they preferred this durable wood for their plank houses and canoes. However, giant cedar trees remained scarce until the climate cooled enough after the Great Drought to allow them to invade Sitka spruce-hemlock forests. So western redcedar did not reach its current abundance on the coast until 3000 years ago. Nevertheless, redcedar represented a significant part of the forest in the northern Puget Trough as early as 5000 years ago, about the same time that woodworking tools appeared among Pacific Northwest Indians. Thus large-scale woodworking projects had to wait until enough cedar trees became available.[205]

A single village of plank houses required many trees and parts of trees to construct. Northwest Indians cut trees of different sizes for posts, beams, totem poles, and canoes. They also pried planks from the sides of living trees, sometimes leaving them alive so that the trees could heal. In addition, they peeled strips of bark from pole size cedar trees to make many household items, including baskets, boxes, canoe bailers, and hats.[206] Dr. J. D. Cooper, who served as the naturalist for an 1853–1855 railroad survey, saw this in western Washington. He said that cedar bark "is thin, coming off in long ribbon-like strings, of which the Indians make bags and articles of dress."[207]

Tree stumps, and wounds on the sides of trees from peeling bark and removing planks, must have been a common sight in the forests around villages, as Alexander Mackenzie noted in British Columbia. "Many of the large cedars appeared to have been examined," Mackenzie wrote in 1793, "as I suppose by the natives, for the purpose of making canoes, but finding them hollow at heart, they were suffered to stand."[208] There is little doubt that the construction of hundreds of villages and dugout canoes, over thousands of years, could easily change the surrounding forest.

Presumably, Pacific Northwest Indians cut single trees and small groups of trees, depending on their needs at the time. Such cutting created small to medium size gaps or patches in the forest, with partial shade or brief sunny periods that resulted from the sun passing overhead. In addition, they scraped away the covering of thick moss on the ground that prevented tree seedlings from becoming established, and they churned the soil as they processed logs within the gaps. As a result, western redcedar or, perhaps, Sitka spruce or western hemlock trees seeded into the gaps and prospered. These species all grow well in such openings. On the other hand, western hemlock continued to invade the understory of the surrounding uncut forests because it also grows well in shade. Here it gradually replaces the overstory trees as they die. As a result, the openings produced by Indian logging probably helped to keep Sitka spruce and western redcedar from being replaced by western hemlock in some coastal forests. Even so, strong winds, Indian-set and lightning fires, and other natural forces played the dominant role in regenerating trees in forest openings.[209]

Indians living in many other parts of North America also created gaps in forests by cutting or burning down trees to build dugout canoes. They were less elaborate in their construction than those made in the Pacific Northwest, but they were "very fine boats," wrote Anthony Parkhurst when he saw them in Virginia in 1564.[210] In southern New England Indians preferred American chestnut for their dugouts because it resisted decay.[211] Chestnut sprouts vigorously, so cutting the trees helped to maintain them within the forest. However, throughout northeastern forests from the Great Lakes to Maine and eastern Canada, where paper birch trees grow, Indians built bark canoes instead of dugouts. Many lakes and rivers cut through this vast region. Therefore, the birch bark canoe served as their primary form of transportation on the northern waterways that Indians called, "Paths that Walk."[212]

These northeastern Indians did not have to cut birch trees to build their canoes. Nonetheless, girdling a tree by stripping the bark will kill it as effectively as felling it with a hatchet. In addition, they peeled bark from birch trees to wrap food and bundles, to cover wigwams, and for containers, arrow quivers, armguards for archers, and for many other items. Besides, northeastern Indians also cut large trees to build dugouts if they needed canoes in the winter when the birch bark was too stiff to shape.[213] They also cut white cedar trees for the gunwales, thwarts, ribs, and deck planking of their bark canoes. Finally, they laced the sheets of bark together with spruce roots and sealed the seams with pine gum. James Rosier, who wrote an account of English Captain George Waymouth's voyage to Maine in 1605, provided one of the earliest descriptions of birch bark canoes. He said that Abnaki canoes "are made ... of the bark of a birch tree, strengthened within with ribs and hoops of wood, in so good fashion, with such excellent ingenious art, as they are able to beare seven or eight persons."[214]

Birch bark canoes also gave northeastern Indians an advantage in war. Father Hennepin noted this advantage after meeting a Sioux war party while paddling with two companions up the Mississippi River in 1680. He said, "we suddenly saw 33 bark canoes, manned by 120 Nadouessioux [Sioux]" who came "down the river with extraordinary speed, to make war on the Miamis, Illinois and Maroa."[215] The Sioux captured them and turned back because they could no longer surprise their enemies. Nonetheless, they could easily escape if seen because, according to Father Hennepin,

> the enemies of the Sioux have only dug-out, wooden canoes, which cannot go as fast as the lighter bark canoes of the Sioux. Only northern tribes have birch trees to make bark canoes, which wonderfully help them in going from lake to lake and by rivers, to attack their enemies. Even when discovered by their enemies, they are safe if they can get into their bark canoes, for their enemies cannot pursue them fast enough.[216]

The lesser known plank canoes built by the Chumash, Fernandeño, and Gabrielino of California represent an equally clever design. They used stone and shell tools to fall trees, split them into boards, and shape them for the sides of their canoes. Then they sewed the boards together with plant fibers and sealed the cracks with hardened asphalt and pine sap to make the canoe watertight. These coastal tribes built their 25-foot-long plank canoes so well that they could live on offshore islands near Santa Barbara and commute 17 miles back and forth across the ocean to the mainland.[217] Nevertheless, bark and plank canoes are the exception because most American Indians made dugouts.

Some dugout canoes in other parts of North America were as large and seaworthy as those found in the Pacific Northwest. William Bartram (the Seminoles called him Puc-Puggy, meaning "Flower Hunter"), the son of Crown Botanist John Bartram of Philadelphia, traveled to the deep south between 1773 and 1778. His detailed account of the journey through northeastern Florida included a remarkable observation on the distance Indians could travel by dugout canoe:

> These Indians have large handsome canoes, which they form out of the trunks of Cypress trees..., some of them commodious enough to accommodate twenty or thirty warriors. In these large canoes they descend the river on trading and hunting expeditions to the sea coast, neighbouring islands and keys, quite to the point of Florida, and sometimes cross the

gulph, extending their navigations to the Bahama islands and even to Cuba: a crew of these adventurers had just arrived, having returned from Cuba but a few days before our arrival....[218]

Captain John Smith also noted the seafaring skill and speed of the Indian's dugout canoe more than a century earlier: "Instead of oars, they use paddles and sticks, with which they will row faster than our barges."[219] He did not know that the birch bark canoes used farther north were even faster.

In the southeast, and presumably elsewhere, American Indians cut trees either with stone tools or fire. Sometimes they cut enough trees to create a large opening in the forest, as Captain John Smith noted in 1608 on the Virginia coast "where we found many trees cut with hatchets."[220] Otherwise they circled a tree with sticks of wood to create a bonfire that burned through the base, as John White's famous 1585 engraving shows in depicting the way Virginia Indians made dugout canoes.[221] Accord-

Fig. 5.2   John White's 1585 engraving of Virginia Indians using fire to cut trees and build a dugout canoe. (Photograph courtesy of the North Carolina Collection, University of North Carolina Library at Chapel Hill)

ing to Captain John Smith's account, and John White's engraving (Fig. 5.2), "they make [the canoe] of one tree by burning and scratching away the coals with stones and shells, till they have made it in [the] form of a trough."[222] Paleoindians built dugout canoes at least 6000 years ago, and probably much earlier, and they undoubtedly used the same methods to construct them as later Indians.[223]

The mammoth hunters found a wild continent when they crossed Beringia and walked southward between the ice sheets. This was the first and only time humans saw true wilderness in North America. Each generation of the mammoth hunters, and the bison hunters and hunter-gatherers who followed them, further modified the land and forests intentionally and inadvertently. Gradually, but irresistibly, they tamed a wilderness and made it home.

SIX

Enhancing Nature's Bounty

They must be very wise, eh? Those people? That time?
Beaver woman, age 69
High Level area, Alberta, Canada
Interviewed 1975–1977[1]

WILD GARDENS

American Indians learned how to manage trees and other plants because they depended on them for construction materials, firewood, weapons, clothing, basketry, cordage, foods, wooden tools, dyes, and medicines. Obtaining what they needed without destroying all the plants required an intimate knowledge of each species and the restraint that comes from a respect for living things. Julia Parker, a California Kashia Pomo, said it best: "We take from the earth and say please. We give back to the earth and say thank you."[2]

Indians throughout the Americas knew what time of year to harvest each type of plant and how much and how frequently to harvest. California Indians waited until after Douglas-fir, ponderosa pine, and Jeffrey pine trees flowered in late spring to harvest their small roots. This ensured that the roots would be tough enough to weave into cups and baskets.[3] In addition, Klikitat basketmakers of southern Washington protected western redcedar by giving the tree a 3-year rest before gathering another bundle of roots.[4]

Today the California Yokut and Mono still avoid digging up all the wild potatoes and onions in a particular area of a ponderosa pine forest. Like their ancestors, they know it would destroy future crops. As a Mono man said in a recent interview:

We gathered Indian potatoes in May or June when the leaves are green and when in flower with digging sticks. They grow in grassy areas near ponderosa pines. You boil them just like a potato and they're eaten plain. We'd go back to the same area and gather them. My mother and grandmother would only take the best and the biggest. They wouldn't harvest the smaller ones. They also gathered two or three kinds of wild onions in the foothills along streams. They never cleaned everything out. They would always leave some behind.[5]

Similarly, the Hualapai of Arizona and the Coahuiltecan of Texas only harvested the fruit, or "tuna," and the young pads of prickly pear cactus leaving the rest of the plant behind to grow the following year.[6] Thus, besides the timing and frequency of harvest, many American Indians knew how many plants to harvest and how much of a plant to take without harming it.

Some Indian tribes used the same care when hunting. For example, Pierre Radisson thought the Cree "are the best hunters in all America." He said in 1660 that when hunting beaver, "they do not kill young beavers, but leave them in the water, because they are sure they will catch them again."[7] However, other Indians hunted with less concern about future harvests. In July of 1835, trapper and explorer Osborne Russell met a small band of Snake Indians in the Lamar Valley of Yellowstone. He called it the "Secluded Valley." Russell wanted beaver pelts, but the Indians had none to trade.

They said there had been a great many beaver on the branches of this stream but they had killed nearly all of them and being ignorant of the value of the fur had singed it off with fire in order to drip the meat more conveniently.[8]

The Northern Paiute, Western Shoshone, Gosiute, Ute, and other tribes living in the Great Basin and the northern Rocky Mountains also took particular care of the trees they needed to make bows. They removed long strips of juniper wood or staves from live trees to make their bows. However, they did not kill the tree because the best bow trees were not abundant. The Indians selected older juniper trees sheltered in canyons or along ledges and large rocks. Here the wood kept a straight grain because the winds were not strong enough to twist the trees. Older trees also have more knot-free wood because they had time to grow around the stubs of small branches that fell off long ago. Then they used a stone chisel to cut a V-shaped notch at the top and bottom of the strip of wood they wanted. Thus cut off from the flow of nutrients, but still held fast to the tree, this strip of wood stayed straight as it dried for one or more years. Then the Indians pried the cured strip of wood or stave from the tree with a wooden lever and fashioned it into a bow. In one case, the Shoshones removed 16 bow staves from a single juniper tree in western Nevada without killing the tree. It died when struck by lightning many years later.[9]

The harvesting of wild plants by hunter-gatherers gradually developed into a form of gardening that bordered on true agriculture. The Nez Perce, Bannocks, Paiutes, and other tribes on the west side of the Rocky Mountains, and in Washington, Oregon, and California, relied on the camas bulb (a member of the lily family that grows in moist meadows) to supplement their diet. Father De Smet called it "the queen root of this clime," which they dig with "long crooked sticks" and "long and painful labor."[10] Each band of the Nez Perce owned a large digging territory that they divided into family plots.[11] Lewis and Clark said that a camas bulb "is round, much like an onion in appearance, and sweet to the taste" and it "is eaten in its natural state, or boiled into a kind of soup, or made into a cake."[12] The Indians also baked the chestnut flavored bulbs in a pit and ate them whole.[13] Father De Smet described the process:

> They make an excavation in the earth from twelve to fifteen inches deep, and of proportional diameter, to contain the roots. They cover the bottom with closely-cemented pavement, which they make red hot by means of a fire. After having carefully withdrawn all the coals, they cover the stones with grass or wet hay; then place a layer of camas, another of wet hay, a third of bark overlaid with mold, whereon is kept a glowing fire for fifty, sixty, and sometimes seventy hours. The camas thus acquires a consistency equal to that of the jujube. It is sometimes made into loaves of various dimensions. It is excellent, especially when boiled with meat; if kept dry, it can be preserved a long time.[14]

Many Indians waited until after the camas plants dropped their seeds before digging up the bulbs. However, the Lummi Straits, Nooksack, and Nuuwhaha of the Pacific Northwest harvested camas while in seed, but they buried the broken seed stalks in the holes left after removing the bulbs.[15] In addition, they usually took only the largest bulbs and left the smaller ones in the soil to grow. Using digging sticks also mimicked what gophers do when they dig with their claws, so it prepared the soil in a natural way that aided seed germination and replenished the plants.[16] Indians certainly knew that they enhanced the growth of plants by harvesting, as Mabel McKay, a Cache Creek Pomo elder in California, explained: "When people don't use the plants they get scarce." She added, "You must use them so they will come up again."[17]

The Assiniboines and other tribes used a similar method for harvesting and reproducing prairie turnips.[18] The careful harvesting and tending of camas and prairie turnips in the same places over thousands of years increased the abundance of the plants, which also created distinct areas not unlike weedy agricultural fields. Lewis and Clark noted such a camas field on their 1814 map of an area now near the Nez Perce Indian Reservation, calling it Quamash Flats.[19] Osborne Russell also commented in his journal that he passed Camas Creek and Camas Lake during his trapping expedition of 1835.[20] Similarly, the Blackfoot called their prairie turnip gathering area near Lethbridge, Alberta, Turnip Butte.[21]

Things changed when European farmers arrived. A treaty guaranteed the Bannocks and Paiutes the right to dig camas bulbs on the prairie southeast of Fort Boise, Idaho. White ranchers did not eat camas, so they released their hogs into the area to root out the bulbs. The dispute started the Indian war of 1878. It began when one angry Bannock wounded two whites. Then the Bannock leader Buffalo Horn assembled 200 warriors to continue defending their right to the camas. He died in a skirmish with volunteers, but the Bannocks then joined with Paiutes and continued the war until the U.S. Cavalry defeated them 2 months later.[22]

American Indians pruned, thinned, weeded, sowed, tilled, and transplanted wild plants long before the introduction of agriculture.[23] They knew that they could promote plant growth this way and increase the production of fruits and nuts. Hence the Italian navigator Giovanni da Verrazzano observed in 1524 that New England Indians stimulated the growth of grapevines climbing tree trunks by trimming shrubs that blocked the sun.[24] The effort they spent cultivating vines was worth it. The Indians ate the grapes fresh and dried, they squeezed them to make grape juice, and they even drank the sap from large vines in the spring.[25]

California Indians pruned oak trees by knocking on branches with sticks to shake loose acorns, which also broke off brittle and dead twigs. In addition, they intentionally broke off dead or diseased limbs to protect the trees and to increase acorn production.[26] They even pulled up cottonwood seedlings from bordering meadows to prevent hot fires from burning into the oak groves and killing the trees. Thousands of years of such careful tending converted many oak woodlands into orchards.[27]

Sometimes California Indians had to quickly gather as many acorns as they could to avoid losing them to birds, but they still tried to ensure the next harvest. They probably knew that knocking on the limbs to harvest the highly desired tanoak and black oak acorns could reduce the following year's crop even though it enhanced the crop of other oaks. These acorns take 2 years to mature, so the buds for the next crop appear while the current crop is still on the tree. Thus knocking on the limbs could easily break off these buds with the acorns. In spite of that, knocking off the acorns before they matured often saved the crop from being devoured by scrub jays. One of these birds can remove as many as 400 acorns from a tree in an hour. In a poor year they can strip a tree clean of acorns. Therefore, California Indians had no choice but to carefully knock the acorns free before they ripened enough to attract the voracious jay.[28] Even so, the Miwok and probably other tribes left some acorns on the ground to grow into new trees.[29]

However, when the choice was between starvation and cutting the trees that provide nuts, like anyone else, Indians chose survival. Ponderosa pine nuts provided starving people with a tasty alternative to boiled lichen. Sometimes, the Nez Perce and Pend d'Oreille in Idaho, like the Ottawa in the Great Lakes region, were desperate enough to scrape lichen from trees and rocks for food, especially during a bitterly cold winter. Father Menard said it "formed a kind of foam or slime" when boiled that "served to nourish their imaginations rather than their bodies." He had to eat the "slime" himself in the winter of 1660–1661.[30] In 1841, Father De Smet did likewise when among the Pend d'Oreilles in northern Idaho. He said that the boiled lichen formed "a thick elastic soup, having the appearance and taste of soap."[31] Thus, in May of 1806, Lewis and Clark reported that the Nez Perce ate "moss" the previous winter because they were "much distressed" for food. They also had to "cut down nearly all the long-leaved pines [ponderosa pine], which we observed on the ground, for the purpose of collecting the seeds."[32]

Indians also found edible mushrooms growing under acorn and nut-bearing trees. Mushrooms are the fruiting bodies of fungi that live on tree roots or rotting wood. The tree's roots provide the fungus with food and the fungus aids the tree in absorbing nutrients. Even today the Karok (which means "upstream"), Yurok (a Karok word for "downstream"), and Hupa ("Trinity River") people of northern California and southern Oregon harvest the tanoak, or matsutake, mushroom for food.[33] The Miwok also harvested mushrooms from beneath pine trees in Sierra Nevada forests. As with harvesting other plants, they took care to ensure future crops of mushrooms. A southern Miwok-Yokut woman recounted the traditions of her people when she said "we were always told in gathering mushrooms to leave the stems of the mushrooms—without that nothing comes back."[34]

Likewise, they pruned shrubs to stimulate the growth of straight young twigs. The Shoshone and Paiute in the Great Basin cut old branches from snowberry bushes so that they could use the new stems for small bird arrows.[35] In California, the Pomo pruned elderberry bushes the same way, but they used the young stems for musical instruments such as whistles and flutes.[36] In the winter, the Yuki and Pomo cut down redbud so that the straight young branches that grew back could be used to make baskets the next fall.[37] Even today, the California Mono cut sourberry for the same reason, as one Mono woman pointed out during a recent interview:

> When you cut sourberry it makes new shoots. They're nice and strong. Every two years I cut it. If it's not cut back it has a lot of little branches coming out of the stem and it's not good for anything.[38]

In the Southwestern deserts and dry plains, American Indians also carefully tended mesquite because these small thorny trees provided many vital resources. They made wooden mortars from mesquite logs, and they used it to make stools. They also used mesquite for poles to build their homes, which they covered with mesquite leaves, and they used it for charcoal and firewood, war clubs, and digging sticks. They also hardened mesquite branches with fire and used them as a foreshafts or points inserted into the mainshafts of their arrows. They even used its sap for food and medicine, and its root and bark fibers to make baskets. Mesquite trees were so valuable that, in some cases, an individual could stake out a grove and own it outright.[39] Not surprisingly, the Shoshone pruned their mesquite trees, removed dead branches, and cleared undergrowth from around the base of the trees.[40]

The most important part of the mesquite tree was its highly nutritious bean. The Indians made flour from the bean for later use, but they also prepared it right after picking, as Cabeza de Vaca explained:

They make a hole of requisite depth in the ground, and throwing in the fruit [bean], pound it with a club the size of a leg . . . until it is well mashed. Besides the earth that comes from the hole, they bring and add some handfuls, then returning to beat it a little while longer. Afterward it is thrown into a jar, like a basket, upon which water is poured. . . . He then beats it, tastes it, and if it appears to him not sweet, he asks for earth to stir in. . . . Then all sit round, and each putting in a hand, takes out as much as he can. Those present, for whom this is a great banquet, have their stomachs greatly distended by the earth and water they swallow. [41]

They paid a price for this kind of cooking. "Many of them have their teeth worn to the gums by the earth and sand they swallow with their nourishment," said Father De Smet.[42] Here he speaks of people in the northern Rocky Mountains, but the same must have been true for the Shoshone.

American Indians also spread some of the seeds they collected, intentionally and sometimes accidentally. For example, Captain John Smith implies that Indians in Virginia may have grown mulberry trees next to their villages. In 1625 he said: "By the dwelling of the savages are some great mulberry trees, and in some parts of the country, they are found growing naturally in pretty groves."[43]

Similarly, groves of American chestnut often occur around village sites in lower Ontario, Canada. Most likely Indians planted some of the trees while others probably grew from chestnuts dropped by accident. The Iroquois also planted Canada plum trees around their villages, and perhaps pawpaw, a small tree with a large edible fruit.[44] Likewise, the Kentucky coffee tree often grows around Indian village sites in Wisconsin, New York, and elsewhere. They concentrate in these places because Indians used the large hard seeds in a dice game and lost them from time to time.[45]

In the Southwest and southern California, many tribes such as the Navajo and Southern Paiute spread grass seeds after the harvest to replenish their gathering sites.[46] The Cahuilla even planted desert fan palm seeds around oases.[47] In New York, the Seneca dug holes to sow seed stalks and fertilized them with leaf mold.[48] The Ojibway, or Chippewa, and Assiniboine of the Great Lakes region also sowed wild rice seed in lakes and marshes. The Chippewa typically sowed one-third of the seed crop, and the Assiniboine not only replanted their wild rice fields but they also weeded out other water grasses.[49] Similarly, California Indians intensively weeded patches of sedge and bracken fern to stimulate the growth of the long rhizomes they needed to lace their baskets.[50]

As American Indians tended their wild crops they gradually began the long process of domestication that ultimately led to intensive agriculture. By the end of the Great Drought eastern Indians already occupied territories. They also started burying their dead in cemeteries about 6000 years ago, and they adopted other religious and ritual practices that helped to reinforce their claim to the land.[51] Their territories surely became more crowded as populations grew. Thus pressure increased to use existing plants and animals more intensively and to find new ways to cultivate wild plants, which aided the spread of agriculture from the Southwest to the East.

AGRICULTURE AND FORESTS IN THE SOUTHWEST

The domestication of squash somewhere in Meso- or Latin America about 10,000 years ago, followed about 4000 years later by beans and maize, made the agricultural revolution possible in North America.[52] When the Great Drought ended, people living on the eastern flanks of the Sierra Madre began to pass maize and squash seeds northward from one band to the next. This hardy small form of maize, or corn, called Chapalote reached the Cochise people of southwestern New Mexico at least 4500 years ago. They readily accepted maize because it grew in the cool highlands where they lived. Furthermore, they could roast the kernels and then use a mano (stone pestle) and metate (flat stone with a depression) to grind them into flour in the same way that they processed wild grass seeds.[53] However, maize and squash remained a minor part of their diet because it furnished too little food. So they had to continue their hunter-gatherer life.[54]

The common kidney bean arrived in southwestern New Mexico about 3000 years ago.[55] What is more important, a major genetic change occurred in maize at about the same time. Through selective breeding, southwestern Indians developed Maiz de Ocho, which probably made widespread maize cultivation possible. This variety has a large kernel and grows well in the hot dry lowlands because it flowers early and matures quickly.[56] Fortunately, they developed this new variety of corn at about the same time that droughts returned and shortened the growing season in the Southwest and the Great Basin.[57]

The Hohokam

About 2300 years ago immigrants from Mexico, the Hohokam (a Pima word meaning "those who have gone"), moved into the Gila River Valley in southern Arizona. They established a village at Snaketown and began to build a vast network of irrigated fields to grow maize, beans, and squash. They also gathered cactus fruit and mesquite pods from the surrounding desert.[58] Within a few hundred years (AD 1) the Anasazis (which means "early ancestors" in Navajo) started farming the canyons and mesa tops of the northern Southwest. Then about AD 250 the Cochise people settled into farming villages to become the Mogollon culture. Their villages spread throughout the mountains of southeastern Arizona and southwestern New Mexico. At first they lived on the ends of mesa tops and farmed the valleys below. Eventually, however, they moved their villages into the valleys nearer their fields.[59] Each of these cultures consisted of many villages, some large and some small, connected to one another by trade and social and religious ties of various kinds.

By AD 500 pottery and new crops that tolerate high temperatures also appeared in the Southwest such as lima beans, cushaw squashes or pumpkins, pigweed, and cotton.[60] Cotton moved into the Southwest along the east side of the Sierra Madre with other crops to become part of Hohokam agriculture. From there it spread throughout the region.[61] Cotton was especially important because it not only provided fiber that the Indians wove into cloth but they also ate the seeds and used them to make vegetable oil. Within a few hundred years southwestern Indians had also domesticated turkeys.[62] In addition, they raised parrots and macaws for their feathers and for ceremonial offerings and trade.[63] Agricultural production also must have improved in the Southwest about this time because rainfall increased.[64] Even so, southwestern Indians continued to exploit wild plant foods such a piñon nuts while they improved the variety and productivity of their crops, and the effectiveness of their farming methods. However, piñon nuts declined in their diet as they became more dependent on corn and beans.[65]

Southwestern Indians also continued hunting wild game such as pronghorn and bison on the prairies, bighorn sheep, mule deer and elk in the mountains, and cottontails in brushy areas. However, cottontails declined near their villages because the shrubby habitat where the rabbits lived became more open as agriculture expanded. By historical times, hunters had also eliminated nearly all the bison from the lands west of the Pecos River, and they severely reduced the number of elk. However, the loss of this game did not jeopardize their survival because they had developed alternative sources of food.[66]

Indian farmers still had their crops, and their agricultural fields provided the perfect habitat for other game. So they shifted to hunting the jackrabbits and deer that thrived around their fields.[67] The same thing happened in the East. Here Indian farmers thought that the deer around their fields were not unlike the livestock of Europeans, so they referred to them as their "sheep."[68] Thus the new agricultural landscapes that Indians carved out of the valleys and mountains provided an abundance of domesticated plant foods and semidomesticated game.

The people of the Southwest prospered in this managed landscape and their populations increased. Consequently, the Mogollon, Anasazis, and Hohokam cultures dominated the region for many centuries. Even after the central organization of these cultures collapsed, their influence and way of life persisted until well after Spanish explorers arrived. Pedro de Castaneda, chronicler of the expedition of Coronado (1540–1542), commented on the large number of settlements they found. He said that while traveling "130 leagues [390 miles]" through one section of Arizona and New Mexico the army saw "sixty-six villages," which is about one village every six miles. Many of these villages also were large because "in all of them" he said, "there may be some 20,000 men [braves and warriors, so perhaps 70,000 people]."[69] In spite of this impressive level of development, the Spanish only saw the last remnants of the three great prehistoric Indian cultures of the Southwest.

The Hohokam and the Anasazis, which gradually absorbed the Mogollon culture, reached a spectacular level of development that was all the more impressive because of the harsh dry environment of the American Southwest. The Hohokam, for instance, built pueblos that covered 500–1000 acres, with ball courts and other structures, all supported by an extensive irrigation system.[70] They located their pueblos about 3 miles apart along the main canals.[71] In 1565, Baltasar de Obregón, chronicler for Francisco de Ibarra, said that "their plantations are well provided with canals used for irrigating them."[72] About two decades later Antonio de Espejo noted that the Rio Grande pueblos had "fields planted with corn, beans, calabashes [gourds], and tobacco in abundance." Furthermore, he said, "these crops are seasonal, dependent on rainfall, or they are irrigated by means of good ditches."[73] Thus canals ensured a crop during periodic droughts. However, in the searing heat of the desert, where the growing season was long, the canals also allowed the Hohokam to grow two crops a year.[74]

Hohokam irrigation canals stretched out from the Salt and Gila Rivers like the branches of a tree.[75] In AD 1200 these canals irrigated fields that covered 58 square miles on just the south side of the Salt River in the Phoenix Basin alone. Farther south Hohokam irrigated fields covered 1200 acres in the foothills of the Tortilla Mountains.[76] They dug more than 500 miles of canals in just the Salt River Valley.[77] Some of these canals were huge. One canal extended 14 miles and measured 30 feet wide and 10 feet deep.[78] The engineering feats of the Hohokam seem even more sophisticated when we realize that the modern city of Phoenix still follows many of the same routes in its water diversion system that the Hohokam used.[79] In addition, the Hohokam irrigation system continued to function until the middle of the fifteenth century.[80]

The Hohokam started to decline about a century before the arrival of Spanish explorers.[81] Nevertheless, their influence spread as far as Owens Valley, California. Here the Paiute imitated their irrigation system successfully until at least 1859. Captain Davidson marched his soldiers from Fort Tejon to Owens Valley that year in search of horse thieves. He recorded what he saw in the valley.

> Large tracts of land are here irrigated by the natives to secure the growth of the grass seeds and grass nuts—a small tuberous root of fine taste and nutritious qualities, which grows here in abundance [wild-hyacinth or blue dicks and yellow nut-grass]. Their ditches for irrigation are in some cases carried for miles, displaying as much accuracy and judgment as if laid out by an engineer, and distributing the water with great regularity over their grounds, and this, too, without the aid of a single agricultural implement.[82]

Thus the Paiutes used the same kind of irrigation system as the Hohokam, but they did not grow corn. Instead they cultivated native plants.

The Anasazis

The Anasazis could not build elaborate canals in the same way as the Hohokam. They lived on mountain plateaus and in canyons rather than the lowlands. Here the land is rugged and streams are small. Nevertheless, they diverted streams, springs, and rainfall runoff where possible. Mostly they concentrated on conserving water and soil because of the steep slopes and narrow canyons.[83] So the Anasazis built terraces and check dams. They stacked several layers of rock on top of each other to build a terrace wall, and then they either filled it with soil or let eroded soil from above fill it.

These terraces covered the steep slopes near their villages like giant staircases. They designed them so that rainwater flowed onto the upper terraces, and the excess flowed to each successively lower terrace. Thus little of the water escaped their fields. Other terraced fields sat in narrow intermittent stream channels to catch greater amounts of rainwater. At Mummy Lake in Colorado they even built a reservoir with canals to irrigate their terraced fields. The Anasazis also built check dams in a way similar to terrace walls, but they served mainly to protect the fields below by slowing the speed of runoff and trapping sediment. Examples of old terraced hillsides still exist at Mesa Verde in Colorado.[84] In addition, the Anasazis intentionally planted their corn on sunny and shady hillsides with different amounts of moisture. They did this as insurance so that some of their crop would still survive during unusually hot or cold, or dry or wet, years.[85] This clever form of intensive agriculture supported a rich and complex society for many centuries.

The Anasazis culture centered in Chaco Canyon in northwest New Mexico. By the time it peaked, they dominated as much as 50,000 square miles within the area where Utah, Colorado, Arizona, and New Mexico come together. Here they built towns of all sizes connected by visual communications from hilltop to hilltop and an extensive road system. The road system follows straight lines with stairways and ramps cut out of steep rock faces. One road extends 50 miles. It took many people and sophisticated organization to build such elaborate structures. However, it ended around AD 1130 because of a 50-year drought that caused widespread crop failures and a reduction in wild plant foods and game. Until that time they enjoyed enhanced rainfall during a wet cycle in the southwestern climate.[86] However, raiding parties from the nonagricultural Athabaskan people who invaded the Southwest from Canada, now known as Navajo and Apache (who called themselves Shis-Inday, "Men of the Woods"),[87] probably contributed to the Anasazis' demise. [88]

The Anasazis culture finally collapsed after a century of decline, but the people survived by scattering here and there where conditions could still support smaller communities or individual families.[89] This loosely linked society of many small villages probably survived more or less intact until the Spanish arrived a few centuries later.

Like other prehistoric and historic peoples, southwestern Indians depended primarily on trees to frame their adobe and stone buildings and pithouses. Most southwestern Indians lived in pithouses before the rise of the Hohokam and Anasazis.[90] Even these subterranean houses needed many large and pole-size logs to support the dirt-covered roof. While traveling with Coronado through New Mexico in 1541, Pedro de Castaneda described a pithouse:

> They [pithouses] are underground, square or round, with pine pillars. Some were seen with twelve pillars and with four in the centre.... They usually had three or four pillars. The floor was made of large smooth stones.... They have a hearth made like the binnacle or compass box of a ship, in which they burn a handful of thyme [sagebrush] at a time to keep up the heat.[91]

After AD 700 they also constructed pueblos or villages composed of single or multistoried stone and adobe houses with interconnecting rooms. Pedro de Castaneda said that the villages are "crowded" and the houses look as if they are "crumpled all up together." He added that they are "terraced" so that the roof of one house forms a patio for the house above. Some pueblos reach "three and four stories high, with the houses small and having only a few rooms" according to Castaneda. However, he did see one pueblo "that has houses with seven stories." Finally he remarked on the lack of doors on the first floor, pointing out that "they generally use ladders to go up to those balconies, since they do not have any doors below."[92] Pueblo Indians did this as a defense against attack by people from other villages.

Southwestern Indians had to cut many trees to clear their fields, build their pueblos, and fuel their fires. Pueblo Bonito alone stood five stories high and contained 800 rooms.[93] Therefore, just 10 major buildings constructed in Chaco Canyon over 1000 years ago by the Anasazi contained 200,000 trees.[94] Each pueblo also had one or more kivas, which are subterranean ceremonial structures.[95] The Great Kiva at Pueblo Bonito was 65 feet in diameter, and the pueblo contained an additional 4 large kivas and 33 smaller ones.[96] It took hundreds of logs to support the roof of just one large kiva.

Even though they only required logs now and then to build houses and kivas, southwestern Indians still needed a continuous supply of fuel. They faced the same problem that many Indians and people throughout the world must deal with, even today—a shortage of firewood. For instance, the Cheyenne had to climb trees and throw down dead branches for firewood after they picked up everything near their camps.[97] The Miwok of California went so far as to make special hooked sticks to knock down branches for firewood.[98] New England Indians even moved their camps to avoid depleting their firewood. Thomas Morton said in 1632: "They use not to winter and summer in one place, for that would be a reason to make fuel scarce."[99] Thus in the Southwest logging for construction and firewood necessarily went on unabated around Anasazis pueblos. Not surprisingly, the archaeological record shows a steady increase in the number of Anasazis stone hatchets.[100]

At first the Anasazi used piñon pine logs to construct their villages and build their fires, which gradually depleted not only the trees but the nuts they needed for food. The piñon trees also failed to regenerate because they need nurse trees such as Gambel oak to shade them during early stages of growth. Thus the cutting of oak for fuelwood further reduced the number of piñon trees. So the Anasazi shifted to juniper because it can grow in the open. Even so, they had to cut smaller and smaller junipers because they did not have enough time to allow this slow-growing tree to become large. Eventually the piñon-juniper woodlands around their villages, including Chaco Canyon, disappeared almost completely.[101] Consequently, the Anasazi began cutting more and more ponderosa pine trees because, like juniper, they grew well in the openings they created.[102] Eventually they cleared so many ponderosa pines from around Chaco Canyon that they finally switched to spruce and fir for beams, and traveled nearly 100 miles round trip to cut and haul them back to their pueblos.[103]

The Anasazi affected the woodlands around the Grass Mesa village in southwestern Colorado similarly. Within less than 200 years they cleared most of the original forest within a 2-mile radius of the village. Little of the original forest remained throughout the region since other nearby Anasazi villages did the same.[104] Thus both firewood and wood for construction became harder to find. At what point they would have exhausted their forests is unknown, given that the Anasazi moved their villages periodically and then returned when the forests grew back.[105] Nonetheless, the Anasazi population could not expand indefinitely without the forests that provided them with logs and fuelwood.

AGRICULTURE AND FORESTS IN THE EAST

The shift to agriculture east of the Great Plains took place about the same time as it did in the Southwest, and extensive forest clearing came with it. However, forests grew thicker and more quickly in the East than in the desert Southwest. Agriculture also developed differently among eastern Indians. Some scientists think the stimulus for agriculture in the East came from the importation of squash and gourds from the Southwest about 4500 years ago. They appeared in Missouri, Kentucky, and elsewhere in the region at about that time.[106] Even so, gourds from local species may already have been in use at the time squash arrived in the East.[107] Unlike the Southwest, however, it took a long time for eastern Indians to shift to intensive agriculture, and adopt corn and beans as their primary crops. Instead they tended small gardens of native plants to supplement their hunter-gatherer life.

The introduction of squash and gourds may have induced eastern Indians to start transplanting wild plants into gardens. Regardless, native wild food plants must already have grown within their villages before domesticated plants arrived. Some of the seeds they gathered inevitably spilled when they brought them home. Others probably ended up in their garbage dumps. Since most of the native plants they ate grew well on disturbed soil, especially when enriched by organic refuse, they had to thrive around villages. It only takes a small step to shift from accidentally planting seeds to doing it intentionally. Just the same, we may never know if eastern Indians planted gardens before or after they imported domesticated crops from the Southwest. However, we do know that they managed wild plants and built impressive civilizations long before adopting intensive agriculture.

Eastern Indians long ago reproduced native plants in areas where they harvested them, and they probably started transplanting them as well. How long ago they began to transplant remains a question. We do know, for instance, that the Iroquois spread groundnuts to more convenient or productive locations by planting the tubers.[108] The Chickasaw also brought sand plum trees to the East from Texas and Oklahoma.[109] Southeastern Indians also transplanted the sacred evergreen shrub known as yaupon from the coastal plain to the Piedmont and the Appalachian highlands where it did not grow naturally.[110] According to William Bartram, "the Indians [Cherokee] call it the beloved tree, and are very careful to keep it pruned and cultivated."[111] They roasted the yaupon leaves to make a strong and frothy, caffeine-rich, tea that the Spanish called the "black drink," even though it had a yellow color. They usually drank it from a conch shell on special occasions.[112] It also could induce vomiting, which the Indian's said made them feel good.[113] "The Indians . . . are extremely fond of it" wrote J. Adair in 1775.[114] Cabeza de Vaca noted the importance of this drink more than two centuries earlier when among the Karankawas who lived on the Texas Gulf Coast.[115] Therefore, agriculture probably seemed reasonable to eastern Indians because they were cultivating and perhaps transplanting wild plants before squash and gourds arrived.

The agricultural revolution in the East began when Indians transplanted small-seeded annuals into garden plots near their villages. In all likelihood they started with starchy plants such as goosefoot, knotweed, and maygrass, and the more oily sumpweed.[116] Paleoindians already harvested sumpweed seeds on a large scale in Illinois as much as 5800 years ago.[117] By 3250 years ago they also transplanted sunflowers from the West into their gardens in Illinois, but it took another 600 years to domesticate them.[118] They probably ate sunflower seeds in much the same way Lewis and Clark described in 1805. According to their account, "the Indians of the Missouri" dried the sunflower seeds and pounded them into meal, and then they used it for "bread, or in thickening their soup." However, Lewis and Clark noted that "sometimes they add a portion of water, and drink it thus diluted; at other times they add a sufficient proportion of marrow-grease to reduce it to the consistency of common dough, and eat it in that manner." Lewis and Clark decided that they preferred "this last composition . . . to all the rest."[119]

The easily stored seeds of native plants grown in gardens provided eastern Indians with a hedge against the uncertainties of finding wild plant foods and game. In addition, gardening naturally led to improving crops by selecting seeds from only the best plants. The resulting increase in the quantity and dependability of food most likely led to a higher growth rate in the Indian population. That, in turn, reinforced the need to further enhance productivity and the size of their gardens. However, small garden plots did not completely support a village. So they continued to gather wild seeds and nuts, hunted game in the surrounding forests and marshlands, and fished in nearby rivers. This combination of small-scale agriculture combined with hunting and gathering eventually led to the development of sophisticated mound-building civilizations.

Early Mound Builders

The oldest known mound-building civilization appeared about 5400 years ago at Watson Brake, on the Ouachita River in Louisiana, before the Egyptians constructed their first pyramid. Here they built a ring of 11 mounds connected by ridges that encircle a level area. The tallest mound rises 25 feet. The people of Watson Brake lived a hunter-gatherer life, so they probably occupied the site from spring to early fall. They lived primarily on fish and mussels, supplementing their diet with other animals, as well as such plants as goosefoot, knotweed, and marshelder. They abandoned Watson Brake about 4800 years ago when a shift in the river decreased the swamp and small stream habitats that provided them with food.[120]

A similar but more advanced civilization developed at Poverty Point 100 miles away from Watson Brake about 3200 years ago. The Poverty Point people built a carefully planned town of mounds and other earthworks that covered a square mile of land on a bluff overlooking the Mississippi River swamplands.[121] Poverty Point consists of one large earthen mound that stands 70 feet tall, three smaller ones, and six sets of low ridges that form five sides of an octagon. The 82-foot-wide ridges stand 10 feet tall. Most of these earthworks remain visible today. A series of waterways linked the largest town of several thousand people to nine smaller towns, so they probably traveled mostly by dugout canoe. Together these towns comprised the cultural center of a society that spread out to include 100 towns in Arkansas, Missouri, Louisiana, and Mississippi.[122]

Poverty Point sat at the hub of a vast trade network in exotic raw stones and finished stone tools, and other objects. They obtained their stones such as flint, quartzite, slate, granite, magnetite, soapstone, quartz, copper, and many others from the Ouachita, Ozark, and Appalachian Mountains, and the upper Mississippi Valley and Great Lakes region. Their artisans carved an enormous number of beautiful stone ornaments and symbolic objects. These include beads made of red jasper, galena, and copper, and pendants shaped like birds, animal claws, and turtle shells. They also shaped stones to look like open clam shells and owls. These and other Poverty Point trade goods found their way as far up the Mississippi River as Tennessee and Missouri, and as far east as Florida and Georgia.[123]

Scientists thought that such an ordered society could not exist without agriculture. Now we know that the people of Poverty Point did not farm. Furthermore, there is little evidence to show that they had many garden plots. However, they did cultivate some native goosefoot and knotweed in their gardens, as well as a small amount of squash for their seeds and bottle gourds for containers. Even so, cultivated seeds represented a minor part of their diet. This highly organized and cosmopolitan culture mostly relied on hunting and gathering for food and clothes. Their diet consisted primarily of acorns and hickory nuts, supplemented by fish, deer, small mammals, waterfowl, and turtles. They also ate snakes, alligators, and frogs, as well as persimmons, wild grapes, hackberries, and seeds from honey locust, goosefoot, knotweed, and doveweed.[124]

The Poverty Point people cooked their food in open hearths and earth ovens. They made their ovens by digging a hole, filling it with food and then placing hot clay or silt balls around the food before covering it. They molded these balls into different shapes and used only one shape to cook a particular food. It turns out that different shaped balls retain heat for different times. Therefore, these ingenious people designed the clay balls to regulate the temperature of their ovens much as people do today with more modern technology.[125]

Unlike most eastern Indians of the time, the people of Poverty Point probably lived in a ranked society ruled by a chief and nobles. This social structure served them well since it lasted for at least 1000 years, a great achievement in itself. Furthermore, the people did not disappear when they abandoned their towns. They became what Europeans know as the Creek, Choctaw, Shawnee, and Natchez.[126]

The trading network of the Poverty Point culture represents an early high point among southeastern Indians. However, trade had gone on for thousands of years before Poverty Point. Denser and denser populations contained within ever shrinking territories made the difference. Now bands lived closer to one another and competition for resources increased. In addition, they became more dependent on each other because some basic resources such as fine-grained rocks, iron, or copper only existed in another tribe's territory, often far away. For example, Arctic people traded pieces of iron, their most precious resource, through an extensive network that originated from one main source, the massive Cape York meteorite that crashed into Greenland. They did not know how to smelt the iron, so they hammered it into tools and weapons.[127] People also depended on their neighbors for food during lean times. So trade became essential.

New ideas also came with trade goods. One of the more important of these involved elaborate ceremonies for burying the dead. Thus, even as the Poverty Point culture disappeared a new mound-building culture—the Adena—emerged farther north in the beech-maple forests of the Ohio River Valley and radiated into Indiana, Kentucky, and West Virginia.

The Adena

The Adena culture began about 2500 years ago.[128] It consisted of a large number of hunter-gatherer bands that logged the surrounding forests and cultivated imported sunflowers, pumpkins, and gourds, with native maygrass, sumpweed, and goosefoot in garden plots near their villages.[129] They lived in unique circular houses that varied from 20 to 80 feet in diameter. They set pairs of posts in the ground and slanted them outward to support the walls. Then they wattled the walls by weaving flexible poles horizontally between the posts. Wooden rafters held up by four large poles in the center of the house supported a mat or thatched roof.[130] Besides the Adena's foraging and gardening way of life, and unique houses, what linked these independent bands were common rituals for commemorating and interring the dead.

The Adena put a person's tools and possessions in the grave during funerals, just as their Paleoindian ancestors had done thousands of years earlier. They also carried on the old practice of adding other valuable objects as grave offerings. By this time such goods might include intricately carved stone and clay figurines, engraved stone tablets, jewelry hammered out of Lake Superior copper, and animal cutouts made from sheets of mica imported from North Carolina.[131] In any case, what set the Adena apart was their mounds.

Funerals and the construction of tombs and other earthworks brought the Adena together now and then. Afterward they returned to their villages and resumed foraging and gardening.[132] At first, Adena burials involved nothing more than a shallow bark-lined pit for one person. They heaped baskets of dirt on top of the pit to make a small mound.[133] Additional burials added to such mounds caused them to grow larger and larger, although some large mounds contained a single grave. Other Adena mounds contain no graves and no one knows what purpose they served, such as the Serpent Mounds in Ohio and Ontario, Canada.[134]

Even today the spectacular quarter-mile-long serpent in Ohio seems to slither northward through the forest. The Adena also built large ceremonial circles composed of an inner trench and an outer earthen ridge pierced by one or more entryways.[135]

The Adena people had to cut more and more trees to build their elaborate log-lined burial chambers, platforms, and circular houses for the dead. They set the houses on fire to cremate the body and then covered the ashes with baskets of dirt in the same way that they covered other graves. Again, the mounds grew in size because the Adena kept burying people on top of each other. As a result, some of these mounds became enormous, such as the Grave Creek Mound in West Virginia that originally stood seven stories tall. Anything approaching this height became unmanageable, so they started another mound, and then another. Thus far we know of as many as 500 Adena mound sites scattered across four states, as well as Ontario, but most of them concentrate in the central Ohio Valley.[136]

Around 2100 years ago, just before the birth of Christ, the Adena culture evolved into the Hopewell culture.[137] Immigrants from Illinois may have stimulated the change. Nonetheless, the Hopewell culture still included many Adena traditions, and it remained centered in the Ohio Valley.[138] The women only wore skirts with belts and the men wore loincloths. But they lavished themselves with decorations, such as freshwater pearl necklaces, bracelets, and anklets, copper-covered buttons, and copper earrings and breast plates. They also sewed pearls onto their garments.[139] Thus, what distinguished the Hopewell from the Adena was the richness of their possessions and the enormous scale of their burials and trade networks.

The Hopewell

The Hopewell culture most likely resulted from the consolidation of villages in river valleys, an increase in the size and importance of gardens, and the emergence of leaders with great wealth and prestige. These leaders, or "Big Men," accumulated their wealth through trade and rose to power because of the need to organize, represent, and defend large villages.[140] At least some of their prestige also came from success as hunters or warriors, as well as from being taller than the average male of the time.[141]

Hopewell people, like the Adena, were hunter-gatherers who supplemented their diet by cultivating native plants and squash in garden plots. Their gardens were larger than those of the Adena, and sometimes they included small amounts of maize, but their gardens were still too little and unproductive to feed people all year. Possibly maize failed to become important at this time because people were growing an old variety that did poorly in the northern climate. Whatever the reason, the Hopewell continued to live primarily on hickory nuts, deer, small mammals, turkeys, and other birds, reptiles, fish, and shellfish instead of agriculture.[142] Thus they supported large towns and a high civilization on nothing more than seasonal foraging and trade, but trading took place on a grand scale.

The Hopewell trade network extended throughout the East and as far west as the Rocky Mountains. People in the Southeast provided the Hopewell with mica, quartz crystals, and chlorite from the Carolinas and Tennessee, and conch and turtle shells and shark and alligator teeth from the Florida Gulf Coast. In exchange, the people in the Southeast received galena from Missouri, flint from Illinois, grizzly bear teeth, obsidian, and chalcedony from the Rockies, and copper from the Great Lakes. Most of the obsidian came from present-day Yellowstone National Park in Wyoming.[143] Many other finished goods also found their way along the network. Ultimately elements of the Hopewell religion and culture spread throughout most of the area in which it conducted trade.

The large scale of Hopewell mound building reflects the scale of their trade. The elite had log-lined tombs like the Adena, but their burials contained a richer collection of finely crafted grave goods made by skilled artisans. Their mounds also covered much larger areas. They cleared 110 acres to build just one mound complex in the Scioto River Valley near Chillicothe, Ohio. It contains 38 mounds enclosed by a rectangular embankment. The Hopewell also cleared most of the area within a 4-square-mile block at Newark, Ohio, to erect their earthworks. These consist of embankments that form giant circles, a square and an octagon, and a set of roads that connect them. The central circle in this complex is nearly a quarter mile in diameter.[144] Even more amazing, a perfectly straight 60-mile-long Hopewell road may tie the Chillicothe and Newark sites together.[145] Other large Hopewell mound complexes dot the landscape in the Midwest and Southeast. Plains Indians as far away as the eastern Dakotas and southern Manitoba, Canada, also built mounds imitating those of the Hopewell.[146]

The Hopewell culture collapsed about AD 400, just 500 years after it began. No one knows why, but the breakdown came at about the time that the bow and arrow replaced the spearthrower in the East. This revolutionary new weapon greatly enhanced their hunting efficiency, which may have temporarily depleted game herds. Just the same, the greater range and accuracy of the bow and arrow also made warfare more efficient and deadly. The construction of earthworks in defensive locations on hilltops, and signs that some villages burned, no doubt show increasing violence among villages.[147] Therefore, increased warfare may also have helped to break up the Hopewell culture.

The Mississippians

As with earlier cultures, the Hopewell people did not disappear, they merely disbanded. Populations even grew during the coming centuries as people became more prosperous. Southeastern Indians also carried on some Hopewell traditions and trade networks well past the time when the culture broke apart in the Ohio Valley. So mound building continued during this time, especially in the lower Mississippi Valley, and along the Gulf coastal plain from Georgia to Florida and Alabama. Some of these mound complexes were equally huge, such as the 297-acre Weeden Island culture site at Kolomoki in southern Georgia. These mounds differed from the Hopewell, however, in that they were platforms for temples and houses of the elite as well as burial places. These differences, along with extensive logging, land clearing, and agricultural fields, became hallmarks of the next development in American Indian culture—the Mississippian.[148]

Maize cultivation erupted in the East about AD 700 and surged across the land, and with it came a new unity among eastern Indians based on intensive agriculture, large towns and cities, and a rigid political hierarchy of chiefdoms.[149] This new Mississippian culture represented a giant leap forward in North American civilizations. To be sure, the earlier southwestern cultures reached a high level of sophistication, but the Mississippian exceeded them in size and complexity. Even the eastern Apaches of New Mexico and West Texas knew about the dense populations of Indians living in the Mississippi Valley, as Coronado's chronicler Pedro de Castaneda documented in 1541:

> They said that there was a very large river over toward where the sun came from, and that one could go along this river through an inhabited region

for ninety days without a break from settlement to settlement. They said that ... the river was more than a league wide and that there were many canoes on it.[150]

Most likely, the introduction of a faster growing and more productive form of maize known as northern flint helped to fuel the advance of Mississippian civilizations in the East.[151] This also accelerated population growth because food production increased far beyond what foraging and small gardens could provide. The introduction of beans from the Southwest to the East sometime between AD 1070 and 1200 further enhanced the spread of agriculture.[152]

Beans may have taken longer than corn to become part of eastern Indian agriculture because people who raise beans must live a more sedentary life. Beans do not grow well unless weeded and protected from animals. Corn, on the other hand, can be planted and left untended until it matures. This form of agriculture fits well with a hunter-gatherer way of life. That may explain why the nomadic Apache grew corn long before they grew beans.[153] However, Robert Stuart, who discovered the Oregon Trail, noted that the northern Plains Indians could grow beans with their corn by living a life that was partially sedentary and seminomadic. During his trek from Fort Astoria at the mouth of the Columbia River to St. Louis in 1812 and 1813, he noted:

> These Indians come to their towns early in April, plant their Corn, Pumpkins and Beans towards the end of May, stay till it is a certain height, when hoeing it, they then abandon it to the benign care of the Allseeing Providence and return to the plains to pursue the humpbacked race [bison]. In August they again revisit their village and after gathering in the harvest depositing safely and secretly in excavations made for the purpose in the earth they once more leave their homes for their favorite pursuit of the Buffaloe at which they employ themselves till the following April.[154]

People usually planted beans and corn together so that their vines grew up the corn stalk. Samuel de Champlain observed this practice at the mouth of the Saco River in southeastern Maine in 1604. He said that

> with this corn they put in each hill three or four Brazilian beans [kidney beans], which are of different colors. When they grow up, they interlace with the corn, which reaches to the height of from five to six feet; and they keep the ground very free from weeds.[155]

This not only produced more food per acre but, since beans are legumes, it also enriched the soil with nitrogen. The Penobscot also built a tepee shaped frame of 10-foot-long poles for the vines to climb. In contrast, low soil moisture forced southwestern Indians to plant their corn and beans separately, so they grew mostly bush beans instead of vining beans.[156]

Beans also added proteins that provided the essential amino acids lysine and tryptophan to the Mississippian diet.[157] People cannot live on corn alone because it lacks these amino acids. Early on, squash contributed proteins containing tryptophan, but this was not enough.[158] Even beans do not furnish enough protein for pregnant women and young children, so meat and fish remained an essential part of the Mississippian diet.[159] Deer meat became especially important because the animals thrived around corn fields.[160] The Mississippian people also supplemented their diet of corn and beans with seeds from native plants such as goosefoot and little barley, which they continued to cultivate as people had done for centuries.[161] Other eastern Indians did the same as late as 1640.[162] Northeastern Indians cultivated other native plants as well, such as Jerusalem artichokes and groundnuts that provided nutritious roots.[163] Not only that, the Mississippians and later Indian peoples kept on collecting nuts in the surrounding oak-hickory and oak-chestnut forests as the Hopewell and Adena did before. Regardless, the nutritious combination of beans and corn still supplied the foundation for the Mississippian and other great civilizations throughout the Americas.

The Mississippian culture first appeared in the valleys of the Mississippi, Tennessee, Cumberland, and lower Ohio Rivers where periodic flooding had built up deep rich soils.[164] Oak-gum-cypress and elm-ash-cottonwood forests grew in these bottomlands surrounded by oak-hickory and oak-chestnut forests on the drier uplands. The Mississippian people girdled the trees to clear the land for their crops. The time when they began clearing their fields in the upper Mississippi basin shows clearly in the fossil pollen record for elm, which grows on bottomlands. Elm pollen dropped abruptly here about AD 1000 and then it nearly disappeared.[165]

The Mississippian culture spread rapidly along other river valleys as well, until it covered most of the Midwest and Southeast. It also extended into the loblolly-shortleaf pine and longleaf-slash pine forests of East Texas where the Caddo Confederacies flourished, and even up the Missouri River and into the Great Plains.[166] Mostly the Mississippian culture spread as ideas passed from one tribe to another while trading. In other cases it expanded by colonization, such as the large colony of hamlets, fortified villages, and even a temple-town that they established in southern Missouri about AD 1275. It failed around AD 1350, so the Mississippian colonists set fire to the settlement and left. The same thing may have happened at Aztalan in southern Wisconsin.[167]

Mississippian societies could be small or large, but they were all unique because of local conditions and constant changes in leadership, alliances, and trade. Just the same, the larger towns and cities followed a standardized plan that included a central plaza surrounded by platform mounds topped by wooden temples and the houses of important people. Commoners lived in thatched houses on the outside of this complex. A log palisade usually protected the town from invasion.[168] Examples include Etowah and Ocmulgee in Georgia, Spiro in Oklahoma, the Angel Site in Indiana, Kincaid in Illinois, and Moundville in Alabama.[169] However, the most spectacular example of a Mississippian community is Cahokia, the largest city ever constructed by North American Indians (Fig. 6.1).

The city of Cahokia sits at the confluence of the Mississippi and Missouri Rivers in Illinois, across from present-day St. Louis. This is an ideal location to conduct trade with other cities using dugout canoes as well as inland trails. Cahokia covers a total of 13 square miles, but the city's influence on surrounding forests and prairies extended even farther. The development consisted of immense agricultural fields, as well as 40 hamlets and farmsteads, 5 small towns, 4 large towns, and the central city. About 20,000 people lived here. The core of the city covered 2000 acres. It contained numerous mounds and plazas surrounded by rows of thatch-covered pole houses with small gardens scattered here and there. These upland gardens gave families some security against the loss of their crops to flooding on the bottomlands. A total of 120 mounds dot the development, mostly in the center of the city. The city also included a large circle composed of 48 equally spaced, red-stained redcedar (a sacred wood) posts that we now call Woodhenge because of its similarity to Stonehenge in Britain. It served as a solar calendar to mark planting times and ceremonies. The ruler of Cahokia lived in a huge wooden house in the center of the city that was 105 feet long, 48 feet wide, and 50 feet tall. What made his home all the more impressive was its location on top of Monks Mound, an earthen pyramid that stood 10 stories high (Fig. 6.2).[170]

We do not know what powers the ruler of Cahokia may have had. Most likely he controlled a number of lower-ranking chiefs that col-

Fig. 6.1 View of Cahokia to the Northeast circa AD 1100–1150. (Photograph courtesy of Cahokia Mounds State Historic Site; painting by William R. Iseminger)

131

Fig. 6.2 View of the Grand Plaza and Monks Mound in Cahokia circa AD 1100–1150. (Photograph courtesy of Cahokia Mounds State Historic Site; painting by Loyd K. Townsend)

lected tribute from people living within certain districts. This paramount chief undoubtedly maintained a stockpile of grain to which all members of the community contributed so that he could help people when their crops failed.[171] He also must have had great social, political, and religious influence, so great that this ruler may have lived on the top of a mound to be closer to his "brother" the sun. The chiefs of the Mississippian's direct descendants—the Natchez—also claimed such kinship.[172] They most likely had unlimited power, as did the later Natchez chief known as "The Great Sun."[173] "He exercises an absolute power over his subjects, whose lives and goods are entirely at his disposal, and they can demand no payment for any labor he requires of them," wrote Father Pierre de Charlevoix in 1721.[174]

The Great Sun's feet never touched the ground. Either his people carried him in a litter or they placed mats on the ground before him when he walked. They killed wives, retainers, guards, and sometimes even close relatives when such rulers died so that they could join him in the afterlife. Likewise, the body of one of the paramount chiefs at Cahokia rested on a blanket of 20,000 shell beads surrounded by 800 arrowheads, mica and copper sheets, and polished stone discs. The disarticulated remains of several other people rested next to him. Buried nearby were the bodies of six of his relatives. A mound near the ruler's tomb contained an even more macabre burial. Here scientists found the skeletons of 50 young women between the ages of 18 and 23, probably the dead ruler's wives. They had all been strangled.[175]

We may not know the details, but what is certain is that warfare increased in the Mississippian culture. Warfare intensified because good farmland was scarce and chiefs were powerful. This inevitably led to attempts at conquest, as well as rivalries and other disputes. Therefore, the Mississippian people fortified their towns with walls of log posts, or palisades, and added skeletons full of arrowheads to their cemeteries. Even their artwork portrays severed trophy heads, and one stone effigy pipe depicts either a beheading or a scalping.[176]

WARFARE AND FORESTS

Warfare also plagued the people who lived before Hopewell times, and surely it occurred to a greater or lesser extent throughout the human history of North America. We know that Paleoindians fought each other at least 9000 years ago. The recently excavated skeleton of a 50-year-old Paleoindian in Kennewick, Washington, shows that this very tough person somehow survived more than one attack. He had a slashed chest that withered his left arm, and a chipped right elbow and shoulder blade. Even more surprising, he walked around with an inch-wide spearpoint embedded in his pelvis. Somehow he survived these wounds only to die later of an infection.[177]

People also fought and killed each other elsewhere in North America during Mississippian times. Evidence pieced together from Robber's Gulch in northwest Wyoming allows us to relive another deadly drama that took place over a thousand years ago. It seems that a young prehistoric Plains Indian failed to outrun his pursuers. They plunged 12 arrows into his back, probably as he ran. Perhaps he kept running, but he finally had to stop at the edge of a gulch. So, with arrows bristling from his back, he most likely turned to face his killers. They shot him two more times in the chest, and he fell face down into the gulch. His killers then heaved several large flat rocks down on his body, crushing him and the arrows embedded in his back. Eventually 25 feet of dirt slumped over the body and hid the gruesome scene until scientists uncovered it a short time ago.[178]

The people of Cahokia must have endured many wars and similar violence because they built a 15-foot-high palisade 2-miles long around the center of their city with guard towers spaced 70 feet apart. They rebuilt this fortification at least four times, and each time it required 15,000–20,000 oak and hickory logs, 1 foot in diameter and 20 feet tall, to construct. The palisade also served as a social barrier to segregate the elite people in the center of the city from the commoners who lived outside.[179] However, smaller palisades also encircled outlying villages, so their main function was undoubtedly defensive.[180] Similar defensive works protected other Mississippian communities throughout the South and Midwest. Finding large numbers of suitable trees to cut for such huge wood structures required logging operations that covered hundreds and perhaps thousands of acres of forest.

Clearing bottomland hardwoods over vast areas to plant crops and build towns certainly destroyed many bottomland forests during the time when Mississippian farmers practiced intensive agriculture. European farmers simply cleared most of the remaining bottomland trees when they displaced the Indians. However, chopping down trees to build homes, palisades, and log stairways up the great mounds, and to fuel fires, may only result in thinning a forest in a way that allows it to regenerate itself. This was likely around Cahokia and other fortified Mississippian settlements since most upland tree species in this region such as oak, hickory, and cedar grow well in openings. The Indians may also have spared many hickories from the ax because they produce valuable nuts, and they surely noticed that hickories produce even more nuts when they grow in open forests. Therefore, other than clearing forests for fields and settlements, logging by Mississippian people probably helped to keep more shade-tolerant hardwoods from displacing the oak-hickory, oak-chestnut, and cedar forests that European explorers saw centuries later.

The Mississippian culture started to unravel during the fourteenth century when construction stopped at Cahokia. By 1500 Cahokia no longer existed as a city, and Moundville, Etoway, and Spiro also collapsed. As in the past, the people did not disappear, only their culture. They carried on in many smaller settlements instead of a few large cities. Perhaps the introduction of beans in the twelfth century allowed maize cultivation to expand into the uplands since people no longer needed fish and waterfowl as a source of protein, and deer were plentiful anyway. This might have undermined the control of paramount chiefs whose districts only covered the river valleys.[181] Maybe the Mississippians destroyed their environment by cutting the forests, which then allowed erosion to strip away the soil.[182] This seems unlikely, however, because their society flourished for over 700 years, and those who followed them continued to prosper using intensive agriculture, as we do even today. Possibly too many people tried living on too little land and they had to spread out to survive. No one knows why the Mississippian culture ended, but any growing and war weary society held together by the absolute authority of a few people must surely be unstable.

Somehow the Natchez managed to retain most of the Mississippian culture long after their great cities fell into ruin. They continued the political and religious traditions of the Mississippians, and they even constructed mounds and temples in the lower Mississippi Valley of Louisiana. Their culture came to end in 1729 when the French governor of Louisiana ordered the evacuation of their city so that he could build his new plantation. The Natchez revolted, but the French sent soldiers out of New Orleans and nearly destroyed the tribe, selling many of the survivors into slavery. However, some of the Natchez escaped and joined neighboring tribes while others continued their hopeless fight against the French.[183]

When the Natchez disappeared, the grand period of the Adena, Hopewell, and Mississippian temple-mound-building cultures closed as well. By the end they may have managed to build several hundred thousand mounds in the South and Midwest, one basketful of dirt at a time.[184] Even more important, they left an enduring legacy of art, technology, and agriculture. Agriculture began in the Southwest, but with the help of the Mississippians, American Indians practiced agriculture within nearly two-thirds of the continental United States in a relatively short time.[185] Other Indians cultivated their wild gardens in a way that came close to modern agriculture.

The agricultural practices that the Mississippians inspired continued in much of eastern North America, although their political and religious traditions and even their history faded away with their great cities. Thus, the Cherokee no longer knew their ancestors when William Bartram visited them in 1773. In his description of their buildings he said:

> The council or town-house is a large rotunda, capable of accommodating several hundred people; it stands on the top of an ancient artificial mount of earth, of about twenty feet perpendicular, and the rotunda on the top of it being above thirty feet more, gives the whole fabric an elevation of about sixty feet from the common surface of the ground. But it may be proper to observe, that this mount, on which the rotunda stands, is of a much ancienter date than the building, and perhaps was raised for another purpose. The Cherokees themselves are as ignorant as we are, by what people or for what purpose these artificial hills were raised.[186]

HISTORICAL TIMES

"All that has been discovered up to the year forty-nine [1549] is full of people," wrote Father Bartolomé de las Casas, "like a hive of bees."[187] Substantial evidence supports the conclusion that 40–80 million people lived in the Americas when Europeans arrived. Population estimates for North America alone range between 3.8 million to as many as 18 million, with 12 million being a widely, if not universally, accepted number. The higher estimates normally account for the 89% of native people who died from Old World diseases soon after exposure to Europeans.[188] Regardless of the population estimates, nothing can change one inescapable truth: Millions of people managed most of North America for a very long time.

Verrazzano remarked on the large number of people who lived in North America in 1524. As he sailed by Block Island while exploring the New England coast, he wrote that the land is "well-peopled, for we saw fires all along the coast."[189] We now know that New England supported from 27 to 109 Indians per square mile at that time.[190] Similar densities probably existed in the South. Garcilaso de la Vega, who marched with Hernando de Soto's (1539–1542) army through Alabama, recorded that the land was "so fertile and thickly populated that on some days the Spaniards passed 10 or 12 towns, not counting those that lay on one side or the other of the road."[191] Another chronicler of the expedition wrote that in Mississippi, "the land was thickly inhabited ... and as it was fertile, the greater part being under cultivation, there was plenty of maize." [192] By this time most American Indians in the East, Midwest, southeastern Canada, and the Southwest, lived in villages surrounded by agricultural fields in which they planted corn, beans, and other crops. They even practiced intensive agriculture in the river valleys on the northern and central Great Plains.

The Arikaras (or "Corn Eaters"), Madans, Hidatsas, Pawnees, and Omahas grew several varieties of corn along the rivers of the Great Plains and, depending on the tribe, pumpkins, beans, squash, gourds, tobacco, and sunflowers as well. They added meat to their diet by hunting bison and other game. These tribes descended from farmers who immigrated to the Plains from the eastern woodlands about AD 900, shortly after the rise of the Mississippian culture.[193] They all built permanent villages, and sometimes they fortified them with palisades and dry moats.

Mandan farmers in particular were an important center of culture and influence on the upper Missouri River. The Teton and other tribes usually mentioned them in ways that showed respect, even when at war, and so did the French fur traders and Lewis and Clark.[194] German Prince Alexander Maximilian traveled up the Missouri River in 1833 with the Swiss artist Karl Bodmer and returned with paintings and written accounts of these people. Maximilian wrote:

> The Mandans and Manitaries [Minnitari is the Dakota name for the Hidatsa] cultivate very fine maize without ever manuring the ground, but their fields are on the low banks of the river ... where the soil is particularly fruitful. They have extremely fine maize of different species. The Indians residing in permanent villages have the advantage of the roving, hunting tribes in that they not only hunt but derive their chief subsistence from

their plantations which afford them a degree of security against distress. The plants which they cultivate are maize, beans, French beans, gourds, sunflowers, and tobacco.[195]

Even such hunting tribes as the Osage, Kansa, Otoe, and Ponka practiced some agriculture. In 1806 the explorer Lieutenant Zebulon Pike said they raised "a sufficiency of corn and pumpkins to afford a little thickening to their soup during the year."[196]

The Midwest and South still contained the largest and most productive agricultural fields. In Rodrigo de Ranjel's diary of Hernando de Soto's (1539–1542) expedition written in 1545, he speaks of cultivated fields in northern Florida "extending over the plain as far as the eye could see."[197] Another recorder of Hernando de Soto's march through the South said that in Arkansas, "the country ... was level and fertile, having rich river margins, on which the Indians made extensive fields."[198] When de Soto's army reached the Little Red River Valley, the recorder also noted with obvious wonder that "the soil was rich, yielding maize in such profusion that the old was thrown out of store to make room for the new grain." He added that "beans and pumpkins were likewise in great plenty."[199]

William Bartram described what remained of the fields around the Mississippian town of Ocmulgee in eastern Georgia 200 years after de Soto landed in Florida. The people themselves had long since disappeared because of the ravages of European diseases when he wrote his account:

> On the east banks of the Oakmulge, this trading road runs nearly two miles through ancient Indian fields, which are called the Oakmulge fields: they are the rich low lands of the river. On the heights of these low grounds are yet visible monuments, or traces, and banks, encircling considerable areas. Their old fields and planting land extend up and down the river, fifteen or twenty miles from this site.[200]

Thus, American Indians cleared the bottomlands along most of the major rivers in the South, Midwest, and much of the East, and converted them to agricultural fields. They did not stop there, however; they planted the best forest lands on the uplands as well.[201] For example, Indians living in the Lake Champlain Valley often located their villages on hilltops and other places far away from large rivers and lakes.[202] William Bartram also noted where forests had reclaimed old Indian villages and agricultural fields on the uplands during his travels through northern Florida in the mid-1700s:.

Passing through a great extent of ancient Indian fields, now grown over with forests of stately trees, Orange groves, and luxuriant herbage, the old trader, my associate, informed me it was the ancient Alachua, the capital of that famous and powerful tribe, who peopled the hills surrounding the savanna, when, in days of old, they could assemble by thousands at ball play and other juvenile diversions and athletic exercises, over those, then happy, fields and green plains. And there is no reason to doubt of his account being true, as almost every step we take over those fertile heights, discovers remains and traces of ancient human habitations and cultivation.[203]

A century and one-half earlier Captain John Smith recorded the Indian villages and fields he saw in the forests of Virginia. He said their villages contained from 2 to 50 houses surrounded by fields that varied in size from 20 to 200 acres or more. [204] Captain Smith also made two voyages farther north in 1614 and 1615. He noted in "the country of Massachusetts" that "here many Iles are planted with corne" and "savage gardens." He added that it "shewes you all along large cornefields, and great troupes of well proportioned people."[205] A few years later the colonists arrived at Plymouth Harbor. They placed their settlement "upon a high ground where there is a great deal of land cleared, and hath been planted with corn three or four years ago." By then many Indians had already died from smallpox or perhaps bubonic plague, so much of their land was unoccupied and their fields were still usable.[206]

Indians cultivated corn, beans, and squash, which some called "the three sisters," throughout the eastern half of the continent to as far north as Canada. Eastern Indians must also have planted many gourds because Adriaen Van der Donck noted in New York in the 1650s that "it [the gourd] is the common water pail of the natives."[207] They even planted crops in Maine, although we do not know how widespread farming might have been there. We do know it extended at least as far north as the Penobscot River where, according to Pownall, there were "old worn-out clear Fields" in 1776 that reached as much as 5 miles down river.[208] Samuel de Champlain drew a map of an Indian village surrounding the mouth of the Saco River in southeastern Maine in 1604. It showed scattered "cabins in the open fields, near which they cultivate the land and plant Indian corn," a "fortress" and "a large point of land all cleared up except some fruit trees and wild vines."[209] Here he also described the crops they planted, saying that in addition to corn and beans, "we saw there many squashes, and pumpkins, and tobacco, which they likewise cultivate."[210] Captain John Smith also recorded cornfields along the Kennebec River in 1614. He said in his descriptions of the area around the river that "I saw nothing but great high cliffs of barren rocks overgrown with wood, but where the savages dwell there the ground is excellent salt [rich] and fertile."[211]

Jacques Cartier documented Indian agriculture even farther north near what is now Montreal, Canada, in 1534. He said it was "fine land with large fields covered with the corn of the country." He added that their village was "near and adjacent to a mountain, the slopes of which are fertile and are cultivated, and from the top of which one can see for a long distance."[212] Similarly, in 1749, the explorer La Vérendrye told Peter Kalm that the land west of Montreal was cultivated long before the French arrived. Kalm wrote the story in his travelog:

> As they [La Vérendrye] came far into the country beyond many nations they sometimes met with large tracts of land, free from wood, but covered with a kind of very tall grass for the space of some days' journey. Many of these fields were everywhere covered with furrows, as if they had once been plowed and sown. . . . In what manner this happened, no one knew. . . .

Others told Kalm much the same thing about the West, so he concluded that these were once "the grain fields of a great village or town of the Indians."[213]

The Indians usually abandoned the fields they carved out of the forest after about 10, 20, or even 30 years and let them grow back to trees.[214] Then they returned to clear and plant the land again. As the Jesuit missionary Francois du Peron noted in 1639: "The land . . . produces for only ten or twelve years at most; and when ten years have expired, they are obliged to move their village to another place."[215] Even this much time required careful tending. So the Huron planted their crops on hills to prevent erosion. Wisconsin Indians planted corn in the same way. Here old Indian corn hills still covered large areas on the Campus of Carroll College in Waukesha as recently as 1902.[216] Indian corn hills also covered Martha's Vineyard, an island off the southern coast of Cape Cod, and hillsides near Fall River, Massachusetts, and Mohegan, Connecticut, as late as the 1930s.[217]

The Huron, like other Indians, also grew beans in the same hill with their corn to maximize production. By doing so they surely noticed that the corn grew better and the soils stayed productive longer. They could not know that beans added nitrogen to the soil. They did know that weeding increased their crops.[218] William Wood noted this about Indians living in New England in 1634. He said they keep "it so cleare with their Clamme shell-hooes, as if it were a garden rather than a corne-field, not suffering a choaking weede to advance his audacious head above their infant corne."[219] A century and one-half later William Bartram also said that Indian fields in Georgia were "kept clean of weeds."[220] Virginia Indians even ridiculed European colonists for not keeping their cornfields weed free.[221]

Captain John Smith described the way Virginia Indians planted and weeded the fields they hacked out of the forests, and then produced three crops a year. His description must be similar to what the Huron and other Indian farmers did when they cultivated the uplands:

> They make a hole in the earth with a stick, and into it they put four grains of wheat [corn] and two of beans. These holes they make four feet one from another; their women and children to continually keep it with weeding, and when it is grown middle high, they hill it about like a hop yard. In April they begin to plant, but their chief plantation is in May, and so they continue till the midst of June. What they plant in April they reap in August, for May in September, for June in October.[222]

The Huron sometimes went to the added trouble of removing tree stumps from their fields before planting, but they waited until the stumps started to rot.[223] Ultimately, the Hurons may have cleared and cultivated more than 23,000 acres of forest.[224] Young forests of eastern white pine, eastern redcedar, or other trees grew on thousands of additional acres that they had abandoned when the soils became unproductive. These second-growth forests growing on old croplands fit perfectly into the ingenious plan the Huron, Iroquois, and probably many other eastern Indians followed to periodically relocate and rebuild the wooden structures in their villages.

The Huron and the Iroquois lived in long houses within well-protected villages placed on hills to improve their defenses.[225] These were sophisticated fortifications. Samuel de Champlain noted in 1615 that there villages were

> enclosed and fortified by palisades of wood in triple rows, bound together, on the top of which are galleries, which they provide with stones and water;

the former to hurl upon their enemies and the latter to extinguish the fire which their enemies may set to the palisades.[226]

Their houses could be as much as 100 feet long. Champlain said that "there may be twelve fires, and twenty-four families" in one house.[227] They built them with a log frame and rafters that supported a gabled or arched roof, and then they covered them with bark.

A village of 1000 people and 36 long houses took 16,000 poles 4–5 inches in diameter, and 250 additional poles 10 inches in diameter, to construct. These houses also required 162,000 square yards of elm or redcedar bark to cover. The palisade took another 3600 poles at least 5 inches in diameter and 15–30 feet long to build. The second-growth forests, probably eastern white pine, that grew on their abandoned croplands provided just the right size trees and in large enough quantities to construct long houses and a palisade. [228] In the 1650s, for example, Van der Donck was surprised when Hudson River Indians told him that a stand of large trees was growing on cropland that they abandoned just 20 years earlier.[229] Therefore, as planned, when the Indians returned to an old village site a decade or two after moving, they had plenty of construction materials available to start again.

The day-to-day activities of Indians produced changes almost as important to America's original forests as the glaciers, dust storms, and floods that scraped and scoured the land. They cut trees, chipped stone tools out of rock, moved basketfuls of dirt, shot bison and other game, cleared campsites, dug holes with clamshell hoes, tended wild gardens, planted and domesticated crops, and set fires. By themselves these little human acts cannot rework the land, but millions of people doing these things over and over again for thousands of years can gradually transform a continent. There can be no doubt that North America would have been a different place when Europeans arrived if American Indians had not lived here.

Fire Masters

Used to be we lived in teepees all year. Moved round a lot. One place in the fall we'd burn, another place in the spring we'd burn that. Country was a lot more open then and wasn't so hard to travel. Not like now. You can hardly travel in the bush and it's not so good for hunting.

Cree, age 70
Fort Vermilion, Canada
Interviewed 1975–1977[1]

Of all the tools carried by Paleoindians into North American nothing came close to matching the power of fire. Sophisticated weapons and clothes and superior intelligence were not enough to allow them to live in the Arctic. Paleoindians had to control fire. They had to use fire to thaw frozen caches of meat or to cook meat stored in the frigid waters at the bottom of ponds.[2] Above all, they had to use fire for warmth. Paleoindians prospered in the Arctic because their skill in the uses of fire represented the accumulated knowledge and experience of 25,000 generations. There can be little doubt that Paleoindians were masters of fire. By mastering fire they also became masters of the New World.

The first true humans, *Homo erectus*, harnessed fire at least one-half million years before their Paleoindian descendants arrived in North America.[3] They were intelligent enough to make stone tools in that distant past; thus they also could easily learn about the effects of fire. They must have seen predatory birds drop down to the charred earth to pick up cooked rodents after a fire. They probably copied these birds and found out that crispy grasshoppers and game animals broiled in the flames tasted good. They surely saw large animals come to burned areas to lick the ashes for salt. They also could not help noticing that grasses and shrubs grew greener after a fire and that herds of game animals came to graze on the

143

lush new growth. The warmth of the glowing embers also must have comforted them on cold nights. It would only be a small step to carry these embers somewhere else to start a fire of their own.

Eventually humans learned to make fire whenever needed. Once tamed, they certainly used fire for cooking and heating, but they also used it to drive mammoths, mastodons, and other game during hunts and to promote the growth of grasses and shrubs for them to eat. Thus long before Paleoindians stepped foot in Beringia, their ancestors had already modified the vegetation of Africa, Eurasia, and Australia by adding their fires to lightning fires. Paleoindians and their descendants would do the same in the Americas.[4]

A WORLD OF FIRE

When Paleoindians emerged from the passageway between the ice sheets and walked into the vast spruce forests beyond, they entered a world of fire.[5] Several million acres of spruce forest probably burned each year, just as it does today, and a single fire most likely burned a million acres or more.[6] Even larger areas burned during periodic dry periods. The Paleoindians who entered such a world must have seen the fires burning every summer, or at least they smelled the smoke and had the horizon obscured by haze. Thus massive wildfires presented a constant danger to Paleoindian explorers.

Hissing, Roaring Flames

Anyone caught in front of a forest fire would have little chance for escape. Prairie fires also posed a serious hazard to Paleoindian explorers. Thousands of years later Lewis and Clark observed a tragic fire while traveling up the middle Missouri River on October 29, 1804. They recorded in their journals what may have been the fate of many Paleoindians:

> In the evening the prairie took fire, either by accident or design, and burned with great fury, the whole plain being enveloped in flames. So rapid was its progress that a man and a woman were burnt to death before they could reach a place of safety; another man with his wife and child were much burnt, and several other persons narrowly escaped destruction.[7]

Even the fast running bison sometimes failed to escape a rushing prairie fire. In 1804, English fur trader Alexander Henry witnessed the plight of bison that had been caught in a prairie fire in southern Canada. He was traveling only a short distance north of where Lewis and Clark saw the fatal fire along the Missouri River, and it was almost exactly one month later. Henry recorded the scene in his journal:

> Plains burned in every direction and blind Buffalo seen every moment wandering about. The poor beasts have all the hair singed off, even the skin in many places is shriveled up and terribly burned, and their eyes are swollen and closed fast. It was really pitiful to see them staggering about, sometimes running afoul of a large stone, and other times tumbling down hill and falling into creeks, not yet dead. . . . The fire raged all night toward the S.W.[8]

The first people to view these wildfires could not record what they saw. We can only imagine their fear and the danger they faced by experiencing wildfire through the eyes of those who followed. Such descriptions abound for one of the largest fires that ever raged through a North American forest in historic times. That fire blackened 3 million acres of northern Idaho and western Montana in August 1910. Excerpts from interviews and written accounts provide just a glimpse of what it must have been like to witness and survive this awesome spectacle.[9]

> The hissing, roaring flames, the terrific crashing and rending of falling timber was deafening, terrifying. The fire had closed in: the heat became intolerable. . . . The men fell as they ran before the merciless fire. (Joe Halm)

> The wind swept up the main valley. . . . It drove the hot flames in searing blasts across the dividing ridges between the creeks, leaping from crest to crest across mile-wide chasms in walls of flame. . . . roaring furnaces of fire, hot ash, smoke and exploding fumes from the thousands of burning trees, shrubs and brush moved with merciless velocity up the narrow gorges, generating infernos of heat and suffocation beyond description. . . . Then the flames sucked down into the depths of the canyon . . . and swept upward . . . in a seething caldron of falling trees, with soot and smoke and flaming branches soaring high in the air. . . . Appalling desolation everywhere. (Orland Scott)

We know from fossil evidence that the Ice Age spruce forest burned in much the same way as a modern spruce forest. Spruce charcoal fossils show that crown fires swept through parts of western Kansas 17,930,

14,450, and 10,245 years ago, and southwestern Nebraska 14,770 years ago. In addition, fossils of land snails that now live in subalpine forests of the Rocky Mountains occur in scattered locations over the Great Plains. These land snails live primarily in the decayed leaves that pile up beneath broad-leaf trees, and they are most abundant and diverse under aspen stands that have spruce growing in the understory. Aspen is a pioneer tree that invades freshy burned areas. So the Ice Age spruce forest must have burned more often than the fossil charcoal evidence alone can show.[10] These forest fires created the diversity of habitats that supported the abundance of animal life that Paleoindians exploited when they first arrived.

Lightning started most fires before Paleoindians entered the spruce forest. We do not know how often lightning burned the forest, but we do know that lightning strikes somewhere in the world at least 8 million times a day.[11] Even so, only one in a hundred to less than one in a thousand lightning strikes starts a fire. First, the lightning discharge must have enough energy and last long enough to ignite the fuel. Then there must be fine fuels available for the fire to start, such as pine needles and duff. These fine fuels must also be dry, which means that lightning cannot ordinarily start a fire in a rainstorm. Nevertheless, 3–4% of lightning flashes can meander through the air for nearly 20 miles before hitting the ground. These strikes can occur far outside the rainstorm that generated them. Therefore, fine fuels can still be dry enough to burn where the lightning strikes.[12] Fires can also start in punky, decaying logs and smolder for weeks, and then spread through the forest when it becomes dry. Even though it is unlikely that any one lightning bolt will start a fire, they still cause hundreds of forest fires each year just as they must have done during the Ice Age.

Fire People

Paleoindians certainly had reason to fear wildfires, but they had no means to stop them. On the contrary, fires burned more often, not less, after they arrived. Whether accidentally or intentionally, they set forests ablaze during any season of the year in which vegetation would burn. That means that their fires burned during the dry season, and even during droughts, not just in the rainy season when lightning fires started and the ground was usually moist. In some places this caused their fires to burn hotter and cover larger areas than lightning fires. In other places their fires broke up the vegetation and reduced fuels, which lessened the effects of wildfires. So, Paleoindians and modern Indians not only increased the frequency of fires, but they also changed the size and behavior of fires, the time of the year when they burned, and even the places they burned.

Paleoindians also knew how to take advantage of the benefits of fire while trying to avoid the dangers. Thick forests and heavy brush offered Paleoindians and modern Indians very little to eat. What they needed were young grasses, forbs, and shrubs to provide food for game and for themselves. Open forests and grassy plains also made hunting easier as well as more productive. Trees that produced acorns and nuts also thrived in open forests, not dense forests. Fire sets back vegetation to the earlier and more productive stages of growth, and it favors the most desired species. Therefore, it is not surprising that fires increased when Paleoindians arrived. By the time Europeans set foot in North America, Paleoindians and their descendants may have as much as doubled the number of fires that would normally burn because of lightning.[13] Such an increase in fires had widespread and lasting effects on vegetation throughout the continent, especially forests and the wildlife that depends on them.

Lightning and human-ignited fires disrupted the movement of trees and other plants as the climate warmed at the close of the Ice Age. Pioneer species led the way and settler species gradually moved in and replaced them. They sorted themselves over the landscape on soils where certain species had an advantage over others. Some trees stayed on the better soils, such as beech, some stayed near streams where moisture was plentiful, such as cottonwood, and others stayed on poor, sandy soils, such as jack pine. Regardless of the soil, fires, as well as hurricanes, tornadoes, floods, and other disturbances, periodically carve openings within forests, or even sweep them aside. These openings allow pioneer species to invade, and future fires and other disturbances retard or prevent the infiltration of pioneer forests by settler species. Thus pioneer forests can remain in the same place indefinitely as long as the disturbances continue. By increasing the number of fires, Paleoindians and modern Indians created even more openings in which pioneer species could grow. As a result, the forests that fire maintains, such as pine, oak, and aspen, covered larger areas of North America than would have been possible in an uninhabited wilderness.

We cannot be certain what caused particular fires in the distant past, but we know that fires increased because of humans. We also know that a warming climate not only increases the chance of lightning fires, but it also increases opportunities for human fires. The size of fires can also markedly increase when the climate warms, as they did in the Northern Rockies during the Great Drought.[14] Similarly, a cooler or wetter climate usually decreases opportunities for both lightning and human-caused fires. Consequently, changes in climate affect, but they cannot fully explain, changes in the frequency and size of fires when people are present. For example, as the climate grew cooler and wetter in the Great Basin at the end of the Great Drought about 4000 years ago, juniper forests expanded and grasses grew thicker. This increased the amount of fuel in the basin for fires to burn. As a result, wildfires increased in part because Paleoindians had been using fire in the region for at least 8000 years. Thus, fires set by Paleoindians created additional openings that favored grasslands and controlled the expansion of more juniper than would have been possible if lightning was the only source of ignition.[15]

Paleoindian fires may also have been responsible for other changes in vegetation that cannot be explained by climate alone. For example, alder abruptly spread through southwestern Alaska about 7500 years ago, even though the climate did not change.[16] Alder is a fire-adapted species, and human-ignited fires may be the most logical explanation for its rapid increase. Similarly, Paleoindian and lightning fires may have inflicted the final blow that caused the dramatic collapse of the white spruce forest in Minnesota about 10,700 years ago. Evidence from fossil charcoal fragments in lake sediments dating back more than 10,000 years support that possibility. The disappearance of the spruce forest allowed jack pine, and later white pine, both fire-adapted species, to move in and replace it.[17]

In north-central Florida, beech, a settler species, and other hardwoods gradually spread over the landscape 13,500 years ago as the climate grew warmer and wetter. However, about 11,200 years ago, close to the time that Paleoindians arrived in Florida, pine and oak displaced most of the hardwoods. Fossil pollen records also show that wildfires repeatedly swept through the pine and oak forest. Consequently, pine increased shortly after the fire and oak declined, and then oak increased between fires as pine declined, and then the next fire started the cycle again.[18] Thus, the combined affects of lightning and Paleoindian burning most likely increased the fire frequency enough to shift the forest from hardwoods to pine and oak.

The same thing may have occurred in the Pacific Northwest when Douglas-fir forests displaced the primeval forests of Englemann spruce, mountain hemlock, and lodgepole pine. Douglas-fir, alder, and bracken fern rapidly spread up the Puget Trough and the west slope of the Cascade Range between 11,000 and 6800 years ago. These are pioneer species that germinate and grow best on burned areas. The climate became warmer and drier about then, which made it possible for lightning and Paleoindians to ignite massive wildfires.[19] We know these fires occurred because charcoal deposits increased in lake sediments about that time.[20] We also know that Paleoindians had already been living in the region for a thousand years.[21] Therefore, the vast area of Pacific Douglas-fir forest that inspired awe among early European explorers probably owed their origin to massive fires set by Paleoindians and lightning at the end of the Ice Age.

The people who greeted the Europeans when they arrived in North America had lived with fire and used it as a tool for thousands of generations. No wonder Father Pere Marquette called the Maskouten Indians, who occupied the lands along the Fox River in Wisconsin, the "Fire People" when he met them in 1673.[22] Even the name Potawatomi, another nation from the Great Lakes region, means "Fire People" or "Fire Nation." The same name could have been applied to nearly all American Indians. They certainly added even more smoke and flames to a landscape already swept by fire.

Country Very Smoky

A smoky haze from Indian fires filled the air when Europeans explored the continent, not every day and not all the time, but frequently enough to be noted in many journals in many regions. In March 1524, Giovanni da Verrazzano remarked when sailing off the coast of North Carolina that, "we smelled the odor a hundred leagues and farther when they burned the cedars and the winds blew from the land."[23] A Dutch traveler made a similar comment while sailing off the coast of Delaware in 1631:

> It comes from the Indians setting fire, at this time of year, to the woods and thickets, in order to hunt. ... When the wind blows out of the northwest, and the smoke is driven to sea, it happens that the land is smelt before it is seen.[24]

Colonel William Byrd, in his history of the 1728–1729 survey of the dividing line between Virginia and North Carolina, also wrote that "the atmosphere was so smoky all around us that the mountains were again grown invisible." He concluded that "this happened not from haziness of the sky, but from the firing of the woods by the Indians."[25]

Comments about the clarity of the air also appear in journals. Father De Smet was traveling west from la Ramée (Fort Laramie, Wyoming) toward the Rocky Mountains in the spring of 1840 when he wrote about his side trips to explore the surrounding country:

> I was often deceived in regard to distances; sometimes I wished to examine more closely a big rock or an odd-looking hill; I started for it expecting to reach it in an hour; and it took me at least two or three hours. This must be due to the great purity of the atmosphere in the prairies of this high region.[26]

The clear air presented other problems as well. In 1810, while trying to elude Blackfoot warriors near Three Forks, Montana, Thomas James and another trapper "ascended a small height to watch the Indians, while the rest went on with the horses." "Here we saw the Indians go up to our deserted camp," he said, "the smoke from which had attracted them thither." James then pointed out in his journal that "smoke in this clear atmosphere is visible to a great distance."[27]

Smoke still obscured the view on many occasions. While traveling along the upper Missouri River in May 1805, Lewis and Clark said, "the air was turbid in the forenoon, and appeared to be filled with smoke; we supposed it to proceed from the burning of the plains, which we are informed are frequently set on fire by the Snake Indians."[28] They had to take precautions because fire represented an ever present danger, as they noted on August 23: "I also laid up the canoes this morning in a pond near the Forks; sunk them in the water and weighted them down with stone . . . hoping by this means to guard against both the effects of high water, and that of Fire, which is Frequently kindled in these plains by the natives."[29]

In the 1870s, John Wesley Powell made a general comment about the effects of Indian fires. He said that "everywhere throughout the Rocky Mountain Region the explorer away from the beaten paths of civilization meets with great areas of dead forests." This was especially true "in seasons of great drought." It was then, he noted, that "the mountaineer sees the heavens filled with clouds of smoke." These were mostly Indian fires, not natural fires, as he verifies: "In the main these fires are set by Indians."[30]

Most explorers would have agreed with Powell. Trapper Osborne Russell said that Fort Hall in Idaho was "the most lonely and dreary place I think I ever saw." He probably felt that way partly because of the heat and smoke from Indian fires that were burning at the time. "The country very smoky and the weather sultry and hot," he noted in his journal in September 1835.[31] On a rainless day in July 1860, near Lookout Pass in northern Idaho, Army topographical engineer P. M. Engle made a notation in his journal that showed fires and smoke were still commonplace:

> Mr. Sohon, myself, and one Indian guide ascended today the mountain, but before reaching the summit we were convinced that we would have no distant view, and therefore retraced our steps. At one pm the smoke of the burning timber enveloped the whole country, and our Indian guide assured that we would have no view until after a heavy rain.[32]

Smoke from Indian fires also spread across the Pacific Northwest, California, and even Alaska. Members of Captain Nathaniel J. Wyeth's expedition spent days in smoke in September 1834 as they passed northeast of the Blue Mountains in Oregon near the Umatilla River. "Every day has been thick smoke like fog," he said, "enveloping the whole country . . . and the whole country burnt as black as my Hat."[33] Similarly, in 1850, writer Frank Marryat traveled through Sonoma County in northern California and wrote:

> At times the Indians would fire the surrounding plains, the long oat-straw of which would ignite for miles. The flames would advance with great rapidity, leaving everything behind them black and charred. At these times a dense smoke would hang over the atmosphere for two or three days, increasing the heat until it became insupportable.[34]

A military reconnaissance party in Alaska had a similar experience with smoky skies. In 1885, Henry Allen recorded "heavy smoke" on the upper Tanana River from a fire set by Athabascan Indians that "obscured the sun the entire day, so that an observation was impossible."[35]

On April 28, 1841, Captain Charles Wilkes and the frigate *Vincennes* arrived at the Columbia River. A crashing surf prevented him from crossing the bar at the mouth of the river that day, so he turned north to explore Puget Sound since he could not enter the harbor. Captain Wilkes was the commander of a flotilla of six ships that made up the 1838–1842 United States Exploring Expedition. The expedition was completing a scientific survey of Antarctica, the South Pacific, Australia, Hawaii, and the West Coast of North America. Eventually Captain Wilkes returned to explore the Columbia before sailing to San Francisco Bay. He dropped off Lieutenant George F. Emmons and a party of officers and scientists while anchored at Fort Vancouver. The assignment he gave Lieutenant Emmons was to travel down Oregon's Willamette Valley and into California to make certain they did not overlook any other navigable river that flowed into the sea north of Spanish California. Emmons and his party began their journey during the peak of the Indian burning season.[36]

The Emmons party spent most of their time in an atmosphere filled with the smoke of Indian fires as they made their way to California. On August 7, 1841, Emmons stood on the Eola Hills and looked down into the Willamette Valley, noting in his journal: "The country becoming smoky from the annual fires of the Indians who burn the prairies" and "the forests." He was justifiably concerned that Indian fires would destroy the grass their horses needed during the expedition. So he added that "these two burnings combined form the greatest obstacle the travelers encountered in this country—one blocking up the way [because of fallen trees]—& the other destroying the food of the animals." The smoke lingered for weeks, leading another member of the party, Midshipman Henry Eld, to record on September 9 that the atmosphere was "filled with smoke" and "consequently unable to see much of the surrounding country." A few days later he added, "it proved to be a thick smoky evening so as to preclude all possibility of getting the North Star." Then six days later, as they left the valley and began their climb into the mountains, Emmons wrote: "Calm, sultry & smoky as ever—the air from the prairies fanning past me—some thing like the heated air from [an] oven."[37]

If sixteenth century and later European explorers had traveled across North America 10,000 years earlier, they most likely would have written the same comments about Paleoindian fires. We cannot be certain if Paleoindians or lightning started a particular forest fire recorded in fossil charcoal or ash. Nevertheless, in many places there is nearly an equal chance that Paleoindians set the fire if it started after they settled the continent more than 12,000 years ago.

THEIR FIRES ARE LEFT BURNING

Even if Paleoindians did not burn vegetation intentionally, and the evidence is overwhelming that they did, they still would have increased wildfires by accident alone. Such accidental fires were unavoidable. Accidental fires could start anywhere along a trail, and by the time Europeans arrived Indian trails laid across North America like an intricate spider web. Indians walked these trails so often over the ages that they became nearly permanent features on the landscape. European travelers used the same trails because they were the most efficient and logical routes. Eventually many of these ancient trails became modern highways and roads. Thus we can gain a sense of the complexity of the network of Indian trails that crisscrossed the continent by looking at modern road maps. In addition, we know the locations of many Indian trails in such places as Ohio, California, Massachusetts, Rhode Island, Connecticut, Missouri, and the Southeast. Maps show that most forests were within a day's walk from an Indian trail.[38] Thus, accidental fires along trails could easily have burned many forests many times, even if Indians did not maintain a permanent or seasonal campsite nearby.

Some accidental fires probably started as embers fell from firebrands, slow matches, or burning wood held in containers, that people carried along trails. Slow matches consisted of a tightly rolled rope of bark that burned slowly at one end. Paleoindians may have carried slow matches because they could not be sure that they could start a fire whenever they needed it. Hence, like modern Indians, they most likely tried to preserve fire rather than make it.

Paleoindians may also have carried firebrands to keep themselves warm as Indians did during more recent times. For example, during Coronado's expedition in Arizona and New Mexico in 1540–1542, Pedro de Castaneda wrote in his chronicle that

> on account of the great cold, they carry a firebrand in the hand when they go from one place to another, with which they warm the other hand and the body as well, and in this way they keep shifting it every now and then.[39]

Embers from such a "firebrand" could easily start a fire on a cold, dry day.

Indians relied primarily on friction and dry fuel to start fires. This method of fire starting is effective but difficult, especially in wet weather. Captain John Smith observed their methods in Virginia in 1607: "Their fire they kindle presently by chafing a dry, pointed stick in a hole of a little square piece of wood, that firing itself, will so fire moss, leaves, or any such like dry thing, that will quickly burn."[40] Indians were still using the same technique in 1835. Trapper Osborne Russell met Snake Indians in the Lamar Valley of Yellowstone and noted that "they produce [fire] by the friction of two pieces of wood which are rubbed together with a quick and steady motion."[41] Bacqueville de la Potherie, a friend of French agent Nicolas Perrot, documented the problem of starting a fire this way in wet fuel when he wrote about Perrot's adventures in Wisconsin in 1668–1670:

> As the Frenchmen reached the bank of the river [the Fox River], a dignified old man appeared with a woman, carrying a clay pot filled with cornmeal porridge. More than 200 strong, young men arrived, their hair adorned with headdresses, their bodies covered with black tattoos. . . . They carried arrows and war clubs. . . . the old man made a harangue, like a prayer, to assure Perrot of their joy at his arrival. One of the men then spread on the grass a large, painted buffalo hide, with hair as soft as silk, on which Perrot sat. The old man rubbed two pieces of wood together to make fire, but because the wood was wet, he could not light it. Perrot then drew forth his firesteel and immediately made fire with tinder. The old man uttered loud cries about the iron, which seemed to him a spirit.[42]

Roger Williams, the founder of Rhode Island, had a similar experience around Narragansett Bay in Massachusetts in 1643. The Indians felt awe when they saw the flint and metal "strike-a-lights" that Europeans used to start fires. They told him that "this fire must be a God, or Divine Power, that out of a stone will arise a Sparke."[43] Many Indians believed fire was "supernatural" and, according to Father De Smet, it was "an emblem of happiness or of good fortune" as well. He said that " 'having extinguished the enemy's fire,' signifies with them to have gained the victory."[44]

Most accidental wildfires caused by Indians probably started from campfires rather than embers dropping from firebrands and slow matches. A pastel drawing by Joseph Drayton, an artist with the United States Exploring Expedition, of "Indians playing the spear game" in 1841 near Mt. Hood illustrates such a fire.[45] The drawing shows Indians camped in a clearing at the base of a mountain covered with a dense Pacific Douglas fir forest. Huge swaths of forest appear to have been swept away by escaped fires that ran up the mountainside from the camp. The burning campfire in the drawing shows the smoke curling up the slope, which also illustrates the probable direction of the wind at the time of the fire. This accidental forest fire would not have endangered the Indians since the wind blew it up the mountain and away from their camp. This kind of Indian-caused forest fire would have been common throughout North America because they were impossible to prevent.

Rarely did Indians put their campfires and cooking fires out before moving to the next resting place or campsite. These abandoned campfires started many more forest fires. In 1658, Pierre Esprit Radisson and his brother-in-law, Medard Chouart de Groseilliers, used an abandoned campfire to find their Indian guides. They decided to disobey the governor of Quebec and explore and trap in the country around Lake Huron. So they left Montreal at midnight and went looking for a party of "wildmen" who they thought might take them up the Ottawa River. "After three days we see the tracks of seven canoes and fires yet burning," Radisson wrote in his journal. Thus the abandoned campfires showed that the Indians were near. "We take no rest until we overtake them," he said.[46] So began a 2-year odyssey in which they survived a fight with an Iroquois war party and nearly starved in the winter. They returned to Quebec with a fortune in beaver pelts, but the governor took them away and put Groseilliers in jail. Radisson's bitterness led him and Groseilliers to defect to the British and help them establish the Hudson's Bay Company.

Paleoindians, and the Indian people who followed them, had no reason to put out a fire. On the contrary, they had good reasons to leave their campfires burning. The slow match they carried could go out if an Indian slipped crossing a stream or was caught in a sudden rainstorm. So they undoubtedly banked some of their campfires with ashes or dirt to keep them burning even longer in case they had to come back and light their match again.[47] This also made it easier to restart campfires.[48] Some of these fires must have escaped their firepits, and a few of these surely tore through the surrounding forests and grasslands.

Colonel Robert Campbell, Jim Bridger, and William Sublette of the Rocky Mountain Fur Company, had a good laugh about one fire abandoned by Indians. They were on a trapping expedition in the fall of 1826 when an "incident occurred ... that created some excitement in the party," wrote Colonel Campbell.

> We came across the tracks of a bear, and while following it, Jim Bridger, who was in advance, saw a smoke on the head waters of the Missouri. We,

Sublette, Bridger and myself determined to see what it was. As we habitually had to be on the alert when hostile Indians were suspected in our vicinity, we dashed along, until we came to a place where the Indians had camped a month before, leaving some burned logs, from which the smoke still issued. The exploit became known for a long time in camp as "the battle of the burned logs."[49]

Father De Smet's experience in the summer of 1840 was less humorous. He and his companions found many campfires left burning while trying to avoid Indian war parties. They journeyed down the Yellowstone River and "espied, upon waking very early in the morning, the smoke of a great fire a quarter of a mile away; only a rocky point separated us from a savage war-party." He said "without losing time, we saddled our horses and started at full gallop; at last we gained the hill" and "reached the top without being perceived." He spent a "sleepless night" and the next day he took "the greatest precautions, because the country we had to traverse was most dangerous." However, he came upon a buffalo that "had been killed, not more than two hours before." This, he said, was "a fresh cause for alarm." He turned and went in the opposite direction and then camped "among the rocks." The next morning, however, "towards ten o'clock," he "came to an abandoned camp of forty lodges; the fires were not yet out; but luckily we saw no one." However, when he reached the Missouri, he arrived at a "place where 100 lodges of Assiniboins had crossed an hour before." When he finally reached safety, Father De Smet related his story to an Indian chief who commented: "The Great Spirit has his manitous (guardian spirits); he sent them out to you." Father De Smet agreed, and made a note in his journal: "I have never seen a plainer instance of the special Providence that protects the poor missionary."[50]

In August 1590, John White, governor of Virginia, also saw several abandoned Indian campfires and what was most likely an accidental forest fire. He stood at the railing of his ship and "saw a great smoke rise in the Ile Roanoke neere the place where I left our colony in the yeere 1587, which smoke put us in good hope that some of the Colony were there expecting my returne out of England." The next day he sat on board one of two small boats and rowed into the shallow waters of Pamlico Sound toward the "smoake." But, he said, "before we were halfe way betweene our ships and the shore we saw another great smoke to the

Southwest of Kindrikars Mountes." The boats turned and headed for the second "smoake," but they found a burning campfire with "no man nore signe that any had bene there lately." A few days later they "espied ... the light of a great fire thorow [sic] the woods, to which [they] rowed." They lowered the boats from the ship and rowed into the sound. They arrived at the north end of Roanoke Island after dark. In the light of the forest fire Governor White said, "we let fall our Grapnel neere the shore, & sounded with a trumpet a Call, & afterwardes many familiar English tunes of Songs, and called to them friendly; but we had no answere." The next morning at daybreak Governor White landed and walked to the fire, where he "found the grasse and sundry rotten trees burning about the place." To this day no one knows the colony's fate.[51]

Colonel Byrd's survey expedition of 1728–1729 passed the famous "Warriors Path" used by the Iroquois in raids and wars with the Cherokee and other southern Indians. It began in western New York, traveled down the Susquehanna River in Pennsylvania, through the foothills of the Appalachians, and into South Carolina. Where the trail crossed the line between Virginia and North Carolina, Byrd saw an accidental fire and recorded the incident:

> We were now near the route the northern savages take when they go out to war with the Catawbas and other southern nations. On their way, the fires they make in their camps are left burning, which catching the dry leaves which lie near, soon put the adjacent woods in a flame.[52]

Similarly, Lewis and Clark mentioned an accidental Indian fire in their journals that started near the lower Missouri River in August 1804:

> In the morning some men were sent to examine the cause of a large smoke from the northeast, which seemed to indicate that some Indians were near; but they found that a small party, who had lately passed that way, had left some trees burning, and that the wind from that quarter blew the smoke directly toward us.[53]

Again, the Indians who "passed that way" cared little about campfires escaping into the forest. This is especially true because, as Roger Williams said in 1643, "burning of the wood to them they count a benefit."[54] Scattered charcoal around an 11,000-year-old Goshen bison kill site in southeastern Montana shows that a fire burned through the meat processing area about the time that these ancient hunters left.[55] These are the people who replaced the Clovis hunters. Therefore, it seems reasonable to assume that Paleoindians felt the same lack of concern as modern Indians about leaving their campfires burning.

THE OMINOUS SMOKE SIGNAL

Smoke billowed from signal fires whenever Europeans entered Indian territory. Eastern Indians used smoke signals in the same way that Indians did in the West, as Samuel de Champlain discovered when he explored the New England coast in 1605. Indians paddled their canoes out to his ship and then, Champlain said, they "went back on shore to give notice to their fellow inhabitants, who caused columns of smoke to arise on our account."[56] This old and effective method of communication was universal among native people in North America and throughout the world. "Smoke," Father Escalante said in 1776 while traveling in northern Utah, "is the first and most common sign which in case of surprise, all the people of this part of America use."[57] Some of these signal fires were small and remained controlled, some of them escaped and became wildfires, and still others resulted from Indians setting wildfires that also served as signals. Regardless, signal fires occurred so often, and in so many places for so long, that, by themselves, they can account for substantial changes in America's ancient forests, brushlands, and grasslands.

Indian signal fires served many purposes. The most obvious was to warn of approaching enemies, which Captain John C. Fremont recorded during his second expedition for the Corp of Topographical Engineers in 1843–1844. Captain Fremont traveled south from Fort Vancouver to California, through western Nevada, where he made the following note in his journal before crossing the Sierra Nevada in the snow:

> I rode out with Mr. Fitzpatrick and Carson to reconnoiter the country, which had evidently been alarmed by the news of our appearance.... Columns of smoke rose over the country at scattered intervals—signals by which the Indians here, as elsewhere, communicate to each other that enemies are in the country.[58]

Hudson's Bay Company trapper, and explorer, Peter Skene Ogden had the same experience about 15 years earlier. Ogden was the son of an American

Tory who fled to Canada during the Revolution. He was the first to travel the West from north to south, and he discovered the Humboldt River and named Mt. Shasta (he called it "Shasty") in northern California. He could also be cruel. Ogden once forced a man to climb a tree and then set fire to it just for fun. Hudson's Bay Company considered him their "ultimate weapon" against American expansion into the West. Nonetheless, David Douglas, a botanist and plant collector for the Royal Horticultural Society and the person for whom the Douglas-fir tree is named, had a pleasant meeting with Ogden in 1826. Douglas said Ogden was "a man of much information and seemingly a very friendly-disposed person."[59]

During his trapping expedition of 1828–1829, Ogden was traveling through northern Nevada where he recorded that "it is very evident from the numbers of fires in all directions that we are discovered by the natives." Shortly afterward his party had a fight with Indians at the Humboldt Sink. Later in their expedition they had to fire a volley into Mojave Indians who charged them with spears. Ogden said, "the first, however, sufficed, for on seeing the number [26] of their fellows who in a single moment were made to lick the dust, the rest ingloriously fled."[60]

The "heavy smoke" that Henry Allen wrote about when troops entered the upper Tanana River area of Alaska in 1885, also "originated from signal fires which were intended [by Athabascan Indians] to give warning of our presence in the country." He described the scene:

> When we first arrived at Nandell's there was only an occasional smoke around, but as his guests departed for their different habitations each marked his trail by a signal fire. The prevailing wind was from the east and carried the smoke along with us. In answer to the fires on the south bank new ones started on the north, so that for nearly two days we barely caught a glimpse of the sun except through the heavy spruce smoke.[61]

Large signal fires were also common on the Great Plains and mountain prairies. Lewis and Clark recorded such a fire in their journal on July 20, 1805. They approached the Rocky Mountains near Helena, Montana, on their way west when "one of Captain Clark's men" fired a gun. Shortly afterward they "discovered a great smoke, as if the whole country had been set on fire." Thus, the Indians "believing that their enemies were approaching had fled into the mountains, first setting fire to the plains as a warning to their countrymen."[62]

Experienced explorers and military leaders knew that not all smoke signals warned of enemies. Even in Apache country, people General Crook called "the tiger of the species," smoke could be a sign of grief, sympathy, or something else rather than a warning.[63] Captain John G. Bourke, an able cavalry soldier who fought the Apache in Arizona under General Crook in the 1870s, wrote in his journal:

> From every peak now curled the ominous smoke signal of the enemy, and no further surprises could be possible. Not all of the smokes were to be taken as signals; many of them might be signs of death, as the Apaches at that time adhered to the old custom of abandoning a village and setting it on fire the moment one of their number died, and as soon as this smoke was seen the adjacent villages would send up answers of sympathy. [64]

On June 22, 1806, Captain Clark recorded an incident in his journal in which Flathead Indians burned a spruce-fir forest just to improve the weather. The party was trying to cross the Bitterroots on their return trip from the Columbia even though snow still covered the "road." They were anxious because, as they said, "we already knew that to wait till the snows of the mountains had dissolved, so as to enable us to distinguish the road, would defeat our design of returning to the United States this season."[65] They camped below the snow line and prepared to make their second attempt over the mountains that separate Idaho and Montana. Then their Indian guides began burning trees. Captain Clark described the event:

> In the evening the Indians, in order as they said to bring fair weather for our journey, set fire to the woods. As these consisted chiefly of tall fir-trees, with very numerous dried branches, the blaze was almost instantaneous; and as the flame mounted to the tops of the highest trees, it resembled a splendid display of fire-works.[66]

The Apaches did the same thing in the Southwest. However, they burned "miles" of mountain lands to bring rain instead of "fair weather."[67]

Indians also used fires to signal their families and tell them they would soon be home. Sometimes these fires were carefully set to avoid accidentally causing a wildfire that could endanger a village. An elderly Cree-Metis who lived on the front Range of the Canadian Rockies, in Alberta, during the late 1970s recalled the way he used to burn signal fires:

> Where the trail came into the valley it was up on a high, rocky ledge. At that place we were still an hour ride from home and if we wanted to let someone

know that we were coming we'd pick out an isolated tree along the ledge and then set it off. The greener the tree the better cause you wanted lots of smoke.... No, it wasn't dangerous; just one, small green tree and nothing around but rock. We didn't burn the whole forest down just to let someone know that we were home. We weren't crazy, you know.[68]

Nonetheless, in August 1832, trapper Warren Ferris noted that the Flatheads could set large fires to announce their arrival. Ferris was in the grasslands and open ponderosa pine forests of the Bitterroot Valley in Montana when he made the following notation in his diary:

On the 13th, we continued down this [Bitterroot] river, till evening and halted on it. The [Flathead] Indians with us, announced our arrival in this country by firing the prairies. The flames ran over the neighboring hills with great violence, sweeping all before them, above the surface of the ground except the rocks, and filling the air with clouds of smoke.[69]

These large signal fires set by Indians were so common that members of General Henry D. Washburn's second expedition into Yellowstone set one to find a lost companion in 1870. Truman C. Everts disappeared somewhere near Yellowstone Lake. The other members of the party spent days searching for him, firing their guns, blazing trees, and even leaving maps and rations behind, but without success. Nathaniel P. Langford and Samuel T. Hauser also climbed "a peak" above Yellowstone Lake on September 10 "and fired the woods, in hope of giving him a point of direction," reported Lieutenant Gustavus C. Doane. Three days later Lieutenant Doane wrote:

The fire kindled on the summit of the mountain has by this time spread to a vast conflagration, before the devouring flames of which tall pine trees shrivel up and are consumed like grass. The whole summit of the mountain sends up a vast column of smoke which reaches to the sky, a pillar of cloud by day and of fire at night.[70]

They never found Everts, but he survived for 37 days before being rescued farther north near the Gallatin River.[71] Ironically, Langford, one of the two people who started the "vast conflagration," became the first superintendent of Yellowstone National Park. He stated in his report for 1872, the first year of the park: "It is especially recommended that a law be passed, punishing, by fine and imprisonment, all persons who leave any fire they may have made, for convenience or otherwise, unextinguished."[72]

Lewis and Clark also adopted the use of large single fires. During the early part of their journey up the Missouri River they made an entry in their journal about it on August 17, 1804:

> In order to bring in any neighboring tribes, we set the surrounding prairies on fire. This is the customary signal made by traders to apprise the Indians of their arrival; it is also used between different nations as an indication of any event which they have previously agreed to announce in that way, and as soon as it is seen collects the neighboring tribes, unless they apprehend that is made by their enemies.[73]

According to Lewis and Clark, the Flatheads and Pend d'Oreille also set "Prairies or open Valies on fire" to assemble different bands. The purpose, they said, was "to go to the Missouri where they intend passing the winter near the Buffalow."[74] Two decades later Indians were still firing the prairies to begin the trek across the Rockies for the annual buffalo hunt. On July 14, 1827, Peter Ogden noted the event in his journal while traveling along the edge of the Snake River Plain in eastern Oregon. In Ogden's words, "the country on all sides is on fire, these are signals for Indians to assemble as they shortly will steer their course to Buffaloe."[75]

FIRING THE FORESTS OF THEIR ENEMIES

American Indians waged "continual warfare" with their neighbors, so said Father De Smet in 1840 about native people living in the Northern Rockies.[76] Centuries earlier, and long before Europeans began encroaching on their lands, Cabeza de Vaca made the same observation about Indians in east Texas. He wrote: "All the nations of the country are their foes" and "they have unceasing war with them."[77] William Bartram tried to explain Indian warfare in the late 1700s, and he succeeded as well as anyone:

> Their motives spring from the same erroneous source as they do in all other nations of mankind; that is, the ambition of exhibiting to their fellows a superior character of personal and national valour, and thereby immortalizing themselves, by transmitting their names with honour and lustre to posterity; or revenge of their enemy, for public or personal insults; or, lastly, to extend the borders and boundaries of their territories.[78]

Indians used many weapons in war, such as the bow and arrow, war club, lance, tomahawk, and knife. But nothing surpassed the reach and power of fire. Such was the conclusion of S. J. Holsinger, an examiner for the General Land Office, who wrote in 1902 that "the most potent and powerful weapon in the hands of these aborigines was the firebrand." Especially, he said, as a weapon to "vanquish the enemy."[79] In 1870, W. A. Bell said much the same thing: "The Apaches also have a very destructive habit amongst their long catalogue of vices [that] of firing the forests of their enemies."[80] The Apaches set so many fires during their wars with the Spanish, and later the Mexicans and Americans, that they dramatically increased the frequency of fire in the Chiricahua Mountains of southeastern Arizona.[81]

Indians used fire as a weapon because it could flush an enemy from cover or stop their attack. It could also block their movements with fallen trees, harass and demoralize them, destroy them, deprive them of game and forage for their horses, or obliterate their villages and crops. In 1853, Dacotahs came to a large Arickara village to trade bison meat and robes for corn and pumpkins. According to a War Department report, the Dacotahs stole everything they could during their visit. They also "set the prairies on fire, in order to prevent the buffalo from visiting the Rees country." "An act of dastardly malignity, as it deprives the Aricarees of the means of support," the report stated.[82] Surely Indians had been burning large areas of forest and prairie during the many wars they fought with one another long before they turned this weapon on Europeans.

The native people of North America used fire against Europeans early in their exploration of the continent. In 1603, Martin Pring and his crew landed on the shore of Plymouth Bay to search for sassafras, which they wanted to collect for medicinal purposes. They took a nap during their search, but were awakened by passing Indians. When they left, Pring said: "The Indians dissembled [in] a jesting manner." Then "not long after, even the day before our departure, they set fire on the woodes where wee wrought, which we did behold to burne for a mile space."[83] Apparently these Indians set several fires over a period of days. They probably burned the forest to harass the Europeans, drive them away by destroying the sassafras trees, or perhaps even kill them.

Over two centuries later, and on the opposite side of the continent, Indians were still burning forests to harass their enemies. The Emmons party climbed the Umpqua Mountains in September 1841 after leaving

Oregon's Willamette Valley when, wrote Lieutenant Emmons, the "atmosphere [was] getting smoky again." He suspected the Indians were burning the forest, in his words, "doubtless to obstruct us," but they may have been burning it for some other reason. Nonetheless, the fires did slow them down because, wrote Emmons, "large trees had fallen over our path, so that we were in many instances obliged to cut our way through or around them."[84] Granville Stuart, who discovered gold in Montana in 1857, also found himself in the midst of Indian fires in July 1861. He was traveling along the Clark Fork River in western Montana when "war parties of Bannocks," he said, "have the mountains on fire in all directions."[85]

Lieutenant Doane wrote about a similar incident. He commanded a tiny force of five troops assigned to protect General Washburn's expedition into Yellowstone. It was late August, and the Crow were passing through the area on their way to the Great Plains to hunt bison. So members of the expedition felt threatened, and for good reason. Doane wrote in his report, "Passing over the high rolling prairie for several miles, we struck at length a heavy Indian trail leading up the river, and finding a small colt abandoned on the range, we knew they were but a short distance ahead of us." Later that day he also wrote: "We passed, a mile before going into camp, near a small lake, the wickey ups of fifteen lodges of Crows, the Indians whose trails we had been following across the plateau." These Indians could be hostile, so Doane posted guards every night. His concern heightened when he met hunters who told him that nearby they found "the skeletons of two hunters murdered by the Indians two years ago." Thus, it is not surprising that Lieutenant Doane suspected that Indians were burning the prairies and forests to harass the expedition. "The great plateau," he said, "had been recently burned off to drive away the game, and the woods were still on fire in every direction." The Crow left them alone throughout the remainder of the expedition.[86]

The Crow may also have set fires in Yellowstone to deprive Washburn's expedition of grass for their horses, as well as game. The Comanche, Dakotas, and many other tribes often harassed their enemies this way in the West. The Dakotas burned the grass along hundreds of miles of cattle trail to stop cattle drives in the 1880s.[87] Cheyenne and Arapaho scouts also fired the prairies to keep cattle, known as "drifters," from wandering into Indian territory around Fort Elliott in Texas. Colonel Homer W. Wheeler, commander of the scouts, ordered them to stop this age-old practice, but not before the prairies burned for several days. It took a rainstorm to extinguish the blaze.[88]

Warfare on the Great Plains and other western prairies frequently involved firing the grasslands. Colonel Richard Dodge, who led the Black Hills scientific expedition of 1875, and his troops found themselves plagued by prairie fires as they crossed the Plains. This time Indians probably set the fires with lethal intent. Colonel Dodge said that "setting fire to the grass in the vicinity of the camp at night was one of the Indian modes of annoying a party too strong for attack and too vigilant for a successful attempt at theft." He recounted such an incident. His troops had, in Colonel Dodge's words, "been followed for several days in succession by a party of Indians, who fired the grass to windward of my camp every night forcing me to burn all around the camp every evening before posting sentinels."[89]

In October 1826, Peter Ogden's trapping party barely escaped a similar prairie fire that Indians lit to drive them out of eastern Oregon.[90] "We had certainly a most providential escape," he said.

> Last night the Indians crossed the river and set fire to the grass within 10 yards of our camp. The watch perceived it and gave the alarm. Had there not been a bunch of willows to arrest it everything would have been lost; a gale blowing at the time.[91]

Similarly, Captain Wyeth wrote in his journal on August 10, 1834, while trapping near the Snake River in Idaho, that he "saw a large fire in the mountains." He said that "Diggers keeping for safety in the hills" set the fire because "the Blackfeet trouble them even here."[92] Likewise, on the northern Great Plains the Slave used fire to defend themselves against the Sioux and Cree.[93]

Indians also used fire to drive their enemies from camp so that they could steal their horses, as British author and adventurer George Ruxton found out when he was hunting near the east side of Pikes Peak in 1847. "I was following a band of deer," he wrote in his account of the incident, when "I came suddenly upon an Indian camp, with the fire still smoldering, and dried meat hanging on the trees." "I saw two Indians, carrying a deer between them, emerge from the timber bordering the creek," he said, "whom I knew at once by their dress to be Arapahos." Ruxton slipped away unnoticed, and the next morning hid his camp in a canyon "entered by a narrow gap." He thought he was safe, but that evening he was "surprised to see a bright light flickering" on the mountainside.

> A glance assured me that the mountain was on fire ... I saw at once the danger of my position. The bottom had been fired about a mile below.... A dense cloud of smoke was hanging over the gorge, and presently ... a mass of flame shot up into the sky and rolled fiercely up the stream ... roaring along the bottom with the speed of a racehorse.... The dry pines and cedars hissed and cracked, as the flame ... ran up their trunks, and spread amongst the limbs, whilst the long waving grass underneath was a sea of fire.

It was night, yet "the whole scenery was illuminated, the peaks and distant ridges being as plainly visible as at noonday." Now Ruxton was "surrounded by fire," so he mounted his horse Panchito and charged through the burning brush, but "Panchito's ... mane and tail were singed." The Indians must have seen him escape. He recalled that, "just as I had charged through the gap I heard a loud yell, which was answered by another at a little distance." "Once in safety," he said, "I turned in my saddle and had leisure to survey the magnificent spectacle." "The fire had extended at least three miles on each side the stream, and the mountain was one sheet of flame." It swept eastward, down the mountain, over the foothills, and on to the prairies, where "for fourteen days its glare was visible on the Arkansas, fifty miles distant." "I had from the first no doubt but that the fire was caused by the Indians," Ruxton noted in his account. He suspected that the Arapahos "had taken advantage of a favorable wind to set fire to the bottom, hoping to secure the horse and mules in the confusion, without the risk of attacking the camp."[94]

Father Louis Hennepin, a Jesuit priest who accompanied Robert Cavelier de La Salle's 1679–1680 expedition, documented an unusual way to harass an enemy with fire. La Salle ordered Father Hennepin and two French companions to go down the Illinois River and then up the Mississippi River to find its headwaters. Unknown to Father Hennepin, a large Sioux war party of 33 bark canoes was paddling down the Mississippi to attack the "Miamis, Illinois and Maroa" as he was traveling upstream. When they met, the Sioux fired arrows at them because they thought they had lost the chance for a surprise attack. Then, Father Hennepin said, "as they drew near our canoe, the old men saw the peace pipe in my hands and prevented the young men from killing us." The Sioux took them prisoner to prevent them from alerting their enemies, and retreated back up the Mississippi. Father Hennepin recorded what happened after the Sioux debated their fate during a frightening 2 weeks on the river:

> They gave each of us to the head of a family—in place of a son who had been killed in a war—and then they broke our canoe to pieces, for fear we would escape to their enemies. Although we could easily have reached their country by water, they forced us to march from daybreak to two hours after nightfall for 150 miles, and to swim across many rivers. On leaving the cold water, I could scarcely stand. We ate a few pieces of meat just once a day. I was so weak that I often lay down to die, rather than follow those Indians, who marched with a speed that surpasses the strength of Europeans. To force us onward, they often set fire to the grass of the prairies, so we had to advance or burn.[95]

The Indians finally allowed Father Hennepin and his companions to return to Green Bay, Wisconsin, the following fall after Sieur Du Lhut [Daniel Greysolon Duluth] and four French soldiers found them with the Sioux.

Forcing an enemy from cover was a common use of fire by Indians. Osborne Russell nearly lost his life at the hands of the Blackfeet because of it. He was among a small party of trappers and hunters who camped along a stream that fed into the Madison River, northwest of Yellowstone, in September 1835. "We had not been encamped about an hour," Russell said, "when fourteen white Trappers came to us in full gallop." They were from Jim Bridger's camp about 20 miles farther south. "The trappers remained with us during the night telling Mountain Yarns and the news from the states," he said. However, "early next morning 8 of them started down the stream to set Traps on the main Fork but returned in about an hour closely pursued by about 80 Blackfeet." The Blackfeet had the high ground because Russell's party camped between two bluffs. The Indians also had lodgepole pine trees on one bluff, and a grove of aspen on the other, to hide behind. Russell gives us a graphic account of what happened next:

> ... the Indians commenced shooting into the camp from both sides. ... from these heights they poured fusee balls without mercy or even damage except killing our animals who were exposed to their fire. In the meantime we concealed ourselves in the thicket around the camp to wait a nearer approach, but they were too much afraid of our rifles to come near enough for us [to] use Ammunition. We lay almost silently about 3 hours when

finding they could not arouse us to action by their long shots they commenced setting fire to the dry grass and rubbish with which we were surrounded: the wind blowing brisk from the South in a few moments the fire
was converted into one circle of flame and smoke which united over our
heads. This was the most horrid position I was ever placed in death seemed
almost inevitable but we did not despair but all hands began immediately
to remove the rubbish around the encampment and setting fire to it to act
against the flames that were hovering over our heads: this plan proved successful beyond our expectations. Scarce half an hour had elapsed when the
fire had passed around us and driven our enemies from their position. At
length we saw an Indian whom we supposed to be the Chief standing on a
high point of rock and give the signal for retiring which was done by taking
hold of the opposite corners of his robe lifting it up and striking it 3 times
on the ground. The cracking of guns then ceased and the party moved off
in silence.[96]

The party lost two horses and a mule, and five other animals suffered
wounds. No one was killed. They packed up and quickly left their camp
and joined Jim Bridger. A few days later a French trapper from their party
"started down the mountain to set his traps for beaver contrary to the
advice and persuasion of his comrades." "He had gone but a few miles,"
said Russell, "when he was fired upon by a party of Blackfeet killed and
scalped."[97]

FIRE HUNTERS

The most common and by far the most widespread use of fire by Indians
was for hunting. Their burning techniques were as varied as the needs of
the animals they hunted. They knew their quarry well, especially white-
tailed deer because it furnished the bulk of their meat throughout the
East. Indians also knew the habitat requirements of bison, elk, moose,
mule deer, black-tailed deer, and many other game animals. Thus, they
burned forests, brushlands, and grasslands to provide each species of game
with the food and cover it needed and to increase their success in the
hunt.

In 1655, Adriaen Van der Donck published a paper in which he
described several of the most important reasons that Indians burned to
improve hunting. Van der Donck was a Dutch colonist who lived on an
island in the Hudson River. The colony began just 32 years earlier when
30 Dutch families landed on Manhattan Island and founded New Nether-

lands; a territory granted by Holland to the Dutch West Indian Company. The colonists spread up the Hudson River Valley to settle Fort Orange (Albany, New York) as well as New Amsterdam at the tip of Manhattan Island. The British took the territory in 1664 and divided it into the colonies of New York and New Jersey.

Van der Donck said that "the Indians have a yearly custom ... of burning the woods, plains and meadows in the fall of the year, when the leaves have fallen, and when the grass and vegetable substances are dry." "Those places which are then passed over are fired in the spring in April," he added.

> This ... is done for several reasons: First to render hunting easier, as the bush and vegetable growth renders the walking difficult for the hunter, and the crackling of the dry substances betrays him and frightens away the game. Secondly, to thin out and clear the woods of all dead substances and grass, which grow better the ensuing spring. Thirdly, to circumscribe and enclose the game within the lines of the fire, when it is more easily taken, and also, because the game is more easily tracked over the burned parts of the woods.[98]

Van der Donck overlooked only one more important reason for fire hunting. That is, depriving game of food in one place so that they must move to another place where they are easier to hunt. Otherwise, his list is reasonably complete.

Van der Donck also made it clear that many of these Indian fires did not destroy the oak-chestnut forest that dominated the area. "On seeing it from without," he said, "we would imagine that ... the whole woods would be consumed where the fire passes, for it frequently spreads and rages with such violence, that it is awful to behold." "Still," he pointed out, "the green trees do not suffer." These must have been surface fires that burn along the ground rather than crown fires that burn through the tops of the trees. Van der Donck confirmed this when he wrote: "The outside bark is scorched three or four feet high, which does them no injury, for the trees are not killed."[99]

Fires stayed on the ground if they burned often enough to keep the forest understory clear of small trees and free of dead wood. However, some patches of forest still escaped the fires long enough to grow thick. Then the next fire could climb from the ground to the tops of the trees and destroy them. Van der Donck saw this happen when an Indian fire entered a thick patch of pine, probably an eastern white pine forest, and began "leaping from treetop to treetop."[100]

Circles of Fire

Nearly all Indians throughout North America used fire to, in Van der Donck's words, "enclose the game." Roland Dixon called them "circles of fire" when he saw California Indians using them to hunt deer in the forests around Mt. Shasta.[101] The Miwok who lived in Yosemite Valley did the same thing, and the Apache used fire to drive game in the Southwest as well.[102] Ethnographer Edward Sapir also reported that Takelma Indians to the North in Oregon set fires in the "mountain forests" around the Rogue River "to facilitate the driving of game."[103] Different tribes invented fire circles that best fit the environments in which they lived. Regardless, these ancient fire hunting methods still served the same purposes whether they used them in forests, brushlands, or grasslands. That is, to "drive" and "enclose" the game so that the animals were easier to kill.

Cabeza de Vaca was the first to report Indian fire circles, or surrounds, when he published his narrative in 1542. On his long and hazardous journey back to Mexico, Cabeza de Vaca met another Spaniard, Figueroa, who was enslaved by a band of Karankawa near the Texas coast. Figueroa told him that the Indians "are accustomed also to kill deer by encircling them with fires."[104] Such fires must have been common in the oak woodlands and prairies of Texas because many early accounts mention them.

Henri Joutel, a personal aide to La Salle, recorded Indian burning several times in his account of La Salle's tragic third expedition to the New World. The explorer sailed from France in 1684 to found a colony at the mouth of the Mississippi. He passed the Louisiana coast and landed by mistake in Matagorda Bay, Texas, in January 1685, where he built Fort St. Louis. A fire set by Karankawa hunters endangered their temporary camp a few weeks after landing. "About an hundred, or an hundred and twenty of the Natives came to our Camp, with their Bows and Arrows," Joutel wrote. Then "when the Indians were about departing," he said, "they made signs to us to go a Hunting with them." La Salle suspected them and declined the invitation. "Besides that," wrote Joutel, "we had enough other Business to do." A few days later one of the Indian's hunting fires threatened their camp. Joutel recorded the incident:

We perceiv'd a Fire in the Country, which spread it self and burnt the dry Weeds, still drawing toward us; whereupon Monsr. *de la Sale* made all the Weeds and Herbs that were about us, be pull'd up, and particularly all about the Place where the Powder was. Being desirous to know the Occasion of that Fire, he took about twenty of us along with him, and we march'd that Way.... We perceiv'd that it run toward *W.S.W.* and judg'd it had begun about our first Camp, and at the [Indian] Village next the Fire.[105]

The settlement at Fort St. Louis faced one disaster after another. Finally in 1687, La Salle decided to go overland to find the Mississippi. As they traveled Joutel reported "plains that had been burnt," and later that spring, near the Trinity River, he said:

We found the Country made up of several little Hills, of an indifferent Height, on which there are Abundance of Wallnut-Trees and Oaks, not so large as what we had seen before, but very agreeable. The Weeds which had been some Time before burnt by the Natives, began to spring up again, and discover'd large green Fields very pleasing to the Sight.[106]

Nearly two centuries later Indians still burned the prairies and oak woodlands in Texas when they hunted deer. However, by then there were probably only a small number of Indians left due to the disastrous effects of European diseases. Herman Ehrenberg, son of a Royal official in Prussia, described such a fire, probably set by Tonkawas. Ehrenberg fought with the volunteer army in the Texas Revolution against Mexico. He was among a few fortunate survivors of the Goliad massacre in 1836. The Indian hunting fire he witnessed burned somewhere between Bastrop and San Antonio in 1835:

The hour was late.... A dusky reddish light which shone dimly somewhere in the distance had attracted our attention.... Suddenly, piercing the unbroken silence of the night, a clear sound like a dog's bark rang out.... Several minutes later thousands of them mingled in a deafening chorus. The performers of this frightful serenade were the coyotes. Soon the wolves took their turn and swelled with their deep howls.... The dark masses of clouds which had by degrees spread over the entire half of the horizon were lit up by a crimson glare ... and the sky was tinged with deep vermilion hues. "The prairie must be burning; it means that the Indians are near," cried someone finally. They have been hunting today, and I guess that many a poor deer, to escape the fire, will run into the muzzles of these scoundrels' guns." Suddenly, fiercely burning flames, like an army, like a torrent, rushed past a hill at some distance from the camp, then leapt forward in our direction. . . . We sprang up, ran after our horses, already frightened by the fire, and drove them into the island of trees. . . . The flames leapt closer and closer; . . . then, abruptly, the whole incandescent horizon, as far as the eye could reach, flickered out . . . soon everything was over; all that remained were the odor and thick columns of smoke mounting into the air.

This "small detachment" of volunteers "had no new excitement during the remainder of the night," wrote Ehrenberg. "But," he said, "how different was the view on this morning from what it had been the day before!" All he saw was "an appalling blackness." [107] That was in October. However, Frederick Law Olmsted wrote in 1857 that, by February, "the dreary, burnt prairies, from repulsive black, changed at once to a vivid green."[108]

Indians who lived in the forests of Virginia and throughout the South used fire to drive and surround deer just as they did in oak woodlands and prairies along the coast of Texas. Botanist William Bartram wrote in his journal for 1773–1778 that Indians set fire in the pine forests of northeastern Florida "for the purpose of rousing the game."[109] Bartram found these Indians friendly, but Captain John Smith had the misfortune of coming upon a large number of hostile Chickahominy Indian hunters in 1608. He survived to tell the story, but his men did not. "In one of these huntings," Captain Smith said, "they found me in the discovery of the head of the river of Chickahominy, where they slew my men, and took me prisoner in a bogmire." Then he "gathered these observations" of fire hunting:

At their huntings they leave their habitations, and reduce themselves into companies . . . and go to the most desert [remote] places with their families, where they spend their time in hunting and fowling up towards the mountains, by the heads of their rivers, where there is plenty of game. . . . Their hunting houses are like unto arbors covered with mats. These their women bear after them, with corn, acorns, mortars, and all bag and baggage they use. . . . At their huntings in the deserts they are commonly two or three hundred together. Having found the deer, they environ [surround] them with many fires, and betwixt the fires they place themselves. And some take their stands in the midsts. The deer being thus feared by the fires, and their voices, they chase them so long within that circle, that many times they kill 6, 8, 10, or 15 at a hunting. They use also to drive them into some narrow point of land, when they find that advantage; and so force them into the river, where with their boats they have ambushes to kill them.[110]

These hunts took place in the fall and winter, according to Robert Beverley. He wrote in his *History of Virginia*, published in 1722, that "they would fire the woods in a circle any time in the winter when the leaves were fallen and so dry that they would burn."[111] John Lawson concurred, saying in 1709 that "these Savages go a hunting" in the Carolinas "when the Leaves are fallen from the Trees." "'Tis then they burn the Woods," he said. Lawson added that they set their fires "with a Match made of the black Moss that hangs on the Trees in Carolina, and is sometimes above six Foot long." However, he said, "in Places, where this Moss is not found, (as towards the Mountains) they make Lintels of the Bark of Cypress beatn, which serves as well."[112]

Indians who lived on the Plains, as well as the oak woodlands and prairies to the south and east, also included fire circles or surrounds among their many techniques for hunting bison. This method became especially important for people without horses, and most tribes probably used fire circles before the introduction of the horse. Like Indians nearly everywhere, they carried out their fire hunts in the fall. Father Hennepin watched the Miamis hunt bison with fire in the oak woodlands and prairies of Illinois. He recorded what he saw during the La Salle expedition of 1679–1680:

> The Miamis hunt the buffalo toward the end of autumn. When they find a herd, they gather in great numbers and set fire to the tall grass around the herd, leaving only a small passage where they take post with their bows and arrows. The buffalo, fleeing the fire, are forced to pass near the Indians, who sometimes kill as many as 120 in a day. The Indian hunters, triumphant at the massacre of so many animals, notify their women, who at once go to bring in the meat.... The Indians distribute their kill according to the needs of each family.[113]

These fire hunts were exciting and special occasions for the Indians, and everyone wanted to participate. So much so that Father Hennepin said: "If our canoe men had found a chance, they would infallibly have all abandoned us, to strike inland and join the Indians whom we discerned by the flames of the prairies to which they had set fire in order to kill the buffalo more easily."[114]

Sometimes fire circles could be enormous, although they may not have covered an entire county as Samuel Clarke reported in Oregon. About 1880, newspaper reporter Clarke and pioneer John Minto interviewed Joseph Hudson, an old Santiam Kalapuyan Indian who lived in the Willamette Valley. Hudson told them how they used fire to hunt deer in the fall before the 1830–1833 malaria epidemic devastated the Indians of California and Oregon. Afterward, too few Indians survived for the "grand hunt" to continue. Samuel Clarke's account of the interview appeared in the *Oregonian*:

> The bands that occupied the region that included the east side of the valley ... all united in this annual roundup. ... Men were placed in position along the rivers ... and including the foothills of the Cascades. The great square encircled all Marion County as constituted today that is not rough mountainous country. ... At a given signal, made by a fire kindled at some point as agreed, they commenced burning off the whole face of the country and driving wild game to a common center. ... When the circle of fire became small enough to hunt to advantage, the best hunters went inside and shot the game.[115]

The Coeur d'Alene in Idaho developed a clever variation of the fire circle for some of their fall hunts. Father DeSmet said in his journal for 1858 that "they wait until the mountains are covered with three to five feet of snow and the deer have been driven down to the valleys, where they pass the winter, feeding on moss off the trees, the tenderer branches of the underbrush and shoots of the herbs and plants." In this case, the Coeur d'Alene only used fire at the ends of a semicircle of hunters:

> They choose by preference the neighborhood of some lake or river which is not yet frozen over; and they determine the extent of the surround, according to the number of hunters of which the band is composed. A hunting-chief is chosen, and all his orders are thereafter executed promptly and punctually. On both ends of their line they light fires, some distance apart, which they feed with old garments and worn-out moccasins. The hunters are now formed in a long curved line, something like a half-moon. At a given signal, they utter the hunting-cry and move forward. The frightened deer rush to right and left to escape. As soon as they smell the smoke of the fires, they turn and run back. Having the fires on both sides of them and the hunters in their rear, they dash toward the lake, and soon they are so closely pressed that they jump into the water, as the only refuge left them. Then everything is easy for the hunters: they let the animals get away from the shore, then pursue them in their light bark-canoes and kill them without trouble or danger.[116]

Sometimes the Coeur d'Alene formed a "complete circle," wrote

Father DeSmet, "burning their old rags in a hundred little fires round about, to prevent the deer from escaping from the circle." Then, he said, "pursued in every direction, the terrified animals flee from one clump of wood or brush to another, until finally enveloped on all sides and finding no issue, they fall into the hands of the hunters." He observed them kill "as many as 200 to 300 ... in a single surround." However, deer were not the only things shot during some of these hunts. The danger increased as the Indians walked forward to tighten the circle. Eventually they faced hunters who were shooting in their direction from the other side of the circle. Not surprisingly, as Father DeSmet observed, "in the eagerness and excitement of such a chase, bullets and arrows are liable to fly wild or glance, and do mischief." [117]

Alaskan Indians found another inventive method of fire hunting. They burned spruce and birch forests to weaken the trees so that they could push them over and build fences. The small size of the trees and the shallowness of the soil made this possible. They used the fences to concentrate the game. Then they torched trees and made banging noises to drive caribou into the fences. The smoke that blew toward the caribou also provided a screen and masked human scent. The Shasta in California, and the Takelma and Umpqua in Oregon, also built fences for some of their deer drives. However, they used brush and rope to build their fences instead of trees. [118]

Indians had to use fences to hunt deer when the weather turned too wet, cold, or snowy for fire hunting. Thus, in 1615, the Hurons decided to replenish their meat supply after an unsuccessful raid on an Iroquois village. Champlain had fought beside them and returned with two wounds in his leg from Iroquois arrows. Nevertheless, he joined the party of Hurons conducting the deer hunt. Other Indians went to hunt bear and beaver, and some went to fish, but Champlain said that "the deer-hunt ... is esteemed by them the greatest and most noble one." [119]

They traveled deep into a forest northeast of Lake Huron to a place the Hurons knew deer congregated during the winter. There was heavy snow, wind, and hail just a few days earlier, and the country was "marshy," so nothing would burn even if the Indians had wanted to use fire. The Indians built three cabins for shelter and then prepared for the hunt. Champlain described the construction of an elaborate trap that took the Indians 10 days to build, and required cutting a considerable number of trees:

> They went into the woods to a small forest of firs, where they made an enclosure in the form of a triangle, closed up on two sides and open on one. The enclosure was made of great stakes of wood closely pressed together, from eight to nine feet high, each of the sides being fifteen hundred paces long. At the extremity of this triangle there was a little enclosure, constantly diminishing in size, covered in part with boughs and with only an opening of five feet, about the width of a medium-sized door, into which the deer were to enter.[120]

Then the Hurons went into the forest "separated from each other some eighty paces," and began walking toward the enclosure while banging two sticks together. The deer ran into the fence and followed it toward the center of the triangular, at which time the Indians howled like wolves to frighten them even more. This forced the deer to crowd into the small inner enclosure where the Indians promptly killed them with "arrow shots." The Hurons repeated the hunt every two days, but it took them over a month to kill 120 deer.[121]

Great Basin Indians also built elaborate fences just to trap rabbits. They did so because the brush did not grow fast enough to burn every year. These fences probably followed a very old design just like other animal traps. Edwin Bryant, a journalist who traveled in a wagon train to California in 1846, the same year the Donner party met disaster in the Sierras, saw such a trap and noted it in his journal. He climbed a ridge above Mary's River in the northeast corner of Nevada while "searching to find a passage presenting the fewest difficulties." Here, he said, "I discovered, at the entrance of one of these gorges, a remarkable picketing or fence, constructed of the dwarf cedars of the mountain, interlocked and bound together in some places by willow withes." The fence was "about half a mile in length, extending along the ridge," he said. Then, "at the foot of the mountain," he found "another picketing of much greater extent, being some four or five miles in length, made of the wild sage." Bryant thought these were defensive works. Later, he wrote, "I have since learned from trappers that these are erected by the Indians for the purpose of intercepting the hares, and other small game of these regions, and assisting in their capture."[122]

The fire circle or fire drive proved effective for hunting smaller game just as it did for large game. Indians from the middle of Baja California northward burned brush to drive rabbits, according to naturalist Jose Longinos Marinez. In 1792 he wrote: "In all of New California from

Fronteras northward the gentiles [Indians] have the custom of burning the brush." They do so, he said, "for hunting rabbits and hares."[123] In addition, a report written by the Spanish in October 1771 described an Indian fire that threatened the Presidio at Monterey in Alta California. It stated: "The heathens are want to cause these fires ... they set fire to the brush ... to catch the rabbits that get confused and overcome by the smoke."[124] The Cahuilla, Cocopa, Mohave, Yuma, Kamia, Diegueño, Paiute, and many other Indians living in the West also set fire to grass and brush to flush, drive, and surround rabbits and hares. They used fire to hunt wood rats and ground squirrels as well.[125]

Their Wings Are Scorched

Indians put fire to use even when hunting insects. The Kalapuya and Takelma of Oregon, the Shasta and Wintu of California, and the Paiute of the Great Basin all used fire to collect grasshoppers and crickets, and even wasp larvae.[126] For example, the Kalapuya and the Takelma set fires on top of yellowjacket nests to cook the larvae while they were underground.[127] California Indians also "burned the grass to enable them to get at ... wasps' nests," wrote Frank Marryat in 1855, because "young wasps" are "a luxury with them."[128] Often they dug the nest from the ground, roasted it on hot coals, and then shook out the cooked larvae to mix with acorn meal or manzanita berries.[129]

More commonly, Indians collected grasshoppers and crickets by driving them toward the center of a circle with fire or by beating the ground with sticks. In July 1854, Lieutenant E. G. Beckwith traveled near the Pitt River in northern California during a Pacific Railroad survey and saw "Indian smokes curled upwards from every part of the mountains." He said the Indians "were engaged in burning the grass to catch grasshoppers, upon which they feed, regarding them as a great delicacy."[130] So many fires were burning that a few days earlier Lieutenant Beckwith could barely see the landscape from high in the Sierra Nevada.

> The view of the mountains for any considerable distance below us was obscured by a smoky atmosphere, and the valley of the Sacramento entirely invisible from the dark cloud of smoke which hung over it, over which, however, as over a blue sea, peaks of the Coast range were occasionally visible.[131]

The Paiutes of the Columbia Basin in eastern Oregon also collected "a great many large black crickets, and grasshoppers by the bushels," wrote A. M. Armstrong in 1857. They found fire an efficient way to catch the insects and to cook them at the same time. Armstrong wrote a description of what he saw: "I have often seen them encircle the grasshoppers in a ring of fire by igniting the grass; their wings are scorched by the blaze, and they fall to the ground, when the Indians gather around, [they] collect them and eat them." He said they also "put [them] into a mortar with acorns or bread root, and pound into a mass which is then kneaded, placed on a board and baked for bread—the legs of the grasshoppers and crickets making a very rough crust."[132]

A Kalapuya interviewed in the 1940s remembered that they once collected grasshoppers the same way in the oak woodlands of the Willamette Valley. The Indian said:

> When it was summertime they burned over the land when they wanted to eat grasshoppers. When they burned the land, they burned the grasshoppers [too]. And then they [women] gathered up the grasshoppers, and they ate those grasshoppers it is said.[133]

Botanist David Douglas traveled through the Willamette Valley in September, 1826, with a Hudson's Bay Company trapping expedition led by Alexander McLeod. He made an entry in his journal that also documents the use of fire by the Kalapuya to collect grasshoppers and probably wasps. The "natives tell me," he wrote, that one reason they burn is "that they might the better find wild honey [wasps?] and grasshoppers, which both serve as articles of winter food."[134]

Many Indians who lived in piñion-juniper and juniper woodlands, and sagebrush-covered steppes, hunted grasshoppers with fire and other methods. They did so on the lower Snake River Plain because few bison could live there, although they thrived on the lush grasslands of the upper Snake River Plain. Big sagebrush dominated the landscape of the lower plain. One traveler who passed that way in 1839 described it as "barren sandy, and level, and produces only prickley pear, sage and occasional scanty tufts of dry grass."[135] Indians often started the few fires that burned there. In 1830, Hudson's Bay Company trapper John Work referred to one of these fires, saying that "the country has recently been overrun by fire." Like most travelers at the time his next entry showed his main concern.

"Scarcely a spot of grass left for the horses to feed," he noted.[136] However, it took many years in this dry land for the sagebrush and grass to grow thick enough to carry another fire, so Indians had to use other methods to hunt grasshoppers in the meantime.

The "very humble people," that Father DeSmet said "roam over the desert and barren districts of Utah and California, and that portion of the Rocky Mountains which branches into Oregon," also hunted grasshoppers. He called these people "Soshocos." "The principal portion of the Soshoco territory," he said, "is covered with wormwood, and other species of artemisia [sagebrush], in which the grasshoppers swarm by myriads." When the Indians could not use fire circles they still surrounded their prey. Father DeSmet described a grasshopper hunt he witnessed in 1858:

> They begin by digging a hole, ten or twelve feet in diameter by four or five deep; then, armed with long branches of artemisia [sagebrush], they surround a field of four or five acres, more or less, according to the number of persons who are engaged in it. They stand about twenty feet apart, and their whole work is to beat the ground, so as to frighten up the grasshoppers and make them bound forward. They chase them toward the centre by degrees—that is, into the hole prepared for their reception. Their number is so considerable that frequently three or four acres furnish grasshoppers sufficient to fill the reservoir or hole.[137]

After the hunt, Father DeSmet said they "eat grasshoppers in soup, or boiled; others crush them, and make a kind of paste from them, which they dry in the sun or before the fire: others ... take pointed rods and string the largest ones on them; afterward these rods are fixed in the ground before the fire." Then they feast, or as Father DeSmet said, they "regale themselves until the whole are devoured."[138]

That Necessity May Drive Them

One of the more popular methods of fire hunting involved depriving game of food in areas where they were difficult or inconvenient to hunt. By this means Indians herded their quarry into places where they were easy to kill. Cabeza de Vaca provides the earliest account of fire herding. He learned from Figueroa that the Plains Indians farther north of where he was on the Texas coast fired the prairies to herd bison. "The pasturage is taken from the cattle [bison] by burning," he wrote in the 1555 edition of his narrative, "that necessity may drive them to seek it in places where it is desired [by the Indians] they should go."[139] This required knowing where to find the animals during certain seasons, why they stayed there, and what would attract them to other areas. This resourceful herding technique surely has its roots in antiquity.

Lewis and Clark recorded an example of fire herding on the Great Plains nearly three centuries after Cabeza de Vaca first heard about it. They prepared to work their way up the Missouri River in the spring of 1805 after spending the winter among the Mandans. They watched an especially dangerous spring hunt on March 29 while they waited for the ice to clear from the river.

> Every spring, as the river is breaking up, the surrounding plains are set on fire, and the buffalo are tempted to cross the river in search of the fresh grass which immediately succeeds the burning. On their way they are often insulated on a large cake or mass of ice, which floats down the river. The Indians now select the most favorable point for attack, and, as the buffalo approaches, dart with astonishing agility across the trembling ice, sometimes pressing lightly a cake of not more than two feet square. The animal is of course unsteady, and his footsteps are insecure on this new element, so that he can make but little resistance; and the hunter, who has given him his death-wound, paddles his icy boat to the shore and secures his prey.[140]

After leaving Fort Mandan Lewis and Clark proceeded up the Missouri toward the Rocky Mountains. They passed the mouth of the Judith River when, on May 28, "an Indian pole for building floated down the river ... which indicates that the Indians are probably at no great distance above us." The low hills around them obscured their view and "the weather was dark and cloudy," they noted. Then Lewis made a separate entry in the journal that documents another example of fire herding:

> The air was turbid in the forenoon, and appeared to be filled with smoke; we supposed it to proceed from the burning of the plains, which we are informed are frequently set on fire by the Snake Indians to compel the antelopes to resort to the woody and mountainous country which they inhabit.[141]

David Douglas also saw the way Kalapuya used fire to herd white-tailed deer in the Willamette Valley. As he traveled through the oak woodlands with Alexander McLeod's trappers in September 1826, he, like earlier explorers, found "the country burned." However, with the keen eye of a botanist he also noted that "only on little patches in the valleys and on the flats near the low hills that verdure is to be seen." He asked why the Indians left these "little patches" unburned. "Some of the natives tell me," he said, "it is done for the purpose of urging the deer to frequent certain parts to feed, which they leave unburned and of course they are easily killed."[142] Indians burned oak forests in New England for a similar reason, but in this case they had to wait until the grass grew back. "The object of these conflagrations," Timothy Dwight noted in 1821, "was to produce fresh and sweet pasture for the prupose of alluring the deer to the spots on which they had been kindled."[143]

On a summer day in 1860, near Lookout Pass in northern Idaho, Army engineer P. M. Engle accidentally witnessed fire herding on a large scale. Engle decided to return to camp before "reaching the summit" because the smoke from forest fires obscured his view. On the way down the Indian guide surprised him when he "set fire to the woods himself." When asked, Engle said the Indian told him "that he did it with the view to destroy a certain kind of long moss ... which is a parasite to the pine trees in this region, and which the deer feed on in the winter season." "By burning this moss," Engle said, "the deer are obligated to descend into the valley for food, and thus, they [the Indians] have a chance to kill them."[144]

Green and Fair Pasturage

Elk, bear, moose, deer, antelope, mountain sheep, caribou, waterfowl, and other game were wild animals to Europeans, but to American Indians they were sources of meat that they could manage. They not only knew how to hunt these animals with fire, but they also knew how to use fire to enrich the habitat on which the game depended. They burned to promote the growth of forage and browse in nearly every part of North America. In 1873, Lieutenant George Wheeler passed through such an area on his trip down the Little Colorado River after climbing Mount Baldy in eastern Arizona. He recognized the beneficial effects of Indian burning in his report: "For a little less than 2 miles the grass is of the old crop, then begins the new and juicy growth of the year subsequent to the burning over by fires set by Indians."[145] However, Indians were unaware that game flourished on burns because grass and shrubs that grow after fire nearly always contain high amounts of protein, calcium, phosphate, and other essential nutrients. As a result, deer produce more fawns on recently burned areas, and even birds such as ducks and sharp-tailed grouse hatch more young in such places. Large game animals such as deer and moose, and even hares and cotton rats, also return to burns soon after a fire or while it is still smoking to nip the salty ash.[146]

In the Willamette Valley the Kalapuya burned just before the fall rains, so grass came back quickly. In some cases it took only "16 or 18 days" to create "green and fair pasturage" with fire, as trapper and pioneer James Clyman reported in 1843. He also saw "greate Quantities of wild geese ... flying and feeding on the young grass of the lately Burned Prairies which are Quite Tame and easily approached on horseback."[147] Large game took advantage of these burns as quickly as waterfowl. A newspaper reported "it was no uncommon sight [in 1822] to behold from four to six hundred deer on the newly burnt patches of the prairie" near the coast of Texas.[148] Elsewhere, Indians burned to provide forage or browse for game the following spring or fall, or even years later. For example, beaver rely on aspen, so it took awhile for the young trees to grow large enough for them to use after a fire. Indians showed patience because they were well aware of the beaver's needs when they burned. "In time many animals go there," an old Slave from the Meander River area of Canada said, "like the beaver, about four to five years after."[149]

In the ponderosa pine and mixed-conifer forests of California and Oregon, sprouting shrubs, such as manzanita and deer brush, produce succulent new leaves shortly after a fire. Deerbrush, the most important deer browse, also stores its seeds in the soil to await the next fire, where they remain viable for many years after the parent plant has died. As a result, in places where no brush seems to grow, many young seedlings can spring from the soil only a year after a fire passes through the forest.[150] Deer rush in to feast on the tender young sprouts and leaves when they appear. The Miwok, Monachi, Yokut, Umpqua, and other Indians surely saw this happen, so they burned these forests frequently to provide more browse for deer. Hector Franco, a Yokut who lives in the southern Sierra Nevada of California, said during an interview, "I remember what the old people told us." "When they burned the Giant Forest," he said, "and they burned even the lowlands and the foothill areas,—it would benefit all the animals, all the creatures, all the plants."[151] Indians usually burned in the fall, and deer brush also grows better after a fall fire than a spring or summer fire. Most likely, they noticed this, and they probably also saw that deer brush thrives in sunny openings that form in the forest when fires kill some of the large trees. Consequently, the Miwok and other Indians concentrated their fire hunts in these patches of brush.[152]

In 1845, Captain John C. Fremont began his third expedition that ultimately involved him in the Bear Flag Revolt in California the following year. While traveling westward in Kansas Captain Fremont wrote that the party passed through "a belt of wood which borders the Kansas [River]." Then, he said "we suddenly emerged on the prairies, which received us at the outset with some of their striking characteristics; for here and there rode an Indian, and but a few miles distant heavy clouds of smoke were rolling before the fire."[153] S. N. Carvalho, "artist to the expedition," wrote his account in a separate journal. He remembered that "the whole horizon now seemed bounded by fire," and a "most disagreeable and suffocating smoke filled the atmosphere."[154]

Painter George Catlin experienced much the same thing near Fort Leavenworth, Kansas. In 1832 he wrote, "Every acre of these vast prairies . . . burns over during the fall or early in the spring, leaving the ground a black and doleful color."[155] Indians set these fires to enhance the grass for game and horses on the Great Plains just as they did elsewhere in North America. They also fired the oak woodlands and prairies toward the East, and the white spruce-aspen parklands that bordered the plains in the North.

Lightning fires swept through parts of the plains every summer as well, but Indian fires usually burned in the spring and fall, especially in the North.[156] Indians who lived on the Canadian Plains set their fall fires on the shortgrass prairies near the edge of the Rocky Mountains. This deprived the bison of winter feed and forced them to seek it toward the East. The bison found fresh grass in the tallgrass prairies on the edge of the white spruce–aspen parklands where the Indians sought shelter for their winter camps. Here the grass was green because the Indians had burned it that spring. They burned it again the next spring. Now bison had to move back toward the West where the shortgrass prairies were turning green early in the places Indians had burned the previous fall. The Indians moved their camps westward trailing after the bison. When fall approached, the Indians again burned the shortgrass prairies and moved East to set up their winter camps where they already knew there would be plenty of bison to hunt.[157] Regardless of the thought that went behind such clever burning methods, all Indians did not burn in the same way or for the same reasons. Thus, in southwestern Alberta, Peter Fidler could record in his diary in 1793 that "these large plains either in one place or another is constantly on fire."[158]

Indian fires also rushed through the wide belt of oak woodlands bordering the eastern edge of the prairies. "The fire spreads everywhere and destroys most of the young trees," wrote Father Vivier when he was on the Ozark Plateaus in 1750.[159] These fires not only renewed the grass between the trees but they also kept the forest open so that the oak trees had plenty of space to grow. This increased the number of acorns that dropped from the trees, which further enhanced the food supply of deer. Lewis and Clark camped in such a place on the lower Missouri River on September 16, 1804. "Our camp is in a beautiful plain, with timber thinly scattered for three-quarters of a mile, and consisting chiefly of elm, cottonwood, some ash of indifferent quality, and a considerable quantity of a small species of white oak," they wrote. The spot where they camped "having been recently burnt by the Indians, is covered with young green grass." They added that the "acorns are now falling, and have probably attracted the number of deer which we saw on this place, as all the animals we have seen are fond of that food."[160]

Indians in eastern Massachusetts burned for the same reasons as Indians in the West. William Wood, who settled in New England in 1629, wrote that it was the "custom of Indians to bourne the wood in November when the grass is withered and leaves dryed." Indians set fires mostly in the fall when forests burned easily, but they probably also were aware that early spring fires endangered young animals. Their fires kept the oak-chestnut and white pine forests open, so "there is no underwood saving in swamps, and low grounds that are wet," wrote William Wood. However, even then Indians were rapidly dying from European diseases and burning declined as a result. Consequently, Wood also commented that "in some places where the Indians dyed of the Plague some fourteen years agoe, is much underwood as in the midway betwixt the Wessaguscus and Plymouth because it hath noth been burned."[161]

Indians not only used fire to keep the forest open so grass would grow, but as Van der Donck pointed out in 1655, the grass also grows "better the ensuing spring."[162] As a result, the swampy thickets that survived the fires, known to Indians as the "abodes of owls," provided hid-ing places for white-tailed deer, and the adjacent open forest furnished an abundance of fresh grass. The game concentrated at the edge between them. That is where Indians hunted.[163]

The Lenni Lenape (which means "Original People"), or Deleware, in New Jersey also burned forests of "stately Oaks." Daniel Denton, a Long Island settler, said they did so "every spring to make way for new [grass]." "They also burned in the fall," he added. That does not mean that the Lenni Lenape burned the same place every year, as fire scars in trees show before 1711. Sometimes they burned more frequently and sometimes less frequently, but on average the Lenni Lenape probably set fire to the same part of the forest once every 14 years.[164] When Daniel Denton described the grass in 1670 as being "high as a mans middle," he was complaining because he felt it was going to waste unless domestic livestock grazed it. The grass, he said, "serves for no other end except to maintain the Elk and Deer."[165] Denton failed to realize that the Lenni Lenape maintained the grass for "Elk and Deer" because they were a form of wild livestock to them.

Indians who lived within the vast longleaf and loblolly pine forests of the Southeast also burned to provide grass for deer. During his 1773–1778 journey through the South, botanist William Bartram made many notations in his journal about these fires. He also saw how quickly the grass recovered afterward. While walking in northeastern Florida, he emerged from a grove of live oak trees and "entered some almost unlimited savannas and plains, which were absolutely enchanting; they had been lately burnt by the [Siminole] Indian hunters, and had just now recovered their vernal verdure and gaiety."[166] In a similar setting, but among "lofty pines," he talked about "being in the midst of plenty and variety, at any time within our reach" with "herds of deer . . . feeding in the green meadows before us [and] flocks of turkeys walking in the groves around us." "How cheerful and gay all nature appears," he said early the next morning, when the "glorious sun gilds the tops of the pines."[167]

To Render Hunting Easier

While climbing the Siskiyou Mountains in southern Oregon on September 28, 1841, Henry Eld, a member of Lieutenant Emmons' party, said that "in passing thro the woods we suddenly came on to an Indian woman who was blowing a brand to set fire to the woods." "We stopped to speak to her," he said, "but she was sullen, & dogged, & made no reply, & we passed." We do not know why she set that particular fire, but on the same day Eld also noted that "we found the woods on fire in several places & at times had some difficulty in passing some of the gullys where it was burning."[168] Six days earlier Lieutenant Emmons thought similar fires in the Umpqua Mountains were set to block their way. However, the Indians may also have been clearing the forest of underbrush, as they often did in the fall. Not only did such fires renew the grass and shrubs for deer and other game, but they also made travel and hunting easier. Many years later, a Klamath Indian from the same region reflected on those old days in an interview: "Now I just hear the deer running through the brush at places we used to kill many deer. When the brush got as thick as it is now, we would burn it off."[169]

Joseph H. Brown, a Salem pioneer interviewed in 1878, also said that Indians burned "to keep down the undergrowth of timber."[170] The Southern Maidu who lived east of the Sacramento Valley in California did the same.[171] Likewise, Indians in Arizona "destroyed the cover in which their quarry took refuge," wrote S. J. Holsinger in 1902.[172]

Indians nearly everywhere set fires in forests, brushlands, and even grasslands to clear obstructions. In 1832, George Catlin personally experienced how difficult it could be to ride through unburned tallgrass prairies in Kansas. Catlin said "the grass is seven or eight feet high, as is often the case for many miles together." "There are many of these meadows," he added, "with a waving grass, so high, that we are obliged to stand erect in our stirrups, in order to look over its waving tops, as we are riding through it." Indian riders had the same problem, so they burned the prairie "for easier traveling during the next summer, when there will be no old grass to lie upon the prairies, entangling the feet of man and horse," wrote Catlin. The thick grass became a more serious problem when Catlin tried to run from a prairie fire during a windstorm that he said "often sweep over the vast prairies of this denuded country." He recalled his narrow escape:

> The fire in these, before such a wind, travels as fast as a horse at full speed, but that the high grass if filled with wild pea-vines and other impediments, which render it necessary for the rider to guide his horse in the zig-zag paths of the deers and the buffaloes, retarding his progress, until he is overtaken by the dense column of smoke that is swept before the fire—alarming the horse, which is wafted in the wind, falls about him, kindling up in a moment

a thousand new fires, which are instantly wrapped in the swelling flood of smoke that is moving on like a black thunder-cloud, rolling on the earth.[173]

In contrast, "the war or hell of fires" in the tall grass became gentle and picturesque on "the elevated lands and prairie bluffs, where the grass is thin and short," said Catlin. Here, he wrote:

> The fire slowly creeps with a feeble flame, which one can easily step over; where the wild animals often rest in their lairs until the flames almost burn their noses, when they will reluctantly rise and leap over it, and trot off amongst the cinders, where the fire has past and left the ground as black as jet. These scenes at night become indescribably beautiful, when their flames are seen at many miles distance, creeping over the sides and tops of the bluffs, appearing to be sparkling and brilliant chains of liquid fire (the hills being lost to the view), hanging suspended in graceful festoons from the skies.[174]

In the 1720s, Le Page du Pratz of Natchez planned his trip northward to avoid the tall grass in the open oak-hickory forest along the Ouachita River. Accompanied by Indian guides, he "set out in the month of September, which," du Pratz said, "is the best season of the year for beginning a journey in this country." He chose this time of year

> ... because, during the summer, the grass is too high for travelling; whereas in the month of September, the meadows, the grass of which is then dry, are set on fire, and the ground becomes smooth, and easy to walk on: and hence it is, that at this time, clouds of smoke are seen for several days together to extend over a long track of country.[175]

The fall season also worked out well since game was more abundant and easier to hunt, as du Pratz pointed out: "By means of the rain, which ordinarily falls after the grass is burnt, the game spread themselves all over the meadows, and delight to feed on the new grass, which is the reason why travelers more easily find provisions at this time than at any other."[176]

The "open wooded country" of the Ouachita Mountains changed during the next century. G. Featherstonhaugh sensed the loss in 1844: "Now that Indians have abandoned the country, the undergrowth is rapidly occupying the ground again."[177] It became worse by 1881, according to F. Gerstacker. He complained that "the forests not having been burnt for many years, were so thickly overgrown with underwood, that it was impossible to find the deer, or to shoot game enough to live upon."[178]

Most Indians who lived east of the Mississippi were farmers, and they used fire to clear underbrush from the forests surrounding their fields and villages in much the same way as other Indians. The religious separatists who colonized the shores of Plymouth Bay in 1620 were especially grateful to the Wampanoag for burning forests to keep them open. The colonists had to clear trees to plant their corn. This proved reasonably easy, as E. Johnson noted in 1654: "The Lord having mitigated their labours by the Indians frequent fiering of the woods, (that they may not be hindered in hunting Venson, and Beares in the Winter season)."[179] They took advantage of fields the Indians had already cleared and abandoned as well.[180]

Thomas Morton and William Wood also commented about the extensive use of fire by Indians. They settled in the Massachusetts Bay area, north of Plymouth Bay, where the Massachuset Indians lived. Morton wrote his observations in 1632:

> The Savages are accustomed, to set fire of the country in all places where they come; and to burne it, twize a yeare, vixe at the Springe, and the fall of the leafe. The reason that mooves them to doe so, is because it would other wise be so overgrowne with underweedes, that it would be all a copice wood, and the people would not be able in any wise to passe through the Country out of a beaten path.[181]

William Wood echoed Morton's comments in 1634, saying "it being the custom of Indians to bourne the wood in November when the grass is withered and leaves dryed, it consumes all the underwood and rubbish, which otherwise would overgrow the country, making it impassable."[182] In addition, Roger Williams commented in 1643 that the Indians thought that "keeping downe the Weeds and thickets" was one of the main benefits of their fires.[183] The Indians still burned the forests over a century later, and "these fires run for many miles," George Loskiel wrote in 1794.[184]

Indians in the Southeast burned vast areas to keep forests clear just as they did in New England and the West. While traveling in northeastern Florida, William Bartram wrote that "the deserts are set on fire ... almost every day throughout the year, in some part or other, by the Indians ... as also by the lightning."[185] These fires burned so often and over such large areas that Bartram commented later in his journal about the soil erosion that came after the fires:

> ... for in all the flat countries of Carolina and Florida, except this isthmus, the waters of the rivers are, in some degree, turgid, and have a dark hue, owing to the annual firing of the forests and plains; and afterwards the heavy rains washing the light surface of the burnt earth into rivulets, which rivulets running rapidly over the surface of the earth, flow into the rivers, and tinge the waters the colour of lye or beer, almost down to the tide near the sea coast.[186]

JUST SET YOUR TEEPEE UP THERE

Indians picked their campsites and village sites carefully. The looked for places that had an attractive view, and could also be defended. They also sought protection from wind by hills or trees, a clear spring or stream, and plenty of firewood. Often they needed to use rivers for transportation, so they looked for places where two or more rivers converged.[187] Just as important as location, Indians knew well that they had to protect their campsites and villages from wildfire. They also knew that dense thickets nearby could hide their enemies and make surprise attacks easier. Fire gave them the means to solve both problems. Fire also cleaned up household refuse and helped to keep mosquitoes, snakes, and other pests out of camp.[188]

Indians everywhere feared wildfires. That is why Indians in the Southeast kept the ground bare around their villages and campsites.[189] They also lowered the fire hazard by simply gathering firewood. Captain John Smith noticed this in Virginia:

> Near their habitations is little small wood or old trees on the ground by reason of their burning of them for fire. So that a man may gallop a horse amongst these woods any way, but where the creeks and Rivers shall hinder.[190]

Indians especially feared the enormous wildfires that swept through the spruce forests of Canada, but fires could become deadly anywhere. So the Cree and other Indians camped in clearings. During an interview in 1975–1977, an elderly Cree from the Trout Lake area of Alberta pointed out the importance of maintaining open areas in the forest when trying to recall the Indian name for these places:

What is that name? *Maskuta? Muskotaw!* Yea, that's a prairie like place. There used to be lots of those places around here. Nobody built their house in the woods like they do now. If we get a forest fire now it could be really bad. All the houses would get burned up. It's a lot safer if you got open places . . . you just set your teepee up there.[191]

Indians in northern California also remembered the "old ways" of preventing wildfires. Yvonne Jolley, a Yurok, said, "in my mom's days they would burn all the brush." "They would burn under the trees," she added, "and that way things came back naturally, and there weren't forest fires."[192] One Mono known only as "J. R." said that his people burned in much the same way in the foothills of the Sierras. During an interview in about 1948, he recalled how they burned during his boyhood. The men started the fire and "it burned the hills, all over, clean through to the next one," he said, and "dead trees and logs were all cleaned up that way."[193] James Mason Hutchings, miner, traveler, and publisher of the *California Magazine*, would probably have agreed with these accounts. He wrote in 1859 that "in the fall season, when the wild oats and dead bushes are perfectly dry, the Indians sometimes set large portions of the surface of the mountains on fire."[194]

The "Poet of the Sierras," Joaquin Miller, visited Yosemite and watched the Miwok burn the valley. He learned that one of the benefits of burning was to prevent dangerous wildfires. He wrote his description of the event in 1887.

> In the spring . . . the old squaws began to look about for the little dry spots of headland or sunny valley and as fast as dry spots appeared, they would be burned. In this way the fire was always the servant, never the master. . . . By this means, the Indians always kept their forests open, pure and fruitful and conflagrations were unknown.[195]

Galen Clark noticed the effects of Indian burning on Yosemite Valley when he first saw it in the summer of 1855. "At that time," he later wrote, "there was no undergrowth of young trees to obstruct clear, open views in any part of the valley." The Miwok told him that they kept it open by setting fires every year "in the dry season," and then they "let them spread over the whole Valley."[196]

Writer Frank Marryat did not like the smoke and heat that came from Indian fires during his travels in Sonoma County, California, in 1850.

However, he liked even less the yellowjackets that plagued him. "The wasps are so numerous here in summer," he said, "as to destroy with rapidity everything they attack." So Marryat thought that Indian fires "have the good effect of destroying immense quantities of snakes and vermin." "One can scarcely imagine the extent to which these might multiply were they not occasionally burnt out," he said.[197]

Indians used fire and smoke to get relief from mosquitoes, gnats, and black flies wherever they camped. Even deer will seek refuge from biting insects in the smoke of a fire.[198] Mosquitoes added to the list of torments endured by Cabeza de Vaca and the other Spaniards lost on the Texas coast in the 1500s. "We found mosquitos of three sorts, and all of them abundant in every part of the country," Figueroa told Cabeza de Vaca when they met. "They poison and inflame, and during the greater part of the summer gave us great annoyance," he remarked. So, he said, "as a protection we made fires, encircling the people with them, burning rotten and wet wood to produce smoke without flame." Even when the Indians hunted bison where there was no wood, Figueroa had to carry wood for "cooking" and "to relieve them of mosquitos." Since Figueroa was a slave of the Karankawa at the time, he had little rest because he was responsible for keeping the fires burning. He recalled his misery:

> The remedy brought another trouble, and the night long we did little else than shed tears from the smoke that came into our eyes, besides feeling intense heat from the many fires, and if at any time we went out for repose to the seaside and fell asleep, we were reminded with blows to make up the fires.[199]

Indians did not limit themselves to bonfires. They also set the countryside on fire to wipe out pests. Thus, Figueroa also told Cabeza de Vaca that

> the Indians of the interior have a different method, as intolerable, and worse even than the one I have spoken of, which is to go with brands in the hand firing the plains and forests within their reach, that the mosquitos may fly away.[200]

Northeastern Indians did the same. In 1643, Roger Williams reported that the Massachuset burned "the Country" around Narragansett Bay partly "for destroying of vermin."[201] In Alaska too, "evidences of conflagration in the dense coniferous forests were everywhere frequent," reported Frederick Schwatka. He wrote that Indians set the fires "with the idea of clearing the district of mosquitoes."[202]

THEY KNEW WHERE TO BURN

Indians cultivated and harvested their wild gardens with wooden digging sticks, clamshells, antlers, stone and bone tools, and fire. However, fire could do many things that other tools could not. Most importantly, it eased the Indian's labor and allowed them to become gardeners of landscapes as well as small plots. Fire served a purpose like any other tool, and Indians understood it well, as Hector Franco, a Yokut, recalled:

> The old people, they knew where to burn. They could feel it. It's just like any farmer nowadays. ... Men, whether they're white men or Indian men. People that work with plants, when they're close to nature they know.[203]

Indians managed their wild gardens with hot fires and light surface fires. They burned large and small areas, and even small clumps of shrubs or individual trees. They set their fires in the spring or fall as their needs required, but in each case they demonstrated their mastery of fire.

Little Hair (Pelillo)

In 1769, the newly appointed governor of California, Gaspar de Portolá, led an expedition from Baja California to Alta California. Portolá was sent to establish a garrison and mission at both San Diego Bay and at Monterey Bay, about two centuries after they were first sighted. Two ships carrying supplies, soldiers, settlers, and a few Franciscan priests sailed up the coast. Junípero Serra, father-president of the Franciscan friars, traveled overland with Portolá to meet them when they landed.[204] About 2 weeks' march south of San Diego, Father Serra entered a valley that he described as "more than a league [3 miles] in width, and in parts so green that, if I did not know in what country I was, I would have taken it, without any hesitation, for land under cultivation."[205] It was June, and what Father Serra saw was the fresh green look of a grassland that Indians had burned the previous fall. He did not know it, but to the Indians this was cultivation. He found signs of Indians, but he did not see them because they probably ran from the leather-armored Spaniards and their horses.

When they arrived in San Diego nearly everyone was hungry and sick, but optimistic. Portolá left Father Serra in San Diego to supervise construction of the mission and led a small party northward to find Monterey Bay. He reached Monterey, but he thought that the bay was too small. Thinking he had made a mistake, he went farther north and discovered San Francisco Bay. He knew he made another mistake and set out on a second expedition in 1770. This time he recognized Monterey Bay and established a settlement and the mission of San Carlos Borromeo.[206] Then Portolá returned to Mexico.

Father Juan Crespí accompanied Portolá on both expeditions and kept extensive diaries of what he saw along the way. As in expeditions by other explorers, he commented often about burnt landscapes and Indian fires. They reached the oak woodlands and grasslands near the Santa Barbara coast in August 1769, which Father Crespí described in his journal:

> We went over land that was ... well covered with fine grasses, and very large clumps of very tall, broad grass, burnt in some spots and not in others. . . .[207]

They marched a little farther and he said: "We ... went up to some low-rolling tablelands that end in high bold cliffs near the sea, but are all very good dark friable soil, well covered with very fine grasses that nearly everywhere had been burnt off by the heathens."[208] Again, on August 29 he wrote:

> We went almost all the way over salt-grass, all very much burnt off by the heathens, with some descents to dry creeks. On going about a league and a half, we reached a stream with a good amount of fresh water emptying into the sea, but no village nor soil of any worth upon it. The soldiers had scouted up to this point, and it was not a full day's march, nor was there grass for the animals, as it had all been burned off. . . . On going about three hours, in which we must have made about two leagues and a half, we came to a hollow where the heathens had said there were some pools of water, and although it had been burned off, there were spots that had not been and where there was good grass for the animals; a halt was ordered here.[209]

Indians burned the grass within oak woodlands in the Central Valley as well as along the coast. The log for one Spanish expedition near the Merced River included the following entry for September 27, 1806:

There are at this spot about sixty oak trees and a few willows in the bed of the stream. The forage was extremely scanty, and that the country appeared to have been burned over by the Indians did not conceal the fact that the land is very poor. Consequently there is little pasturage.[210]

These fires burned uncontrolled over the countryside and up the sides of the adjacent hills and mountains "according to the winds and calm," wrote military governor Fernándo Rivera y Moncada in 1776.[211] Indian fires usually burned harmlessly under the oaks that were sprinkled across the grasslands, but they surely became torrents of flame when they reached the coastal sage and chaparral. These shrubs will die if they are burned every few years, but they thrive if burned once in a decade or more. Thus the fires set by Indians not only renewed the plants they wanted, but by burning the nearby brush they expanded their wild gardens as well.

Like other Spaniards, governor Moncada complained about the lack of grass, but he recognized that the Indians waited to burn until after they finished harvesting. He wrote in 1776:

In the countryside [there is] extreme need of pasture for the animals ... all occasioned by the great fires of the gentiles [Indians], who, not having to care for more than their own bellies, burn the fields as soon as they gather up the seeds, and that is universal.[211]

However, the blackened landscape that deprived the Spaniards of forage for their horses became lush in the spring. Father Crespí noticed this change from the previous summer when he left with Portolá in May on his second expedition to Monterey Bay:

At once after setting out, we commenced to find the fields all abloom as many as were the flowers we had been meeting all along the way and on the Channel, it was not in such plenty as here, for it is all one mass of blossom, great quantities of white, yellow, red, purple, and blue ones: many yellow violets or gilly-flowers of the sort that are planted in gardens, a great deal of larkspur, poppy and sage in bloom, and what graced the fields most of all was the sight of all the different sorts of colors together. ... On this whole march ... we have seen not a bush.[212]

By late summer many of these plants produced the seeds sought by Indians, including such annual herbs as chia, spikeweed, redmaid, tarweed, and many others. The seeds from such grasses as California brome, rye-grass, and needlegrasses were equally important, as were bulbs from bro-diaea and mariposa lily. All of these plants are known as "fire followers" because fire stimulates their growth.[213]

Indians used seedbeaters to knock seeds into conical baskets at harvest time, which also dispersed the seeds that missed the baskets. They burned and harvested so regularly that the Spanish referred to grasslands as "fields." Father Francisco Palóu recognized the importance of these "fields" to the Diequeño. He wrote in an official report for Mission San Diego in 1772 that the "savages subsist on seeds of the *zacate* [wild grass] which they harvest in the season."[214] That year, Father Jayme also complained in a letter that soldiers let their animals graze in their "fields" near the mission and that "they ate up their crops."[214]

Fall was a time of plenty, but winter and spring could bring starvation if stored foods were insufficient. Berries, nuts, acorns, and seeds were not mature by spring, although game remained abundant. Even fish such as salmon and steelhead were not yet running in the streams. However, fire helped even at this time of year. Naturalist Jose Marinez noticed that California Indians used fire to create extensive fields of grass, not to feed game but to eat themselves. They burned the brush, he wrote in 1792, "so that with the first light rain or dew the shoots will come up which they call *pelillo* (little hair) and upon which they feed like cattle when the weather does not permit them to seek other food." "Little hair" included the green shoots of many plants, including clover.[215]

To Dry and Cook

Indians found that the fires they set could also cook their food before they gathered it. Indians often cooked grasshoppers this way, but they also cooked small animals and even plants. No doubt this was common among Indians in North America because they surely saw birds gorge themselves on animals broiled in wildfires. William Bartram watched a King Vulture in northeastern Florida do this after a fire burned through a southern pine forest in 1774 or 1775. This large colorful vulture was sacred to the Creeks, and they used its tail feathers on their royal standard to signify war or peace, depending on the color they displayed. "These birds," Bartram wrote in his journal, "seldom appear but when the deserts are set on fire." Then, he said,

... they are seen at a distance soaring on the wing, gathering from every quarter, and gradually approaching the burnt plains, where they alight upon the ground yet smoking with hot embers: they gather up the roasted serpents, frogs, and lizards, filling their sacks with them: at this time a person may shoot them at pleasure, they not being willing to quit the feast, and indeed seeming to brave all danger.[216]

The Caracara, a dark hawk about the size of an Osprey, also swoops down to feast on small animals caught by fires in Florida and south Texas.[217] Texas Indians imitated this behavior by using fire to drive and cook small game where they found them. Most other Indians likely did the same. Thus Figueroa told Cabeza de Vaca in the early 1500s that Indians used fire "to drive out lizards and other like things from the earth for them to eat."[218]

Tarweed, an annual herb in the aster family with a yellow flower, has a seed that Indians throughout the West used as food. Some species of tarweed grow only in small local areas while others have wide distributions in North America. They also grow in a variety of habitats, including open ponderosa pine forests, red fir forests, lodgepole pine forests, prairies, and many other plant communities. Lieutenant Emmons observed tarweed collecting while traveling through Kalapuya territory in the Willamette Valley. He wrote in his journal on August 7, 1841, that the Indians "burn the Prairies to dry & partially cook a sunflower seed [tarweed]." He added that it "abounds throughout this portion of the country & is afterwards collected by them in considerable quantities & kept for their winter's stock of food."[219]

Fire not only stimulated the growth of tarweed and cooked the seeds, it also made gathering much easier. Pioneer George Riddle described the way the Takelma used fire to ease tarweed seed harvesting in the Rogue River Valley of Oregon in 1851. The Kalapuya, Umpqua, Hupa, and many other tribes in Oregon and California used similar methods.[220] Riddle said that tarweed "was very abundant on the bench lands of the valley and was a great nuisance at maturity." The reason, he said: "It would be covered with globules of clear tarry substance that would coat the head and legs of stock as if they had been coated with tar." No doubt Indians also found this tarry substance inconvenient when they collected the seeds. So they solved their problem with a light fire that left the plant standing erect, as Riddle explains:

When the seeds were ripe the country was burned off. This left the plant standing with the tar burned off and the seeds left in the pods. Immediately after the fire there would be an army of squaws armed with an implement made of twigs shaped like a tennis racket with their basket slung in front they would beat the seeds from the pod into the basket. This seed gathering would last only a few days and every squaw in the tribe seemed to be doing her level best to make all the noise she could, beating her racket against the tip of her basket. All seeds were ground into meal with a mortar and pestle.[221]

The ground tarweed seeds "resembled pepper in appearance, but was sweet tasting," settler Horace Lyman remembered.[222] Sometimes the Kalapuya mixed ground hazelnuts and camas with the tarweed seed meal before they stored it for the winter.[223]

Wherever Indians gathered acorns, especially in California and Oregon, they cleared debris with fire. This kept oak woodlands open and productive. In addition, Indians who lived in the coastal mountains sometimes set their fires before gathering the acorns to roast them where they lay.[224] In the 1930s, a Karok woman told an interviewer that her people burned oak woodlands in northern California for all of these reasons. Furthermore, she recalled, "long ago where they saw lots of acorns on the ground, in a tanbark oak grove they made roasted unshelled acorn."[225] Mamie Offield, a Yurok, told an interviewer in the 1950s that her people used to do the same thing. "The trees [tanoak] are better if they are scorched by fire each year," because, she said, burning "kills disease and pests" and it "leaves the ground underneath the trees bare and clean and it is easier to pick up the acorns."[226]

Indians living in the Sierra Nevada usually burned oak woodlands after gathering acorns rather than using fire to cook them on the ground. Otherwise, they burned for most of the same reasons as Indians living in the coastal mountains and valleys. Black oak acorns were an especially important food for the Miwok, Monachi, and many other Indian peoples who lived in the Sierras. So important that they often located their villages near oak groves. Holes where they ground their acorns also riddle granite boulders near these groves. However, black oak groves within ponderosa pine and mixed-conifer forests quickly disappear without fire. Brush can overtop oak seedlings and shade them out, and conifers can shade out mature oaks and gradually replace them. Thus Indians set light surface fires to keep conifers and brush from growing among the oaks.

They also knew that black oaks produce larger crops of acorns when they have more space within which to grow.[227]

Besides knocking acorns from limbs and collecting them on the ground, Indians in Oregon, California, and the Southwest, often found it convenient to let the acorn woodpecker gather the crop. The acorn woodpecker lives in oak woodlands, as well as redwood, pine, and mixed-conifer forests that include oak trees. It collects acorns and then stores them in the bark of particular trees. One tree can contain hundreds or thousands of acorns because many birds contribute to the cache.[228] This concentrated supply of food could not help tempt Indians, so they burned these trees and took the acorns. In 1850, Frank Marryat either watched them do it or heard about it from someone else during his travels through the redwood forests of northern California. "The Digger Indian," he said, "will light a fire at the root of a well-stocked red-wood tree until it falls; they then extract the carpentaro's acorns and fill many baskets full, which they carry away."[229]

The Takelma and other Indians burned trees to collect sap, especially sugar pine, because they used the sap for medicine and food.[230] On September 25, 1841, while traveling through the Rogue River Valley in southern Oregon, Henry Eld, a member of Lieutenant Emmons' party, reported their first sighting of the sugar pine tree. He said that the sap "forms in little lumps on the outside where the bark has been removed." "The Indian manner of gathering it," he wrote, "is to set fire to the tree and save the juice as it runs out, in this way they get it in large quantities."[231]

Straight and Slender

California Indians are renowned for their baskets. Baskets served many uses, including cradleboards, winnowers, sifters, boiling containers, conical seed collectors, and as traps for catching fish. Indian basketweavers used many plants and plant parts for their baskets. Sedge root baskets were among the finest made. However, they used many other plants as well, such as deer brush, redbud, hazel, willow, black oak, blue oak, squawbush or sourberry (a member of the sumac family with an acid berry), and bracken fern. Most of their baskets, as well as arrow shafts made of buttonwillow or snowberry, required straight young shoots to make. These shoots had to come from the sprouts of shrubs and trees that were burned or pruned.[232] A Mono elder made the point:

A burn brings the sourberry and the redbud up real nice for baskets or when we prune, it comes up nice every year. If we don't there's nothing there to use for baskets.[233]

Indians pruned when necessary, but burning took less effort and produced far more sprouts. Norma Turner, a Mono, told an interviewer that they "preferred the shoots after a burn because these are the ones that grow right from the ground and they're straight and slender." Sometimes they burned small patches, but mostly they burned large areas. "They'd light the whole hillside on fire," recalled Dan McSwain, a Mono. Indians set most of their fires in the fall so that they could harvest shoots the following spring, and they often burned the same area every year.[234]

BURNED PLACES IN THE FOREST *(Go-ley-day)*

Indians who lived in places where forests grew thick, and edible or useful plants were few because of heavy shade, often used fire to clear patches of forest and narrow corridors or trails to connect them. They could not rely on the uncertainties of lightning fires for their survival. They had to clear parts of the forest themselves. An elderly Slave trapper from the Hay Lakes region of Canada said that "our people have a name for those burned places in the forest [they are] called *go-ley-day*." "They tell one another about those places and when to hunt there," the trapper added.[235] Indians burned holes in forests because sunny meadows provided more food for game than a dense forest, and they concentrated the animals so that they were easier to hunt. Clearings also improved the growth of berries and other edible plants, and they provided Indians with some safety from the wildfires that often threatened their camps. Thus, scattering meadows within a sea of trees became an important method of Indian burning in Canada, the Pacific Northwest, and the East.

Keeping the Country Open

Indians who lived in the vast spruce forests of Canada often burned their clearings in "deadfalls," places where the trees had already died because of lightning fires, insects, or diseases. Deadfalls provided nothing that Indians needed and they increased the danger of wildfire. Indians also felt that deadfalls would burn anyway, most likely in the summer when fires could become huge. So they burned them in the spring when it was safer. An older Cree from the Chipewyan Lake area explained:

> When you burn the deadfall places it burns for a long time, not like the meadows—they burn out fast. Because there's all those dead trees. Maybe the next year you come back, burn it some more and then pretty soon it's all open and the moose really like those places.[236]

Moose thrived in these burned clearings because they are generalists. They eat hundreds of different plants worldwide and 25–30 plants in one area, including grasses, forbs, shrubs, trees, mushrooms, and lichen. However, they prefer the young shoots of trees and shrubs that sprout after fires, and the grasses and forbs that they eat also require fire. Furthermore, the clearings that Indians created favored moose because the thick forests that surrounded them furnished the animals with cover when winter snows became deep and hard.[237]

Indians only burned green trees if they needed a meadow where there were no dead trees. Even then, they chose the site carefully. They did not burn jack pine because the soils were too poor to support a meadow. They usually avoided black spruce for the same reason. They preferred white spruce and aspen groves because these forests grew on soils that produced better grasslands. According to an elderly Cree-Metis from the Lac La Biche area, it must be "the right kind of place or the grass and brushes don't grow." "You can burn places like that but sometimes the fire burns more, more than you want," he cautioned.[238]

Indians took many precautions when burning, but their fires continued to push forests northward in some places and expand the grasslands, which also allowed grazing animals to move in and replace the woodland animals. Alexander Mackenzie inadvertently recorded this change when he asked an old Beaver Indian his age while traveling through Alberta, Canada, in 1793. The Indian did not know his age, but he knew how the landscape had changed during his lifetime. Mackenzie recorded what the old Indian said in his journal:

> An Indian in some measure explained his age to me, by relating that he remembered the opposite hills and plains, now interspersed with groves of poplars, when they were covered with moss, and without any animal inhabitant but the rein-deer. By degrees, he said, the face of the country changed to its present appearance, when the elk came from the East, and was fol-

lowed by the buffalo; the rein-deer then retired to the long range of high lands that, at a considerable distance, run parallel, with this river.[239]

Short-term fluctuations in weather brought on by the Little Ice Age probably contributed to the change.[240] Fires burned less easily during excessively wet periods, so forests most likely expanded. They would contract during droughts, in part, because of an increase in fires. Thus, shortly after talking to the old Indian, Mackenzie noted in his journal that "the whole plain was on fire." Thus the Indians were still burning and forests were still shifting around the landscape.[241] Dr. George M. Dawson, a biologist with a Canadian Pacific Railway survey party, also noticed this in 1879:

> The origin of the prairies of the Peace River is sufficiently obvious. There can be no doubt that they have been produced and are maintained by fires. The country is naturally a wooded one, and where fires have not run for a few years, young trees begin rapidly to spring up. The fires are, of course, ultimately attributable to ... the Indians.[242]

After creating grasslands and meadows, Indians still relied on fire to keep them open. Springtime marked the beginning of the Indian burning season in the vast spruce forests of Canada. "In the spring when there is still some snow in the bush that's the only time most people could burn the open places," explained an elderly Slave from the Meander River area.[243] Sometimes they burned in the fall as well, especially if the spring was too wet to burn large areas. These fall fires were less effective, however, so they kept them small. Furthermore, a Slave from the Hay Lake area warned that fall burns have "to be done at the right time because [the fire] might get away." A Cree gave a similar warning:

> You couldn't just start a fire anywhere, anytime. Fire can do a lot of harm or a lot of good. You have to know how to control it.[244]

Whenever they burned, the one common requirement was a wet forest that would stop the fire at the edge of the clearing. "You wait until there's snow in the bush and it's kind of wet there," a Slave explained, adding that the "grass burns real good but it stops when it gets to the bush."[245]

Still, according to an old Indian from the Beaverlodge area, "fires would always kill some trees around that meadow." This too was a benefit of spring burning. "You had good, dry wood, lots of firewood," the Beaver Indian said. [246] The fires they set each spring gradually expanded the meadows, and the shrubs that came up where the trees used to grow made spring burns even more useful. "There were always berries round a meadow or places where there has been a forest fire, and those are good places to hunt bears," said a Slave from the Meander River area. Nevertheless, the Slave also knew that "burning is good for some berries but not others," and they burned accordingly. They also knew that a "hard" burn delayed the berry crop for a year, while a "light" spring burn produced a good berry crop that summer.

In some years, the grass within clearings was still too wet to burn at the time the Indians left their hunting and trapping grounds. They solved the problem by building fires that smoldered for days or weeks, and then spread to the grass when it became dry. An older Cree described how they did it:

> So we'd just build a big campfire and leave it. Maybe couple weeks later even, when the grass is really dry, the grasses would all get burned up, but the fire would't go anywhere because it was still damp in the bush. . . . We didn't worry about campfires in the springtime; but you had to be careful during the summer cause you could start a . . . big fire.[247]

Indians had to use fire differently when they kept meadows open on the eastern slope of the Rocky Mountains. Here winds normally blow uphill during the day as the air warms in the sun, which is a dangerous time for fires. They knew this and burned in the late afternoon when the cool mountain breezes moved downhill. An older Cree-Metis from the Grande Cache area explained:

> We'd always wait until late afternoon and the fire was set at the upper end [of the meadow]. It would burn down to the low, damp places where the really wet grasses grow. That's the way we burned mountain meadows. See, you have to know the wind; you have to know how to use it.[248]

One witness remembered the early years when Indians could still burn: "You could see the smokes, a string of fires, that the Indian trappers were setting as they came in the spring." "You knew they were on their way home then; it was a grand sight," he said.[249] However, an old settler from England had a different experience. "I remember being scared that first spring when the Indians set fires in all the meadows," he said. "There were smokes everywhere," but, he added, "that's the way they kept this country open, you know."250

Indians who lived in the spruce forests also set fire to marshlands around lakes and streams. They burned to increase waterfowl, muskrats, and many other useful animals. "Burning sloughs was … good for foxes," an older Slave from northwestern Alberta remembered. Then he explained why:

> When you get thick grass all bent over it's hard for the fox to dig down to get the mice, and the foxes are really hungry that time of year [spring]. When you burn it off, well sure you kill lots of mice—maybe hundreds; you can hear them in the fire, but they come back by the thousands. Then it's easy for the fox to get the mice and we get lots of foxes too.251

What the Indians may have heard was the squeaking of adult mice as they gathered their young and herded or carried them away from the fire. This is a common sound in southern pine forests when cotton rats move their young to safety as the flames approach their nests.252

Ducks were an important food source for Indians living in the Canadian spruce forests. They ate nearly three-quarters of a pound of duck meat a day during the duck season. A Chipewyan from the Cold Lake area remembered how they used fire to increase the duck population in the old days:

> A lot of the burning was around the lakes, especially to make it better for the ducks. The ducks like the fresh roots that come in after a fire. … There's still lots of ducks around, but not like they were before. … There was lots more of them then.253

Indians relied on the ducks, so they took care to protect them when they set fire to the marshlands. "We always burned just before all of the snow melted, before the ducks had started to nest," said one elderly Slave. "We didn't burn after the ducks had nested," he said, "we knew when to burn and when not to burn."254 Like ducks, muskrats also depended on fire for their food. "Muskrats like to eat the grass roots, like the reeds that grow round a lake or pond," said an elderly Cree-Metis from the Utikuma Lake area. "If it don't burn regular then it don't grow," he said, "it gets all choked up, dead stuff—and not so many muskrats."255

"There are many places people know to burn," said an old Beaver woman. Likewise, "there are a lot of places they don't burn," an old Slave remarked.[256] In particular, the Beaver woman said, "where there's timber and moss they don't burn, of course, because the fire lasts and lasts." Furthermore, she said, "it's not good for fur bearing animals like pine martens and mink and lynx; that's where they mostly stay in winter, so they don't want to destroy that."[257] Thus, Indians wanted the forest, the meadows and the marshlands, but they had to be productive. They needed a variety of plants and animals to survive, and that meant furnishing each of them with the habitat they required. A burned meadow provided Indians with hay for their horses, plenty of game, berries and firewood, as well as protection from summer wildfires. Similarly, a burned marsh provided them with an abundance of waterfowl and other game. By understanding and managing their world, Indians could harvest their wild gardens, and hunt their wild livestock and wild fowl, with confidence.

A Pleasant Meadow

The red spruce forests of the Northeast confronted Indians with many of the same problems as people faced in the white spruce forests of Canada. So they also burned "open places in the forest." Thus in 1749, while camped at Crown Point on the southern end of Lake Champlain, Swedish botanist Peter Kalm wrote: "The country hereabout, it is said, contains vast forests of firs of the white, black and red varieties, which formerly had been still more extensive." "One of the chief reasons for their decrease," he concluded, "is the numerous fires which happen every year in the woods, through the carelessness of the Indians, who frequently make great fires when they are hunting, which spread over the fir woods when everything is dry."[258] In 1772, surveyor A. Campbell wrote that the plains in the Adirondacks were "almost clear" and "burnt land."[259] Likewise, surveyor Duncan McMartin also recorded in his "survey minutes" of 1823 that Indians burned the plains every year to hunt game.[260] Most likely, Indians cleared the sprunce-fir forests to provide forage and browse for game and then kept them open with hunting fires. Some of their fires undoubtedly escaped and expanded the plains by sweeping through the adjacent forest.

Indians did not just carve large clearings from dense spruce forests; they also made clearings in oak and pine forests. Some of these clearings were no doubt created intentionally. Others probably started as abandoned cornfields that they maintained as grasslands with fire after the soil became depleted. Still others developed from forests that they burned too often or too hot so that trees ceased to grow there. Regardless of the cause, these vast open grasslands became productive hunting grounds for the Indians. In 1805, T. Bigelow described one of these grasslands as he traveled through northwestern New York on his way to Niagara Falls:

> Hundreds of acres may be seen together, on which there is scarce a single tree, there being at most but an oak or a poplar or two, scattered at great distances. The earth here is covered with small willow bushes, brakes, butterfly plant ... wild grass and strawberry vines, with very young trees not more than knee-high. In many of these open grounds, a man may be seen at a distance of two miles. There are patches of trees interspersed among these open grounds. They are of the same kind as are to be met with in the neighboring country.... Various conjectures are indulged as to the scarcity of trees; but the most probable is that it has been occasioned by the Indians repeatedly firing whatever would burn here.... What serves to confirm this opinion is the frequent appearance of charcoal and burnt sticks, and the abundance of young trees which are now shooting up. Wherever groves of trees are yet standing, it may be seen that they were probably protected by the interposition of a stream of water, or by the dampness of the soil where they grow.[261]

Europeans called these open hunting grounds "savannas," "plains," "barrens," "prairies," or "meadows," and most Indians who lived within forests made them.[262] For example, sometime before 1626 two lost colonists discovered a "place where the Savages had burnt the space of five miles in length" in the forests near Plymouth Harbor.[263] Thomas Morton also noted in 1632 that constant burning in the same place by Indians "so scorcheth the elder trees, that is shrinks them, and hinders their growth very much." Some of these open areas must have been large around Massachusetts Bay because Morton also said that

> ... if he would endeavor to find out any goodly cedars, he must not seek for them on higher grounds, but make his inquest for them in the valleys, for the savages by this custom of theirs have spoiled all the rest. For when the fire is once kindled, it dilates and spreads it selfe as well against, as with the winde; burning continually night and day, untill a shower of raine falls to quench it.[264]

Indian hunting grounds might be near or far from a village, some as much as 20 miles away, depending on how the clearing first developed.[265] Given the choice, however, they probably preferred to hunt nearby. Thus, Samuel Purchas, an English compiler of travel books, noticed that Indians living near the Saco River in Maine cleared forests next to their villages for their hunting reserves. In 1602–1609, he wrote:

> Neere to the north of this River ... are three townes. These two last townes are opposite one to the other, the river dividing them both.... Upon both sides of this river up to the very lake, for a good distance the ground is plaine, without trees or bushes, but full of long grasse, like unto a pleasant meadow, which the inhabitants doe burne once a yeere to have fresh food for their deere.[266]

Indian fires also kept the grasslands open in the vast Shenandoah Valley, and other treeless valleys west of the Blue Ridge. Moreover, they created large clearings near the headwaters of the Rappahannock, a river in northeast Virginia that flows from the Blue Ridge into Chesapeake Bay.[267] Some of the grasslands that Indians carved out of forests in other parts of the Southeast were large as well.[268]

Prairies and Open Grounds along the Coast

Indians who lived in the dense coastal forests of the Pacific Northwest and California, where sunlight rarely reached the ground, faced the same challenges as those who lived within eastern spruce forests. So they managed their forests in much the same way. Here Indians relied heavily on shellfish and fish, particularly salmon, but they also traveled inland in the fall where the men hunted elk, deer, and bear, while women harvested roots, seeds, and berries.[269] People who cultivated the open oak woodlands and grasslands of the Willamette Valley had plenty of these foods available. But game and plant foods were less abundant within forests of huge Pacific Douglas-fir, hemlock, Sitka spruce, and redcedar trees. So Indians such as the Tillamook had to burn openings in forests to make room for plant foods and to feed the game they hunted.[270] No doubt their fires occasionally went out of control, particularly during droughts, and became conflagrations that swept over vast areas of forest.[271]

The Tillamook, Kawakiutl, and other native coastal people pushed the thick forests back from the edge of the sea with fire to create prairies. They harvested many plants from these coastal prairies, especially camas, which once decorated the hillsides with patches of blue flowers.[272] They also grew Indian consumption plant or wild celery in burned clearings. Indians used the seeds of this medicinal plant as an analgesic, antirheumatic, cough and cold remedy, and laxative.[273] Without their annual fires the coastal prairies, and the wild celery, camas, and other plants that Indians relied upon, would have quickly disappeared as the forest reclaimed the land.

Coastal Indians also burned large openings in forests to create berry patches, especially huckleberries, salmonberries, and salal.[274] Berries were an important food for Northwest Indians, and dried salmonberries added vitamin C to their diet in the winter. They even ate the young sprouts of salmonberry in the spring, raw or cooked, and the bark and leaves provided them with medicines as well. The bark was particularly important as a remedy for stomach ailments caused by eating too much fish. Huckleberries and salmonberries produce abundant seed, and birds and small mammals carried them into the newly burned clearings where they could germinate. Buried seeds dropped in a forest years before the Indians burned it also remained viable and often sprouted as well.[275] Therefore, berry patches were easy to make.

Sometimes fires escaped and berry patches became much bigger than intended. Charles Nordhoff saw the aftermath of a huge fire that Indians may have set while he was sailing up the Oregon coast from San Francisco. It presented "a barren view" he wrote in 1875, "which you owe to the noble red man, who, it is said, sets fire to these great woods in order to produce for himself a good crop of blueberries."[276] Indians may or may not have started this fire. However, Nordhoff and everyone else knew that Indians set many fires in the forest, and that lightning fires were rare, so they probably set this fire as well. Indians may also have set the fire that burned the forest seen by Lewis and Clark on April 9, 1806, since there were no Europeans traveling inland at the time. "The wood in this neighborhood has lately been on fire," they wrote, "and the firs have discharged considerable quantities of pitch, which we collected for some of our boats."[277]

Besides the large fires that most likely started when Indians created new prairies and berry patches, some escaped fires may have come from drying berries. Indians built campfires to dry their berries, and they even set rotten logs on fire to dry them.[278] Similarly, fires probably escaped into the forest now and then when Indians burned their little tobacco plantations. In 1825, David Douglas met the owner of one of these private gardens when he found it hidden among the trees in the Willamette Valley. He described the meeting in his journal:

> The natives cultivate it [tobacco] here, and although I made diligent search for it, it never came under my notice until now. They do not cultivate it near camps or lodges, lest it should be taken for use before maturity. An open place in the wood is chosen where there is dead wood, which they burn, and sow the seed in the ashes. Fortunately I met with one of the little plantations and supplied myself with seeds and specimens without delay. On my way home I met the owner, who, seeing it under my arm, appeared to be much displeased; but by presenting him with two finger-lengths of tobacco from Europe his wrath was appeased and we became good friends. He then gave me the above description of cultivating it. He told me that wood ashes made it grow very large. . . . Thus we see that even the savages on the Columbia know the good effects produced on vegetation by the use of carbon [burning].[279]

The Indian David Douglas met in the forests of Oregon raised his tobacco the same way that most Indians did.[280] Thus, in 1833, Maximilian, Prince of Wied, commented: "The Blackfeet like most tribes of the Upper Missouri, sow the seeds of the *Nicotiana* [tobacco] . . . having first burnt the place where they intend it to grow."[281] Some of these fires certainly burned nearby forests in many parts of North America.

The fires Indians set to maintain their prairies and berry patches probably escaped and spread into the forest from time to time as well. Huckleberries and salmonberries tolerate some shade, so they will grow in a relatively open forest. Even so, they become low and straggly in dense old forests. Indians knew that they grew best and produced more berries in the open, especially after fire. Consequently, they burned their berry patches every few years, and they may also have set fire within the forest. Fire kills the tops of the shrubs, but they sprout vigorously and grow rapidly after the fire. Thus repeated fires made the shrubs grow thicker and reduced the likelihood that trees would invade the berry patch. However, there would be no berries in the burned patch the next season, but in 2 years the berry crop would be larger than before.[282] This delay in berry production meant that the Indians had to have several berry patches scattered in the forest so that they could pick berries in one patch when they burned another.

Game could be hard to find in the dense forests of the Pacific Northwest, as Captain Clark noted in his journal on January 10, 1806. He had just returned to Fort Clatsop from the coast when he remarked with disappointment that the hunters "were not very successful" in the forests around the fort. "The deer have become scarce," he said, because they are "seen chiefly near the prairies and open grounds along the coast."[283] The Indians knew this, and it was another reason why they used fire to carve prairies and berry patches from the forest.[284]

Mule deer, such as the black-tailed deer that live in these forests, are primarily browsers that feed on the leaves, stems, and shoots of woody plants. They also eat grasses and forbs, especially in the spring. They prefer to browse on the young plants that grow in burned areas rather than the older and less palatable plants found in unburned forests. Elk share similar preferences with deer, and they both relish the leaves and stems of huckleberries and salmonberries.[285] Lewis and Clark noticed this, as they wrote in their journal on February 4, 1806:

> The elk are in much better order in the point near the prairies than they are in the woody country about us.... In the prairies they feed on grass and rushes, considerable quantities of which are still green and succulent. In the woody country, their food is huckleberry bushes, ferns, and an evergreen shrub which resembles the laurel in some measure.[286]

Likewise, both deer and elk use the dense forest around openings as cover. Thus the scattered prairies and berry patches created by Indians also became excellent habitats for game, so that is where they hunted. Robert Stuart, who discovered the Oregon Trail, recorded such a hunt during his trek from Fort Astoria to St. Louis in 1812 and 1813. He said:

> Their mode of hunting Elk and Deer is with the Bow and Arrow, very few possessing or knowing the use of Fire Arms; they frequently go in large parties, surround the game while grazing in a favorable place, such as a small prairie or meadow environed by Wood; they plant themselves in the different avenues, or paths leading to this spot, then set in their dogs, which throws the affrighted animals in such confusion as to scatter in every direction, thereby giving the most or all a chance of exercising their skill, for let the consternation of these poor creatures be ever so great, they can only escape by those leading paths.[287]

Little Knots of Deer

Much like other dense forests, coast redwood forests in northern California could not produce food for native people unless they used fire. So the Wiyot, Tolowa, Tututni, Yurok, and Karok created clearings, known as "prairies," in scattered locations throughout the forest, especially along ridges. Most of their prairies were large, up to $\frac{1}{4}$-mile wide and $\frac{3}{4}$-mile long. Like other Indians, they created these prairies to stimulate the growth of berries and edible plants, and to provide food for elk, deer, and bear, including California grizzly bears. One of the more important plants they cultivated on their prairies was mule ears, a "fire follower." They ate the leaves and stems, and they ground the seeds and mixed them with other seeds in their pinole.[288]

Indians used these prairies as oases not unlike those found in deserts. Thus a network of prairies increased their chance of finding game and other food as they passed through an otherwise sterile forest. An early expedition from the Sacramento Valley to the coast where the Wiyot lived demonstrated how important these prairies could be to people who traveled through the redwoods. During the trip, the party had nothing to eat for three days, and two mules starved to death. They noted in their journal that they rode for 10 days without "the sight of any living thing that could be made available or useful for food." Then they found a prairie with "little knots of deer," on one side, "on another and nearer ... a large herd of elk, and still in another direction both." However, they did not know where to find other prairies, or the network of narrow footpaths Indians made to connect them, so one of the men and several more mules starved before the expedition ended.[289]

After clearing the forest, Indians burned the prairies nearly every year to keep them open and productive. Ridge-top prairies worked well because the fires Indians set had to burn down the slope against the wind when they reached the forest since heated air usually blew upslope during the day. These are called backing fires because they back into the wind, and the short flames lap uphill into areas already burned by the fire. Backing fires move slowly and rarely flare up enough to cause a crown fire. Even so, Indian fires certainly scorched some redwoods on the edge of the prairies just as ridge-top fires do today.

Indians also set surface fires underneath redwood trees to stimulate the growth of young shoots of such plants as five-finger ferns and hazel, which they needed for weaving baskets. A Karok woman recalled that "when they burn ... for hazel sticks, they pick them [in] two years, then they are good."[290] Tanoak grows with redwood on the lower and middle slopes of the mountains, so Indians burned there as well. They burned the litter and duff under the tanoak nearly every year to keep the ground clear for gathering acorns. Even more important, these frequent light surface fires prevented twigs, leaves, logs, and small trees from accumulating and becoming dense enough to support hot fires that would kill the tops of the tanoak.[291] "They do not want it to burn too hard," a Karok woman cautioned when interviewed in the 1930s, because "they fear that the oak trees might burn."[292] Even though tanoak quickly sprouts after a fire, it would still take 5–30 years before the trees would produce acorns again. Therefore, the frequent surface fires set by Indians allowed tanoak to grow to a large size so that they would continue producing acorns, sometimes for as long as 300–400 years.[293] Indians certainly knew this, as the Karok woman said: "An old tree bears better, too."[294]

To Prepare the Ground

By far the largest and most widespread clearings were those Indians made to plant corn, beans, squash, and other domesticated plants. From the Mississippi River to the shores of the Atlantic and the Gulf Coast, they gradually removed forests on most of the river floodplains and even large areas on the uplands. Indians became farmers and they saw the forest as an obstacle to cultivation in the same way that European colonists did when they arrived. Even though the great Mississippian cities were already collapsing, Indians were still clearing land for crops and hunting reserves when Cabeza de Vaca began his remarkable and perilous journey from Florida to Mexico in 1528.

As with other clearings, Indians used fire to cut down the forest for their crops. Captain John Smith described the way most Indians made their fields. He said:

> The greatest labor they take, is in planting their corn, for the country naturally is overgrown with wood. To prepare the ground they bruise the bark of the trees near the root, then do they scorch the roots with fire [so] that they grow no more. The next year with a crooked piece of wood they beat up the weeds by the roots, and in that mold they plant their corn.[295]

In a few years the dead trees would topple and pile up on the ground. This provided enough dry fuel to burn the field again and clear the remaining trees.[296] These could be hot fires so some of them probably burned into the surrounding forest. This was especially likely along the Ottawa River in Canada. Champlain noted this in his narrative after traveling up the river by canoe in 1613: "When they wish to make a piece of land arable, they burn down the trees, which is very easily done, as they are all pines, and filled with rosin."[297]

Indians kept planting their fields until the soil became exhausted, or until they had to abandon them and move their village because of the lack of firewood, game, or war. However, they often made permanent fields on floodplains and other places with fertile soil. Here they might leave the field fallow if productivity declined, and then reburn it and begin planting again before the trees and shrubs invaded.[298] Champlain saw the Indians do this around Boston Bay in 1605. He said:

> There were also several fields entirely uncultivated, the land being allowed to remain fallow. When they wish to plant it, they set fire to the weeds, and then work it over with their wooden spades.[299]

George Percy, one of the Jamestown settlers, may have seen Indians doing the same thing in Virginia in 1607. He reported that "wee marched to those smoakes and found that the Savages had beene there burning downe the grasse, as wee thought either to make their plantations there, or else to give signes to bring their forces together, to give us battell."[300]

Eastern Indians cleared and burned so much land that they changed the historic range of the bison. Cabeza de Vaca did not see a single bison during his 8-year trek across the South. Nevertheless, his companion Esquivel saw them three times when he traveled into northern Texas with the Indians as their slave. He described them to Figueroa, who told Cabeza de Vaca, who then wrote the first description of these "cattle." "They have small horns," he said, "the hair is very long and flocky ... [and] some are tawny, others black."[301] Things changed by the time La Salle mistakenly landed on the coast of Texas in 1685. Another century and one-half of Indian clearing created enough additional grassland to attract bison down to the Texas coast and into Georgia and Florida. They had not roamed there since the end of the Ice Age. [302] As a result, La Salle's personal aide, Henri Joutel, commented in his journal that they named the river where they built Fort St. Louis "*la rivere aux Baufs*," the river of the Bullocks [bison]. He said they named it that "by reason of the great Number of them there was about it." "These Bullocks are very like ours," Joutel added, "there are Thousands of them, but instead of Hair they have a very long curl'd Sort of Wool."[303]

The Spaniards inadvertently contributed to the southern and eastern spread of bison by introducing European diseases. They were not the first, of course, John Cabot and Giovanni da Verrazzano came more than a decade earlier, and "Skrellings," the name Norsemen used for Indians, killed Thorvald, Leif Ericsson's brother, in 1006.[304] However, the Spanish stayed longer and wandered through more of North America than anyone else during these early times. Soon the Indians began dying in great numbers from disease. So bison moved south and into the many fields that lay abandoned and covered with grass.

Cabeza de Vaca may have been one of the first to start this pandemic among the Indians that depopulated large areas. He and his companions built rickety boats that they tried to sail from Florida to Mexico. After sailing for some time along the Gulf Coast, he said that "near the dawn of the day, it seemed to me I heard the tumbling of the sea." Then, he said:

> Near the shore a wave took us, that knocked the boat out of water ... and from the violence with which she struck, nearly all the people who were in here like dead, were roused to consciousness. Finding themselves near the shore, they began to move on hands and feet, crawling to land into some ravines. ... The day on which we arrived was the sixth of November [1528].[305]

They had most likely washed ashore on Galveston Island in Texas. After finding them, Cabeza de Vaca wrote that the Indians "made a house for us with many fires in it." Then, "in the morning," he said, "they again gave us fish and roots, showing us such hospitality that we were reassured." Soon thereafter it grew cold, since the Spaniards were mostly naked and food became scarce, many of them began to die. Eventually only 15 of 80 men remained alive. "After this," Cabeza de Vaca wrote, "the natives were visited by a disease of the bowels, of which half their number died." "They conceived that we had destroyed them," he said.[306] The Indians were probably right.

THEY CLEARED THE WAY WITH FIRE

Indians needed a well-maintained network of trails to take advantage of the patchwork of clearings they created within forests. They relied on their trail systems no less than we rely on our highways. Trails were even essential in regions with many rivers and lakes. Indians needed them to move back and forth between their villages, camps, hunting grounds, gathering places, and fields, and to portage their canoes and trade with one another. Thick forests full of fallen trees presented a formidable obstacle that the Indians had to overcome.

In 1853, the Stevens expedition followed an Indian trail through the Bitterroot Mountains while exploring routes for the Pacific Railroad. Their report commented about the poor condition of one section of the trail:

> The Indian trail ... leads mostly through dense forests, and over irregular ground, and is, moreover, obstructed by great quantities of fallen timber. These last obstructions would probably have to be removed yearly, as there is reason to believe that the timber, in consequence of the great height of the trees—it may be from winds, and from the forests being occassionally set on fire—is constantly falling.[307]

Indians usually found a way around obstacles such as fallen trees, but they still had to keep parts of their trails open, particularly those who rode horses. Stone hatchets could not do the job. Fire was the only tool powerful enough to make a forest passable, and Indians used it often.[308]

According to Bert Davis, an early forest ranger who worked in northwestern Montana, if "trails became blocked by blow-down or the wreckage left by snowslides, the natives simply cleared the way with fire." Furthermore, Indians often did not worry about fires that might burn more than the trail. "If the fire went wild and destroyed a few million trees," Davis said, "the Indians knew that the burn would soon grow up to grass and small brush, which made good grazing and browsing for deer and elk, and in later times for horses."[309] However, unlike most Indians, the Cree who lived in the spruce forests of Canada took care to keep their fires confined to the trail. They were aware that a fire that escaped into the forests could burn for miles over this rolling country and endanger many villages. A story told by the former white leader of an Indian trail crew, who worked for the Alberta Forest Service in the 1920s, illustrates the Indian method of trail building:

This [Cree] Indian crew had gone ahead of me ... and when I come along the trail, all of a sudden, there was fire right on the trail. Well, I put it out. Then I come on another, and then another! Well, of course, I had a pretty good idea who was starting them fires. When I caught up with the crew they were having lunch and just starting another ... fire! ... Later, after I'd been up here a few years, I realized that those [trailside] fires weren't going to go anywhere. They only did it in the spring and the fire burned to the wetter stuff and went out ... and that's the way they kept trails open through the muskeg areas.[310]

An elderly Indian from the same region remembered the old days when they used fire to improve travel and to concentrate game along trails. He said: "We used to make fire where there is too much brush to walk easily." "Then, in the fall time, we would go there to hunt moose," he added.[311]

BECAUSE THE WOODS WERE NOT BURNT

To Indians, an unburned forest was an unproductive forest. Certainly they needed fallen limbs for firewood, large trees for dugout canoes and plank houses, and places for game to hide. Forests furnished them with many of the things they needed, but a burned forest provided even more, and a greater variety as well. Even so, some forests remained unburned for relatively long periods. Indians moved villages to new and more attractive locations, populations expanded and contracted due to disease and war, and some areas remained isolated because they were not close to important trade networks. Still, these areas did not stay vacant. The Indians eventually returned. The first thing most Indians probably did when they moved into such a place was to burn the forest and make it productive. These are known as "corrective fires" in Australia where the Aborigines have been "cleaning up the country" after long absences, including thick rainforests, for tens of thousands of years.[312] Such fires would undoubtedly burn very hot, often becoming raging crown fires.

Captain John Smith captured an Indian warrior after a skirmish near Chesapeake Bay in 1608. He interrogated the warrior and "asked him how many worlds he did know." The warrior replied that "he knew no more but that which was under the sky that covered him," including some "that were higher up in the mountains." "Then we asked him what was beyond the mountains, he answered," said Captain Smith, "the sun: but of anything else he knew nothing; because the woods were not burnt."[313]

This would be "rubbish country" to Australian Aborigines,[314] and American Indians most likely felt the same way.

Indians burned these areas wherever they found them, usually when they moved there. However, sometimes they burned the places they were leaving so that the trees and other plants would grow to the right size by the time they returned. At other times they just burned because they thought they might want to move into the area later.[315] Colonel William Byrd may have seen a corrective fire while traveling through southwestern Virginia during his surveying expedition of 1728 to 1729. This fire, which he recorded in his journal, may also have been the start of a hunting reserve.

> As we marched along we were alarmed at the sight of a great fire which showed itself to the northward. . . . We could not see a tree of any bigness standing within our prospect, and the reason why fire makes such a havoc in these lonely parts is this: The woods are not there burnt every year, as they generally are among the inhabitants, but the dead leaves and trash of many years are generally heaped up together, which being at length kindled by the Indians that happen to pass that way, furnish fuel for a conflagration that carries all before it."[316]

Setting corrective fires must have been a common practice among North America Indians. For example, the Miwok moved into California's Yosemite Valley about 650 years ago. They may have returned from a long absence or, perhaps, another tribe abandoned the valley much earlier. Regardless, the Miwok found the dense mixed-conifer forest that grew there unsuitable for their needs, so they burned it. Charcoal deposits in the soil jumped up abruptly at that time, and fossil pine and fir pollen taken from a pond declined rapidly.[317] This clearly shows that these were corrective fires. The Miwok had burned most of the trees and opened the forest. They continued to burn the valley with management fires until historic times. Gradually they transformed it into an open oak woodland that provided them with an abundance of acorns, seeds, roots, berries, game, basketmaking materials, and the numerous other things they needed to survive. Thus the Miwok's mastery of fire helped to create such a productive and picturesque landscape that Europeans set it aside as a national park.

PART TWO

Forests at Discovery

*The spacious land ... full of the largest forests, some thin and
some dense, clothed with various sorts of trees, with as much
beauty and delectable appearance as it would be possible to express.*
Giovanni da Verrazzano (1480–1527), Italian navigator
Exploring the East Coast of North America for France in 1524[1]

The age of discovery lasted several centuries because North America is a vast continent. The Spanish conquistadors became the first Europeans to wander deep into America's ancient forests. Spanish soldiers boldly marched through the South, the Southwest, California, and the Great Plains of Kansas. English and Dutch colonists came next, but they stayed close to the eastern seaboard. French soldiers of fortune, fur traders, and Jesuit priests saving souls followed closely behind the colonists. They paddled and trudged their way into the forests of the "Up Country," the land around the Great Lakes and the interior of Canada. They also were the first to explore the Mississippi River. Finally, it was America's turn to send Lewis and Clark's Corps of Discovery, and the waves of trappers, scientists, and settlers that followed them, into the ancient forests of the West.[2]

America's ancient forests inspired deep feelings among the Europeans who saw them for the first time. They often used words such as "enchanting," "stately," "noble," "majestic," "parklike," "spacious," and "picturesque" to portray forests and trees. Of course, not all of their descriptions were favorable. Some Europeans explained their feelings about these ancient forests with words such as "daunting," "terrible," "solemn," "dreary," and "gloomy" as well. They had nothing in their experience upon which to base their impressions of America's ancient forests. Most forests back home in Europe were cut long ago, but here in North America forests covered roughly 45 percent of the continent.[3]

What the first European explorers saw were forests of amazing diversity and awe-inspiring vastness. They felt especially drawn to trees of immense size and great age that no longer existed in Europe. Such trees grew everywhere in North America. Europeans also seemed surprised to find that most of America's ancient forests were open rather than dark and dense. They soon realized that Indians used fire to help keep them that way. So they had little trouble traveling within most forests unless they followed streams or crossed marshes where thickets grew. Even here travel was sometimes easy because Indians cleared trees from many floodplains for cornfields and hunting grounds. Occasionally, European explorers had to hack through forests blown down by winds or burned by wildfires. The difficulty they faced when cutting a path through the jumble of fallen trees was unusual enough to end up as stories in some of their journals.

Timeless Qualities of Ancient Forests

Science! True daughter of Old Time thou art! Who alterest all things with
thy peering eyes.
"To Science"
Edgar Allan Poe (1809–1849)

Even though America's forests seem complex, and in many ways they are, they all share a few timeless qualities, and ancient forests were no different (Fig. 8.1). First among these is the patch, which is a relatively uniform group of plants. Second is succession, or the way a patch of forest advances from one stage of development to another as it ages. Third is the shifting mosaic, or the way different stages of development shift from place to place in the mosaic as each patch of forest changes. Fourth is mutual dependence, or the way a shifting mosaic and the forces that shape it interact to create an entire forest.

PATCHES

Patches are the fundamental units of vegetation, so forests consist of patches. The size of a patch of forest depends on how much land is cleared or made suitable for plants to grow. Likewise, the dominant trees in a patch are roughly the same age because they seeded in or sprouted at about the same time. Even though each plant is an individual that is capable of behaving independently, plants still tend to grow in patches. This is so because the conditions that favor or inhibit growth are also patchy, so they usually affect more than one plant at a time. A tornado, fire, lightning strike, insect infestation, landslide, or even a single large tree that falls clears an opening

Fig. 8.1 America's ancient forests during the age of discovery.

in a forest in the same way that a gardener tills the soil. The plants simply respond by growing in the opening and forming a patch.

Forest patches come in all sizes and shapes, depending on the tree species that grow in the forest and the forces at work there. Individual trees are also usually scattered among the patches. A single patch can cover an entire hillside or it can be the size of a few trees. It can also change in shape or size over time. Sometimes a fallen tree, fire, or landslide will cut through a patch and split it into smaller pieces. Others may become larger as adjacent patches take on similar characteristics and merge. This can happen when a clearing fills with seedlings, and another clearing opens

Northern Region

- Tundra
- White spruce–black spruce parkland
- White spruce–aspen–birch forest
- White spruce–balsam fir–jack pine forest
- Aspen parkland
- Mixed pine–spruce–hardwood forest

Western Region

- Sitka spruce–hemlock forest
- Sitka spruce–hemlock–redcedar forest
- Pacific Douglas-fir forest
- Pacific oak woodlands
- Coast redwood forest
- Mixed-conifer forests
- Grasslands and prairies
- Chaparral
- Spruce–fir, hemlock–fir, and fir forests
- Lodgepole pine forest
- Ponderosa pine and Douglas-fir forests
- Larch–fir–white pine–cedar forests
- Aspen forest

Western Region

- Piñon–juniper and juniper woodlands
- Sagebrush and sagebrush–juniper
- Oak–juniper–mesquite savanna
- Oak–mesquite savanna
- Southwestern shrub-steppe
- Desert shrub

Eastern Region

- White spruce–balsam fir forest
- Red spruce–fir and balsam fir forests
- White–red–jack pine forests
- Pitch pine forest
- Oak–gum–cypress forest
- Beech–maple forest
- Mangrove forest
- Oak–chestnut forest
- Elm–ash–cottonwood forest
- Loblolly–shortleaf pine–oak forest
- Longleaf–slash pine forest
- Wet grassland
- Maple-basswood forest
- Oak-hickory forest, woodlands, and savannas

Fig. 8.1 (*Continued.*)

next to it and fills with seedlings just a few years later. They can also appear to merge when adjacent patches grow old and look much the same, even though one patch of scraggly old trees became decadent long before the other.

SUCCESSION

The patches that form a forest share similar qualities, regardless of the forest type. They each progress through a series of successional stages that continues for as long as the patch escapes destruction. When Indian fires, wind, old age, or something else finally destroys a patch of trees, it progresses through these stages again, cycling endlessly in a process of renewal, aging, and destruction. How old a patch becomes depends on the life span of the dominant trees, and the frequency and severity of disturbances.

The successional process begins when a disturbance opens a space within a forest where a new patch of young plants can grow. As long as the patch escapes another major disturbance, it will progress through a somewhat predictable series of stages. The first stage usually includes some combination of grasses, forbs, and shrubs that fills the fresh clearing. A single generation of young pioneer trees that require bright sunlight and bare soil may join the other plants or move into the patch a little later. Settler trees that tolerate shade and deep litter may occasionally move into a patch of young pioneer forest, but they usually grow more slowly than the pioneers and seldom share the canopy. The young pioneer forest advances to the second stage as the trees grow to pole size. Now it is a patch of middle-aged pioneer forest.

The trees often stand so close to one another during the young and middle-aged pioneer stages that they have to grow straight and tall to reach the light. The thick canopy casts such a deep shade over the ground that most smaller plants, and even settler trees, cannot invade the forest. It becomes more open over the years as the weaker trees die and the survivors mature. The forest has now advanced to the third stage of development; it is an open old pioneer forest.

Eventually the old pioneer forest starts to break apart as the aging trees crash to the ground. This may not happen for a thousand years in some forests because the trees live so long. These trees are long-duration pioneers. Trees that live a few hundred years or less are short-duration pioneers. The trees may die in little groups or one at a time, but a small gap or hole opens in the forest canopy each time an old tree falls. The tall trees surrounding these small gaps keep them shady most of the day, but there is usually enough light for settler trees to grow. More gaps open and fill with young settler trees as the old pioneers die. So little patches of settler trees keep popping up here and there within the large patch of old pioneers. Now the old pioneer forest has entered the fourth stage of development. It has become a dense old transitional forest. Many of the old pioneers still tower over the younger trees, but the forest is gradually dissolving into smaller patches of settler trees.

When the last old pioneer topples, all that remains of the original large patch is a mosaic of very small patches of settler trees. Sometimes only one settler tree in a patch will make it into the canopy, so single trees also grow here and there within the mosaic. The old transitional forest has now become a self-replacing or late-successional forest. This is the fifth and final stage of development. Thus, a self-replacing forest is a collection of small patches of settler trees that sustain themselves by filling gaps that open when old trees die. Therefore, only a portion of these smaller patches will contain large old trees. For example, two acres of self-replacing maple-basswood forest can contain as many as 50 small patches. Large old trees occupy about 63% of these little patches. The rest of them consist of young and middle-aged trees.[1]

America's ancient forests seldom had a chance to reach the old transitional stage of development, and fewer still became self-replacing forests. If they had, the explorers who saw them would have felt more anxious when traveling through such dense forests. The shadowy understory of tangled growth within old transitional and self-replacing forests would not have inspired the use of such words as "open" and "spacious" that occur so often in early accounts. Notable exceptions existed, such as self-replacing beech-maple forests in the Northeast, and long-duration pioneer Pacific Douglas-fir forests filled with fallen trees in the West. Nonetheless, pioneer species dominated the majority of America's ancient forests, not settler species, and the forests were generally open and sunny.

SHIFTING MOSAICS

A forest mosaic consists of adjoining patches that have something in common, such as sharing a similar set of dominant species. Mosaics usually cover landscapes, but they can cover smaller areas as well if the patches are very small. The size of the majority of patches within a forest mosaic defines it. Patches that cover thousands of acres form a very large patch mosaic. A small-patch mosaic consists of a great number of small patches, each covering about an acre or less. Nevertheless, a very large patch mosaic can still contain many small patches, and a small-patch mosaic can include some large patches.[2]

How often disturbances occur, and their severity, determines the way a forest mosaic develops. A very large patch mosaic usually came about because of severe but infrequent disturbances, such as crown fires that only burn every few centuries. On the other hand, a small-patch mosaic develops in a forest when minor disturbances occur on a regular basis. Frequent light surface fires often produce such forests. Finally, a very small patch mosaic develops when one or a few trees fall here and there and open tiny gaps in a forest. Therefore, much can be learned about the history of a forest by looking at the size of its patches, and the proportions of its mosaic that are in different stages of development.

The patches in a forest mosaic are all changing together, but each is at a different stage of recovery from a destructive event. It is like turning a kaleidoscope and watching shapes and colors shift from place to place to form new patterns. The various kinds of patches that make up a particular forest mosaic seem to shift around the landscape as they disappear and reappear in different places over time. As long as the climate and human activities remain relatively constant, the forest mosaic will look much the same even while the patches within it change. Therefore, no single patch within a forest mosaic can be stable, no matter how large it is. It is either recovering or degenerating.[3] Only the forest mosaic can be stable, and even that depends on maintaining a balanced mix of patches in different stages of development. This rarely happened in an ancient forest.

MUTUAL DEPENDENCE

The structure of a forest and the forces that shape it work together; they are inseparable and mutually dependent parts of a single whole. Change one part and the other changes. It becomes a different forest. For example, the plants and game upon which American Indians depended thrived in forest mosaics that included a variety of successional stages. Most animals require two or more kinds of vegetation for their habitat, such as openings where they can find food and closed forest for cover.[4] Indians knew this from experience, so they burned forests to enhance the abundance and diversity of game, as well as plant foods. Thus the forests and the Indians sustained one another. Remove the Indians and the forest and the wildlife must change. They were inseparable. Captain Pages sensed this while traveling through the southern pine forests of East Texas in 1767. "It would be difficult for me to recount the sweet sensations ... I felt," he wrote, "both at the sight of simple, original Nature and at the rapport between the land and its inhabitants."[5] There is no doubt that American Indians were an integral part of America's ancient forests.

NINE

The Spanish
Explorer's Forests

I wandered lost and in privation through many and remote lands.
Alvar Nuñez Cabeza de Vaca, 1537[1]

With armor gleaming, swords rattling, hooves thudding, and feet shuffling through dirt, armies of Spanish conquistadors tramped a dusty and noisy path across the southern end of North America. Just as Paleoindians were the first people to cast their eyes on America's Ice Age forests, the Spanish were the first Europeans to see the interior of America's ancient forests. But unlike the empty land that lay before Paleoindians, the conquistadors saw a densely populated land where the forests and the people had developed together over millennia. It took the Spanish over two centuries to travel from the Atlantic to the Pacific, but during that period they explored some of America's most distinctive ancient forests. These include the vast pine forests of the South, the piñon-juniper woodlands of the Southwest, and the oak woodlands and towering coast redwoods of California.

The first descriptions of southern forests came from Alvar Nuñez Cabeza de Vaca's account of his journey in 1528. Juan Ponce de León touched land even earlier, naming the place *La Florida* in 1513, but we know little of what he saw. "It appeared very delightful, having many and fresh groves, and it was all level," writes the late-sixteenth-century historian Antonio de Herrera.[2] The anonymous "Gentleman of Elvas," who traveled with Hernando de Soto's expedition from 1539 to 1543, wrote another important narrative describing the South. The narrative of Pedro de Castañeda, written in 1540, provides equally valuable insights into the people and forests of the Southwest. Two years later Juan Rodriquez Cabrillo sailed north along the California shoreline, but he did not travel inland. That had to wait until 1769 when Gaspar de Portolá, and Fathers Junípero Serra, Juan Crespí, Francisco Garcés, and Pedro Font wrote the first descriptions of some of California's ancient forests.[3] These Spanish accounts, and the descriptions of those who followed them, as well as scientists who studied the Spanish explorer's forests, give us a rare glimpse into a time that was really not that long ago.

SOUTHERN PINE FORESTS

Throughout are immense trees and open woods.
Alvar Nuñez Cabeza de Vaca, Spanish explorer, 1528[4]

The southern pine forests the Spanish saw first were among the most picturesque ancient forests in North America (Fig. 9.1). The anonymous Gentleman from Elvas, who traveled with Hernando de Soto, wrote in 1547 that the "country was delightful and fertile," and covered with "tall pine-trees."[5] These were also old forests, reaching their historic condition about 5000 years ago.[6] Southern pine forests spread across 125 million acres of the South.[7] They grew on a series of terraces that looked like steps leading from the sea to the Appalachian Mountains, as well as on the Ozark Plateaus and the Ouachita Mountains. By far the largest was the longleaf-slash pine forest. It covered about 74 million acres from southeastern Virginia down into the Florida peninsula and westward to East Texas.[8] This forest grew in a broad curve along the low flatlands of the

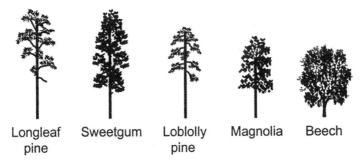

| Longleaf pine | Sweetgum | Loblolly pine | Magnolia | Beech |

Fig. 9.1   Trees of ancient southern pine forests.

Atlantic and Gulf Coastal Plain and the hills farther inland. It also grew on the Piedmont Plateau above the plain, as well as the interior uplands of Alabama and northwest Georgia. Most of the loblolly-shortleaf pine-oak forest grew above it on the Piedmont and Cumberland Plateaus, and parts of the Appalachian Highlands.[9]

William Bartram described the southern pine forests that grew on these terraces while traveling from Savannah to Augusta, Georgia, in 1773–1778. Bartram was on a plant collecting expedition through the south when he wrote his account.

> First, from the sea coast, fifty miles back, is a level plain, generally of a loose sandy soil, producing spacious high forests, of [pine]. . . . Nearly one-third of this vast plain is what the inhabitants call swamps. . . . We now rise a bank of considerable height, which runs nearly parallel to the coast, through Carolina and Georgia: the ascent is gradual by several flights or steps for eight or ten miles . . . when we find ourselves on the entrance of a vast plain, generally level. . . . This plain is mostly a forest of the great long-leaved pine . . . the earth covered with grass, interspersed with an infinite variety of herbaceous plants, and embellished with extensive savannas, always green, sparkling with ponds of water, and ornamented with clumps of evergreen, and other trees and shrubs. . . . The next ascent, or flight, is of much greater and more abrupt elevation, and continues rising by broken ridges and narrow levels, or vales, for ten or fifteen miles, when we rest again on another extensive nearly level plain of pine forests, mixed with various other forest trees.[10]

Southern pine forests, like other forests, consisted of different mixtures of trees that grew where they did because of past disturbances and their adaptations to fire, light, moisture, and soil. Thus vast forests of nearly pure longleaf pine usually covered the drier soils on ridges, plateaus, and south-facing slopes (Fig. 9.2). Bartram described them as "high pine forest," and he often rode through them for "miles," before entering another type of forest. He wrote that they were "open forests, consisting of exceedingly tall straight Pines . . . that stood at a considerable distance from each other." "The surface of the ground covered with grass, herbage, and some shrubbery," he said.[11] Slash pine stayed around streams and wetlands near longleaf pine forests, but they did not range as widely in the South as longleaf pine.[12]

Loblolly-shortleaf pine-oak forests not only grew farther inland than longleaf pine, but they also occupied the moist lower slopes between the longleaf pine forests and the wet bottomlands where hardwoods grew.

Fig. 9.2 Longleaf pine, Summerville, South Carolina. (Courtesy of M.E.I. Productions, Bend, Oregon; photograph by Mike McMurray)

Loblolly pine dominated this forest on moist soils, along with sweetgum, a tree with leaves that look like five-pointed stars, water oak, red maple, and other hardwoods. In places where fires had not burned for many years, a few old loblolly pine trees that came in after the last fire might poke through the canopy of invading hardwoods. In other places where light fires kept burning a few old hardwoods that had escaped the flames might stand among the pines. Scattered pockets of loblolly pine with an understory of yaupon and shade-tolerant hardwoods also grew within the forest, while other parts of the forest had open grassy understories. Shortleaf pine, occasionally mixed with some blackjack oak, southern red oak, and post oak, became dominant on drier soils in the forest. Thick patches of self-replacing hardwood forest of settler species such as beech and magnolia also grew here and there, usually in places seldom reached by fires, such as on lower slopes and in ravines.[13]

A sprawling grid of Indian trails connecting large treeless hunting grounds, cornfields, villages, and towns crisscrossed ancient southern pine forests and added even more to their diversity.[14] In the late 1700s, William Bartram rode along many Indian trails while making notes about this ancient mosaic of pines, hardwoods, and Indian villages. After passing through "delightful plains and meadows," Bartram wrote:

> We next entered a vast forest of the most stately Pine trees that can be imagined, planted by nature, at a moderate distance, on a level, grassy plain, enameled with a variety of flowering shrubs. . . . This sublime forest continued five or six miles, when we came to dark groves of oaks, Magnolias, Red bays, Mulberries, &c.[15]

A Portuguese gentleman from the town of Elvas, whose name was probably Alvaro Fernandez, made a similar comment in his eyewitness account of Hernando de Soto's march through Florida in 1539. He wrote that, "the land is level, the forest open, and in places the fields very fertile and inviting."[16] The pine forests of East Texas looked much the same as the rest of the South. In 1719, Father Francisco Celiz recorded his impression of the country between the Angelina River and Nacogdoches, saying:

> We travelled in the direction of the northeast, passing several villages of Asinai Indians that are intermediate, and passed through country that consisted of open woods of pecans, some pines and oaks, and in places through clearings and ravines.[17]

The openness of the ancient southern pine forests impressed most of those who saw them. Even so, thick patches of trees grew on the wet bottomlands within these pine forests, and in spots that escaped the many fires that burned throughout the South. These were the areas that the "Gentleman of Elvas" was referring to when he said "the land" he saw with Hernando de Soto had "in places . . . high and dense forests, into which the Indians that were hostile betook themselves, where they could not be found; nor could horses enter there."[18]

Piles of fallen pine trees also cluttered a few parts of this ancient forest in places that had suffered the ravages of insect attacks, hurricanes, tornadoes, and lightning. Cabeza de Vaca walked into such an area in 1528 while being guided by Florida Indians:

> They conducted us through a country very difficult to travel and wonderful to look upon. In it are vast forests, the trees being astonishingly high. So many were fallen on the ground as to obstruct our way in such a manner that we could not advance without much going about and a considerable increase of toil. Many of the standing trees were riven from top to bot-

tom by bolts of lightning which fall in that country of frequent storms and tempests.[19]

Such areas of broken and shattered trees were uncommon in ancient southern pine forests. For example, hurricanes only move inland and destroy large areas of southern pine forest every few years, and then they usually affect a different area each time. Blowdowns are patchy because short gusts of wind do much of the damage to forests during a hurricane. Nevertheless, hurricanes can flatten enormous numbers of trees when they strike. In 1875 a hurricane nearly destroyed 2000 square miles of pine forest in Texas. By 1989, nineteen more hurricanes had seriously damaged large areas of pine forest in the South. Erratically spinning tornadoes create additional havoc as they cut swaths through these forests.[20] Still, such areas are small relative to the size of the forests.

The great height of the pine trees and the immense size and straightness of the hardwood trees impressed the explorers who saw the ancient forests of the south. "The trees growing about over the country, without planting or pruning, of the size and luxuriance they would have were they cultivated in orchards, by hoeing and irrigation," wrote the "Gentleman of Elvas" during his travels in 1539.[21] William Bartram seemed equally affected when he wrote his account of the hardwood trees in eastern Georgia over two centuries later:

> We rose gradually a sloping bank ... and immediately entered this sublime forest. The ground is perfectly a level green plain, thinly planted by nature with the most stately forest trees, such as the gigantic black oak, ... whose mighty trunks, seemingly of an equal height, appeared like superb columns ... many of the black oaks measured eight, nine, ten, and eleven feet diameter five feet above the ground, ... and from hence they ascend perfectly straight, with a gradual taper, forty or fifty feet to the limbs ... the tulip tree, liquidambar [sweetgum], and beech, were equally stately.[22]

This ancient forest of immense black oak trees followed a low terrace above a stream for nearly 7 miles. These trees might have been up to 150 feet tall and 8 feet in diameter because they were nearing the end of their lifespan.[23]

Looming near the black oak forest was "a stupendous conical pyramid, or artificial mount of earth," said Bartram. Numerous other large mounds and earthworks showed this to be a long abandoned Mississippian Indian town.[24] Thus, a plausible explanation exists for the history of this forest. About two centuries earlier, these oak trees most likely grew in a large orchardlike forest around the town, just as they did at other Indian towns. Frequent light surface fires set by the Indians kept most other trees out of the forest. Then beech, a settler species, and a few tuliptrees and sweetgum invaded the spaces between the oaks and grew to a great size after the Indians deserted the town. Occasional surface fires that did little damage to the mature oaks and tuliptrees still burned through the forest, but they probably thinned the sweetgum and beech trees because they are more susceptible to fire damage. The widely spaced oaks in the original forest, and the longer time between surface fires caused by the departure of the Indians, probably created the imposing mixed-hardwood forest described by Bartram.

The striking beauty of an old pioneer longleaf pine forest resting in a carpet of grass and flowers could not help attract the attention of explorers. However, the Spanish and most other early travelers overlooked a bird that later came to symbolize this forest—the red-cockaded woodpecker. It prefers to peck nesting holes in living old pine trees that have a soft center due to red heart fungus. An open understory helps to prevent the nest from being invaded by predators. A patch of pole-size pine trees near its nest completes its habitat by providing a greater variety of insects and other foods.[25] So red-cockaded woodpeckers must have flourished in the ancient forest because Indian and lightning fires created the ideal combination of open old pioneer and young and middle-aged pioneer forests to serve their needs. Even though unnoticed, these little birds no doubt flitted around Spanish explorers on many occasions as they tramped by their nests. The longleaf pine forest was also ideal habitat for 85 other birds and 36 mammals, including the red wolf and the now extinct ivory-billed woodpecker.[26]

The largest area of longleaf pine forest contained an understory dominated by wiregrass, especially in the East. Wiregrass is a perennial bunchgrass that invades burned areas. It provides forage for grazing animals in the spring, but it becomes tough and wiry by early summer. Bluestem grasses also grew in the East, but they were more prominent in the understory farther west. Even so, more than 200 other herbaceous plants, including over 50 grass species, lived under these forests as well. Not only that, the number of different kinds of plants growing in the understory of ancient longleaf pine forests increased as the frequency of fires increased, reaching maximum diversity with annual light surface fires.[27]

Indian and lightning fires, more than any other natural forces, were responsible for the openness, patchiness, and diversity of plant and animal life that characterized southern pine forests. Light surface fires swept through ancient longleaf pine forests nearly every year, but they entered the slightly moister shortleaf pine forests only about once in 5 years. Hot surface fires burned loblolly pine forests on the wetter slopes above and below longleaf pine about once in 10 years. Fires seldom crept into damp bottomlands. Even though surface fires did not burn all parts of a forest in the same way, they still burned often enough to keep oak, and settler species like magnolia and beech, from spreading among the pines and replacing them. [28]

William Bartram often wrote notes in his journal describing recently burned forests as he wandered over Indian trails during the late 1700s. One such account links open forests and savannas to surface fires and the presence of Indians:

> We kept no road or pathway constantly, but as Indian hunting tracks by chance suited our course, riding through high open, pine forests, green lawns and flowery savannas in youthful verdure and gaity, having been lately burnt, but now overrun with a green enamelled carpet, chequered with hommocks of trees of dark green foliage, intersected with serpentine rivulets, their banks adorned with shrubberies of various tribes. [29]

Sometimes light surface fires created patches in southern pine forests by burning into piles of fallen trees and flaring up and destroying the trees above. Young pine trees quickly invaded such openings. Lightning strikes that killed trees but did not start fires, and insect attacks, also created clearings that helped keep southern pine forests patchy. [30] In addition, the size of the clearing often depended on the size and spacing of the trees that died. The loss of large widely spaced trees usually created a bigger clearing and a larger patch than the loss of smaller, more densely packed trees. [31] Large patches also developed where adjacent groups of old trees died at the same time because they were equally susceptible to insect attack.

Clearings also formed slowly as trees in a patch gradually fell victim to repeated surface fires. In 1767, Pierre Marie Francois Pages saw how fire thinned overstory pine trees so that they "stood at a considerable distance from each other," as Bartram had observed. [32] Pages made the following note in his journal while traveling through a southern pine forest somewhere between the Sabine River and Nacogdoches, in East Texas:

Where there is moisture, the hills are covered by woods consisting of diverse types of trees and by very tall pines and other large trees in dry places. I was surprised to see a large quantity of these pines lying on the ground, blackened and with a type of carbon powder at the base; one would have said that they had been set on fire. I noticed the same thing in those which were very old and still standing. At the ground level, the base of the tree becomes blackened and is reduced to powder, and little by little, because of the missing foundation, the trunk falls.[33]

Pages probably saw the effects of a surface fire that burned through a thick litter layer, so it moved slow enough to severely scorch or girdle the trees. Such fires often result in the loss of some large pines, especially since they become weakened and vulnerable to insect and disease attack. Even so, lightning and wind topple more longleaf pine trees than fire, and southern pine beetles can devastate loblolly and shortleaf pine trees.[34]

Longleaf pine is a fire species like no other. Longleaf pine seeds germinate best on bare soil because the large wings on the seeds prevent them from working their way through dead leaves and twigs, and into the moist soil below. Fire clears this litter and prepares the seedbed. However, the large seeds lay exposed to seed-eating animals when they fall on bare soil, so they normally germinate within a week to lessen the chance of being eaten. The dry soil on which longleaf pine grows also requires the seedling to quickly send its carrotlike taproot deep into the ground to find moisture. So the stem grows slowly and allows the tree's energy to be directed into the roots. Seedlings stay in this grasslike stage for several years, which makes them vulnerable to competition from faster growing trees. However, frequent fires keep other trees cleared away. The longleaf pine seedling survives these fires because a tuft of needles protects the bud. Even if fire kills the top of the seedling, it can still sprout from the root collar. A disease called brown-spot needle blight attacks the longleaf pine seedling when it is in the grass stage, but surface fires also cleanse the seedlings of this disease. A thick fire-resistant bark quickly develops on the stem once longleaf pine begins height growth. This protects the tree from light surface fires, but the bark is less effective against hot surface fires. Nevertheless, fires seldom become hot since the pine-needle-draped grass that covers the forest floor burns quickly, and it takes little time for this fuel to recover and become flammable again.[35]

Longleaf, slash, loblolly, and shortleaf pine are all intolerant of shade.[36] That means they need openings larger than mere gaps in order for young trees to get enough sunlight to grow. Nitrogen also increases in the soil toward the center of larger openings, so that is where the trees grow best.[37] In addition, each opening usually contains trees only a few years apart in age because they invade it at about the same time. Thus ancient southern pine forests consisted of groups of trees of about the same age, but some groups were older and some were younger because clearings formed at different times. Doubtless individual pine trees were also sprinkled through the forest because they survived the last disturbance, they were the only trees left in former patches, or their seeds dropped into small openings with just enough sunlight for them to survive.

Herman Chapman provided one of the few early scientific accounts of the patchy structure of ancient longleaf pine forests. He wrote in 1909:

> The longleaf pine is found in pure stands . . . the natural form of this forest constantly trends toward small, even-aged groups of a few hundred square feet. Being naturally resistant to fire, large clearings never occur from this cause. In regions of severe winds, or tornadoes, larger even-aged patches and strips are found, sometimes one-quarter to one-half mile in width, which have come in after blowdown. These are pretty well interspersed with patches or single survivors of the old forest which have acted as seed trees.[38]

Surely some of the larger patches Chapman noticed also became established on abandoned Indian farmlands, hunting grounds, and towns. Likewise, some of the smaller patches probably formed in openings where the cavities created by a cluster of red-cockaded woodpeckers had weakened groups of old trees and made them vulnerable to wind and attack by southern pine beetles. Therefore, patches ranged from less than a tenth of an acre to several hundred acres. However, most of them were small, and the proportion of the mosaic in larger patches decreased as the patch size increased. Consequently, ancient longleaf pine forests probably consisted of a mosaic of small patches spread over an open grassy understory because light surface fires burned so frequently. That also means that young, middle-aged, and open old pioneer forests were more plentiful than dense, old transitional and self-replacing forests.

Loblolly pine forests also consisted of a mosaic of patches.[39] However, the patches were probably larger than those found in longleaf pine forests because fires were less frequent, which gave shrubs and hardwoods more time to grow in the understory. This increase in fuel produced more severe fires, which most likely burned bigger areas.

The tall pines that so impressed early explorers could not have stood everywhere in ancient southern pine forests, even though explorers often implied that they did. Patches of old pioneer forest, and a few old transitional forests, could have covered up to 50% of the longleaf pine mosaic since this tree often reached 300 years, and matured at half that age.[40] This assumes that southern pine forests remained stable because patches of trees, each of a different age, occupied about the same amount of space. Therefore, a patch of old trees that died would always be replaced by an equivalent area of young trees. This rarely happened in the ancient forest. The proportions of the mosaic in different stages of development varied from time to time because disturbances were sporadic and variable in their effects. Nevertheless, open old pioneer forests dominated the ancient landscape.

A few very large patches of older trees helped to make the ancient forest appear uniformly tall to early explorers. The boundaries between small patches could also go unnoticed because middle-aged trees look nearly as tall as old pioneers. This probably reinforced the appearance of "a vast forest of the most stately Pine trees" that Bartram and other European explorers described in their accounts of ancient southern pine forests.

PIÑON-JUNIPER AND JUNIPER WOODLANDS

In that country are small pine trees.
Alvar Nuñez Cabeza de Vaca, Spanish explorer, 1536[41]

The first written account of America's ancient piñon-juniper woodlands came from West Texas, near the Pecos River (Fig. 9.3). The year was 1536, and Cabeza de Vaca and his three companions were nearing the end of their harrowing trek across the South and Southwest. Naked and bare-

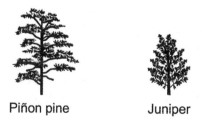

Piñon pine   Juniper

Fig. 9.3   Trees of ancient piñon-juniper and juniper woodlands.

foot they stumbled from one Indian village to the next. Not surprisingly, Cabeza de Vaca had little time or energy to make detailed observations, so his account was brief. He said: "In that country are small pine trees, the cones like little eggs." Then, thinking mainly of food, he went on to describe piñon nuts and the way the Indians prepared them. [42] Shortly afterward, Cabeza de Vaca's 8-year ordeal among the Indians ended when 20 Spanish soldiers found him and his companions and escorted them to Mexico.[43]

The ancient piñon-juniper woodlands Cabeza de Vaca saw covered about 57 million acres, much of it in the Southwest.[44] They also snaked along the lower east slope of the Sierra Nevada, the mountains of the Great Basin, and both sides of the Rocky Mountains. Piñon pine dropped out of these woodlands north of Colorado and Utah, but juniper woodlands continued into eastern Oregon and parts of Idaho and Wyoming.[45]

Southwestern piñon-juniper woodlands developed at the end of the Ice Age, about 11,000–8000 years ago. So they are among the oldest of America's ancient forests.[46] The western juniper woodlands of northeastern California and eastern Oregon developed about 7000–4000 years ago.[47] Piñon pine and juniper moved off the southwestern lowlands and up the mountain slopes as the climate warmed. Desert scrub moved in to replace them, and the forests above provided room for the invading woodlands by moving even higher.[48] Thus, the ancient piñon-juniper woodlands seen by European explorers covered the foothills and lower slopes of the mountains, wedged between grassy brushlands or shrub-steppes on the deserts and ponderosa pine forests on the higher slopes.[49] Within this band of woodland, piñon pine probably became more abundant than juniper at higher elevations and on north slopes, and juniper became more abundant at lower elevations.[50]

Piñon-juniper woodlands look much the same wherever they occur, partly because they consist of small evergreen trees about 20–60 feet tall. Furthermore, different species of piñon pine and juniper look similar. Thus a woodland in the southern Rocky Mountains made up of Colorado piñon pine and one-seeded juniper still looks like a woodland in the Great Basin that consists of single-needle piñon pine and Utah juniper.[51]

Piñon pine and juniper also share many of the same traits. They both grow on lower slopes, ridges, and mesas, with shallow, rocky soils of low fertility. In addition, they both have heavy seeds that drop near the tree, or are dispersed by birds and small mammals. However, slight differences can give one species a competitive advantage over the other. So piñon pine is the settler species in this forest. The seedlings of piñon pine and juniper grow best under the canopy of larger trees that protect them from the hot sun, even though they are both intolerant of shade as they grow larger. Still, piñon pine is slightly more tolerant of shade than juniper. Likewise, both trees can grow in warm climates with low rainfall, but juniper requires less moisture than piñon pine. Similarly, they both grow slowly and can live to be at least 300 years old. However, piñon pine grows faster than juniper. Therefore, after several centuries without a fire or other disturbance, piñon pine has a slight advantage and will gradually replace juniper on all but the driest sites.[52]

Indian and lightning fires swept over the dry landscape too often for a dense self-replacing forest of piñon pine to dominate anything but a minor fraction of the ancient piñon-juniper woodlands. Fast-moving grass fires pushed piñon and juniper up into the foothills and kept the trees from invading the lowlands. Since these fires moved uphill with the wind, they usually burned hotter than the fires that backed downhill into the woodlands from the ponderosa pine forests above. These fires further increased the fire frequency within the woodlands and helped to keep them open and patchy. Forest assistant A. Ringland wrote a report for the Lincoln National Forest in New Mexico that depicted these woodlands as they looked in 1905:

> The wide stretch of country lying between the White Mountains and the Capitans, may be referred to as the woodland–pastoral zone. This zone is characteristically rolling, with many small rounded hills and knolls, drained by numerous arroyos of intermittent flow. The hills and knolls are sparely covered here and there with a scrubby growth of pinion . . . and one-seeded juniper . . . or perhaps the low matting of the scrub oaks.[53]

Ringland saw these woodlands after the Indians had left. So he and other observers from this time could not include the effects of Indian clearing in their descriptions. However, we know that great gaps also occurred in ancient piñon-juniper woodlands near villages where Indians cleared them for building materials and firewood. They also burned the woodlands on plateaus to prepare the ground for crops. Indians most likely planted their crops at the base of the charred trees to take advantage of rainwater that flowed down the stem. Even so, the crops depleted the soil of nutrients within a few years and the Indians had to move on and burn another woodland. They returned to burn and plant again after the woodland grew back. [54] Cabeza de Vaca even noticed in 1536 that piñon-juniper woodlands in southwestern New Mexico reinvaded cleared areas around villages shortly after the Indians moved, either to avoid the Spanish or to seek better soil for their crops. Thus, he said, "from abandonment the country had already grown up thickly in trees."[55]

Indian and lightning fires burned through ancient piñon-juniper woodlands about once in 10–30 years.[56] The grasses and shrubs that grew under the trees carried the fires, destroying most of the trees under 4 feet tall. These surface fires rarely killed the larger trees unless tumbleweeds or other debris piled up beneath them so that the fire could climb into the crowns. Thus frequent surface fires created and sustained the open character of the woodlands. Even so, some patches of woodland that grew in protected and rocky areas, or just by chance, still managed to escape the fires long enough for the trees to grow large and dense. Stringers of these old piñon and juniper trees curved through the woodlands along rocky ridges and rimrocks. These patches of closed forest created enough shade and litter to prevent grasses and shrubs from becoming thick enough to carry fires, which made them difficult to burn. So they could escape even more fires. However, on hot dry days with high winds a fire could still reach the crowns of the trees. Since the trees grew close to one another, the fire could jump from crown to crown and destroy whole patches. Sometimes, under extreme conditions, a fire would also flare up and kill a patch of open woodland. These patches of charred trees, or skeleton forests, gradually filled in with grasses and forbs, then shrubs, and eventually some trees, so that they again became part of the open piñon-juniper woodland.[57]

Repeated fires made the ancient piñon-juniper woodlands look like patchy savannas. Grasses and shrubs grew in the understory, and D. M. Lang and S. S. Stewart, who conducted a forest survey in Arizona in 1909, said that "brushy [and] crooked" trees made up the overstory. They also said that this woodland "is very irregular [and] patchy and consists principally of clumps of pinion pines ... and juniper."[58] A 1904 report by a U.S. Geological Survey team working in Arizona added more detail, saying that the woodlands consisted of "a multitude of scattering trees, small groups, copses, and stands of medium density." Some of these stands covered "100 to 300 acres," and a few of them were even larger. The report also said that the woodlands were "interrupted by numerous tortuous lanes of bare ground, varying in width from 15 to 150 feet, and irregular tracts containing 20 to 100 acres."[59] However, H. Calkins called them "park-like openings" rather than bare ground during his 1909 forest survey in New Mexico.[60]

America's ancient piñon-juniper woodlands were open, patchy, and diverse. Each part of the mosaic represented a different stage of successional development, depending on the severity of the last disturbance. Grassy and shrubby openings, land cleared or abandoned by Indians, skeleton forests, savanna-like old pioneer forests, sparse groups of trees, and a few stringers and blocks of dense self-replacing forest were all present in the mosaic. This great variety of habitats supported a large number of animals, including deer, elk, bison, mountain sheep, and desert bighorn sheep. Hundreds of small animals and birds, such as the pinyon jay that helped to distribute piñon seeds, also lived here.[61] The trees were small but these ancient woodlands were productive and vast. They gave color and life to the sun-baked landscapes of the West.

PACIFIC OAK WOODLANDS

They look like old orchards.
Joel Palmer, Oregon pioneer, 1847[62]

The Spanish marched into California's ancient forests more than two centuries after they had first entered the ancient forests of the Southeast. In California, just as elsewhere, they concentrated their explorations on the lowlands and foothills. The Spanish had little interest in the mountains because the soils were unsuitable for agriculture. Nevertheless, the forests they saw still impressed them and those who followed. The Pacific oak woodlands were among the finest, not just for the lush grass that cattle and horses could graze beneath the trees but also because of the beauty of the landscape (Fig. 9.4).

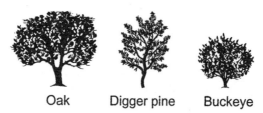

Oak   Digger pine   Buckeye

Fig. 9.4   Trees of ancient Pacific oak woodlands.

Pacific oak woodlands consisted of four different woodlands that covered over 10 million acres in California, and they also extended northward between the Coast Range and the Cascade Range to southern Canada. The coastal segment of the woodlands spread over the rolling hills and intermountain valleys of the central and southern California coast. Coast live oak dominated most of these coastal woodlands. In the interior of California, valley oak dominated the valley woodlands that followed the rivers in the Great Central Valley. Blue oak dominated the foothill woodlands that stood above the Central Valley, encircling it like a huge picture frame. Farther north, higher rainfall allowed the Oregon white-oak-dominated northern woodlands to spread over the broad valleys and foothills of northwestern California and western Oregon and Washington. Various combinations of interior live oak, canyon live oak, California black oak, and tanoak also grew in parts of different woodlands.[63]

The soldiers and priests in Gaspar de Portolá's expedition were the first Europeans to see the ancient Pacific oak woodlands of Alta California.[64] Father Crespí wrote a brief description of the coastal woodlands as the expedition traveled near Santa Barbara in August 1769:

> We went over land that was ... well covered with fine grasses, and very large clumps of very tall, broad grass, burnt in some spots and not in others. ... All about are large tablelands with big tall live-oaks.[65]

The expedition stayed close to the coast because Portolá was searching for Monterey Bay. However, in 1776, Father Francisco Garcés wandered into the valley woodlands of California's Great Central Valley. He recorded in his diary that he traveled 6 miles along the north bank of the Kern River near Bakersfield "wondering again at the extent of woodland."[66]

These fragmentary images became clearer when later travelers wrote their descriptions. One of these was Captain John C. Fremont. He and his weary companions peered into the Sacramento Valley while still standing in deep snow high in the Sierra Nevada. It was the spring of 1844, and they were on their way from Nevada to California. A thunderstorm loomed over the valley and the rivers were flooding, but "for us," he said, "as connected with the idea of summer, it had a singular charm." They struggled down the mountain until they finally entered the oak-covered foothills on the west slope. Here, where "the mere traveling and breathing the delightful air being a positive enjoyment," Captain Fremont wrote a description that fits most ancient Pacific oak woodlands. "The country is smooth and grassy," he noted, "the forest has no undergrowth; and in the open valleys ... the low groves of live-oak give the appearance of orchards in an old cultivated country."[67]

Even though the principal oak species changed in different regions, the woodlands all looked very much alike because they shared a long history of Indian use. Lightning fires were rare, usually many decades apart. Indians set most of the fires that burned here nearly every year.[68] Tens of thousands of Indians made their home in the Pacific oak woodlands of California, Oregon, and Washington. These woodlands provided Indians with an abundance of acorns, grass seeds and other plant foods, as well as deer and small game. Consequently, Pacific oak woodlands were among the most intensively managed ancient forests in North America when the Spanish first saw them. Not only that, but Indians had probably lived in the woodlands since they first took on a modern appearance at the end of the Ice Age.[69] It is no wonder that Captain Fremont and other early explorers often described these oak woodlands as "orchards." That is what they were to Indians.

Valley Woodlands

The ancient valley woodlands of California's Great Central Valley may have been the most spectacular of the Pacific oak woodlands. The setting in which they grew contributed to the wonder of the scene. This huge valley, 60 miles wide and 400 miles long, lies in a basin formed by the curving mass of the Sierra Nevada to the east and the Coast Ranges to the west. Many large rivers drain into the valley from these mountains. Their waters collect to form the Sacramento River, which flows southward, and the San Joaquin River, which flows northward. Then these two great rivers join in a vast marshland near the center of the valley and turn westward to empty into San Francisco Bay. These woodlands followed the rivers, and they spread along the west side of the bay and into the valleys of the Coast Range.[70]

The first Europeans to see the Great Central Valley beheld a vast grassland that lapped up the sides of the bowl-shaped valley and into the foothills of the surrounding mountains. Trapper Zenas Leonard entered the valley in 1833, and said that "the plains or prairies ... stretched as far as the eye can reach."[71] These ancient grasslands consisted of perennial bunchgrasses, such as purple and nodding needlegrasses that stood 2–6 feet high. Hundreds of other grasses and forbs intermixed with the tall needlegrasses.[72]

Valley woodlands curved back and forth over the grasslands in belts that varied from $\frac{1}{2}$ to 6 miles wide along rivers.[73] Edwin Bryant described the scene during his tour of California in 1846–1847:

> The spacious valley of the Sacramento suddenly burst upon my view, at an apparent distance of fifteen miles. A broad line of timber running through the centre of the valley indicated the course of the main river, and small and fainter lines on either side of this, winding through the broad and flat plain, marked the channels of its tributaries.[74]

These woodlands became thicker and ladened with grape vines near the edge of the rivers, and trees such as ash, sycamore, walnut, cottonwood, and willow grew among the oaks. Away from the rivers, they became open oak woodlands, then widely dispersed patches, and finally single oaks that rose from the landscape like giant umbrellas as the woodlands blended into the surrounding grasslands.[75] "Sign of Indians," also appeared nearly everywhere in the Great Valley "as moccasin tracks, and smoke rising from the prairies in different places," said Zenas Leonard.[76]

These were patchy forests of young, middle-aged, and old pioneer valley oaks. Dense, old transitional and self-replacing forests did not exist in the Great Central Valley. However, California live oak, canyon live oak, and tanoak, all of which are shade-tolerant settler species, did invade valley woodlands that grew on moist sites within the Coast Range. Still, such late stages of development were extremely rare even in the Coast Range because of frequent Indian burning.[77]

In the fall of 1837, the British ship *Sulphur* sent a party to explore the lower Sacramento River. They traveled northward following the meandering course of the river for 150 miles. The party then returned to the ship and wrote the following description of the patchy character of ancient valley woodlands in their report:

> Oaks of immense size were plentiful. These appeared to form a band on each side, about three hundred yards in depth, and within (in the immense park-like extent, which we generally explored when landing for positions) they were seen to be disposed in clumps, which served to relieve the eye, wandering over what might otherwise be described as one level plain or sea of grass.[78]

An 1854 survey party for a railroad route also noted in their report that these "magnificent oaks ... grow scattered about in groups or singly, with open grass-covered glades between them." In addition, the surveyors gave us a glimpse of the interior of this ancient forest: "There is no undergrowth beneath them, and as far as the eye can reach, when standing among them, an unending series of great trunks is seen rising from the lawn-like surface."[79] British navigator George Vancouver saw the valley woodland near Palo Alto during his expedition to the Pacific Northwest in 1791–1794. It reminded him of "a park which had originally been planted with the true old English oak."[80]

The massive trunks and broad crowns of the valley oaks, made all the larger by the openness of the scene, gave these woodlands a memorable appearance. This is the largest North American oak, sometimes reaching a height of 120 feet or more with trunks up to 9 feet in diameter. This mighty oak withstands flooding, drought, and fire. It reproduces primarily from acorns that squirrels store underground. The nearly annual fires that burned through the woodland usually destroyed the acorns that fell on the ground, but buried acorns survived the flames. The young seedlings also took advantage of the extra moisture no longer used by the burnt grasses, and they could grow in the shade of the parent tree. Young trees also sprouted after a fire killed the top.[81] Thus frequent light surface fires set by Indians and lightning kept valley woodlands open, patchy, and diverse, which also gave the oaks enough room to grow to great size.

The grasslands and valley oak woodlands of the Great Central Valley supported an incredible abundance of wildlife, comparable in many ways to the Great Plains, although bison did not venture this far west. Ducks and geese abounded in the marshlands, and large bands of the small tan-colored tule elk numbering in the hundreds, along with deer and antelope, looked at a distance like herds of cattle.[82] This led trapper Zenas Leonard to comment that, "we had reached a country thickly filled with almost all kinds of game."[83] About one-half million tule elk probably roamed the valley at the time.[84] Wolves and coyotes stalked the herds, and troops of the now extinct California grizzly bear, or "golden bear," wandered through the open woodlands.[85] Meadowlarks and killdeer fluttered above the grasslands while red-tailed hawks swooped down on squirrels and rabbits. At least 67 species of birds nested in the valley woodlands near the rivers, including snowy and American egrets, which often crowded the tops of isolated groups of trees.[86] Few ancient forests could rival the beauty and magnificence of California's valley woodlands, with their spectacular display of wildlife, great oaks standing among tall grasses, and the Sierras towering in the background.

Foothill Woodlands

Oaks gradually increased as the woodlands climbed out of the Great Valley and into the foothills of the Coast Range and the Sierra Nevada (Fig. 9.5). The smaller blue oaks began replacing the valley oaks as well, which marked the beginning of the foothill woodlands. Many of the same animals that lived in the valley woodlands also lived here, but the majestic California condor probably soared overhead more often.[87] This was also the land of the roadrunner, a little bird that runs as fast as a horse. People seldom see it today, but it was plentiful in ancient foothill woodlands.[88]

In the spring, the live oaks and leafless blue oaks seemed to blend into the green lawn that covered the slopes and valleys of the foothills. The white and pale-rose flowers of little buckeye trees added spots of color

Fig. 9.5   Blue oak, California buckeye, and chaparral in the Sierra Nevada foothill woodlands of California. (Photograph by author)

here and there, but it was the wildflowers that transformed this canvas into art. One traveler remembered such a scene as it was in the late 1800s:

> The open spaces among the foot-hills, and more especially the prairies that skirt them, bloom in spring time with fields of wild flowers of every hue—all exceedingly brilliant and graceful, though generally deficient in odor. Sometimes a single variety will occupy several acres, to be followed by another patch equally extensive, covered by a different kind.[89]

The dry heat of summer changed everything. The grasses shriveled and dried to a yellowish brown, wildflowers faded, and the little buckeyes dropped their leaves. But now the broad, dark green oaks stood out, looking like emeralds set in gold.

Fire was the brush that painted this landscape. It surged into the foothills from the valley grasslands below. More fires crept into the foothills from the ponderosa pine forests on the mountains above, and others started within the foothills. So fires burned nearly as often here as they did in the valley woodlands. These fires maintained the woodland's grassy understory, and kept it open, patchy, and colorful. Such flashy, fast-moving fires favored blue oak because they quickly passed the tree's thin bark without causing serious damage. Slower, hotter fires could easily kill the trees, although some grew back from sprouts. Young blue oaks added to the patchiness of the landscape by filling the sunny openings that fires created.[90]

The hilly country, the young, middle-aged, and old pioneer oak forests, and various mixtures of trees, shrubs, and grassy openings gave the ancient foothill woodlands a more diverse look than valley woodlands. Nevertheless, they still had the appearance of an open oak savanna. The ancient foothill woodlands ranged from sparsely scattered trees to a few patches of nearly closed forest.

Digger pine, a pioneer species, mingled with blue oak on some steeper, drier slopes with shallower soils. A scrubby growth of interior live oak and shrubs also mixed with blue oak where digger pine grew. Digger pine rises above blue oak, but it does not live as long. This relatively small pine has one to several trunks that tend to lean, with limbs topped by tufts of long, stiff, gray-green needles that form a thin, airy canopy. The digger pine's large heavy cones also produce nutritious nuts that Indians relished nearly as much as acorns.[91]

Shrubs such as coffeeberry and wedgeleaf ceanothus grew beneath blue oaks in places that fires skipped for a while.[92] A light surface fire that entered a grove with an understory of shrubs could easily become hot enough to kill the overstory oaks. This would start a new patch that could contain any one of a number of different kinds of plants. For example, young blue oaks might sprout back along with some of the shrubs, but older oaks could not.[93] A few digger pine seeds might also germinate among the sprouting oaks and shrubs. Acorns stored in the ground under the trees would sometimes produce enough seedlings to fill the opening as well. If not, the loss of the overstory oaks might allow the patch to revert to grasses and flowers.

Chaparral usually occupied the dry hilltops and south-facing slopes that were less suitable for oak woodlands and grasslands, and more prone to hot fires. Chaparral consists of brushy thickets, or a shrubby growth of live oaks. It burns less often than oak woodlands because grass cannot grow underneath the thick canopy, making fires more intense when they do burn. Chaparral shrubs such as coffeeberry also sprout vigorously and quickly renew their dominance after a fire, which prevents blue oak from invading. Thus, thick patches of chaparral also grew within the ancient foothill woodland mosaic.[94]

The ancient foothill woodland was diverse and old, but it did not include very much dense self-replacing forest. Frequent fires were an integral part of this forest and few places escaped the flames for long. So old transitional and self-replacing forests probably existed only briefly. Here patches of self-replacing forest may have consisted of interior live oak and, in some northern areas, perhaps California laurel, or bay, since they are shade-tolerant settler species.[95] Like the valley woodlands, this was an open patchy forest, but instead of vast expanses greeting the eye it was a flowery parkland of grasses, oaks, and shrubs spread over rolling hills.

Coastal Woodlands

Ancient coastal woodlands covered the wetter ocean-facing side of the Coast Range, and they looked much the same as foothill woodlands.[96] The coast live oaks that dominated them also stood nearly as tall as blue oaks, but their hollylike leaves stayed green all year, unlike the blue oak's blue-green leaves that dropped in the fall. Moreover, coast live oaks have short trunks with large limbs, and rounded, wide-spreading crowns, while blue oaks stand straighter.[97] The somewhat shorter, evergreen Englemann oak became important in these woodlands on drier slopes south of Los Angeles, but it did not change the open, parklike appearance of the coastal woodlands.[98]

Chaparral also grew on the dry ridges above the coastal woodlands, although the less dense sage scrub replaced it near the coast. Frequent Indian fires on the lowlands kept sage scrub from spreading downslope. So grasslands usually occupied the terracelike bluffs overlooking the ocean.[99] Nevertheless, this "impenetrable thicket," as one traveler called chaparral in 1860, covered much greater areas in coastal woodlands than foothill woodlands. This was particularly true in the mountains of southern California where chamise chaparral dominated vast areas.[100]

Chamise chaparral may have looked like a continuous covering of thick shrubs, but like forests it consisted of a mosaic, each part in a different stage of recovery from the last Indian or lightning fire. John Leiberg described it in 1900 as "a growth which varies from extremely dense to thin or open, but rarely forms large uninterrupted patches."[101] Patches of young shrubs in the mosaic acted as fuel breaks that slowed or blocked the movement of fires that burned in the more flammable thick old shrubs. This helped to perpetuate the mosaic by making fires burn in a patchy manner. Nevertheless, some ancient chaparral fires could still be huge, especially when fanned by Santa Ana winds.[102]

Coastal woodlands huddled in the valleys away from the fires that swept over the dry chaparral-covered slopes of southern California. No doubt these valleys helped to protect the oaks, partly because of the frequent light surface fires Indians set to keep them open and free of brush. Indian fires not only reduced fuels in the valleys, but they also must have created fire breaks above the valleys by burning upslope into the chaparral.[103] Not only that, the flush of grasses, forbs, and young shrubs that grew back on the slopes that surrounded the oak woodlands surely improved hunting because they provided nutritious forage and browse for deer.[104] Moreover, Indians normally set their fires in the fall, which further improved deer habitat the following year by promoting the growth of important shrubs such as *Ceanothus* that rely on seeds to reproduce. Perhaps the Indians knew that spring and summer fires could reduce or eliminate this deer browse from chaparral.[105]

The ancient coastal woodlands had their own distinct character. Still, like foothill woodlands, they formed a mosaic of trees, grasses, and shrubs.

They also shared the open, grassy look of a well-manicured park with other Pacific oak woodlands.[106] No doubt a small portion of the ancient mosaic contained self-replacing forest, particularly in areas where fires were infrequent, such as ravines and north slopes. Like foothill woodlands, interior live oak and California laurel, or bay, probably dominated these few patches of self-replacing forest that existed in the ancient coastal woodlands.[107]

Northern Woodlands

The ancient northern woodlands displaced coastal woodlands in the cooler temperatures and higher rainfall of northwestern California. The drumming sound of the ruffed grouse also marked the beginning of the northern woodlands.[108] These woodlands grew in valleys and around small prairies that sat on hills above coast redwood forests. The chronicler of the 1851 expedition of Indian agent Colonel Redick M'Kee stated that "almost the only open country was upon these high slopes."[109] Northern woodlands also covered parts of valleys within the Douglas-fir forests that grew along the eastern border of the redwoods. Colonel M'Kee's journal described this country as "well timbered with oak and fir . . . and interspersed with fields of bunch grass and little valleys affording good pasturage."[110] Even larger areas of northern woodland occupied the interior valleys of the Umpqua and Rogue Rivers, the Willamette Valley, and other valleys in western Oregon and Washington. They grew within parts of the Columbia River Gorge as well. Besides valleys, bands of ancient northern woodlands often grew in the foothills along the lower edge of ponderosa pine and Douglas-fir forests, separating them from the dry, treeless grasslands and shrub steppes on the lowlands.[111]

Regardless of where they grew, the ancient northern woodlands looked like other Pacific oak woodlands. In 1847, pioneer Joel Palmer even described the northern woodlands in the Willamette Valley with the same image that Captain Fremont used for the foothill woodlands of California:

> Upon the slopes of these hills are several thousand acres of white oak from six to twenty feet in height, some of them large diameter, and all with large bushy tops: the ground being covered with grass, at a distance they look like old orchards.[112]

Midshipman Henry Eld, from the U.S. Exploring Expedition, had a similar impression when he traveled through the Willamette Valley in 1841:

> Our route has been through what might be called a hilly prairie country, the grass mostly burnt off by recent fires, and the whole country sprinkled with oaks, so regularly dispersed as to have the appearance of a continued orchard of oak trees.[113]

Light surface fires set by Indians almost every year, along with a few lightning fires, kept large areas of ancient northern woodlands open, patchy, and orchardlike in appearance. This happened because light fires seldom kill the mature Oregon white oaks that dominate northern woodlands. Young oaks also sprout vigorously if a fire scorches the top of the tree. These sprouts grow fast enough to overtop shrubs within just a few years. However, they grow slowly afterward, usually only reaching heights of about 70 feet, and seldom attaining diameters of more than 3 feet in 250 years.[114] Oak seedlings also came up after these fires. They could not grow in thick grass, or even grass that was just coming back after a fire, but they did well on small bare spots where logs and other debris burned intensely.[115]

Northern woodlands contained few, if any, dense self-replacing forests. They consisted of mostly young, middle-aged, and open old pioneer forests, and some old transitional forests. The few self-replacing forests that grew here consisted of settler species such as western hemlock and bay, or in some areas Douglas-fir, which is a pioneer species that can replace oaks. Oregon white oak grows best with frequent light surface fires, but settler trees and Douglas-fir do not. Douglas-fir cannot sprout, and fire easily kills the young trees. So surface fires that burned nearly every year also drove back the young Douglas-fir, western hemlock, and bay trees that tried to invade northern woodlands. If the fires became less frequent or stopped, Douglas-fir would quickly move into the understory of the oaks. Then they would overtop the oaks and cast enough shade to kill them. Eventually the oak woodland would disappear and a dense Douglas-fir forest would stand in its place. Settler species such as hemlock or bay might then replace the Douglas-fir forests and convert them to self-replacing forests. However, Indian and lightning fires kept burning, so the ancient northern woodlands flourished within the great forests of redwood and Douglas-fir that covered northwestern California and the Pacific Northwest.[116]

COAST REDWOOD FOREST

Rising like a great tower.
Father Pedro Font, 1776[117]

Coast redwood forests were among the most spectacular ancient forests in North America, and members of Portolá's expedition were the first Europeans to see them (Fig. 9.6). The Spanish mostly stayed near the sea as they ventured northward from San Diego in search of Monterey Bay. They went too far north and became the discoverers of San Francisco Bay instead.[118] This mistake also led them to the discovery of the ancient redwood forest. They missed the redwoods near Monterey Bay, but finally saw them in the Santa Cruz Mountains farther north. We can barely imagine their feelings as they gazed up the great trunk of one of these colossal trees and into its tapering crown high above. Father Crespí described them for the first time on October 9, 1769:

> Here begins a large mountain range covered with a tree very like the pine in its leaf ... it has some small sharp-pointed cones ... the heartwood is red, very handsome wood, handsomer than cedar. ... There are great numbers of this tree here, of all sizes of thickness, most of them exceedingly high and straight like so many candles: what a pleasure to see this blessing of timber.

He added that "some of them are extremely thick, the soldiers measured

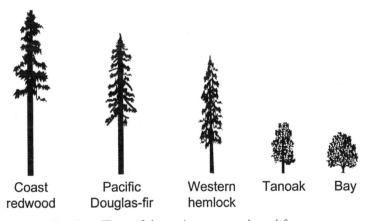

| Coast redwood | Pacific Douglas-fir | Western hemlock | Tanoak | Bay |

Fig. 9.6   Trees of the ancient coast redwood forest.

one that was three musket-lengths in diameter." In March 1772, Father Crespí named the tree when he called it "*palo colorado* [red tree]" in his journal. The English translation became redwood.[119]

The Spanish saw the redwood forest again just a few years later. In the late summer of 1776, Juan Bautista de Anza led a party of colonists from Monterey to San Francisco Bay where they intended to build Mission Dolores.[120] They saw an unusual twin-trunked redwood tree on their way up the western shore of the bay. Father Pedro Font, chronicler of the expedition, described it as "rising like a great tower." He called it "pal o alto."[121] The tree still stands in the city of Palo Alto, although one of its trunks fell long ago. Shortly after arriving at San Francisco Bay, Father Font etched his name permanently into the California landscape. He looked eastward from a hill at the mouth of the Sacramento River and drew the first map showing the Sierra Nevada. Father Font described them as "a great snow-covered range [*una gran sierra nevada*]," hence their name.[122]

Coast redwoods over 200 feet high and 5 feet in diameter are common, and many reach a height of 300 feet and a diameter of 16 feet, but 20 foot diameter trees are rare (Fig. 9.7). The largest trees grow on the deep soils deposited by flooding rivers, including the tallest tree known today that is over 367 feet high. Up to 12 inches of deeply furrowed, reddish brown bark protects the gigantic trunks of these trees.[123] The cones of redwoods are small and egg-shaped, and their leaves form flat sprays that fall each 3–5 years to create a deep layer of litter on the forest floor.[124] Even the crowns of the largest redwoods seem surprisingly narrow and short, yet they appear perfectly proportioned when the trees stand together in the forest.

Ancient redwood forests wound their way 450 miles up the foggy coast of northern California in a narrow band, from 5 to 35 miles in width. They began in southern Monterey County and extended into the southwest corner of Oregon, covering an area just under 2 million acres. Therefore, coast redwoods were one of the smallest of America's ancient forests. Rainfall varies between 25 and 122 inches a year in this area. Still, redwoods need fog to reduce the rate at which they lose water through their leaves, and fog adds additional moisture to the soil. Thus, redwood grows within the summer fog belt near the ocean. However, grasslands usually grow along the sandy shoreline between the forest and the ocean because redwoods are sensitive to salt spray.[125]

Fig. 9.7 Coast redwood, Del Norte State Park, California. (Courtesy of M.E.I. Productions, Bend, Oregon; photograph by Mike McMurray)

Open grasslands also dotted the ridges above the ancient redwood forests. "Prairies of rich grass lie on their southern slopes, and especially on their tops," wrote the chronicler of Colonel M'Kee's 1851 expedition to the redwoods.[126] These were the hunting and gathering places that Indians maintained as oases along their trails.[127] So little herds of elk and deer were often seen sprinkled over the grasslands. This must have been a welcome sight to any Indian that emerged from the depths of the forest after a long hike. Surely they also appreciated the grand views their little prairies provided of the fog drenched redwood forests that blanketed distant mountains. An outer fringe of Oregon white oak and Pacific madrone, an evergreen tree with a twisted red trunk, served as a decorative frame that separated these picturesque openings from the darkness of the surrounding redwoods.[128]

The coast redwoods of today are probably among the most ancient forests in North America, partly because the trees can live to be 2200 years old, perhaps older.[129] Furthermore, redwoods, including some of their associates such as Sitka spruce and western hemlock, may have moved on to the exposed coastal shelf during the Ice Age where the air was warmer and foggier. Then they moved back when the warming climate poured glacial meltwater into the oceans. However, at the end of the Ice Age redwood forests must have looked different from the ones the Spanish saw because fires burned less often. Even today, lightning fires seldom occur in coast redwood forests.[130]

Indians did not settle among the redwoods until about 12,000–11,000 years ago, and when they did they started fires. Their fires added to those that already burned because of lightning.[131] Furthermore, both Indian and lightning fires generally started in the fall when the leaves, twigs, and logs were dry enough to burn.[132] So the change was dramatic. Fires swept through redwood forests about once in 135 years before Indians began burning.[133] These had to be mostly hot surface fires and crown fires because fuels had time to accumulate. Such fires create large patches. Consequently, at the end of the Ice Age, redwood forests most likely formed a large patch mosaic, with each patch representing a different stage of recovery from fire. After the Indians started burning, only about 17–82 years passed between each fire, and in some areas it dropped to 8 years.[134] Nevertheless, fires usually occurred less often within the fog belt near the coast than farther inland, and a few long periods between fires intermixed with short periods in both places.[135] Therefore, ancient redwood forests were subjected to frequent light surface fires, as well as some infrequent hot surface fires. These more frequent fires changed the ancient forest into a mosaic of mostly small patches.

Coast redwoods do well in a forest where surface fires burn, and they prospered in the more severe fires at the end of the Ice Age. The thick bark of redwoods insulates the trunk from most light fires, and their great height helps to keep flames from scorching the foliage. Dormant buds along the trunk also produce sprays of new foliage if a hot fire destroys the crown. These sprays create "fire columns," tall trees with short feathery foliage, but the long life of the tree gives them enough time to grow new branches. Likewise, if fire kills the top of the tree, it quickly sends up sprouts from the base of the trunk, although young trees sprout more vigorously than old trees. Some of these sprouts can grow 7 feet tall the first year. Redwood seedlings also grow well on bare soil in sunny openings that fire or other disturbances clear within the forest. On the best of these sites, redwoods can reach heights of 150 feet within 50 years.[136]

Fire, wind, and flooding were the principal forces responsible for perpetuating ancient redwood forests and creating their patchy structure. Fires or windstorms cleared openings of various sizes in a forest that quickly filled in with stump sprouts and redwood seedlings. Most of these young pioneer redwood forests were small, but some were large. The ages of the dominant trees in different patches varied because they opened at different times. A large patch of old pioneer redwoods that came in at one time gradually dissolved into smaller patches as light surface fires eroded the overstory trees and replaced them one by one with little groups of young redwoods and settler trees. This was now an old transitional forest.

Repeated light fires created gaps in a forest by gradually burning away the thick bark at the base of a few scattered redwood trees, which eventually killed the tops and caused the base to sprout. The many sprouts that encircled each stump grew their own roots and formed a ring of trees where the old tree used to stand. When another fire killed the tops of the trees in these little groups, they sprouted again, widening the ring and enlarging the patch. Additional fires expanded these circular patches until single trees died within them and formed new patches of sprouts. Thus, the original rings of redwoods gradually blended into the surrounding forest and the process started again.[137]

Sometimes fire and wind worked together to form these patches. A redwood tree cannot sprout and replace itself if wind blows it over and tears the roots from the ground. The seedlings that grow on the mound of dirt left by the fallen tree may survive, or they may not. The problem is that a gap must be about as wide as half the height of surrounding trees for the seedlings of even many tolerant trees to receive enough sunlight to grow.[138] That means the gap would have to be about one-quarter to one-half acre because redwoods are so tall. However, most redwood gaps are only a few hundredths of an acre.[139] If redwood seedlings manage to survive in the understory or a small gap, they may stay stunted or suppressed for decades. Furthermore, the next light surface fire would normally kill many of these suppressed trees, but a few might survive and grow into the canopy.

The chance of a redwood replacing itself increases if wind snaps it off above the base so that it can sprout. This usually happens when heart rot infects the inside of the redwood and weakens its trunk. It becomes susceptible to infection when repeated fires burn through the protective bark of the tree. Thus, fire also worked indirectly through a fungus and strong winds to help create the patchiness of ancient redwood forests.[140]

Surface fires often left most of the large redwoods in the overstory unharmed when they burned through a patch of ancient forest, leaving only a black char on the outside of the bark. Even so, these fires could profoundly affect understory trees, and even many of the other large trees that grew among the redwoods, especially western hemlock. Such fires simply cleaned thick litter from the ground and exposed the mineral soil underneath. Western hemlock seedlings grow well on bare soil, although the seedlings of this settler species also grow in thick litter. Moreover, western hemlock grows well in dense shade. So conditions would be ideal for a flush of hemlock seedlings to pop up under the redwoods. Eventually the patch would consist of an overstory of redwoods and an understory of western hemlock.[141]

Another surface fire could eliminate most of the western hemlock, even if they were large trees, because their thin bark and shallow roots make them vulnerable to fire. A second flush of young hemlock seedlings would most likely appear after such a fire.[142] Thus, generations of hemlock would cycle in the understory of the towering redwoods because they can only live 400–500 years at most, while the redwoods live more than 2000 years. This means a patch of redwood forest could cycle back and forth between the open old pioneer and dense old transitional stages of development. Not only that, a few redwoods might still sprout here and there because of the fires, renewing the overstory redwoods and prolonging the cycles.

Hotter surface fires that removed large redwoods from various parts of the overstory produced conditions that no longer favored western hemlock. The openings left by the blackened redwoods allowed more sunlight to shine on the bare soil than before, making it ideal for Douglas-fir. Consequently, a flush of Douglas-fir seedlings often sprang up along with the sprouts of redwoods burned by the fire. Tanoak thrived in the understory of such a forest as well, eventually creating a layer of hardwoods below the Douglas-fir trees. Additional light surface fires further stimulated the growth of tanoak because it does well when burned in shady places. Douglas-fir lives to be nearly twice as old as western hemlock, and it is more resistant to fire, so it stayed in the canopy longer and cycled less often.[143]

Since fires burned each part of the mosaic differently, an ancient redwood forest consisted of patches with many different combinations of trees and layers. Even so, redwoods persisted as long as the fires kept burning. Without fire, fewer redwoods would sprout because the roots pull out of the ground when a tree topples. Still, some redwood seedlings would find places to grow on the bare soil left behind by fallen trees. Douglas-fir, western hemlock, and other trees such as California laurel, or bay, which is a hardwood and a settler species, would occupy many others. Western hemlock and bay grow better in deep litter and dense shade than redwood and Douglas-fir. That means they are the ultimate, self-sustaining settler species, and Douglas-fir is an interim settler species. So fewer and fewer young redwoods and Douglas-fir would appear in the understory as western hemlock and bay became more abundant.[144]

Eventually, the last redwood in a patch of old transitional forest would fall. Then a self-replacing forest dominated by western hemlock, with the shorter bay trees in the understory, would grow where the ancient pioneer redwoods once stood. Bay would probably replace redwood in the southern part of its range where western hemlock does not grow. The entire redwood forest would gradually shrink if this happened in many patches, but it would probably not disappear.

Occasional windstorms knock down pockets of trees on hillsides, and periodic floods sweep away trees in river valleys and bury their roots in silt. Soil exposed by falling trees tearing up the ground, and sediment deposited from floods, provides the bare soil redwood seeds need to germinate. Flooding also thins the forest enough so that redwood seedlings have sufficient light to grow. Furthermore, unlike most trees, large redwoods survive flooding because they can grow new roots to replace those that were buried. Therefore, windstorms and floods would probably create enough clearings and bare soil to maintain some redwood groves. Individual redwood trees would probably survive in scattered locations as well. Even so, only remnants of the overpowering forests of redwoods that the Spanish described would remain if the fires stopped.[145]

Ancient redwood forests did not shrink or disappear. They thrived along California's foggy coastline because the fires kept burning. However, not until the Indians came and started more fires did redwood forests take on their modern character and diversity. Botanist David Douglas probably felt like the Spanish explorers, and most of the early travelers who followed them, when he first saw the ancient redwoods. Douglas said in 1831 that these forests were "the great beauty of Californian vegetation ... which give the mountains a most peculiar, I was almost going to say awful, appearance."[146]

TEN

Forests of the Colonies

Heaven and earth never agreed better to frame a place for man's
habitation.
Captain John Smith, 1624[1]

The arrival of Captain John Smith at Jamestown, Virginia, in 1607, marked the beginning of English colonization of North America. However, Italian navigator John Cabot (Giovanni Caboto) had probed the coast for Henry VII of England more than a century earlier. He planted his banners in Newfoundland's soil on June 24, 1497, near where Leif Erics-son probably landed about 500 years earlier. Even so, the Spanish became the first successful colonists when Pedro Menendez de Aviles established St. Augustine, Florida, in 1565. The English followed in 1587, but their first attempt failed when Governor John White's Roanoke Colony disappeared in Virginia. Captain Smith finally succeeded and began making detailed descriptions of the ancient forests of the colonies.[2]

The Italian navigator Giovanni da Verrazzano glimpsed the North American forests much earlier than Captain Smith. He left a brief, but enduring, image of them while exploring the East Coast for France in 1524. The landscape 15 or more miles inland from Narragansett Bay consisted of "open plains twenty-five or thirty leagues in extent, entirely free from trees or other hindrances," he said. That would make these "open plains" 75–90 miles long. A colonist in 1647 estimated that Indian agricultural fields in the Hudson River Valley were 21–24 miles long. Verrazzano added that the forests that surrounded the openings "might all be traversed by an army ever so numerous."[3] What Verrazzano saw resulted from widespread Indian burning to keep the forests free of undergrowth and to clear them for crops and hunting grounds. These clearings helped

258

the colonists settle the land a century later. When the Pilgrims landed in 1620 they just planted their corn in fields left abandoned by the Indians because "great multitudes" had died from European diseases just 4 years earlier. "God," said John Winthrop, "hath hereby cleared our title to this place."[4] Thus, Indians had colonized and altered these ancient forests long before the English came and took advantage of their labor.

Oak-chestnut, eastern white pine, beech-maple, and red spruce–fir forests dominated the colonial landscape. Most of them clustered in the northeast, but the oak-chestnut forest extended far to the south and the west. Forests also sorted themselves on the landscape by soil type. The first English colonists knew this, as Jeremy Belknap wrote in his *History of New Hampshire* in 1792:

> In the new and uncultivated parts, the soil is distinguished by the various kinds of woods which grow upon it. Thus: white oak land is hard and stony. . . . The same may be said of chestnut land. Pitch pine land is dry and sandy. . . . White pine land is also light and dry, but has a deeper soil. . . . Spruce and hemlock, in the eastern parts of the state, denote a thin, cold soil. . . . Beech and maple land is generally esteemed the most easy and advantageous for cultivation, as it is a warm, rich, loamy soil.[5]

Like other forests, the ancient forests of the colonies consisted of mosaics of irregularly shaped patches, some no larger than a single large tree, and others that covered hundreds or thousands of acres. Likewise, the frequency and severity of disturbances that created these mosaics differed from place to place. So each forest of the colonies had a distinctive character that resulted from a web of relationships among the trees, the climate, the land, and the Indians.

OAK-CHESTNUT FOREST

An ocean of woods.
Thomas Pownall, former governor of Massachusetts, 1776[6]

The ancient oak-chestnut forest covered over 84 million acres, the largest forest of the colonies (Fig. 10.1). It extended from southern New England southward along the Appalachian Mountains to as far south as northern Alabama. From there it extended southwestward through Pennsylvania, eastern Ohio, Kentucky, and Tennessee to the Mississippi Valley.[7]

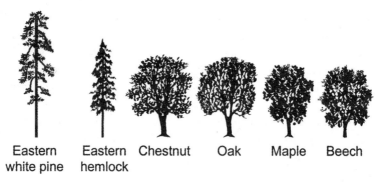

Eastern   Eastern   Chestnut   Oak   Maple   Beech
white pine   hemlock

Fig. 10.1   Trees of the ancient oak-chestnut forest.

One of the first Europeans to see this ancient forest was Henry Hudson, an English navigator and explorer. It was 1609, and he was on an expedition for the British East Indian Company when he described the oak-chestnut forest in his journal. He sailed 12 days up the Hudson River and "rode still and went on land to walke of the west side of the river." Here Hudson "found goode grounde for corn and other garden herbs, with great store of goodly oakes and walnut-trees and Chestnut trees, ewe trees and trees of sweetwood in great abundance."[8] Even though the oak-chestnut forest that Hudson saw covered a large area, it did not grow everywhere. Other types of forest prospered within its boundaries in places less favorable to chestnut.

The Appalachian and Allegheny Mountains in eastern Pennsylvania fell within the oak-chestnut forest. These mountains form a seemingly endless series of long parallel ridges that stretch across the landscape in broad curves. Thomas Pownall, former governor of Massachusetts, stood on a ridge in 1776 and gazed into the distance. He said that the forest-clad mountains had the "general appearance" of "an Ocean of Woods swelled and depressed with a waving Surface." Pownall declared it one of "the most picturesque Landscapes that Imagination can conceive."[9]

Anyone who crossed these mountains would find that forests in valleys differed from those on the slopes and mountain tops. Lt. Col. Adam Hubley documented some of these differences when he wrote his account of the campaign against the Iroquois in 1779. Repeated Iroquois raids on European settlers led Congress to direct General Washington to retaliate. He sent three large expeditions, one under Major General Sullivan, to destroy all Iroquois villages and crops, engage Tory and Indian marauders, and drive them out of the country. General Sullivan's army marched northwestward through the Allegheny Mountains in eastern Pennsylvania and up into the Finger Lakes region of New York.[10] Excerpts from Lt. Col. Hubley's journal describe how some of the forests they saw changed during their march.[11]

Lt. Col. Hubley began his account in a "delightful ... valley," a "mere garden," he said, "bounded by large chains of mountains." Then the troops marched up a mountain where the "wood [is] chiefly low, and composed of pine only [pitch pine]." A few days later they climbed another mountain, finding it "exceedingly level for at least six miles" and covered with "pine and white oak, chiefly large." "We then descended the mountain," Hubley wrote, and found the "land ... well timbered, chiefly with black walnut, which are remarkably large, some not less than six feet over, and excessively high." They dropped into a narrow valley on the other side of yet another mountain and entered "a dark, difficult swamp ... covered with weeds and considerable underwood, interspersed with large timber, chiefly buttonwood [American sycamore]." Again they climbed "with the greatest difficulty," and stood on the top of a mountain covered "with large timber, chiefly oak, interspersed with underwood and excellent grass." "We had a view of the country of at least twenty miles round," wrote Hubley, "the fine, extensive plains, interspersed with streams of water, made the prospect pleasing and elegant." It was August 28, and they finally saw their quarry. Hubley said, "We observed, at some considerable distance, a number of clouds of smoke arising, where we concluded the enemy to be encamped." The next day General Sullivan's army fought in the Battle of Newtown, routed the Iroquois, and pushed onward through America's ancient forests.[12]

This is the northern Ridge and Valley Province. During the age of discovery the upper slopes and ridges usually supported open forests of American chestnut and chestnut oak, an oak tree with dentated leaves that look similar to those of the American chestnut. Scarlet oak, black oak, eastern white pine, and pitch pine also grew on the ridges. The sheltered mountain coves also contained chestnut, as well as hemlock, black oak, beech, ash, and some pine. Northern red oak stayed in these coves as well, even though it could also have grown on ridges, slopes, and valley floors. However, frequent Indian fires forced it to take refuge there because it is more sensitive to fire than most oaks. White oak grew in many places, but primarily in scattered locations within grasslands on valley floors, along with black oak, hickory, and some white pine. This is where Indians burned most frequently and white oak is moderately fire resistant. Hickory is less fire resistant, but it prospered since injured trees sprout vigorously. Ravines that cut through the valleys normally contained settler species such as maple, basswood, beech, elm, and hemlock because the soils were moist and fires rare. Hickory grew in the ravines as well. Black walnut could withstand the fires but it needed moist soils, so it also grew in the ravines.[13]

This arrangement of trees held true elsewhere in the oak-chestnut forest. Areas with deep moist soils supported a wide variety of hardwoods, including some chestnut and such settler species as maple and beech. Oaks tended to dominate upper slopes with drier soils, and chestnut also grew well in these areas. Peter Kalm said that he "found chestnut trees in some spots in the forests," but that "the soil in which they grew was in some places rather poor."[14] Species such as chestnut oak and black oak grew on the driest soils, and white oak and northern red oak grew on the moister soils. Open forests with a combination of oaks and pines, or just pines, dominated dry ridges, rock outcrops, and sandy soils. The large number of species that grew within the ancient oak-chestnut forest, together with the effects of periodic disturbances, created many combinations of trees, even on a particular kind of soil. Still, oaks and the American chestnut defined the forest because they were widespread and prominent.[15]

The American chestnut was abundant in the ancient oak-chestnut forest, constituting 25–40% of the trees.[16] Patches of chestnut were scattered throughout the forest, but it also grew in nearly pure groves in the southern Appalachian and northern Blue Ridge Mountains.[17] These groves made a spectacular display in early summer when the long, creamy-yellow flowers of the chestnuts gave hillsides the look of a recent snowfall.[18] Chestnut trees also produced a heavy crop of nuts every fall that furnished a dependable food supply for bear, deer, wild turkey, squirrels, passenger pigeons, and Indians.[19] Henry David Thoreau looked forward to fall. "Now is the time for chestnuts," he wrote in October of 1855, "a stone cast against the trees shakes them down in showers upon one's head and shoulders."[20] E. Johnson wrote in 1654 that to the Indians "boiled chestnuts is their white bread."[21] They also became an important food for the colonists who thought they were "very sweet in taste."[22] Captain John Smith said that America's chestnuts "equalize the best in France, Spain,

Fig. 10.2 American chestnut in the Great Smoky Mountains of North Carolina (Photograph courtesy of the Forest History Society, Durham, North Carolina)

Germany, or Italy."[23] The other nuts in the forest were less tasty and crops were less dependable.

The American chestnut had a commanding presence in the ancient forest (Fig. 10.2). It gained the title of the "Redwood of the East" because of its great size, dark brown fissured bark, and resistance to decay.[24] The largest specimens stood as much as 130 feet tall with trunks up to 10 feet thick, although most mature trees were about 100 feet tall and 4–5 feet thick.[25] Their great size, easily worked wood, and resistance to decay made them one of the Indian's favorite trees for building dugout canoes. One Indian could fell a large chestnut and build a canoe in 12 days using only stone tools, clamshells, and fire.[26] The colonists also found many uses for chestnut, especially split-rail fences.[27]

To be sure, other trees that grew among the chestnuts could also be huge. The northern red oak can reach a height of 160 feet and a diameter of 5 feet. The chestnut oak does not grow so tall, only 100 feet, but it can be 6 feet thick. Even the tupliptree can be 150 feet tall and 6 feet thick.[28] No doubt grand trees such as these grew in parts of the ancient oak-chestnut forest. The scarlet oak was a little smaller, but it made up for its size by standing out when the leaves turned color in the fall. To Henry David Thoreau in 1858, the "intense, burning red" leaves made the scarlet oak look "like huge roses." "The whole forest is a flower-garden," said Thoreau.[29]

Even though oak-chestnut forests had many distinctive features, they shared two qualities that could apply to most of America's ancient forests that Indians burned regularly. The dominant trees were often large and few trees and shrubs grew in the understory. In short, they fit Verrazzano's original descriptions of the forests in Rhode Island and Massachusetts. These qualities applied as well to the forests that Captain Smith described in Virginia in 1631:

> The woods for many a hundred mile for the most part grow slight; ... neither grow they so thick together ... and much good ground between them without shrubs, and the best is ever known by the greatness of the trees and the vesture it beareth.[30]

Captain Smith also noted that the forests consisted of "many" trees "so tall and straight, that they will ber two feet and a half square of good timber for 20 yards long."[31] All the trees were not huge, just "the best," nor were they all tall, just "many." Similarly, these ancient forests were not entirely open or composed of widely spaced trees, just "for the most part ... slight." However, open forests must have been more common than dense forests because so many early travelers made the same comments, such as Andrew White's account of the forests that grew along the Potomac River in 1633:

> On each bank of solid earth rise beautiful groves of trees, not choked up with an undergrowth of brambles and bushes, but as if laid out by hand, in a manner so open that you might drive a four-horse chariot in the midst of the trees.[32]

Johnson concluded in 1633 that "the Indians frequent fiering of the woods," made the forests around Plymouth colony "thin of Timber in many places, like our Parkes in England."[33] Thomas Morton added that it also made "the Country very beautiful, and commodious."[34]

The same thing held true far to the west, on the sandy soils along the southern shore of Lake Ontario, which was still within the oak-chestnut forest.[35] French missionary Galinee wrote that the land there was "for the most part beautiful, broad meadows, on which the grass is as tall as myself." "In the spots where there are woods, these are oak plains, so that one could easily run through them on horseback," he said.[36] It is not surprising that bison wandered into western Pennsylvania and New York.[37] Bison also roamed in West Virginia. No doubt because grasslands increased when disease, war with Iroquois invaders, and colonial expansion forced large numbers of Indians to abandon their lands. They left behind many "brave meadows and old fields," noted Robert Fallows in 1671.[38]

The oak-chestnut forests that so impressed early explorers and colonists would have looked less majestic a few thousand years earlier. The American chestnut only represented 7–8% of the trees back then. However, by the time the colonists arrived, it represented up to 40% of the ancient oak-chestnut forest.[39] Something happened about 2000 years ago to cause the dramatic expansion of chestnut. Scientists believe that increased burning by Indians to clear land for crops may have triggered the change. This seems reasonable since chestnut trees clustered around stone walls, roadsides, and woodlots in the colonies, and it also invaded forests that grew on abandoned fields and pasturelands.[40] Chestnut sprouts vigorously, and an increase in fires would have helped it to spread, just as fuelwood cutting did during the colonial period.[41] Furthermore, Indians most likely planted groves of chestnuts around some of their villages.[42] Whatever the cause, the American chestnut became one of the dominant trees in the ancient forest.

Chestnuts do not grow well under the thick canopy of a closed forest. Thoreau even noticed the absence of chestnut seedlings within chestnut groves. However, too few light surface fires burned through northeastern forests in his time to keep them thinned enough for chestnut. Thoreau did find chestnut seedlings under the sparser canopies of pine and birch trees that grew on old pastures.[43] Consequently, chestnut most likely reproduced and grew in a manner similar to oak.[44]

Oak and chestnut hold an intermediate position in succession. They both act like settler species because their seedlings can grow in light shade under larger trees, but they are not self-replacing. Likewise, they both act like pioneers because they need fire to reduce competition from other plants. However, seedlings that start from seeds do not grow as often on bare soil in freshly cleared openings as they do in light litter and shade. Their sprouts do better in clearings and in the understory than their seedlings. Therefore, oak and chestnut rely primarily on sprouting to replace themselves, and they often follow their parent trees or other pioneers such as pine. However, they cannot sustain themselves without fire. They are pioneers that can become interim settlers. They are not the ultimate settlers. That means it took a special combination of fire and shade to maintain the ancient oak-chestnut forest.

The relatively thick bark of the older oaks and chestnuts allowed them to resist serious damage from light surface fires, but some large trees still died. Even so, some oaks such as black oak, white oak, and chestnut oak, resist fires better than others, such as red oak. Light surface fires also helped the oaks by clearing litter from the forest floor. Squirrels and blue jays prefer to bury acorns where the litter is thin, and oak seedlings also grow best on such soils. Fire also reduced the insects that live in litter and prey on acorns. Moreover, frequent fires thinned overstory trees so that there was enough light for the seedlings to survive. However, these fires also killed the tops of oak and chestnut seedlings, but the roots of many remained alive. The survivors quickly sprouted and kept pushing their roots deeper into the soil. Some settler species such as maple also sprouted, but they had a higher mortality rate from fire than oak and chestnut seedlings. So oak and chestnut gradually increased while maple and other settler species declined. By the time the big old pioneers started falling there were usually thousands of deeply rooted oak and chestnut seedlings ready to replace them. However, chestnut grew faster than oak when exposed to full sunlight. So chestnut either reached the overstory first and dominated the forest or it shared dominance with the oaks. Once established, oak and the chestnut kept replacing themselves because surface fires continued to burn.[45]

If the fires stopped for a time, debris would accumulate, and the trees would become thicker. Then the next fire could be intense, climbing into the crowns of the tallest trees and killing more than 80% of their tops. However, the roots of most of the sprouting species could survive such a fire and sprouts would emerge to replenish the forest with oak and chestnut. Even so, after a hot fire some species such as chestnut oak would sometimes form dense stands that choked out the other species.[46] The "mighty thicket, at least three miles long" that Colonel William Byrd "Scuffled thro" near the boundary between Virginia and North Carolina in 1728 represented such an oak forest. He said it was "rich high land . . . overgrown with Saplings of Oak, Hiccory, and Locust" that had "been burnt not long before."[47]

Frequent Indian fires certainly played an important role in maintaining the ancient oak-chestnut forest. Timothy Dwight saw this in New England in 1821, stating that these forests were "usually subjected to an annual conflagration." They "were selected [by the Indians] for this purpose," he said, "because they alone were, in ordinary years, sufficiently dry."[48] Lightning fires contributed as well, but they happened too rarely to account for the widespread occurrence of oak and chestnut. Only about two to six lightning fires burn per million acres of oak forest in any one year. It took many more fires than that to sustain the ancient forest and keep it open. Furthermore, the ancient oak-chestnut forest could only prosper in the presence of both light surface fires and crown fires or hot surface fires. Light surface fires burn where fuels are low, and that only happens where fires are frequent. Sporadic lightning fires could probably not do it alone. Therefore, Indians were the only source of fire that could make up the shortfall.[49]

If the Indians abandoned an area for a long time, as they sometimes did, the oak-chestnut forest changed dramatically. The elimination of Indian fires from the Ridge and Valley region would cause a decline of oak, chestnut, and eastern white pine. The more shade-tolerant red maple would replace them on drier sites and sugar maple on moister sites. Hemlock or a beech-maple forest would replace the oak-chestnut forest on moist, fertile soils outside the Ridge and Valley region, but red maple would still replace it on drier soils.[50] Thus, large parts of the oak-chestnut

forest would change to thick self-replacing forests of shade-tolerant settler species if Indian burning ceased. Nevertheless, remnants of the open
oak-chestnut forest would still exist in a few places because of occasional
lightning fires.

Oak and chestnut seedlings rarely grew tall enough to reach the upper
canopy when they started within the very small gaps created by the death
of a single large tree. These tiny openings favored settler species such as
maple and beech because they grow faster in low light than less tolerant
species such as oak and chestnut.[51] Oak and chestnut are intermediate in
tolerance to shade.[52] So they needed a larger hole in the canopy, usually
from one quarter to one acre or more in size.[53] Therefore, oak and chestnut had to grow in large patches. Still, they did not fill all the openings
because disturbances were sporadic, and old seedlings were not always
there to take advantage of the opportunity.[54]

The ancient oak-chestnut forest was a dynamic and diverse landscape
of people and forests. It consisted of a mosaic of young, middle-aged,
and open old pioneer forests, and dense old transitional and self-replacing forests, in a wide range of sizes. Large patches of trees that covered
hundreds of acres were probably rare because crown fires were rare. Those
covering more than an acre occurred more often, and those covering an
acre or less were probably the most numerous. Very small, single-tree size
openings were also numerous, but they favored the more shade-tolerant settler species rather than oak and chestnut. Thus, the bulk of the
ancient oak-chestnut forest formed a small to large patch mosaic, with
very small patches and single large trees sprinkled throughout the forest.
Most groups of middle-aged and old trees were relatively open, except for
tree seedlings, sprouts, and small plants. The dominant trees were often
large. Old transitional and self-replacing forests probably became more
numerous at greater distances from Indian villages. Indian villages, croplands, and hunting grounds were also part of the ancient forests mosaics,
especially along rivers and coastlines, and a vast network of trails also cut
through them.

Oak still thrives in the oak-chestnut forest, although it is declining in
abundance because of the lack of Indian fires. Sadly, the American chestnut tree that helped define this forest has nearly disappeared. The victim of a fungus that entered the United States before 1900. By 1950
nearly all the chestnut trees were dead. What remains are sprouts that
keep growing from old seedlings and the stumps of giants, only to be

struck down as saplings. There is hope. Even now scientists are working to breed resistance into the American chestnut.[55] Perhaps someday the chestnut will resume its rightful place as one of the hallmarks of America's ancient forests.

EASTERN WHITE PINE FOREST

The most noble is the mast pine.
Jeremy Belknap, parson, 1792[56]

Ancient eastern white pine forests (Fig. 10.3), mixed with some red pine and jack pine forests, covered nearly 16 million acres in the United States, but only about a third of it existed in the colonies of New England.[57] White pine represented the largest proportion of these forests, and it was also an important part of the 68-million-acre pine-hardwood forest in southern Ontario, Quebec, and New Brunswick.[58] The eastern white pine forest concentrated in central New England. It occurred in southern New Hampshire and Maine, eastern Massachusetts, the Champlain Valley in Vermont, the Connecticut River Valley, and the eastern slopes of the Adirondacks in New York. Lesser amounts of white pine extended southward into Pennsylvania and the southern Appalachian Mountains, where it grew at higher elevations. White pine forests also grew far to the west in Michigan, Wisconsin, Minnesota, and southwestern Ontario, but they were different in many ways from colonial forests.[59]

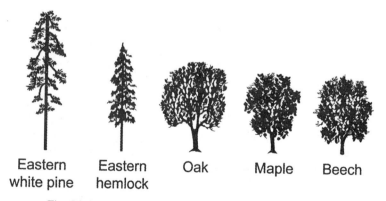

| Eastern | Eastern | Oak | Maple | Beech |
| white pine | hemlock | | | |

Fig. 10.3   Trees of the ancient eastern white pine forest.

This is an old forest that most likely survived the Ice Age some-where on the eastern continental shelf. However, the eastern white pine forest probably developed its modern appearance about 6000 years ago, although white pine arrived in the colonies over 6000 years earlier.[60] Even so, white pine never detached itself completely from the Ice Age. It still prefers young soils left behind by retreating ice sheets, such as glacial tills. These consist of unsorted mixtures of clay, sand, gravel, and boulders. This kind of soil is usually dry, and it has a lower nutrient content than an older, finer soil. Such soils are found on outwash from the front of glaciers and on eskers, which are ridges of debris that meander over the landscape where rivers used to flow beneath ice sheets. Hardwoods are less able to compete with white pine on these soils. Eastern white pine also grows on better soils when they can find room among the hardwoods.[61]

Mature eastern white pines are impressive trees. Many colonists thought so when they saw them in the ancient forest, including parson Jeremy Belknap who described eastern white pine in 1792:

> Another thing, worthy of observation, is the aged and majestic appearance of the trees, of which the most noble is the mast pine. This tree often grows to the height of one hundred and fifty, and sometimes two hundred feet. It is straight as an arrow, and has no branches but very near the top. It is from twenty to forty inches in diameter at its base, and appears like a stately pillar, adorned with a verdant capital, in [the] form of a cone.[62]

Belknap called them "mast pine" because a century earlier King William III issued Massachusetts a new charter that reserved the largest white pine to the crown to provide masts for the Royal Navy. No wonder, it was the tallest tree in the colonies. In addition, old white pine growing within a forest only had limbs on the top one-third of the tree, which kept most of their great brown trunks free of knots. One mast shipped to England was 154 feet long and $4\frac{1}{2}$ feet in diameter. Timothy Dwight measured an even larger white pine around 1800. It was 6 feet in diameter and 250 feet tall.[63] This must have been a very old tree because a 170-foot giant was 460 years old when its rings were counted in 1899.[64]

The ancient eastern white pine forests grew primarily on river terraces, lowlands, and mountain slopes below beech-maple forests.[65] Sometimes white pine grew in extensive and nearly pure groves, especially on river terraces. In 1629–1633, William Wood saw these spectacular old pine forests "shooting up exceeding high." He seemed impressed, but like most colonists of his time his interest was primarily commercial. "I have seene these stately high growne trees, ten miles together, close by the river side," he wrote, "from whence by shipping they might be conveyed to any desired Port."[66] White pine also grew among other trees, but even here "its summit is seen at an immense distance, aspiring to heaven, far above the heads of the surrounding trees," said Francois Michaux in 1803.[67] Such a forest seemed "gloomy," to Jeremy Belknap when he walked into one in 1792, but he also appreciated "the silence which reigns through it." "On a calm day," he said, "no sound is heard but that of running water, or perhaps the chirping of a squirrel or the squalling of a jay."[68]

Eastern white pine is a pioneer, although it is intermediate in its tolerance to shade. Like other pioneers, white pine must have places with relatively bare and moist soil, adequate sunlight, and little competition from other trees for its seedlings to survive and grow. Therefore, large groves of nearly pure eastern white pine only grew where they did in the ancient forest because of the abandonment of Indian croplands and hunting grounds, or wildfires.[69] Indians often used river terraces to grow maize, and they also burned them more frequently than other areas. So many of the spectacular white pine forests along rivers in the northeast were undoubtedly the result of Indian cultivation and burning. White pine responded the same way when the colonists set fires and let their pastures revert to forest.[70] Therefore, large patches of young, middle-aged, and old white pine would have stood along the banks of many rivers that flowed through the ancient forest. However, patches of small, young white pine trees would have been much less common than patches of large trees because the mature part of a tree's life span lasts so much longer than the youthful part.

Crown fires and hot surface fires cleared large openings more frequently in some regions of the white pine forest than in others. In southwestern Quebec such fires probably burned the same area an average of once every 230 years. That is less than the life span of the tree. Light surface fires burned more often and kept many of these forests open. Indians probably set many of the fires because in eastern Canada only about 1.6 lightning fires start per million acres of land each year.[71]

Even some hot fires failed to regenerate eastern white pine. White pine only became established on about one-fourth of new clearings because it does not produce a good seed crop more than once in 3–5 years.[72] Not only that, some patches of white pine disappeared just because fires are unpredictable and can skip areas long enough for self-replacing forests to replace them. Balsam fir, black spruce, or maple might dominate these self-replacing forests in southern Ontario.[73] In Pennsylvania, hemlock eventually takes over white pine forests on stream terraces, but red maple gradually replaces them on uplands.[74] In other parts of New England settler species such as beech, hemlock, and maple often replace white pine.[75] Even so, hot fires burned the ancient forest often enough to consistently make room in the mosaic for many large patches of eastern white pine.[76]

In New England, large nearly pure groves of eastern white pine were less plentiful on uplands than on river terraces. This means that on parts of the uplands the period between large fires must have been longer than the 230-year average for white pine. No one knows how much longer, but it probably varied a great deal. However, large fires most likely burned in old transitional forests before the older first-generation white pines died, which would be about 500 years. Most of them surely burned before the small number of scattered second-generation white pine disappeared from the forest, or about 1000 years. Otherwise, too few pine trees would be near enough to a fresh burn to provide seed for the next forest. White pine seeds are mainly dispersed by wind, and they only travel about 200 feet from the tree within a forest, and about 700 feet in the open. Gray squirrels also bury white pine seeds, but they travel a limited distance as well.[77] Therefore, the hot fires that renewed white pine had to burn every few centuries. However, hot fires were less frequent in areas remote from Indian villages and hunting grounds.

A Mr. Pringle described an old transitional white pine–hemlock forest during his trip through northwestern Pennsylvania about 1879. The area he visited was said to be "little known ... scarce explored, and ... uninhabitable" by T. F. Gordon only 47 years earlier.[78] Mr. Pringle wrote:

> We saw much original forest still standing and composed principally of hemlock. Some white pine appeared as scattering trees, or in groves, and some hardwood.[79]

The events that took place in the forest before Mr. Pringle's arrival can be inferred from later descriptions and similar forests.[80] The forest he saw most likely began a few centuries earlier when a hot fire burned the area.

Seneca Indians may have started the fire. Cornplanter, a chief who lived near there earlier in the century, said the Seneca often burned along the Allegheny River below the forest to kill rattlesnakes.[81] Doubtless some of their fires crept up the slopes and into the forest above. Most of them probably went out before causing much damage, while others became conflagrations. We do not know, but the fire that established Mr. Pringle's forest may have covered thousands of acres. Since it was probably a dry year, there may have been several large fires.[82] This kind of forest is difficult to burn, so the fire probably gained momentum in a jumble of dry fallen trees that blew down in a windstorm.[83] Then the forest remained unburned until Mr. Pringle arrived.

Normally, most white pine seedlings invade a burned area within the first decade, but they may not grow thick enough to stop other trees from gaining a foothold. So, eastern hemlock, a settler species that also does well on bare soil, probably joined white pine in the clearing where Mr. Pringle's forest started, but it kept invading long after white pine seedlings could no longer tolerate the shade. Beech and maple seedlings gradually invaded as well. The understory usually becomes too shady for even these settler species before the end of the first century. However, the young settler trees held on until small gaps opened in the canopy, which happened on about 0.5% of the forest each year.[84]

By the time Mr. Pringle passed through the forest only 3% of the living trees were white pine, and they stood above the much shorter hemlock trees. These white pines were almost all mature trees that came in after the fire. Doubtless, many more white pines used to grow in the forest, but they dwindled in number during the following centuries. Some of their bleached skeletons still stood in the forest, while others lay decaying on the forest floor. Beech represented about 43% of the trees in Mr. Pringle's forest, 20% was hemlock, and 5% was sugar maple, all settler species. Unlike white pine, these trees were in seedling, sapling, pole, and mature sizes because they kept replacing themselves. Therefore, the white pine forest Mr. Pringle saw was a dense old transitional forest well on its way to becoming just a memory. A self-replacing forest of beech, hemlock, and maple would soon take its place.[85]

Pockets of the original white pine and even a few second-generation pines might still exist in Mr. Pringle's forest far into the future, but there would be too few of them to call it a white pine forest. Timothy Dwight saw such a rare grove of great white pine in a nearly identical beech-hemlock-maple forest in New York. It was 1822, and this forest was older than Mr. Pringle's because the white pines had nearly disappeared. "This cluster," of white pines he said, "is the only considerable one composed of full grown trees, of this kind, which I had seen." The grove Dwight wandered into contained 250-foot-tall veterans that were probably part of the original forest.[86]

White pine can survive in scattered places within a self-replacing forest. They normally grow in groups of two to five trees near charred stumps, places where a small surface fire flared up and cleared a patch of bare ground. However, clearings sufficiently large to regenerate white pine, about one-quarter acre or more, are still limited in such a forest. Even more unusual are patches of white pine that started on eroded stream banks or slopes, or on soil exposed by trees uprooted in windstorms, although these sometimes occur on the windy shores of lakes.[87] Single white pines also find a few places to grow in a forest. Nonetheless, the large patches of nearly pure eastern white pine that grew in the ancient forest would not have existed unless lightning fires and Indians had cleared places for them to grow.

Oaks were perhaps the most common associates of eastern white pine in the ancient forests. "To speak from recollection of pines and oaks, I should say that our woods were chiefly pine and oak mixed, but we have also ... pure pine and pure oak woods," wrote Henry David Thoreau in 1860. However, Thoreau, an astute observer of nature, also noticed that "pines are continually stealing into oaks, and oaks into pines, where respectively they are not too dense ... and so mixed woods may arise."[88] This would help explain the widespread occurrence of old white pines towering above what Jeremy Belknap called in 1792 "the common forest trees."[89] Peter Kalm noticed the same thing in northeastern New York while traveling on a river in 1749:

> In the beginning, and almost along the whole way, we had mostly pine woods around us, though here and there was a clump of oaks. ... Both white and red oak grew abundantly among the pines, but they were small.[90]

White pines that poked through the forest canopy were usually older than the other trees because they filled a clearing first, but they could be younger. This normally occurred when white pine grew among widely spaced oaks and chestnuts. Fires burned pine-oak forests often enough to keep the ground underneath the trees free of deep litter. White pine can germinate and survive in grass and light litter, and a little shade also protects the seedlings, but they still need full sunlight for maximum height growth after becoming established.[91] As long as the overstory trees are not too dense, white pine can invade the understory and slowly grow up through the canopy until they eventually stand above the rest of the trees. Many of the ancient mixed forests of white pine and oak started this way in the colonies.[92]

More often, white pine invaded a new clearing, and then the oaks followed, as Thoreau observed, "The pine is the pioneer, the oak the more permanent settler." "It is evident to any who attend to the matter that pines are here the natural nurses of the oaks, and therefore they grow together," he said.[93] Even when the smaller but more aggressive pitch pine arrives in a clearing first, white pine and oak will follow. Again, Thoreau said "to my surprise, I find that in the pretty dense pitch pine wood ... where there are only several white pines old enough to bear ... yet there are countless white pines springing up under the pitch pines (as well as many oaks), and very few or scarcely any little pitch pines, and they sickly."[94] Therefore, in this part of the white pine forest the successional sequence may begin with pitch pine, which is not tolerant of shade. Then white pine will invade because the pitch pine canopy is airy and lets in a little sunlight. Oak will usually follow the white pine and, if the soil is moist, settler species such as beech, hemlock, or maple will finally move in and dominate the forest.

Ancient white pine forests seldom reached this last, self-replacing stage of development. Fire stood in their way. The many surface fires that burned through a white pine forest did nothing more than keep it open and clear of undergrowth. They seldom eliminated the oak sprouts growing in the understory. However, occasionally a fire would destroy the white pines when they were still young, which allowed the oaks to replace them much sooner than usual. Thoreau saw this happen near his home in Concord, Massachusetts, and wrote about it in his journal on October 17, 1860:

> It is surprising how many accidents these seedling oaks will survive. . . . when a fire runs over the lot and kills pines and birches and maples, and oaks twenty feet high, these little oaks are scarcely injured at all, and they will still be just as high the next year.[95]

Thoreau also noted that "if in the natural course of events a fire does not occur nor a hurricane, the soil may at last be exhausted for pines, but there are always the oaks ready to take advantage of the least feebleness and yielding of the pines."[96] Even here Thoreau seems to know that strong winds from a hurricane or severe thunderstorm can accelerate the replacement of a mature pine forest by settler trees. These winds can topple the overstory pine long before they begin to die of old age, which bathes the young settler trees waiting in the understory in light and stimulates their growth.[97] No doubt this happened fairly often since hurricanes tore through parts of southern and central New England every 20–40 years.[98]

The ancient white pine forest consisted of two mosaics. One nested inside the other. The large patches of white pine that made up the first mosaic developed after severe fires. A smaller mosaic gradually developed within it because of frequent surface fires, windstorms, and the loss of individual trees. Some of the large patches in the first mosaic consisted of pure white pine, while many more contained a mixture of pine and oak. Of these, some had dense pine with an understory of oak seedlings and grass, and some had scattered pines emerging from a canopy of oaks. However, frequent surface fires usually kept the ground clear of undergrowth in both the white pine groves and mixed forests of white pine and oak. In the more remote places where fires were infrequent, giant white pine often stood above a canopy of mature hemlock, and a thick growth of beech, hemlock, and maple grew underneath. Some of these dense old transitional white pine forests escaped fire long enough to become self-replacing forests, and eastern white pine disappeared.

The number of small patches and scattered individual settler trees that appeared within the larger patches increased as ancient white pine forests grew older and degenerated. Each great white pine that died probably left a big enough opening to stimulate the growth of some of the young oaks or young beech or other settler trees that grew underneath. This allowed them to grow upward and form a new smaller patch. However, oaks became scarce in the understory as the forest became thicker. Smaller openings favored the most shade tolerant settler species, especially hemlock and beech. When a tree fell and created an opening, a group of saplings grew a little, then the overstory closed in and they stopped growing. Another overstory tree near the opening might fall a few years or decades later, and the young trees would have enough sunlight and space to grow a little more. This opening and closing cycle repeated many times in a gap before a small group of hemlock or beech trees, or a single survivor, finally reached the canopy.[99]

Trees that blew over instead of snapping or dying produced a mound of dirt near the upturned roots, and a shallow pit where the roots pulled out of the soil. Small black or sweet birch trees took advantage of the mounds and formed more tiny patches in the forest. On the other hand, red maple grew in tiny patches within the pits.[100] Yellow birch also grew on mounds as well as on rotten logs in openings of about one-tenth acre. When a log decayed, it left the yellow birch trees standing on their roots where they used to wrap around the log. The less frequent the fires the greater the number of these small groups of settler trees that would pop up throughout the forest. Single trees would also squeeze their way in among the larger trees in many places. As the settler trees increased, the oaks and white pines would gradually dwindle to a few patches and single old veterans.[101] Toward the end of the life of the original white pine forest, isolated groves of giant pines would still be standing here and there waiting for the next great fire to help them recapture the land. They would also be an awe-inspiring surprise to travelers such as Timothy Dwight who ventured into the depths of the forest.

Jeremy Belknap's "noble" white pine fell to the colonist's ax too quickly for observers to write many accounts of the ancient forest. Even in Thoreau's day the second-growth forest that came in afterward was itself rare, and most forests consisted of a third generation of trees that followed a second cutting.[102] Nevertheless, a few remnants of the white pine forest remain, although they have changed substantially. However, we can also peer into the past and see the ancient forest in our mind's eye. Picturing for a moment what Captain George Waymouth saw in 1605 when he wrote his description: "And surely it did all resemble a stately Parke, wherein appear some old trees with high withered tops, and others flourishing with living greene boughs."[103] Ancient eastern white pine forests were spectacular, to be sure, but they were transient. However, the loss of giant white pines in one area made little difference. As long as the fires burned, and Indians cleared and then abandoned land, more giant trees replaced them in another area. It was still the same forest, an ever-changing mosaic of patches that, using Thoreau's words, "appeared to be all in motion."[104] An active landscape interwoven with the lives of the native peoples who called it home.

BEECH-MAPLE FOREST

Overpowering and awful is the solemn gloom.
James Buchanan, Esq., His Majesty's Consul for the State of New York, 1824[105]

The ancient beech-maple forest grew mainly in central Pennsylvania and New York (Fig. 10.4). It also grew in a belt along the middle slopes of the mountains wedged between the eastern white pine forests in the valleys and the spruce-fir forests above. This belt of beech-maple forest intermingled with other forests as it wound its way northeast through New York, western Massachusetts, Vermont, New Hampshire, and southern Maine. Arms of this forest also extended westward across southern Ontario, Michigan, and northeastern Wisconsin, as well as down through the Ohio Valley and into southern Indiana. Here its lower border followed a path that marks the southern extent of Ice Age glaciers. Parts of it also found their way along the upper slopes of the southern Appalachians.[106] The ancient beech-maple forest, and the closely related maple-basswood forest in Wisconsin and Minnesota, covered nearly 57 million acres in Canada and the United States.[107]

This is a very old forest. It has been in the Northeast for about 10,000 years and southern Michigan for 8000 years.[108] However, eastern hemlock did not become part of the beech-maple forest in New England until about 8000 years ago, and it joined the forest in southern Michigan 1000 years later.[109] The barrier created by Lake Michigan, and the Great Drought that followed the Ice Age, slowed the westward expansion of

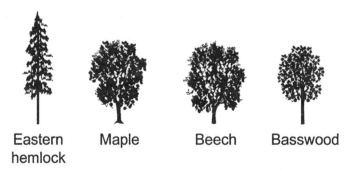

Eastern   Maple   Beech   Basswood
hemlock

Fig. 10.4   Trees of the ancient beech-maple forest.

beech and hemlock, so they came late to the Upper Peninsula of Michigan and northern Wisconsin. Beech is not plentiful there even today.[110]

The great age of this forest is fitting because it is also a self-replacing forest of mostly long-lived trees. Unlike ancient forests in North America that were reborn and revitalized by fire, the beech-maple forest went on with little change for millennia. Fires burned the forest, but only rarely. Light surface fires crept through the forest about once in 600 years, and severe fires only burned it once in 3000 years.[111] This was a wind forest not a fire forest. Even light surface fires can kill beech and maple trees. The beech-maple forest did not need fire to sustain itself, and it shrugged off the hurricanes and tornadoes that battered its trees.[112] It was truly a forest for the ages.

Dense forests such as this did not inspire romanticism among early travelers. A British officer summed up the general feeling in 1777 when he said that "one prodigious forest, bottomed in swamps and morasses, covered the whole face of the country."[113] These forests seemed even less attractive to James Buchanan, Esq., who was His Majesty's Consul for the State of New York in 1824. He thought they were "overpowering and awful." He said the "solemn gloom" gave him "a strange sensation of loneliness and inability to move in any direction without being immediately bewildered."[114]

The beech-maple forest was thick and dark during the summer when the trees and shrubs were in full leaf. The canopy blocked the sun, giving it the look of a dull green ceiling. Still, little openings here and there allowed flecks of light to shine through to the forest floor, like tiny spotlights. Thickets of wispy young trees barred the traveler's way, slapping and tearing at clothes; and logs and other debris cluttered the ground. During the winter, after the leaves had fallen, the sun washed the forest floor with light, but the trees stood naked in the snow, as if dead rather than sleeping. This was not the forest of poets, at least not during these seasons. Only in the brisk air of autumn did the beech-maple forest announce its presence. Flashes of color painted whole hillsides. Beech leaves turned yellow and the maples orange and red, but only briefly. They quickly changed to brown and gradually fluttered to earth, and the forest fell into the anonymity and silence of winter.

Many explorers and colonists knew the beech-maple forest well. However, the forest failed to produce as much admiration as the trees. Indians and Europeans alike used beech nuts for food and medicine.[115] Beech nuts fed the game they relied on as well, including black bear and deer, and many other animals such as ruffed grouse, ducks, and small mammals.[116] Indians also prized sugar maples for their sap. Paul Dudley of Massachusetts was the first to describe maple sugar in New England, although Indians taught the Canadians how to make it decades earlier. It was 1720, and Dudley wrote that "maple sugar is made from the juice of . . . maple trees that grow in the highlands." Reverend Samuel Hopkins, a missionary trying to save Indian souls in western Massachusetts, commented in 1753 that "the molasses that is made of this Sap is exceedingly good, and considerably resembles honey." Indians valued maple groves so highly that they belonged to families, handed down from generation to generation. They used maple sugar for many things, including seasoning their food like salt. The Iroquois even poured hot maple syrup over popcorn and called it "snow food."[117] In 1763, English fur trader Alexander Henry said that "I have known Indians to live wholly upon the same and become fat."[118] He was speaking of the Chippewa who gathered maple sap in the spring near Sault Ste. Marie, on the border of upper Michigan and Ontario.

The trees were useful, but beech and maple only reached a modest size, and hemlock and American basswood were just a little bigger. Sugar maple commonly grew 90–120 feet tall and 3 feet in diameter in the ancient forest, although some were larger. Old beech trees were about the same size, but their trunks did not grow as large as the biggest maples. These trees stood nearly straight in a dense forest, like pillars holding a roof. Still, a traveler could easily see the difference between them. Sugar maple bark is gray-brown, furrowed, and rough, while beech bark is blue-gray and smooth, like metal.[119] This difference hints at their individuality.

Beech and maple do not always live side by side even within the beech-maple forest. American beech can grow on poorer and wetter soils than sugar maple. So beech occurs in more places than maple.[120] They also go their separate ways in the South and the Great Lakes region. Beech drops out of the forest in eastern Wisconsin, where it becomes the maple-basswood forest that extends into central and western Wisconsin, and Minnesota. Likewise, sugar maple drops out on the lowlands in the South, and southern magnolia replaces it. This forms the beech-magnolia forest, which intermingles with southern pine forests.[121] Nevertheless, beech and sugar maple have grown together for thousands of years as if they were inseparable partners.

The partnership between beech and sugar maple endures because each tree can take advantage of slightly different conditions in a forest, and each creates a different understory environment for the other. Beech needs less light to grow than sugar maple, but both trees are very tolerant of shade. However, sugar maple blocks more sunlight than beech. Therefore, young beech trees grow taller than maple in the dense shade underneath sugar maple. That means that beech usually replaces large maple trees that fall. Young maple trees grow as well as beech under the less shady canopy of beech trees, and sugar maple grows faster than beech when light increases. So, sugar maple usually gets to the canopy first when beech trees fall, especially in the larger openings. Therefore, sugar maple usually replaces beech, and beech replaces maple, in a cycle that only ends when the next tornado or fire levels the forest.[122]

Eastern hemlock also grows within some parts of the beech-maple forest, particularly on sandy soils.[123] However, hemlock grows poorly under maple, but it does well under beech. Maple may exude a chemical that inhibits the growth of hemlock, but hemlock seeds are so small that they also have difficulty reaching moist soil underneath a layer of maple leaves. A dense blanket of maple seedlings also crowds out hemlock seedlings. Hemlock seedlings even have difficulty getting their roots to penetrate a thick layer of hemlock needles. So they tend to grow on rotten logs, and on mounds of bare soil where wind uproots trees.[124] Therefore, hemlock can replace hemlock, and it can replace beech, but it is not as successful at replacing maple.

Once hemlock invades a forest, or becomes established in a burned area, it usually takes over and keeps most other trees out, although it might take centuries.[125] Hemlock saplings can wait longer than any other tree in the forest for a larger tree to fall and open the canopy. Not only that, mature hemlock inhibits the growth of both maple and beech seedlings, most likely because hemlock litter contains acids and aluminum that leaches into the soil.[126] However, yellow birch is unaffected, but it is not tolerant of shade. So it often grows in the larger gaps that open among the hemlocks. In addition, hemlock seedlings can grow under hemlock.[127] Therefore, it is likely that hemlock will continue to occupy the same place in the forest until a windstorm or fire removes it and gives other trees a chance to grow.[128] Even then hemlock has a good chance of replacing itself because its seeds can germinate and survive on fresh burns.[129] Some hemlock forests have already stayed in the same place for 2000 years.[130] Nevertheless, heavy browsing by white-tailed deer, snowshoe hares, and New England cottontails can eliminate hemlock seedlings and give beech and maple a chance to replace the overstory.[131] They just have to wait for the overstory hemlock trees to die of old age, which can take 800 years.[132]

Understory seedlings and saplings seldom had to wait that long for at least some overstory trees to fall and release them from shade. It only took an average of 69 years for a windstorm to topple 10% or more of the overstory trees, and 30–50% of the trees probably fell at one time in stronger winds every 300–400 years. So few patches of trees escaped the effects of light to moderate windstorms during their lifetime. Still, things could become dramatically worse in 1500 years. By then, the roaring of a tornado or thunderstorm downburst was likely to brake the silence of the forest, as well as the trees.[133]

The ancient beech-maple forest was not a quiet and serene place where things changed only slowly. On the contrary, its peacefulness hid the brief but violent windstorms that pounded its trees. The sky changed from blue to black in an instant as a line of dark clouds slid over the forest like a shade. Trees began to sway and leaves rustled as in any storm, but then, without warning, a cold wind blasted straight down from a thundercloud and hit the ground at 160 miles an hour. It crushed the forest underneath, and splashed outward like water emptied from a giant bucket, flattening trees and throwing limbs over a wide area. Then the downburst ended. Peace returned, and the shattered forest slowly recovered from the onslaught.

Only two or three generations of trees had a chance to grow in the same place before being destroyed by a windstorm. These powerful winds tore through huge areas when they hit, wrecking as much as 15,000 acres of forest, although the average was about 400 acres.[134] Up to a third of the trees snapped off instead of falling.[135] This left splintered and decapitated trunks standing among the rubble of twisted and uprooted trees. A few scattered old trees from the former forest also stood upright within the debris, bruised and alone. Unless a fire cleaned up the mess, the landscape could look as if it had endured a heavy artillery bombardment for many years. R. D. Irving, a geologist, recalled the devastation that such a storm created in 1872:

> The traveling is rendered yet more difficult by the frequent areas of fallen timber, known as "windfalls." ... They are like great swaths having been prostrated by hurricanes, and are often sharply defined from the standing timber around. The trees at times lie all in one direction, but in other cases

are wholly without such arrangement, having been prostrated by a tornado. The great windfall of September, 1872 ... is as much as forty miles in length, though it is reported to have a greater length than this. It crosses the Chippewa river ... with a width of about one and one half miles ... when last traversed in 1877, this windfall had been partly burned over, but was, for the most part, made more impenetrable than ever on account of the new and dense growth of bushes and small trees. In crossing it ... it was found necessary to cut the way with an axe for nearly a mile and a half.[136]

Many small trees from the original beech-maple forest escape the destructive force of windstorms and remain alive among the jumble of fallen trees. However, dry dead trees burn more easily than an intact forest, so occasionally fire kills the young trees and converts a windfall to bare soil. Then the forest must begin again with pioneer species such as pine, aspen, birch, and basswood. Patches of hemlock may also appear in a burn.[137] It takes many centuries without fire or severe windstorms for a self-replacing beech-maple forest to reclaim the land.

If a windfall does not burn, the young trees that survived the storm become part of the new forest that replaces the old. Patches of maple saplings will spring upward between the fallen trunks of dead trees, and beech and hemlock saplings do the same. Clumps of basswood sprouts grow from the base of their parent trees that snapped in the wind. New seedlings of pioneers such as quaking aspen, paper birch, and yellow birch also find a few spots of bare soil from uprooted trees upon which to grow. However, quaking aspen and paper birch live only a short time, and they need plenty of light. So they gradually disappear as other trees grow taller and block the sun. Yellow birch declines in abundance as well, but it persists in gaps within the maturing forest. In short, the same trees come back, so the forest soon comes back about as it was.[138]

The ancient beech-maple forest did not burn easily, and wind created only temporary setbacks, so it held on to the land tenaciously. From a distance the forest looked like one vast blanket of old trees, but its patchy or mosaic structure revealed itself when seen up close. The patches of eastern hemlock scattered within the northern part of the forest became especially noticeable in the winter when the leafless hardwoods exposed their dark green silhouettes to view.[139] Small groups of yellow birch grew in some of the gaps that formed among the hemlocks. Elsewhere, sugar maple saplings grew in gaps, but many died by the time they reached the canopy. So old patches of maple were less conspicuous than the thicker young patches. The same was true of beech. American basswood formed tight clumps in parts of the forest because it grew primarily by sprouting from the base of large trees. A few isolated pockets of aspen and birch also existed in the forest where creeping surface fires burned.[140] Overall, the beech-maple forest formed a mosaic of very small patches, most covering only a fraction of an acre.[141] Scattered within it were a few very large openings filled with piles of dead trees where downbursts or tornadoes had ripped through the forest.

The patches that made up the ancient beech-maple forest mosaic also represented different stages of recovery from the last windstorm or occasional fire. Most likely it was similar to a maple-basswood forest where about 2% of the patches contained saplings, often with old wind-broken trees looming above them. Pole size trees dominated a little less than 11% of the patches, and mature trees dominated about 24%. Moreover, most of the patches of pole size and mature trees had seedlings or saplings growing underneath, although the hemlocks were often free of dense undergrowth. The majority of the patches, about 63%, consisted of large old trees, most of which had one or more layers of younger trees growing in the understory.[142]

The ancient beech-maple forest was exceptional. North America's ancient forests usually contained some patches of self-replacing forest, but few were dominated by them. Fewer still escaped the ravages of fire for thousands of years.[143] Still, the ancient beech-maple forest was not free of disturbance. Wind smashed its trees as often as fires blackened other forests, but it endured. Unlike the effects of severe fires, violent windstorms left enough young trees unharmed, even in very large patches, so that they could quickly replace those that fell. Less severe winds did the same thing on smaller areas. Even so, wind did little to change the forest. Old trees would die anyway and release understory trees from the shade that kept them small. Wind just made this happen faster. Europeans could not stop windstorms in this forest as they stopped fires from burning elsewhere. Therefore, some of the beech-maple forests that exist today still look like the ancient forests.

RED SPRUCE–FIR AND BALSAM FIR FORESTS

Land poor and timber small.
John Pierce, surveyor, 1775

Italian explorer Verrazzano's ship *La Dauphine*, alone and with only 50 crew members, bobbed in a calm sea and clear skies off the Maine coast in

May 1524. *La Normande*, the ship that was to accompany *La Dauphine* to America, had returned to France. Verrazzano was there on behalf of the king, Francis I, but French bankers paid for his voyage because they wanted to find a shorter route to Asia to lower shipping costs. The king only loaned him the ship. Verrazzano began his voyage in January 1524. He had already explored most of the coast from North Carolina northward, and he spent 2 weeks ashore with the Wampanoag Indians while anchored in Narragansett Bay. He said they were "the goodliest people, and of the fairest conditions, that we have found in this our voyage." He thought the Abnaki who lived in Maine would be the same. They were not. They shouted insults, displayed "all signs of discourtesy and disdain," and fired arrows at the explorers. This led Verrazzano to call Maine the "Land of Bad People." The Abnaki kept him from venturing inland, so Verrazzano only saw the coast. He thought it was bare, but "pleasant to the view," and the numerous small islands made "many fair harbors and channels." [144] "A high land," Verrazzano said, "full of very thick forests, the trees of which were pines, cypresses and such as grow in cold regions."[145]

What Verrazzano saw was a finger of the ancient red spruce–fir forest (Fig. 10.5) that extended along the coast of Maine. It separated the white pine and beech-maple forests that grew farther inland from the cold winds of the north Atlantic. This narrow band of coastal forest connected to the red spruce–fir forests in northern Maine, New Brunswick, and Nova Scotia.[146] It also covered the mountainous regions of western Maine, as well as southern Quebec and southeastern Ontario. It also grew high in the mountains of New York and eastern Massachusetts, and

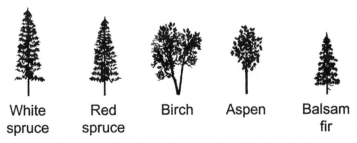

| White spruce | Red spruce | Birch | Aspen | Balsam fir |

Fig. 10.5   Trees of ancient red spruce–fir and balsam fir forests.

on a few mountain peaks in the southern Appalachians in southwestern Virginia, eastern Tennessee, and western North Carolina.[147] The ancient red spruce–fir forest covered approximately 60 million acres in Canada and the United States.[148]

The dark green appearance of ancient red spruce–fir forests, the mosses and lichens that covered the lower tree trunks and the ground, and the abundance of ferns made these forests look old. However, they were relatively young. Red spruce only started spreading from rocky areas along the northeast coast about 2000 years ago. This was several thousand years after the end of the Great Drought that followed the Ice Age. The climate became cooler and wetter during this period, which created the humidity that allowed red spruce to expand its range. Red spruce pushed aside the less shade-tolerant white spruce as it spread, and it mixed with balsam fir in the North and Fraser fir in the southern Appalachians.[149] However, red spruce made up a smaller proportion of the forest as it climbed higher into the mountains. Then it dropped out completely, and the forest changed to nearly pure fir. In New England, balsam fir trees became dwarfed in the harsh conditions near the summits. They finally dropped out as well, and alpine tundra capped the highest peaks. Thus, ancient red spruce–fir forests grew over large areas in northern Maine and southeastern Canada, and in belts along mountain slopes above beech-maple forests and below fir forests in New England and the Southeast.[150]

The red spruce–fir forest, and the fir forests above it, did not inspire awe among those who first saw them. On the contrary, early Vermont surveyors called it "black growth."[151] John Pierce, surveyor for General Arnold's military expedition to Quebec in 1775, described the "timber" in Maine as "not very good." To him the land was "poor" and the "timber small."[152] John Josselyn had an even less favorable view of the red spruce–fir forest. He said it was "daunting terrible," the land "being full of rocky hills, as thick as molehills in a meadow, and clothed with infinite thick woods."[153]

The trees were indeed small, and the forest was thick. In the southern red spruce–fir forest, Fraser fir rarely grows more that 80 feet tall and 2 feet in diameter. More often it is less than 60 feet tall and 1 foot in diameter, and it only lives about 150 years.[154] Balsam fir grows slightly larger in the northern forest, and it lives a little longer.[155] Red spruce is not much larger, although it can live 400 years. However, conditions are more favorable for the growth of red spruce in the southern Appalachians than in the North. There it can reach a height of 115 feet and a diameter of 4 feet.[156]

Ancient red spruce–fir forests depended on windstorms and severe fires to renew themselves in the Northeast, so the forest mosaic consisted mainly of large to very large patches.[157] Windstorms, along with insects, diseases, and landslides, were the primary forces that renewed red spruce–fir forests in the southern Appalachians because fires were rare.[158] Indian fires surely occurred in Appalachian forests as well, but they were probably less important here than elsewhere. Wind and ice substituted for fire in high-elevation balsam fir forests in New England.[159]

Southern Red Spruce–Fir Forest

Small windfalls were common in the southern Appalachians because spruce and fir are shallow rooted, and the soils are thin. However, only a few trees fell at a time, so the gaps were very small. Still, at any one time about 6% to as much as 17% of the ancient forest consisted of trees under 10 years old growing in gaps. Only about 1% of the southern red spruce–fir forest consisted of large patches.[160] Thus, unlike the red spruce–balsam fir forests of the North, the red spruce–Fraser fir forests in the southern Appalachians formed a mosaic of very small patches.

Red spruce, Fraser fir, and yellow birch share a partnership in the southern Appalachians that is similar in some ways to the partnership among beech, sugar maple, and yellow birch in the beech-maple forest. They can live together because of their differences. Red spruce and Fraser fir are tolerant of shade, but Fraser fir is slightly more tolerant.[161] In addition, Fraser fir produces more seedlings than red spruce. Consequently, Fraser fir makes up nearly 80% of understory seedlings and saplings. To compensate, young red spruce suffers less from insects and diseases than Fraser fir, so its mortality is lower and it can linger longer in the understory waiting for an opening in the canopy. Not only that, its life span is nearly twice as long as Fraser fir, so it can hold on to a patch much longer. Unlike either spruce or fir, yellow birch is intolerant of shade, but it grows faster in openings. Therefore, each species has relative advantages and disadvantages that balance each other.

Since gaps can form anywhere, and combinations of understory trees of various ages and densities change from place to place, all three species had a chance to grow in the ancient forest. This allowed them to coexist as long as small gaps continued to open, and a severe fire or windstorm did not destroy the forest. As a result, they formed an ever-changing yet stable forest mosaic. Roughly 56% of the patches were dominated by Fraser fir, 30% by red spruce, and 14% by yellow birch.[162] However, less than 23% of the ancient red spruce–Fraser fir forest consisted of patches in which the dominant trees were at least 100 years old.[163]

Northern Red Spruce–Fir Forest

The ancient red spruce–fir forests of New England and southeastern Canada differed dramatically from those in the southern Appalachian Mountains. Here fires were often big, and windstorms and insect outbreaks ferocious, so patches in the forest mosaic were large and sometimes very large. Surveyor John Richards worked in a forest of the "finest spruce, and yellow birch" in the Adirondaks in 1811. He said, "with very few exceptions, the wind has made great havoc among the timber in many places."[164] Samuel de Champlain experienced such a windstorm in a Canadian forest in February 1606. "The wind was so violent," he wrote, "that it blew over a large number of trees, roots and all, and broke off many others." "It was a remarkable sight."[165]

Severe windstorms and fires did not take their toll very often. The average area in northern Maine burned every 200–400 years, although some areas escaped a major fire for as long as 800 years. This includes Indian fires. Adding hurricanes and other large windfalls drops the average down to 244 years between severe disturbances. Regardless, many smaller fires and windfalls cleared parts of the forest during the quiet periods between these upheavals.[166]

Indians probably started some of these fires, even though they did not build permanent settlements in red spruce–fir forests high in the mountains. Peter Kalm called the Adirondack and Green Mountains of New York in 1749 "waste regions" where "no Indian villages are found." Indians lived mostly in farming villages along rivers and the coast, although they also cultivated the lowlands far up the Saco River in Maine. Even so, Kalm also pointed out that "in the autumn" the Indians "live here [in the mountains] for several months by hunting alone."[167] This means that some escaped campfires were inevitable, and hunting fires were likely. No doubt other Indian fires occasionally entered the forest from villages and fields on the lowlands, and from nearby pine and oak forests. Therefore, Indians surely increased the total number of fires, however slightly, by adding their fires to those that normally started by lightning.[168]

Some wildfires swept through the crowns of live trees, but most fires probably increased in intensity when they entered windfalls filled with piles of dead trees.[169] For example, in 1803 the Katahdin fire devastated over 200,000 acres of eastern white pine, beech-maple, and red spruce–fir forests in Maine.[170] J. Morse wrote in 1819 that these forests were "set on fire . . . by lightning, or by the carelessness of the Indian hunters." Regardless of how it started, the fire became a conflagration when it burned into a jumble of dead trees knocked down in a severe windstorm. J. Morse documented this when he wrote that "the trees on this extensive tract were first prostrated by some violent tempest, which happened about the year 1795."[171] Two more huge fires burned nearby in 1825. Together they blackened over 100,000 acres, but this time they consumed standing trees rather than windfalls.[172] Such mammoth fires were infrequent, and they tended to burn in the lowlands rather than the mountains. Even so, smaller fires also burned in the mountains, and they started more often. Still, many of them covered thousands of acres.[173]

Paper birch, aspen, and the small pin cherry, all pioneer species, were usually the first trees to invade a burn in this ancient forest. Red spruce and balsam fir soon followed. Both trees do equally well in large clearings, but those under 2 acres tend to favor balsam fir. These shade-tolerant settler trees stayed in the understory for as long as 50 years because birch and aspen grow faster. By then pin cherry disappeared, and birch and aspen were rapidly declining. Spruce and fir grew slowly during this period, and many of them were dying as well. The survivors were still only 30–45 feet tall and 4–8 inches in diameter when the forest reached 50 years of age. They formed a nearly impenetrable thicket of stems festooned with sharply pointed dead branches. Leaning dead trees made the forest seem even thicker and more unpleasant, and a jumble of fallen trees and branches cluttered the ground. It took 30 years more for enough trees to die so that the canopy opened a little. This gave the spruce and fir trees room to grow, and it let in just enough sunlight for a sparse ground cover of mosses, lichens, shrubs, and other small plants.[174]

The red spruce–fir forest took on a more mature look about 100 years after fire. Until this time it formed a huge and nearly uniform patch in which all the overstory trees were about the same age. Now windstorms were opening gaps in the forest where a second generation of young spruce and fir could grow, and perhaps yellow birch. Thus it began dissolving into a mosaic of small patches. The overstory trees were also much larger by now and mosses and other plants increased. Finally, it became an old transitional forest well on its way to becoming a self-replacing forest when it reached about 150 years of age. The trees were mature, the older balsam fir were declining, and thickets of young spruce and fir trees crowded numerous small openings. Spruce and fir seedlings also carpeted the ground under the larger trees, intermixed with shrubs and small plants. Mosses and lichens also became more evident. Still, pieces of birch bark littered the ground as silent testimony to the forest's fiery past.[175]

Fire, wind, and insects such as the spruce budworm, created "havoc" in ancient red spruce–balsam fir forests, but these forces also sustained a high diversity of plant and animal life. Moose, white-tailed deer, ruffed grouse, beaver, snowshoe hares, and other animals depend on aspen, birch, and other plants that invade fresh clearings. The gray or timber wolf and Canada lynx indirectly depend on such areas as well because they prey on these animals. On the other hand, martens prey on red squirrels and den in hollow trees or logs, so they need older forests in which to live. It takes a variety of habitats to support a variety of wildlife. Ancient red spruce–fir forest mosaics furnished the many habitats that animals required because they consisted of patches in nearly all possible stages of development, from freshly cleared or burned areas to self-replacing forest. The mosaic also helped to contain insect outbreaks by confining them primarily to those parts of a forest that contained the older and weaker trees.[176]

About 14% of the ancient red spruce–fir forest mosaic was composed of large or very large burns dominated by birch, aspen, and pin cherry. Over 1% was recent windfalls that might have burned a few years later. Thick young and middle-aged forests of red spruce and balsam fir, intermixed with dead trees and aging aspen and birch, dominated approximately 25–30% of the mosaic. The remainder, about 55–60% of the mosaic, contained old transitional and self-replacing forests in which the dominant red spruce and balsam fir trees were at least 150 years old.[177] These older forests were also riddled with gaps filled with trees of various ages and mixtures of species. Most of these older forests lasted only about 100 years before fire or wind swept them away to make room for new forests and more wildlife.[178]

High Mountain Balsam Fir Forest

The balsam fir forests that live near the tops of the mountains in New England share almost nothing in common with the red spruce–fir forests below. Fires very rarely burned these forests. Few people venture into these remote regions even today, and Indians probably shunned them as well. So these forests depend on wind and ice to renew themselves rather than fire, although debris avalanches and snow breakage also play a role on steep slopes in some areas.[179]

These mountains are cold, windy, and cloudy places that give forests a primeval look. This seems proper because balsam fir forests have existed here for over 10,000 years.[180] The interior of a balsam fir forest is dark, and the blistered and scaly tree trunks bristle with dead limbs. Adding to their primordial appearance are large crescent-shaped gashes in the forest filled with the scraggly, bleached skeletons of dead trees (Fig. 10.6). From a distance these openings look like gaping ashy-white wounds that slice through the dark green canopy in nearly parallel lines.

Fig. 10.6 Crescent-shaped waves of regeneration in the high mountain balsam fir forests of New England. (Courtesy Blackwell Science Ltd.; photograph by Douglas G. Sprugel)

The openings that cut through the balsam fir forest do not stay in one place. They usually march upslope with the prevailing winds, like ranks of soldiers slashing their way through the trees above. Each wave begins when a few trees die on the mountainside. This opens a hole in the forest canopy that allows wind to enter and start the process of expanding the hole and moving it uphill. It gradually takes on a crescent shape as wind funnels into the corners of the hole. Old balsam fir trees, weakened by age and disease, provide a poor defense against the wind on the uphill side of an opening. Strong gusts of wind knock down some of them. Others die when winter winds dry their needles, and their roots fail to replenish the moisture because of the frozen soil. These winds also quickly freeze water droplets that accumulate on the upper branches. This builds up a heavy layer of ice that bends and breaks the trees. Eventually all the old trees on the uphill side of the opening topple to the ground. This exposes the weak old trees on the uphill side to the full force of the wind and ice, so they succumb as well. This moving wave of destruction continues upward until it reaches the crest of the mountain.[181]

A wave of young balsam fir trees follows the wave of destruction that advances up the mountain. As an opening spreads uphill, a new generation of fir seedlings emerges from the jumble of dead trees left behind. Saplings grow in the older opening below it, and pole size trees grow in the even older opening below the saplings. The oldest trees grow even farther downhill. Eventually these old veterans fall victim to the destructive front of the next wave, and the cycle begins again.[182]

The balsam fir forests in the mountains of New England are relatively stable mosaics of moving waves or strips. The front of the wave destroys the forest ahead, and each strip behind the front is in a different stage of recovery. The farther back the strip, the more advanced is its recovery. The strips usually become larger on their way up the mountains because they merge toward the top. Hence, balsam fir forest mosaics consist of smaller patches on the lower slopes and larger ones above. However, unlike other forest mosaics, the patches are long and narrow, and they move and change at a relatively constant rate. In addition, the trees in this mosaic rarely become old. The waves move too quickly. Most patches of trees reach an age of only 50–60 years before they are destroyed by another wave.[183] Nevertheless, high mountain balsam fir forests are very old self-replacing forests. They have probably remained much the same for thousands of years.

ELEVEN

Forests of the Fathers

*And in regard to temporal provisions very little was needed to satisfy
a man who demands nothing but perpetual poverty, and who seeks
for nothing but heaven, not only for himself but also for his brethren.*
Samuel de Champlain, in reference to Father Joseph Le Caron's mission
to the Hurons
Quebec, 1615[1]

Italian navigator John Cabot was the first European to set foot in
New France, now Canada, other than possibly Leif Ericsson. It was
1497, just 5 years after Christopher Columbus' first voyage. However,
Cabot only saw the shores of Newfoundland, and he knew nothing of the
vast continent that lay beyond. The Portuguese explorers Gaspar Corte
Real and Joao Alvares Fagundes who followed him, as well as the Italian explorer Giovanni da Verrazzano, saw little more than Cabot. It took
another 37 years before Jacques Cartier's two French ships anchored off
the coast of Newfoundland. He scouted most of the shoreline and then
sailed westward into the Gulf of St. Lawrence.[2]

Cartier explored the Gulf of St. Lawrence during the summer of 1534
and traded with the Indians, who he said were "untamed and savage,"
but friendly. He found that some parts of the coast were "composed of
stones and frightful rocks and . . . not one cartload of earth."[3] Elsewhere
he said the trees were "as excellent for making masts for ships of three
hundred tons and more, as it is possible to find."[4] Cartier's ships were
much smaller, only about 60 tons.[5] Cartier returned the next year and
sailed even farther up the St. Lawrence River, where he found Indian villages on sites that later became Quebec and Montreal.

Explorations into the interior of Canada, the Great Lakes, and the

Mississippi River Valley started after Samuel de Champlain founded Quebec in 1608. Just 2 years later Champlain's lieutenant Etienne Brûlé became the first European to paddle a canoe across the waters of Lakes Ontario, Huron, and Superior.[6] Soldiers, explorers, and the colorful *coureurs des bois* and *voyageurs* seeking furs and adventure, arrived soon afterward. Many of their names, such as Cadillac, Du Lhut, Joliet, La Salle, Mackenzie, McLean, Nicolet, Nicollet, Henry, Radisson, Schoolcraft, Thompson, Tonty, and Varennes, as well as Champlain and Brûlé, are permanently etched into the early history of America's northern and central forests. However, no less important, and far less celebrated, were the "black robes," the Jesuit priests who often gave their lives to save native souls.

The first black robes arrived in Quebec in 1615. Among these, Father Joseph Le Caron was the first to venture into America's ancient forests as a missionary.[7] Many priests followed him deep into the interior of North America. One of the most revered black robes was Father Jean de Brébeuf who went to live among the Huron in 1625. The black robes suffered hardships that few people today could imagine, such as Father Brébeuf who became a martyred saint when the Iroquois burned him at the stake in 1649. These Jesuit priests, and those of other Catholic orders who joined them, were devout, humane, and literate. They wrote mostly about the Indians and less about forests because that was their work. However, due to their courage and sacrifice we have enduring images of America's ancient forests and the people who lived there. We should remember at least some of their names because they are now part of the forests of the fathers. Besides Fathers Brébeuf and Le Caron, they include Fathers Allouez, Charlevoix, Dablon, Dreuillettes, Garreau, Hennepin, Jogues, Marquette, Membré, Menard, Rasles, Raymbault, and Silvy.[8]

These explorers, soldiers, traders, and priests wandered through ancient forests that stretched across central Canada, surrounded the Upper Great Lakes, and spread southward along the western edge of the Mississippi River. This vast territory encompassed America's largest ancient forest, the white spruce forest of Canada and Alaska. It also included the Great Lakes pine forests and the oak-hickory forest. The beech-maple and maple-basswood forests were also part of the forests of the fathers, but they were an extension of colonial forests and mostly avoided by early explorers.

WHITE SPRUCE FOREST

We are here surrounded with the vastest woods in the whole world; in all appearance, they are as ancient as the world itself.
Father Pierre de Charlevoix, explorer and scholar, 1721[9]

The ancient white spruce forest dwarfed all other forests in North America (Fig. 11.1). Not because the trees were huge but because the forest was vast. It spread across Canada and Alaska in a wide curving band that stretched from the Atlantic Ocean to the Pacific Ocean. Even so, it was only part of an even larger belt of northern or boreal forest that encircled the arctic. Today this mammoth forest covers over 500 million acres in Canada alone. The open white spruce–black spruce parkland that separates it from the arctic tundra to the north covers an additional 253 million acres in Canada. Including Alaska, these two forests cover over 900 million acres.[10]

This is a very old forest because the trees assembled while the continental ice sheets were still retreating. White spruce trees led the way northward, but black spruce, jack pine, balsam fir, tamarack, aspen, and birch quickly joined them to form a nearly modern forest. This forest appeared in southern Canada soon after the ice sheets crossed the border 9500 years ago. It expanded steadily until it reached its full extent about 4000 years ago.[11] Therefore, the first European explorers saw a white spruce forest that looked much like the forest that existed when Ice Age glaciers still covered much of Canada.

The ancient white spruce forest probably seemed nearly as primitive and forbidding to European explorers as the Ice Age spruce forest would have seemed. To the Jesuit fathers it was "a region horrible with forests."[12] However, some of the trees of the early Ice Age were missing, such as limber pine, but jack pine took its place. Giant short-faced bears and other menacing Ice Age beasts no longer prowled the forest either, but grizzly bears were nearly as ferocious and packs of wolves still skulked in the blackness beyond campfires. In 1793, Alexander Mackenzie camped on the east slope of the Canadian Rockies when a wolf crept into a nearby Indian village. Mackenzie wrote a brief account of the incident: "On the 22d [of February] a wolf was so bold as to venture among the Indian lodges, and was very near carrying off a child."[13] The swarms of mosquitoes, black flies, and moose and deer flies that made life miserable

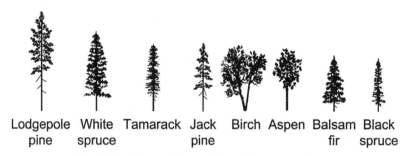

| Lodgepole pine | White spruce | Tamarack | Jack pine | Birch | Aspen | Balsam fir | Black spruce |

Fig. 11.1   Trees of the ancient white spruce forest.

for Paleoindians also plagued modern Indians and European explorers. Fur trader and explorer John McLean wrote that "the natives complain as loudly of the mosquitoes as the whites."[14] Surely the snow and advancing mountain glaciers of the Little Ice Age that was going on at the time enhanced the primeval character of the scene.[15] The ancient white spruce forest that Europeans first entered was a true remnant of an Ice Age world.

This is a forest of cold regions, and Champlain thought it was "far from attractive" when he saw it in 1615.[16] Here in the subarctic, rainfall is low and the growing season is short, even though the day is longer than it is farther south. Ice Age glaciers scraped away much of the land, so soils are generally thin and low in nutrients, and permafrost lies just below the surface in some areas. The white spruce forest seems to fit this rigorous environment. Trees do not reach the grand size that was common in many ancient forests. Most of the conifers are small, dark green, and cone-shaped, which helps them to shed snow and ice in the winter. Tamarack is also a small conifer, but tufts of blue-green needles give it a wispy look, and its needles change to bright yellow and drop in the fall. Aspen, birch, and balsam poplar, the three main hardwoods, are also small trees. Trees grow thick in the southern part of the forest where it is warmer and the soils are deeper. Still, white spruce only averages about 80 feet in height throughout Canada. The trees spread out and grow progressively smaller toward the north. Low evergreen shrubs, herbs, mosses, and lichens dominate the forest understory.

A dense network of rivers and lakes dissect the land, and it stays wet and marshy despite the low rainfall. It remains wet because of the generally level terrain, and the cool summers that reduce evaporation.[17] About 1856, Charles Lanman, an adventurer, described the land of the Chippewas around Lake Winnipeg in Manitoba. He said it was "covered with a stunted growth of trees, where the birch, the elm, the pine, and the spruce mostly predominate." "The surrounding country is a dead level," Lanman added, and "composed of continuous woods, which are everywhere interspersed with lakes and rice swamps."[18]

Botanist David Douglas traveled through the white spruce forest in June 1827. "The journey admits of little variety," he wrote in his journal. He could only say the forest consisted of "low thick marshy woods."[19] Frederick Graham felt the same way as he traveled along the Winnipeg River with a Hudson's Bay Company expedition in 1847. "Trees, trees, trees," he wrote in his diary, "I shall hate the sight of a wood for the future."[20]

Douglas dealt with boredom by listing the various tree species he saw along the way. Champlain and most other European explorers found nothing very remarkable about the white spruce forest either, so they also contented themselves with listing tree species. However, they spent most of their time writing about Indians and the rigors of travel rather than the forest. Occasionally they would add a little detail to their descriptions. McLean wrote that parts of Labrador were "remarkably well wooded." He also said the forests "abound in martens," which made them attractive to McLean because he was a fur trader. Black bear, timber wolf, lynx, and wolverine also roamed these ancient forests.[21]

To be sure, the ancient white spruce forest was somewhat simple. It contained only a handful of tree species and the landscape was relatively flat, except where the forest climbed the slopes of the Rocky Mountains in the West and the Laurentian Mountains in the East. Ice Age glaciers had leveled the land, but they also piled up low ridges of debris and carved out depressions that later became lakes and bogs. This added variety to the landscape by providing different conditions in which trees could grow.

The ancient white spruce forest was far more diverse than the casual observer might have thought. Father Pierre Biard said in 1612 that the "country ... [was] covered on every side by one continuous forest."[22] He overlooked how slight variations in the land and differences among tree species provided diversity by contributing to the mosaic structure of the forest. Here, as elsewhere, trees sorted themselves into areas that gave them the best opportunity for growth. White spruce grew mostly along river terraces and on uplands where the soils were not too wet. Jack pine grew on the driest soils, while black spruce and tamarack occupied the cooler north-facing slopes, and wetter and more poorly drained areas. These patches of black spruce–tamarack forest were often thick, in part because the lower limbs of black spruce can take root and grow into new trees when they touch the ground. Still, white and black spruce could also form open, parklike forests on dry upland sites. Paper birch and quaking aspen grew throughout the white spruce forest because they thrive on a wide variety of soils. Balsam poplar also grows on a variety of soils, but it tended to concentrate on bottomlands near streams, and along the shores of swamps and lakes.[23]

Lodgepole pine mixed with the white spruce forest when it approached the Rocky Mountains, while jack pine and balsam fir gradually dropped out. Jack pine and balsam fir failed to reach the Yukon Territory and Alaska because they could not migrate over the mountains at the end of the Ice Age.[24] The forest became more open as it approached the Canadian Plains toward the south in Manitoba, Saskatchewan, and Alberta. Likewise, it became more open as it blended into the white spruce–black spruce parkland to the north, and then the arctic tundra. McLean described the scene on the northern edge of the forest in 1838. "The country [is] flat," he said, "yielding dwarf pine intermixed with larch." Then the trees dwindled away as he traveled farther northward, just as they do at timberline in the mountains. McLean wrote of the landscape beyond the forest:

> Pursued our route over extensive swamps and small lakes, where there is scarcely any wood to be seen. The face of the surrounding country being level, the least elevation commands a most extensive view; but the eye turns away ... from the cheerless prospect which the desolate flats present.[25]

The landscape was more inviting on the southern side of the forest. Mackenzie followed the edge between the ancient white spruce forest and the aspen parkland that separated it from the plains as he traveled by canoe through Alberta in 1793. "The East side of the river consists of a range of high land covered with the white spruce and the soft birch," he wrote, "while the banks abound with the alder and the willow."[26] When he turned toward "the West side of the river" the land changed dramatically. Here was "the most beautiful scenery I had ever beheld," wrote Mackenzie:

> This magnificent theatre of nature has all the decorations which the trees

and animals of the country can afford it: groves of poplars in every shape vary the scene; and their intervals are enlivened with vast herds of elks and buffaloes: the former choosing the steeps and uplands, and the latter preferring the plains. At this time the buffaloes were attended with their young ones who were frisking about them: and it appeared that the elks would soon exhibit the same enlivening circumstance. The whole country displayed an exuberant verdure.[27]

Indian and lightning fires played an important role in maintaining the diversity of the white spruce forest.[28] As early as 1661, Jesuit priests were reporting huge fires in these forests, and muddy rivers that carried eroding soil from the burns.[29] Mackenzie also saw a large fire the day after he described his "magnificent theatre of nature."[30] Even today a single wildfire can sweep across thousands and sometimes over a million acres of Canadian spruce forest.[31] In 1855, a fire burned about 1.3 million acres between Lake Timiskaming in southeastern Ontario and Michipicoten on Lake Superior.[32] Smoke from a fire like this could blot out the sun. On one occasion in 1785, smoke darkened the sky over a 58,000 square mile area east of Montreal. People called these "Dark Days."[33]

David Thompson, a fur trader who discovered a navigable route from the Canadian Rockies to the Pacific, described the patchwork of burned and unburned forest that these fires created. He was traveling along the eastern foothills of the Rocky Mountains in Alberta. It was early winter in 1810, and he was heading north from the Saskatchewan to the Athabaska River. Thompson made his usual terse journal notes during the next 4 weeks of his journey:

> This Course had several banks to go up & down, but the Road on the whole tolerable for the Country. Woods green Pine & Aspin—latter part burnt Woods. We then turned to the left to avoid the marshes & went to a Hill of green Woods ... then over the Hill ... to a Brook, up along which we went, all burnt Woods ... in a wet willow Plain. ... We set off and cleared the Road, in doing which we had much trouble from the closeness of the Woods. ... Baptiste went off a hunting & to see how the Land lies before us ... the appearance of the Country is against us. He returned in the Evening hav killed a Bull, thank Heaven. ... We went West ... to a Rill of Water in a great Morass with burnt Woods every where as usual. ... A Part of the Rocky Mountains SSE of us. ... We have a hill in front of us all green Woods, to the left pt [point] of which we bend our Co [course] ... thro' mossy Wood & Morass. ... In the last Co we crossed & re-crossed the Brook often as the burnt Woods & Ground form a bad

Country. . . . We then held on over Land—hav at times much fallen Wood &
at times small Plains to the river, which we again left & held on thro' thick
Woods. . . . There are no animals of any Kind, tho' the Woods have never
been burnt. . . . We have now much burnt Woods in Places, but also much
green Woods. . . . We followed down the Current . . . where we camped on
the Banks of the Brook . . . from hence, on climbing a Tree, we see the
Athabaska River.[34]

Ancient white spruce forests were ideal places for fire. Thick, lichen-
draped tree branches hung near the ground, and shrubs and debris
clogged the understory. Red squirrels also tended to pile cone scales at the
base of the trees after opening them for their seeds. Such fuel produced
raging crown fires and hot surface fires that burned with explosive force,
especially when they passed through balsam fir trees that died in a spruce
budworm outbreak.[35] In a strong wind the flames traveled faster than a
person could run. Lightning probably started the majority of these fires,
but Indians increased the number of fires and they changed the time of
year in which some of them burned.[36]

Ancient white spruce forests usually burned before they reached 100
years of age, and very few patches of trees escaped the flames for another
century. Fires occurred even more frequently in the dry interior of Alaska
and northwest Canada. They were less frequent in eastern Ontario and
Quebec where rainfall was higher. Still, it took only a few massive fires to
burn several million acres each year. The largest fires swept across the level
terrain east of the Rocky Mountains. Fires did not reach such a massive
size in the mountains because they tended to move upslope into rocky
ridges. However, most fires burned 100 acres or less, even on level terrain. So, the chaotic behavior of fire created an ancient white spruce forest
mosaic that consisted of very large patches intermixed with many smaller
ones. These patches also had sharply defined boundaries because of the
intense heat of the fires. In keeping with the great size of this forest, some
of its patches were probably the largest in North America.[37]

Each patch in the ancient forest mosaic represented a different stage
of recovery after the last fire, from recent burns to old trees approaching the end of their life span. Blackened dead trees stood among charred
logs and ash in newly burned areas. However, fires burn unevenly because
the heat they produce creates strong winds that swirl flames erratically
through the forest. Occasionally, flames will spin like a tornado, shatter-

ing trees and scattering them in huge circles. Strips of trees with brown scorched needles also wound through the dead trees, although most of them would soon die. Islands of green trees also rose above the carnage here and there, perhaps spared from the searing heat by a shift in the wind, or the intervention of a lake or a ridge.[38]

Fresh burns only remained barren for a short time even though they changed in slightly different ways. White spruce might even replace itself in places where a fire burned hot enough to kill the roots of sprouting shrubs and hardwoods.[39] However, patches of spruce usually went through a similar sequence of stages when they recovered from fire. Grasses, sedges, and other herbs quickly sprang up from the ashes. Just a few weeks after a fire new shoots sprouted from roots that survived the heat, such as aspen, as well as shrubs such as willow, highbush cranberry, and blueberry. Fireweed and birch seedlings appeared a little later. White spruce or balsam fir seedlings only showed up in fresh burns if some of the trees survived nearby, and then only if a good seed crop occurred that year. Balsam fir invaded spruce forests more toward the east, particularly in Quebec. Regardless, by the end of the first year the land turned green again, and the scorched trunks of the former forest poked through a canopy of 10-foot-tall aspen sprouts.[40]

Thickets of young aspen and birch saplings flourished on 1- to 5-year-old burns. The pink blossoms of fireweed added color to these young pioneer forests, raspberries and highbush cranberries grew thick, and mosses and reindeer lichen already covered about a third of the ground. Before the fire, a dark, quiet forest covered this patch of ground, but now it was bustling with life. Even the trees seemed more alive with their whitish trunks and yellow fall color. Snowshoe hares thrived. They found plenty of succulent grasses, forbs, and small shrubs to eat during the summer, and thickets of saplings provided protection from predators. Lynx benefited as well since they prey almost exclusively on snowshoe hares. Small numbers of woodland caribou also moved into these patches to take advantage of the increased food, as did moose, bears, and sharp-tailed grouse.[41] The high concentration of game and berries in these recent burns made them especially attractive to Indians.[42]

This changed when the fur traders arrived. Indians found an adequate amount of game and other food in these fresh burns when they lived alone. Starvation still followed them everywhere, especially in the winter. However, it worsened when Europeans brought their guns and began to share in the limited amount of food that this cold country could provide.[43] Fur traders even ate the snowshoe hares upon which some Indians relied. "These tribes clothe themselves with the skins of rabbits and feed on their flesh," McLean wrote while traveling along the east side of the Rockies in northern Alberta. However, he said, "when the rabbits fail, they are reduced to the greatest distress both for food and raiment."[44] Hare populations failed now and then even before the traders came, but European hunters reduced them further during low years. Still, McLean said, the region around Fort Good Hope near Great Bear Lake was "by far the richest in furs of any in the country" in 1844, but game was less plentiful. He knew why:

> At the first arrival of the Europeans, large animals, especially moose and wood rein-deer, were abundant everywhere. In those times the resources of the district were adequate to the supply of provisions for every purpose; whereas, of late years, we have been under the necessity of applying for assistance to other districts.[45]

Fireweed and other herbs, willows, and reindeer lichen reached their peak in areas that burned about 30 years earlier. Then they started declining. The moose population also peaked about this time, and beavers trickled in to use the young aspens and willows for building materials and food in places that were near streams and lakes.[46]

Places that burned 50 years earlier began to look like forests again because the aspen and birch trees reached a height of about 30 feet. Berries increased as the low, creeping mountain cranberry steadily spread along the forest floor. This was now a middle-aged pioneer forest even though it was young in years. It would become an old pioneer forest and then an old transitional forest just as quickly. Thus, aspen declined as a patch became more mature due to its short life span. Reindeer lichen diminished as well, at least in denser forests. It kept increasing in more open forests. Birch still prospered, and young white spruce trees began to invade the patch and push their way into the canopy. Balsam fir often invaded the understory with white spruce in the East.[47]

Places that burned 150 or more years earlier usually consisted of a few old paper birch trees that managed to survive among the crowded trunks of spruce and fir. Soon the pioneer birch trees would disappear, although some might keeping sprouting. This was now a self-replacing forest. The decaying logs that crisscrossed the forest floor rested on a mat of feathermosses, and mountain cranberries lay over much of the ground. Balsam fir seedlings might also carpet the ground in eastern forests. Tree lichens hung from spruce limbs again, making this ideal winter habitat for caribou. Snowshoe hares also sought protection here in the winter. Gone were the many plants and animals that thrived in younger forests, but these patches of self-replacing forest still provided a home for martens and spruce grouse.[48] Even so, the forest started to look old and ragged. The heavy ground cover and dense shrubs made it difficult for new trees to enter the forest by this time. So it became more open as old trees died and shrubs grew thicker. It would take another fire to restore the forest's youthful vigor and abundance.[49]

Black spruce forests progressed through a similar series of successional stages. However, black spruce differs from white spruce because it has an abundance of seed available every year. Sap tightly seals some of their cones, which protects the seeds from fire and keeps them viable for many years. They normally stay sealed until the heat of a fire melts the sap and opens the cones. Therefore, black spruce often invaded fresh burns with aspen and birch. Even so, the faster growing hardwoods could quickly overtop them. Nevertheless, black spruce can survive in shade, and it lives much longer than aspen and birch. Therefore, it eventually became the dominant tree in self-replacing forests.[50]

Patches of self-replacing spruce lived only a fleeting existence in the ancient forest. Most of them never had a chance to become old before they burned again.[51] Therefore, only a small part of the ancient white spruce mosaic consisted of self-replacing forests, probably less than 25%. This was mostly a forest of pioneers.[52]

Frequent hot fires set by lightning and Indians kept the ancient white spruce forest in a constant state of renewal. This made forests more diverse than they would have been, given that only a few species of trees and shrubs grew there. The mix of different patches also remained relatively stable, even though they shifted around the forest mosaic as fires destroyed them in some places while they recovered in others. These shifting mosaics also provided the variety of habitats that wildlife requires. Caribou could graze and browse in the openings during summer and take refuge in an older forest in winter. Snowshoe hares and many other animals did the same. Each animal had different needs, but nearly all of them required more than one stage of forest development to survive.

They had to have a mosaic of different forests. The many fires that burned the ancient white spruce forest every year sustained that mosaic, and the wildlife and the Indians that depended on it.[53]

GREAT LAKES PINE FORESTS

The timber land has all the beauty of a sylvan grove.
C. C. Andrews, a traveler, 1856[54]

Ancient Great Lakes pine forests concentrated in the northern half of Michigan, Wisconsin, Minnesota, and southwestern Ontario (Fig. 11.2). They probably represented about two-thirds of the 16 million acres of eastern white pine, red pine, and jack pine forests in the United States.[55] The rest were in the Northeast. Oak, hemlock, and beech, were not as prominent here as they were in northeastern pine forests. Hemlock and beech grew around the Great Lakes, but they were on the western edge of their range. Beech dropped out of the Great Lakes pine forests beyond the eastern edge of Wisconsin, and hemlock became increasingly scarce beyond western Wisconsin.[56] Therefore, few of these pine forests resembled the ancient white pine forests in the Northeast. Those that did were mainly in northern lower Michigan. Therefore, ancient Great Lakes pine forests had a distinctive character molded by the species present, the effects of Ice Age glaciers, and thousands of years of Indian and lightning fires.[57]

The mammoth glaciers of the Ice Age sculpted the land of the Great Lakes pine forests. This country lay buried beneath a mile high sheet of ice for tens of thousands of years. However, the ice sheet never ceased moving. It scraped back and forth constantly, pushing away the soil and grinding the rock below. The glacier left nothing behind but a boulder-strewn landscape intersected by sandy ridges and exposed bedrock when it finally slid back into Canada 9500 years ago.[58]

Great floods of meltwater poured from the retreating glacier as it shrank in the hot sun, which washed away soil, and spread sand and gravel over the land. The meltwater gradually merged to become a raging torrent that gouged out the Mississippi River Valley. Floodwaters also formed gigantic Lake Agassiz below the ice wall at the front of the glacier.[59] The sediments that gradually filled in the lake formed a boggy soil over much of northeastern Minnesota and southern Manitoba.[60] Water also filled

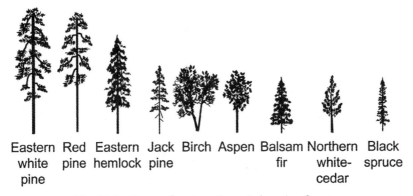

| Eastern white pine | Red pine | Eastern hemlock | Jack pine | Birch | Aspen | Balsam fir | Northern white-cedar | Black spruce |

Fig. 11.2 Trees of ancient Great Lakes pine forests.

the countless channels and basins carved by the glaciers. This created the thousands of crystal lakes, rivers, and wetlands that would one day characterize the ancient Great Lakes pine forests, and influence the cultures of the people who lived there.

Ancient Great Lakes pine forests went through a series of changes as they assembled at the end of the Ice Age. Jack pine and red pine came to this region from the Southeast soon after the glaciers retreated. Balsam fir was there even earlier, and some spruce and tamarack also remained behind while the Ice Age spruce forest migrated northward.[61] Aspen and birch came later.[62] Eastern white pine joined them in Minnesota about 7200 years ago, and eastern hemlock appeared about 1000 years later. Then a disease devastated the hemlocks, and they did not recover until about 2800 years ago. When hemlock, a settler species, spread through parts of the Great Lakes pine forests the second time, it must have caused a major change in the places where it grew.[63] No doubt a similar disruption occurred in the white pine forests farther east. Not only that, white pine was still advancing westward through Minnesota when European explorers first arrived.[64] Even so, the immense fires that swept the Great Plains every year would have kept white pine forests from expanding much beyond where they were when Europeans first saw them.

Eastern white pine is the largest tree in the Great Lakes region. This is "the monarch of our forests," wrote J. W. Foster in his 1869 geography of the Mississippi Valley.[65] However, the white pines he saw growing in the thin, rocky soil of northern Minnesota were not even as large as those that grew in the deep, sandy soil of Michigan and the Northeast.[66] Red pines were nearly as large as white pines, but they commonly grew 80 feet tall with a diameter of 3 feet. Still, red pine reached a height of 150 feet and a diameter of 5 feet in some places. Red pine could also live nearly as long as eastern white pine, achieving ages of up to 400 years.[67]

Jack pine is a scraggly tree that only reaches a height of 100 feet and a diameter of 2 feet, but most trees are much smaller. It also shows signs of age at a mere 75 years, and it seldom lives longer than 200 years. Except for eastern hemlock, most of the other trees in the ancient Great Lakes pine forest are also small when compared to red and white pine, such as spruce, tamarack, balsam fir, and northern white-cedar.[68]

The three main pine trees in these forests are all pioneers that do well on a glacially scarred landscape. Nevertheless, they tend to sort themselves in their tolerance to shade and their preference for soils in a way that mirrors their size. The smallest tree, jack pine, is the least tolerant of shade and grows on the driest, sandiest soils. Jack pine even grows on sand dunes. The larger red pine is a little more tolerant of shade, and it grows on slightly better soils that have more moisture. Eastern white pine, the largest tree, is the most tolerant of shade, although it still requires a considerable amount of sunlight to grow well, and it prefers the better soils. Nevertheless, it is most abundant on sandy soils where it has less competition from other trees. Hemlock prefers about the same soils as eastern white pine. Northern trees such as white spruce, black spruce, tamarack, and balsam fir grow on the wetter soils because the Great Lakes pine forests lie at the southern end of their range.[69]

The many lakes and streams that intersected the land made this canoe country, but Indians had another name for this area. They called it "Mosquito Country." John Carver, a New Englander looking for adventure in the late 1700s, said, "It was most justly named." "I never saw or felt so many of those insects in my life," he noted. Perhaps that is why Carver wrote so little about the forests he saw. He only commented that, "the country . . . is very uneven and thickly covered with woods."[70]

Hills, valleys, river floodplains, swamps, and other features of the landscape, together with fires and windstorms, created a variety of places for trees to grow. This expressed itself in the diversity of forests that European explorers described. Alexander Henry was the first English fur trader to enter the Great Lakes pine forests on the border of Canada and the United States in northern Minnesota. He jotted comments in his journal about some of these forests as he paddled and carried his canoe westward from lake to lake on a fur trading expedition.

Henry began his canoe trip through the region in June 1766. He entered by way of the Grand Portage on the western shore of Lake Superior, near Isle Royal. It connected the lake with the Pigeon River in southwestern Ontario. "The grand portage, or great carrying-place," Henry wrote in his account of the adventure, "consists in two ridges of land, between which is a deep glen or valley with good meadow lands, and a broad stream of water." The "lowlands are covered chiefly with birch and poplar, and the high with pine," he said.[71] The Grand Portage had 16 pauses, or resting places, spaced about a half mile apart. It took Henry "seven days of severe and dangerous exertion" to make the portage.[72]

Henry and his men traveled up Pigeon River, known as "River aux Groseilles" in his day, to another portage. "We left the Groseilles," he said, "and carrying our canoes and merchandise for three miles over a mountain [we] came at length to a small lake." Henry then described the region:

> This was the beginning of a chain of lakes extending for fifteen leagues and separated by carrying-places of from half a mile to three miles in length. . . . The region of the lakes is called the Hauteur de Terre, or Land's Height. It is an elevated tract of country, not inclining in any direction, and diversified on its surface with small hills. The wood is abundant but consists principally in birch, pine, spruce, fir and a small quantity of maple.[73]

Another young Englishman, Frederick Graham, traveled this route with a Hudson's Bay Company expedition 60 years later. He wrote an account of how Henry's *voyageurs* must have looked as they paddled their heavily ladened canoes 16 hours a day in the heat of summer. They seemed like "fiends," Graham said, "their shirts off, their skin like heated copper, and their long black hair all loose, with their wild black eyes glowing like hot coals."[74]

These hard-working *voyageurs* navigated their canoes through a labyrinth of lakes and rivers by following a route marked with "lob trees." Traders on earlier voyages usually chose tall pine trees that stood on points of land or islands along the canoe trail for their landmarks. Then they climbed them and lopped off the branches in the middle of the crown to make a knob of foliage at the top. This gave lob trees a distinctive shape that travelers could see from a great distance, and the trees kept this shape for many years. Some of these lob trees still stand along the canoe route Henry traveled.[75]

Henry paddled a "distant sixty leagues from the Grand Portage" before reaching "Lake Sagunac, or Saginaga [Nequaquon Lake]." "There was formerly a large village of the Chipewa here, now destroyed by the Nadowessies," he wrote. Even at this early date, fur traders and explorers often commented on the decline of Indian populations due to disease, liquor, and war with tribes that colonists had pushed westward. "When populous," he said, "this village used to be troublesome to the traders, obstructing their voyages and extorting liquor and other articles." Then Henry briefly mentioned the landscape: "The lands, which are everywhere covered with spruce, are hilly on the southwest but on the northeast more level."[76]

"We now entered Lake à la Pluie [Rainy Lake]," Henry recorded, "its banks are covered with maple and birch."[77] The river he traveled from Rainy Lake was "forty leagues long, of a gentle current, and broken only by one rapid." "Its banks are level to a great distance," Henry noted, "which was covered with luxuriant grass." This was most likely wild rice. The scene along the river moved Henry to make a comment not typical of his usually dispassionate accounts. He said "they were perfect solitudes ... I was greatly struck with the beauty of the stream." No doubt the laughing and yodeling calls of the common loon intensified his feeling of solitude, especially in the blackness of night when it made a haunting *ha-oo-oo* sound.[78]

Alexander Henry only skirted the upper edge of the Great Lakes pine forests in Minnesota. Henry Schoolcraft penetrated deep into the interior. Schoolcraft was a geologist who accompanied Lewis Cass, governor of the Michigan Territory, on an expedition into the region in July 1820. Naturally Schoolcraft focused on geology, but he recorded a few notes on forests as well. Governor Cass planned to reach the source of the Mississippi River during this part of the expedition. They traveled along the southern side of Lake Superior and pulled their birch bark canoes ashore at the western tip of the lake. The St. Louis River empties into the lake here, and it is now the site of two cities that straddle the river—Duluth, Minnesota, and Superior, Wisconsin.

Schoolcraft said that the river became "one continued chain of rapids and falls," as the party paddled west.[79] So they had to make numerous portages. During one of them they lugged their canoes through an ancient white pine-hemlock forest. This was probably an old pioneer forest that developed when both white pine and hemlock invaded the area after fire destroyed an earlier forest. No doubt it was also thick since hemlock can continue to invade a forest even when there is too little light for white pine. Schoolcraft tried to describe this forest, but fatigue and bad weather colored his impressions:

> Every thing around us wears a wild and sterile aspect, and the extreme ruggedness of the country—the succession of swampy grounds, and rocky precipices—the dark forest of hemlock and pines which overshadow the soil—and the distant roaring of the river ... render it a gloomy and dismal scene.[80]

Governor Cass decided that they had to leave the St. Louis River. So their Chippewa guides led him and half of the party overland to Sandy Lake while the rest continued up the river to meet them later. From there Governor Cass planned to travel by canoe up the Mississippi River to its source. Schoolcraft's account of the country they hiked through typified the diverse landscape of the Great Lakes pine forests:

> Our guides taking their course by the sun, immediately struck into a close matted forest of pine and hemlock, through which we urged our way with some difficulty. On traveling two miles we fell into an Indian path. ... After pursuing it two miles, we passed through a succession of ponds and marshes, where the mud and water were in some places half leg deep. These marshes continued four miles, and were succeeded by a strip of three miles of open dry sandy barren, covered with shrubbery, and occasionally clumps of pitch pines [jack pines]. This terminated in a thick forest of hemlock and spruce, of a young growth, which continued two miles and brought us to the banks of a small lake, with clear water and a pebbly shore. Having no canoe to cross, we took a circuitous route around its southern shore. ... We now again fell into the Indian path. ... We now entered the great tamarack swamp, in which we progressed about eight miles.[81]

This was a land of frequent and hot fires. The "close matted" and "thick" patches of "pine and hemlock" and "hemlock and spruce" forest that Schoolcraft described were both relatively young and no doubt recovering from fires. One patch probably burned more recently than the other. Similarly, the "shrubbery" and "clumps of pitch pines [jack pine]" on the "open dry sandy barren" he saw also had their origin in fires. Only "the great tamarack swamp" escaped the flames. However, even swamps can burn. Schoolcraft noted that a portage in another area went through a burnt swamp:

> Old voyageurs say, that this part of the portage was formerly covered with a heavy bog, or a kind of peat, upon which the walking was very good, but that during a dry season, it accidentally caught fire and burnt over the surface of the earth so as to lower its level two or three feet when it became mirey, and subject to inundation from the ... river.[82]

Windstorms also knocked down these forests from time to time. Schoolcraft was still hiking to Sandy Lake when he struggled through a windfall:

> The dreadful storms which prevail here at certain seasons, are indicated by the prostration of entire forests, and the up-rooting of the firmest trees. These lie invariably pointing towards the southeast, indicating the strongest winds to prevail from the opposite point. It is one of the most fatiguing labours of the route, to cross these immense windfalls.[83]

Schoolcraft finally reached Sandy Lake after crossing "a succession of sandy ridges, covered with white and yellow pine, with some poplar and thickets of underbrush in the valleys, and altogether of a barren appearance." Here he stood on "a lofty pine ridge, which forms its southern barrier, and commands one of the most charming views of this romantic little lake." "The adjoining lands" were also "hilly and covered with pine."[84]

They fired "a volley of musketry" to alert the Indians of their arrival at the lake, but they misunderstood the party's signal. Schoolcraft wrote that "the Indians of this region being at war with the Sioux, had mistaken the firing for an attack of that nation ... and were thrown into the utmost consternation." Soon they met the other half of their party and continued by canoe to another village. Schoolcraft felt no less anxious than the Indians at Sandy Lake when the Indians at this village saluted him:

> We were received with a salute from the Indians ... with balls. The custom of firing salutes was introduced into this region by the North West Company. ... But the Indians never use blank cartridges on these occasions. ... The balls dropped in the water all around us, and it would seem as if they were apparently trying how near they could strike to the canoes without endangering our lives.[85]

About a week later they paddled across a lake and entered the last stretch of the Mississippi River. The river wound "through a prairie-valley, a mile in width, which is bounded by ridges of sandy land covered with yellow and white pine." "Its banks," Schoolcraft said, "are overgrown with wild oats [rice], rushes, and grass." They traveled up the river "fifty miles to its origin." Schoolcraft wrote in his journal that the lake from which it flowed "may be considered the true source of the Mississippi River." Then he described the lake and the surrounding countryside:

> Cassina Lake [Lake Itasca] is about eight miles long by six in width, and presents to the eye a beautiful sheet of transparent water. Its banks are overshadowed by elm, maple, and pine. Along its margin there are some fields of Indian rice, rushes and reeds: in other places, there is an open beach of clean pebbles, driven up by the waves. . . . The pike, carp, trout, and catfish are caught in its waters. . . . On the north shore of this lake, on a cleared eminence, is a village of Chippeways.[86]

Members of Governor Cass's expedition were not the first Europeans to see this spot. Two French traders with the American Fur Company were already living among the Chippewas there.[87] However, these fur traders left no record of what the area was like, and Schoolcraft did not list their names in his journal.

Schoolcraft noted that the Chippewas at Lake Itasca "received the party with every mark of friendship." He also said they "presented us an abundance of the most delicious red raspberries, and a quantity of pemmican, or pounded moose meat."[88] The Indians who lived in this part of the Great Lakes pine forests did not grow corn. The shallow, sandy soils were unsuitable for agriculture. Deer were not plentiful either.

The Chippewa could usually depend on seasonal harvests of fish, waterfowl, and wild rice. What happened in the surrounding forests rarely changed that. They also stored rice for the rest of the year. However, they could not survive without supplementing their diet with other edible plants and game. These foods came largely from the forest, and they were most abundant where forests had recently burned.

Severe fires that destroyed the dominant trees burned through the Great Lakes pine forests an average of once every 100 years.[89] Schoolcraft did not see such a fire when he visited Lake Itasca. However, fires had burned in these forests at least six times during the 100 years before his arrival. One fire blackened about 32,000 acres, and the others burned areas ranging from 5000 acres to nearly 22,000 acres. Several groves of old pioneer red pine forest that we can still see today around Lake Itasca date back to these fires. They were not all crown fires. Some were hot surface fires and others were only light surface fires, but they covered large areas nonetheless. If Schoolcraft had stayed at Lake Itasca a little longer he would have found himself in the middle of a fire, and we would have an eyewitness account. But he left too soon. A fire scorched nearly 23,000 acres around the lake that year. Schoolcraft did not mention it, so the fire most likely started shortly after he paddled his canoe back down the Mississippi.[90]

Some fires were even bigger than the one at Lake Itasca. Alexander Henry did not record a fire as he traveled through the chain of lakes along the Minnesota-Ontario border. However, 7 years before he passed that way a fire swept over 200,000 acres of forest on the south side of Lac La Croix.[91] Dead trees would still have been standing in much of this area when Henry was there. Unlike Henry, Father Aulneau did see a fire as he paddled his canoe through these lakes in 1735. He commented in his journal that the smoke prevented him from "even once catching a glimpse of the sun."[92]

Lightning ignited most fires in the Great Lakes pine forests, and Indians started the rest. Like Indians everywhere, the Chippewa used fire because they needed younger forests for berries, especially blueberries, and game, as well as building materials such as birch bark. No doubt Indians also started many fires by accident.[93] Not only that, they had been burning these forests from their very beginning. We know this is likely because Paleoindians killed a now extinct species of bison in the forest near Lake Itasca about 8000 years ago.[94] Regardless, fires were frequent in the Great Lakes pine forests, even though the number of fires declined slightly during the Little Ice Age.[95] Still, a fire occurred somewhere around Lake Itasca every 9 years since the middle of the seventeenth century.[96] However, they burned more often in some places than in others. The many lakes and ridges that cut through the landscape played a key role in determining how often, and even how severely, a patch of forest would burn. This, in turn, helped to control the kind of trees that grew there.[97]

Storms that swept over the Great Lakes pine forests usually came from the northwest and west. Any Indian fire that was burning at the time, or a lightning fire that started during the storm, would rush eastward with the wind. Long, narrow dry ridges oriented in a west to east direction served as highways for fire. They carried tongues of flame far ahead of the fire front. This allowed the fire to expand even more quickly. The flames would leap from tree to tree and destroy everything in their path. Nothing could stop such fires except a rainstorm, or a big lake. Even then, the lake had to have only a few widely spaced islands or, better yet, no islands. Otherwise the fire could jump the lake as the wind carried burning sheets of birch bark and other firebrands from island to island until they reached the opposite shore.[98]

Flying firebrands started spot fires that could become conflagrations during windstorms, although they generally burned as light surface fires. A swamp, valley, ravine, small lake, or even a steep ridge that crossed its path, might stop a fire if the winds were not too strong and there was no drought. Such barriers easily stopped most fires that came from the south, north, and east where the winds were usually weaker. However, they were less effective in stopping flames fanned by strong winds, and such windstorms normally blew from the west.[99]

When Schoolcraft arrived at Lake Itasca in 1820, the ancient forest he found was highly diverse. Fires were so frequent that only a small portion of the mosaic supported maple-basswood forests. An even smaller portion of the mosaic contained white spruce–balsam fir forests, which mostly grew in wet areas, and black spruce–tamarack forests, which grew in bogs. Fires also drove northern white-cedar to the lakeshores where they were relatively safe. This settler species spread into the understory of surrounding forests between fires, but the next fire quickly forced it back. Therefore, pioneer species that thrived after fire, especially pine, as well as aspen and paper birch, dominated the largest portion of the ancient forest mosaic.[100] This was especially true in northern Minnesota. Here white pine and red pine averaged about 55% of the patches in the mosaic, and jack pine averaged about 23%. Aspen, paper birch, spruce, balsam fir, and northern white-cedar, made up the remaining 22% of the mosaic. Thus, the many fires that burned in ancient Great Lakes pine forests added considerably to their diversity.[101]

Jack Pine Forest

Jack pine grows best where the soils are dry and fires are frequent and severe. Sandy ridges and uplands were ideal for jack pine because hot fires burned them every 50–100 years. Other trees that benefit from hot fires also grew on these sites, including aspen, paper birch, and even red pine. However, they were less abundant than jack pine. Reindeer lichen covered the ground beneath ancient jack pine forests on drier sites and feathermosses grew on moister sites.[102]

The ancient jack pine forest was patchy, and the patches were usually large. Most of them covered tens of acres to hundreds of acres.[103] Some were even larger. William J. Beal, a reporter who accompanied a botanical expedition in 1888, described some of the ancient jack pine forests he saw in Michigan:

> Fires often sweep over the plains, destroying all vegetation. Young pines soon spring up in these burnt areas, ... These groves of young trees are often miles in extent, and are so dense that the traveler is completely hidden from view at the distance of a few paces."[104]

The seeds of jack pine, like any other tree, did not spread uniformly over the ground when they fell. Small mammals such as red squirrels, chipmunks, mice, and voles ate the seeds in some places and left them alone in others, and the quality of the seedbed varied from place to place. Therefore, ancient jack pine forests contained openings where other pioneer trees could grow, particularly quaking aspen and paper birch. Bigtooth aspen also grew within these forests as scattered trees or in clumps.[105]

The sandy pine barrens of northern Wisconsin also contained jack pine forests. Fires most likely swept these relatively level areas even more often than they did the sandy ridges in Minnesota. Even so, the Wisconsin barrens came close to fitting the description of an "open dry sandy barren, covered with shrubbery, and occasionally clumps of pitch pines [jack pine]" written by Schoolcraft in Minnesota in 1820.[106] Jack pine grew in widely spaced, dense patches in the ancient Wisconsin barrens, and grasses, sedges, and other herbs, bracken ferns, shrubs such as blueberries and blackberries, and scrubby oaks covered the ground between them. Sprinkled here and there on the better soils, and in moist depressions, were aspen groves. Scattered old red pines towered over these grassy barrens, and open groves of bur oak crowned the hilltops and ridges.[107]

Jack pine is perfectly adapted to frequent hot fires. Most of its cones stay sealed with sap until a fire opens them, and the ash and mineral soil left behind by fires makes an ideal seedbed. One tree can hold as many as 1000 cones containing over 50,000 seeds, so there are also plenty of jack pine seeds available to restock the forest. They produce the most seed when the trees are 40–50 years old, which is about the time the next fire destroyed them. Not only that, 5- to 10-year-old trees can produce cones with viable seeds. This provides jack pine with insurance against periods of unusually frequent fires.[108]

What remained after a jack pine forest burned was a tangle of skinny blackened tree trunks, standing, leaning, and fallen, in a fluffy carpet of ash. However, signs of life appeared almost immediately. Woodpeckers flew in from surrounding forests to peck for the larvae of wood-boring beetles and bark beetles that multiplied in the dead and dying trees. It still looked like devastation, but it was really the beginning of a new forest. Hanging from the dead branches were clusters of scorched cones. The sap that held the scales together had melted and they were slowly opening. Some jack pine seeds already began dropping from the cones even as stumps continued to smolder. Before long the jack pine seedlings, and aspen and birch seedlings and sprouts, that would become the new forest emerged from the piles of dead trees.[109]

Within 5 years a severely burned jack pine forest in northern Minnesota became a mosaic of young pine, aspen, birch, pin cherry, and alder thickets. Fireweed, and scattered shrubs such as blueberry and hazel, grew in the spaces between thickets. Within just 40 years it would be a crowded middle-aged jack pine forest broken here and there by separate pockets of aspen and birch. A few groups of aging shrubs also grew among the trees, and reindeer lichen started to cover the rocky openings where other plants were sparse. It became an old pioneer forest with a feathermoss ground layer by the time it reached 50–60 years of age.[110]

Ruffed grouse, a plump game bird about the size of a chicken, were plentiful in ancient jack pine forests. The young aspen thickets that grew among jack pine shortly after a fire made perfect cover for ruffed grouse to raise their young. The aspen saplings grew so close together that hawks and other winged predators could not swoop in and take their chicks. The dense aspen canopy also kept the ground too shady for most plants, so there were few places where predators could hide in ambush. However, the aspen thickets became too thin for rearing chicks when they started dying at 10 years of age. Even so, now they provided excellent breeding cover. By the time aspen thickets reached 30 years of age, they were no longer dense enough for breeding either, but the trees were larger and more productive. This was the time they started producing an abundance of flower buds and catkins, which ruffed grouse relied on for food in the winter and spring. Therefore, ruffed grouse thrived in forest mosaics that included young, middle-aged, and old forests of aspen growing near one another.[111] Fires burned ancient jack pine forests at just the right frequency to provide ruffed grouse with this ideal mix of habitats.

If fires skipped a forest for a long time, settler species took over and jack pine gradually disappeared, along with most of the aspen and the ruffed grouse that depended on them. This replacement by settler species normally followed a relatively predictable sequence of changes.[112] Within 100 years it would be an old transitional jack pine forest. The dominant trees were showing their age, and many were dying and falling. Sporadic windstorms and heavy snows knocked down other trees. Seedlings and saplings of black spruce, and sometimes balsam fir, as well as paper birch, filled the gaps left by the dead trees. Spruce and fir also invaded the understory in places where the canopy was less dense. However, they stayed small because of a lack of sunlight. Then they quickly filled the sunny gaps that opened when overstory trees fell. The very shade tolerant balsam fir also invaded the understory of clumps of paper birch because their airy canopies blocked less sunlight than jack pines.[113]

The original patch of jack pine forest that might have covered hundreds of acres was now breaking down into a mosaic of smaller patches. These included the remaining old jack pines and the settler trees that were invading the forest. They ranged in size from half an acre to a few hundredths of an acre. However, the smallest patches could still contain nine or more mature trees, and the largest held thousands because the trees were small and their crowns were narrow.[114]

Within another 100 years only a few scattered old jack pines and aspens stood above the invading young settler trees. A thick mat of feathermosses covered the ground, and mosses and herbs became more abundant. Hazel shrubs also grew in the understory among a jumble of fallen jack pine trees.[115] This would soon become a self-replacing forest of black spruce and balsam fir, or just balsam fir, growing in very small patches. Paper birch, and even some aspen, also grew here and there where larger openings formed within the mosaic. Regardless of the mix of trees, these mosaics of self-replacing forest persisted as long as wind, snow, and insects such as the spruce budworm, continued to open gaps in the canopy, and fires stayed away.[116]

The conversion of a young jack pine forest to self-replacing forest could speed up or slow down, depending on what tree species invaded a fresh burn with jack pine. Black spruce or red pine seedlings often emerged in a burn a few years after jack pine. Unlike jack pine, the seedlings of these two trees do not grow well in fresh ash, so they had to wait to invade until rain and snow washed the ash away. If red pine came in, it eventually replaced jack pine because it lives nearly twice as long. This could delay the development of a self-replacing forest by as much as 200 years. It did so because the forest went through two old pioneer stages; one with jack pine and red pine in the overstory, and one with just red pine. Eventually the old pioneer red pine forest also disappeared as settler species filled gaps and transformed it into a self-replacing forest.[117]

Black spruce had the opposite effect because it is a settler species. It grows more slowly than jack pine, but it can tolerate shade. So black spruce fell back into the understory and stayed there when it joined jack pine seedlings in a fresh burn. This gave it a head start. The burn quickly converted to a self-replacing forest as black spruce saplings grew into the gaps left by jack pine trees that died.[118]

Settler species rarely had a chance to displace jack pine in the ancient forest because crown fires were too frequent, especially on sandy ridges and uplands. The majority of jack pine forests died in a crown fire after they grew to about 60 years of age. As a result, jack pine normally dominated the same areas. Only a small number of patches reached the old transitional stage of development, and there were very few self-replacing forests. Even if self-replacing forests did exist in the ancient forest, they had a very short life expectancy.[119]

Red and White Pine Forests

Red pine and eastern white pine forests were the most impressive of the ancient Great Lakes pine forests (Fig. 11.3). Red and white pine were not the tallest or the largest trees in North America. Still, they were very big trees.[120] An old pioneer forest must have been spectacular. Such open forests of tall pines were a common sight around the Great Lakes because light surface fires burned them repeatedly. These pine forests also burned more often than white pine forests in the northeast, except for those near Indian villages, hunting grounds, and heavily used trails. Nevertheless, Minnesota and southwestern Ontario's forests were unusual even among

Fig. 11.3   Red and eastern white pine, Itasca State Park, Minnesota. (Courtesy of M.E.I. Productions, Bend, Oregon; photograph by Mike McMurray)

Great Lakes pine forests. They not only lacked beech, but hemlock was rare or absent. They also were more extensive and burned more frequently than other Great Lakes pine forests.[121]

C. Andrews, like many adventurers of his day, wanted to visit the West before it became fully developed. So he came to Minnesota in 1856 where, he said, "I was not prepared to see the pine timber so valuable and heavy as it is above and about here." The scene inspired him, so he tried to capture an image of this ancient pine forest with words:

The trees are of large growth, straight and smooth. ... With the exception of swamps, which are few and far between, the timber land has all the beauty of a sylvan grove. The entire absence of underbrush and decayed logs lends ornament and attraction to the woods. They are more like the groves around a mansion in their neat and cheerful appearance; and awaken reflection on the Muses and the dialogues of philosophers rather than apprehension of wild beasts and serpents.[122]

Frequent light surface fires kept most of northern Minnesota's red and white pine forests clear of small trees, shrubs, and debris. This made them accessible and inviting to early travelers. Tree trunks were free of branches to a great height, and the orange-red bark of old red pines was striking in a forest where colors were mostly browns and greens. The wispy foliage of these tall pines also allowed sunlight to filter down to the cushion of needles that carpeted the forest floor. This enhanced the open, airy look of the forests, which made them even more picturesque.

White pine and red pine seldom grew on sandy ridges because they could not survive where hot fires burned too often. Unlike jack pine and black spruce, these pines only needed a hot fire to clear an opening for their seedlings every few hundred years. Then they needed light surface fires every few decades to keep invading trees, shrubs, and logs under control so that there was less chance of catastrophic fire. Thick bark helped to protect the mature pines from light surface fires, but it could not withstand hot fires. Consequently, ancient white and red pine forests tended to grow in protected places, such as islands. The prevailing winds in northern Minnesota and Michigan came from the west, so they also grew on the eastern and northern side of large lakes and streams, swamps, and high ridges. Red and white pine forests prospered in these locations because they provided just the right combination of infrequent crown fires or hot surface fires followed by frequent light surface fires.[123]

Lakes, streams, swamps, and north-south oriented ridges regularly stopped the many crown fires that came from the west. Even so, crown fires were unpredictable. They twisted and turned in any direction that fickle winds blew. Flaming pieces of birch bark, twigs, and even tree lichens, also blew across barriers and started spot fires on the other side. These little fires could grow into crown fires or hot surface fires when conditions were dry and windy. Hot surface fires could be just as deadly as crown fires. They did not climb into the canopy, yet they cooked surface roots and living tissue underneath bark. These fires could also scorch foliage and buds in the overstory canopy if there were small trees and logs in the understory to feed the flames. As a result, killing fires still ravaged red and white pine forests about once in 100–300 years.[124]

Red and white pine forests were easy to burn. The thick, fluffy layer of pine needles that covered the ground provided the perfect fuel to start fires. Unlike large logs that may take an entire summer to dry enough to burn, thin pine needles dry so quickly that they can burn soon after the snows melt in the spring. All it took to set them off was a spark from something like a lightning strike, an abandoned Indian campfire, or a piece of burning bark blown in from a distant crown fire. As a result, light surface fires burned within red and white pine forests about once in 20–40 years during periods between crown fires and hot surface fires.[125]

Fires are not independent of the forests they burn. On the contrary, a forest provides the fuel for a fire, and fire alters that fuel. They interact with one another to produce a particular pattern of fires and a particular forest structure. Weather and landscape features such as hills and lakes add even more variation to this relationship. Thus, frequent light surface fires in ancient red and white pine forests helped to ensure that future fires were also light. They did this by keeping small trees, shrubs, and logs from building up in the understory.[126] This, in turn, kept flames from growing tall enough to reach the forest canopy. Even if a crown fire entered such a forest from outside it would usually collapse. Most crown fires cannot sustain themselves without the heat and burning gases that flow up into the canopy from intense surface fires fed by heavy ground fuels. Open forests lacked the fuel to support such fires.[127]

Unusually severe weather substantially altered the size and behavior of fires. During the drought of 1863–1864, several fires of all levels of severity affected 44% of the ancient forests on the northern border of Minnesota. Extensive fires also ravaged Wisconsin's forests during that period. This was an extremely rare event.[128] Equally rare are strong gusts of wind that bend the flames of a crown fire nearly horizontal. This allows them to heat the crowns of trees in front of the flames enough to dry them out and ignite them, even without the help of an intense fire burning below the canopy. Such a crown fire could roar over the top of a forest with little undergrowth and destroy it. However, these crown fires seldom traveled far. Just the same, they probably accounted for some of the fires that destroyed Great Lakes pine forests, especially those that were dense and middle-aged or younger.[129]

Red pine and white pine are fire species. They need fire to open clearings where their seedlings have a chance to grow. Red pine needs clearings even more the white pine because it is less tolerant of shade. Regardless, not all crown fires and hot surface fires result in the establishment of pine forests. Even the destruction of a red or white pine forest does not ensure its renewal. Still, pine became established on 10 of the 12 major burns that occurred around Lake Itasca between 1712 and 1885.[130]

Red pine does not always come back after a fire because it only has a good seed crop once in 5–7 years. White pine has the same problem since it produces a good seed crop about once in 3–5 years. Furthermore, some mature trees have to survive a fire to produce seed. Even then, a second fire may burn through the debris left by the first fire and destroy any seed trees that remain standing. This would quickly convert a forest to aspen, birch, or brush, or even black spruce or jack pine, if their seeds were available. Therefore, the establishment of a red or white pine forest requires that a fire open a clearing while sparing enough seed trees to restock the forest, and it must burn during a good seed year. Such a lucky coincidence only happened now and then in ancient red and white pine forests.[131]

Just having seed available is not enough to ensure that a fresh burn will become a red or white pine forest. Red pine is particularly exacting in its requirements. Its seedlings grow well on mineral soil, but not on fresh ash. The flush of fireweed and other herbs, shrubs, and aspen sprouts that covers burns also inhibits the growth of its seedlings. Red pine seedlings cannot grow as fast as these other plants, so they suffer from a lack of moisture and gradually die in their shade. However, red pine can restock a burn if a good seed crop occurs a few years after a fire so that the ash has washed away. It also helps if aspen, birch, and shrubs are sparse, and a layer of moss is keeping the ground moist and preventing herbs from growing too thick. Then the burn might become a red pine forest or a mixture of red pine and white pine. A large windfall can also provide an opportunity for the development of a red pine forest, as long as there is plenty of sunlight, a mineral soil seedbed, and little competition from other plants.[132]

Eastern white pine was more successful than red pine because it is far less demanding in its requirements. Its seedlings can grow in ash, bare soil, moss, or light litter. White pine also grows on relatively dry, sandy soils, or rich moist soils, although it does not do as well as red pine on nutrient-poor soils. Like red pine, white pine grows best in large clearings, but it is more tolerant of shade than red pine. Therefore, its seedlings can survive in the understory of other plants and gradually rise above them. That is, if the canopy is not too dense.[133]

In the ancient Great Lakes pine forests, white pine often invaded aspen shortly after it took over a fresh burn. Aspen grows in clones, or clumps, of identical trees connected by a dense network of roots. An aspen grove can contain one or more of these clones.[134] Their roots inhibit the growth of other trees, including white pine. Therefore, only a few white pines successfully penetrated clones. They usually invaded the border areas between clones where the aspen canopy was thinner and the roots less dense. Even so, white pine still grew slowly. They grew more quickly after the aspen started dying at approximately 60 years of age. As a result, a large patch of white pine forest that emerged above aspen was itself patchy, even though it looked like one uninterrupted forest. It consisted of scattered trees that started within clones, and many denser clumps that started between clones. These denser patches of white pine usually covered less than a tenth of an acre.[135]

Once a red and white pine forest became established, it went through a series of changes not unlike those that took place in a jack pine forest. The changes just took longer. A jack pine forest normally died in a fire within 60 years. Those that avoided a killing fire became self-replacing forests composed of settler species within two centuries or less. A red or white pine forest could last two or three centuries before being destroyed by fire. It might take them three or four centuries to convert to self-replacing forests.

During that time, red and white pines, either mixed or growing separately, progressed through the young, middle-aged, and open old pioneer stages. They grew in large patches, most likely covering tens or hundreds of acres.[136] They eventually entered the old transitional stage and began dissolving into mosaics of very small patches that included young settler species. In the western part of the Great Lakes pine forests, settler species such as spruce, balsam fir, or northern white-cedar, along with shrubs such as hazel, invaded the understory and gaps. Paper birch invaded larger openings.[137] In the eastern part of the Great Lakes pine forests, hemlock or maple invaded the understory and gaps, and yellow birch invaded larger openings.[138] The small patches that formed within old transitional red and white pine forests were about the same size as those found in jack pine forests.[139] However, unlike jack pine forests, light surface fires normally beat back the invaders before they could become thick enough to fuel a severe fire. This happened repeatedly in many areas, especially in Minnesota and southwestern Ontario, which prolonged the open, old pioneer stage of ancient red and white pine forests.

Fire is unpredictable. During any given period, fire may skip a patch of forest several times in a row before burning it, or even longer, or it may burn it every time it passes. Thus, fire kept most patches of red and white pine forest open, but some became choked with young settler trees and shrubs as they reached the later stages of disintegration. A few may have even eluded fire long enough to convert to self-replacing forests. Even so, patches of self-replacing settler species were a temporary part of ancient red and white pine forests. They seldom lasted very long because fires were just too frequent.

On the other hand, farther east around the Great Lakes, white pine forests were often a temporary part of self-replacing beech-maple and maple-basswood forests. A large patch of white pine would emerge in a fresh clearing somewhere in these forests and then disappear, only to reappear somewhere else. The clearing white pine invaded was usually a large windfall, especially if the dead trees burned and exposed the soil. White pine could also invade aspen if it arrived in a clearing first, and then take over when the aspen died. Small groves of white pine also grew in places where soils were disturbed often, such as along streams and lakes, and on rock outcrops. So, there was always a reasonable chance that white pine seeds would be available when a fire occurred because they grew in scattered locations throughout these self-replacing forests.[140]

White pine forests that started in windfalls could not sustain themselves without fire, and fires were very rare in beech-maple and maple-basswood forests. Furthermore, it might take 1500 years or longer for a major windstorm to strike the same place, and then it might not burn afterward.[141] That is nearly four times the life span of a white pine. So pioneer white pine forests advanced to the old transitional stage of development and slowly dissolved into the surrounding self-replacing forest. This might take centuries, but the original settler species ultimately reclaimed the land. No doubt many of the ancient white pine forests that early travelers marveled at in Michigan and northern Wisconsin originated this way, and shared a similar fate.

The largest portion of ancient red and white pine forests probably consisted of the more open old pioneer and denser old transitional stages of development. These are the stately forests that impressed early explorers and travelers. They dominated the landscape because the trees lived so long, and because these two stages cycled back and forth. That is, young settler trees and logs constantly built up in the understory of the large old pines, and light surface fires kept clearing them away. Lesser amounts of young and middle-aged pioneer forests of pine, aspen, and birch grew where hot fires had created new openings. So these were mostly a patchwork of pioneer forests created and sustained by Indian and lightning fires. Only a few patches of thick self-replacing forest survived the flames in ancient Great Lakes pine forests.[142]

OAK-HICKORY FOREST

Like islands in the sea.
Colonel George Rogers Clark, 1778

The ancient oak-hickory forest was immense and spectacular (Fig. 11.4). This was not a forest of lofty trees with great trunks. Some trees were indeed large, but it was mostly a forest of relatively small trees in open woodlands or scattered through grasslands. This was the second largest forest in North America, covering 132 million acres of the center of the continent.[143] Only the spruce forests far to the north were larger. However, the great size of this forest was not its most striking attribute. It was a forest you could see. It was spacious and sunny because the trees stood apart, and it seemed as big as it was. Early travelers who saw the

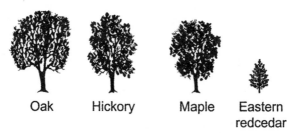

Oak   Hickory   Maple   Eastern
redcedar

Fig. 11.4   Trees of the ancient oak-hickory forest.

ancient oak-hickory forest for the first time stood breathless at the scene. Not only because it looked like an artist's rendition of the perfect landscape, but also because of the animals that teemed within it and gave it life. Herds of elk, bison, and deer grazing among the trees was a common site, and little packs of gray wolves often sat in the tall grass nearby watching their prey. Black bears, coyotes, foxes, and wild turkeys also wandered here and there. In the spring you could see prairie chickens strutting on their booming grounds, beaver splashed in the streams all summer, and great flocks of passenger pigeons swooped in to pick at acorns in the fall.[144] The traveler could see everything in this ancient forest. It was incomparable.

Father Claude Dablon, a Jesuit priest, was among the first to write a description of the oak-hickory forest. He was paddling a canoe through an arm of the forest that poked up into east-central Wisconsin toward Green Bay. The year was 1670. "This country has the beauty of an earthly paradise," wrote Farther Dablon. However, he said, "the route to this paradise is as trying as the way to heaven provided by our Lord." The problem Farther Dablon faced was 10 miles of rapids. Already in poor humor, he found "an idol honored by the savages" near the rapids. It was just a rock shaped like a man, but the Indians made sacrifices to it and asked it for help. So Father Dablon felt compelled to "remove this cause of idolatry," and threw the rock "into the depths of the river."[145] This done, and the rapids behind him, Father Dablon relaxed to enjoy the countryside:

> As a reward for passing this way, rough and dangerous, we enter into the most beautiful country ever seen—prairies on all sides, as far as the eye can reach separated by this river flowing gently through them. We see small hills with groves of trees here and there, as if planted to shade the traveler from the ardent heat of the sun. To paddle on this river is to repose oneself. . . . Here are elms, oaks and similar trees. . . . Among these rich pastures are found buffalo, which the savages call *pisikiou*. The bear and the raccoon fill the country. These animals and the deer are easily hunted in the woods, which are very clear and contain large prairies.[146]

Three years later Father Marquette described the view from a Maskouten Indian village on the Fox River near the portage to the Wisconsin River. He said it "is beautiful and very picturesque, for from the eminence on which it [the village] is perched, the eye discovers on every side prairies spreading away beyond its reach, interspersed with thickets or groves of lofty trees."[147]

The Great Plains bounded the ancient oak-hickory forest in the west, and the beech-maple, oak-chestnut, and southern pine forests bounded it in the east. The oak-hickory forest curved eastward from the southeastern corner of Minnesota, through southern Wisconsin, Iowa, and Illinois, to Ohio. Then it curved toward the Southwest, following the west side of the Ohio and Mississippi Rivers, through Missouri, eastern Kansas, and northern Arkansas, until it bent westward even more and entered Oklahoma and Texas. Thus, the oak-hickory forest formed a wide belt—a transition zone—between the arid grasslands of the Great Plains and the more humid eastern forests.[148]

The oak-hickory forest was very old. Oak and hickory spread from their Ice Age refuge in Tennessee as the climate began to warm. They took over the land left behind by the dying spruce forest as they moved up the Mississippi River Valley and onto the Great Plains. Blue jays helped to spread oaks northward, so they moved fast, and hickory followed closely behind them.[149] Prairie grasses moved northward and eastward from their refuge in west Texas at about the same time and occupied the ground beneath the trees.[150] As a result, the southern part of the oak-hickory forest probably assembled about 12,000 years ago and the northern forest about 10,000 years ago.[151] The melting ice sheets had not even retreated into Canada by this time. Paleoindians were also hunting mastodons in the oak-hickory forest when it first developed in eastern Missouri.[152] So they and their Indian descendants were part of this ancient forest from its earliest beginnings.

The trees differed somewhat in the northern and southern parts of the ancient oak-hickory forest because of the way they migrated in the warming climate at the end of the Ice Age. For example, northern pin oak made it to Minnesota and southern Canada, but it died out below northern Illinois. Northern red oak and bur oak migrated far to the north as well, but they managed to survive in the warmer climate near their Ice Age refuges in the South. However, northern red oak could not survive below Oklahoma and Arkansas, while bur oak made it into south Texas. Other species spread almost as far north, and they grew in south Texas as well, such as white oak, black oak, shagbark hickory, bitternut hickory, and eastern redcedar. Post oak, blackjack oak, and mockernut hickory could only spread northward to central Illinois, but they too survived in the South. Likewise, black hickory only made it as far north as Missouri.[153] However, just because a species could live in the North or the South did not necessarily mean that it would dominate that part of the ancient oak-hickory forest. That depended mostly on their relative ability to endure periodic droughts, and Indian and lightning fires.

This ancient forest grew in a place where rainfall was uncertain and fires were frequent, especially near the edge of the prairies. This was a fire forest where pioneer species thrived and settler species retreated into the background. From the beginning of European exploration, no one ever doubted that Indian fires were largely responsible for the way the oak-hickory forest looked. The role of Indian fires was most obvious in the ancient oak savannas that bordered the Great Plains and midwestern prairies. Peter Pond, a fur trader, even stated in 1773 that it was "proverbial that the fires which run over these prairies destroy the small trees and stop the spread of the woods."[154] Pond was traveling along the Fox River in Wisconsin at the time. In 1819, Ferdinand Ernst, a wealthy German looking for farmland, was farther south in Illinois when he learned firsthand how Indian fires affected both the savannas and the Europeans who tried to live there:

> Every autumn the Indians ... hold a grand hunt. They then set fire to the dry grass of the prairie, and the flame with incredible rapidity spreads over all the country. ... This destructive custom of burning off the prairies is the reason that timber is confined to the banks of streams and a few other places. The heat of the fire not only prevents entirely further extension of the forests but even diminishes their area. Upon these annual hunts the Indians forcibly eject all white settlers from their territory.[155]

The land between the prairies and the hilly country on the Ozark Plateaus in Missouri was no different. Father Vivier drew this conclusion when he traveled thourgh the area in 1750. He said that the "trees are almost as thinly scattered as in our public promenades ... partly due to the fact that the savages set fire to the prairies toward the end of autumn."[156] Indians burned the lowlands and savannas more often than anywhere else. But they still burned everywhere. It was the combination of Indian fires and lightning fires that favored oaks and hickories, and sustained the mosaic character and openness of the ancient forest.

Oaks and hickories dominated the ancient oak-hickory forest because, as Elmer Baldwin pointed out in 1877, they "are the most hardy and least injured by fire, consequently were the only varieties on the bluffs." Many other trees also grew in this forest. "Black and white walnut, linden [basswood], elms, sycamore, ash, maples, etc., were found in abundance," wrote Baldwin, "but were not found on the bluffs, as they would be killed by a fire."[157]

Oaks are drought-hardy, resistant to fire, and vigorous sprouters, so they were more plentiful and widespread than any other tree in this forest. Nevertheless, some oaks are hardier and more fire-resistant than others. Consequently, post oak and blackjack oak dominated the drier more frequently burned areas in the southern half of the ancient oak-hickory forest.[158] Bur oak became the dominant tree on frequently burned lowlands in the northern half of the forest. White oak and black oak are also moderately tolerant of fire, so they grew mostly on upper slopes and even ridges, but they also occurred on lowlands in some areas.[159]

Hickories were less plentiful than oaks in the ancient forest because their thin bark made them vulnerable to fire. Still, they could sprout after being burned like the oaks, so they survived and prospered. Bitternut, shagbark, and mockernut hickory grew throughout the oak-hickory forest. However, these hickories were more widespread in the North than in the South where it was hotter and drier. Therefore, they stayed mostly on river bottoms and moist slopes in the southern part of the forest, and black hickory took their place on the drier sites.[160]

Most of the trees in the ancient oak-hickory forest were less than 100 feet tall, and the largest blackjack oak was half that height. Still, a post oak could be 4 feet thick, and a black oak could be just as wide and stand 150 feet tall. However, white oak, bur oak, and shagbark hickory were the most impressive trees in the ancient forest. The shaggy strips of bark that hung on the thick trunk of a shagbark hickory made this tree as distinctive as a woolly mammoth. Likewise, the tall straight trunk of a white oak made it stand out in a forest, especially those that reached a height of 150 feet and a diameter of 8 feet. Some giant white oaks lived 600 years, which is twice as old as the other forest veterans. Still, a bur oak could grow even taller and nearly as thick as a white oak on rich bottomlands. The corky wings that garnish the twigs of the bur oak, and the fuzzy fringe around the cup of its acorn, seem unnecessarily decorative on so majestic a tree.[161]

Oaks, like other trees, did not reach their maximum height in open grasslands. Rather than stretching upward to gather sunlight as they would in a dense forest, they spread outward. So they had wide crowns, stocky trunks, and heavy branches. These massive old oaks must have been an imposing site when standing alone on the prairies. At least Father Louis Hennepin seemed to think so when he saw them in 1679. He was paddling a canoe down the Illinois River with Robert Cavelier de La Salle when he made a note about them in his journal. "There are boundless prairies interspersed with forests of tall trees," Father Hennepin wrote, "excellent oak . . . of prodigious girth and height."[162] Then they reached the Mississippi River and completed their historic trip to its mouth in the Gulf of Mexico. Fur trader Louis Joliet and Father Marquette had paddled down the Mississippi 6 years earlier, but La Salle traveled farther and took possession of the whole Mississippi Valley for France. He named it Louisiana.[163]

Oak Savannas

Oak savannas, like the one described by Father Hennepin, were prairies sprinkled with isolated patches of woodland. They only made up about one quarter of the ancient oak-hickory forest.[164] However, this varied from place to place. For example, about 45% of southern Wisconsin consisted of oak savanna in the early 1800s. Prairies made up 20% of the area, and oak forest or woodland only covered 10%. The remaining 25% consisted of maple and floodplain forests.[165] Regardless, savannas extended along the entire length of the western side of the oak-hickory forest, from Minnesota to Texas, and they may have stretched as far west as eastern Nebraska.[166] They also extended eastward up to Detroit, Michigan, and into Ohio. A long belt of post oak savanna also existed along the east side of the Blackland Prairies of east-central Texas. Therefore, no one could enter the midwestern prairies, the Great Plains, or even most of Texas, without crossing them. So nearly everyone who traveled west had a chance to see these ancient oak savannas and marvel at their beauty.

Early explorers and travelers seemed to like oak savannas better than any other part of the ancient oak-hickory forest, and for good reason. They had the great open vistas and grand displays of bison and other wildlife offered on the Great Plains. "The deer and elk abound in this quarter," Henry Schoolcraft wrote when traveling through western Missouri in 1818, "and the buffaloe is occasionally seen in droves upon the prairies, and in the open high-land woods."[167] Even something as ordinary as grass was impressive here. This was the edge of the true, or tallgrass, prairie that covered the eastern half of the plains.[168] Big and little bluestem and Indian grasses grew here. This grass, Henry School-craft noted, "attains so great a height that it completely hides a man on horseback."[169] Like most early travelers, the tall prairie grasses also left an impression on Frank Edwards, a volunteer with General Stephen W. Kearny's detachment during the Mexican War. Edwards was just begin-ning his march along the Santa Fe Trail from Fort Leavenworth, Kansas, when he wrote his description in 1846:

> When we emerged from this belt of trees, the first prairies met our view. . . .
> Perhaps it is one of the most beautiful sights in nature to see a puff of wind
> sweep over these grassy plains, turning the glistening sides of the grass to
> the sun, and seeming to spread a stream of light along the surface of the
> wave-like expanse.[170]

Add fields of wildflowers and a sprinkling of massive old oaks to this spec-tacle and the scene must have been overwhelming. No wonder the savan-nas received more attention than any other part of the ancient oak-hickory forest.

The most hardened of soldiers had to pause and look at the oak savan-nas even as they marched toward battle so that they could write lyrical descriptions in their journals when they rested. George Rogers Clark was one such soldier. It was 1778, and he was leading a small band of volun-teers through southern Illinois to seize British forts. Clark interrupted his notes on the military campaign to write about the savannas:

> The country is more beautiful than any I could have imagined. Extend-
> ing beyond eyesight are large prairies covered with buffalo and other game,
> varied by groves of trees that appear like islands in the sea.[171]

Clark's portrayal was suitably brief and he soon returned to writing about the dangerous and bloody work of a soldier. He and his men marched onward and captured three British outposts.

Ancient oak savannas were like ocean shorelines. A traveler could sit under the shade of an old oak on this shoreline and peer out at a sea of grass. In some places they would see rolling green hills that looked like the swells of a stormy sea frozen in time. In level places it would appear like a calm sea where occasional gusts of wind bent the tall grasses in waves that rippled over its surface. Islands of old oaks would also be visible a short way out from the wooded shore. Prairie fires were the storms that battered oak islands and wore them down. These fire storms also swept inland and thinned trees along the shore. Sometimes tsunami-like fingers of flame rushed far into the forest and destroyed everything in their path, leaving unburned oak peninsulas jutting into the prairies. At other times, trees moved into the prairies. Occasionally, these invaders even managed to encircle a piece of prairie and pinch it off so that it became a grassy island within the forest. Some of the invading trees survived, especially in hilly places or behind ponds where fires burned erratically. Others became oak islands when fires encircled them and cut them off from the forests along the shore. But most of the invaders died. Still, trees kept spreading and fires kept burning. So the shoreline flexed back-and-forth over time depending on where trees invaded the grassy seas and fire storms hurled them back. Like a wave-beaten coast, nothing stayed the same.

Light is not limiting in savannas, yet the trees still grew in patches. Thick grasses, and the timing of wet and dry periods, and fires, is more critical to young trees in savannas than light. During dry periods, fires cleared the ground of tall grass and debris so that oak seedlings had plenty of room to grow. However, these fires also kept the new generation of oaks from becoming trees. In order for oaks to grow the fires had to stop for at least 10–20 years. This was unlikely during dry periods, especially when Indians regularly burned the prairies. Occasionally, Indians would abandon an area long enough for patches of seedlings and sprouts to grow into trees. Even so, lightning fires still killed many of them. It took a wet period to reduce fires enough to add a large number of patches of trees to the oak savanna. The coincidence of a wet period, with little or no fire, following a dry period with many fires and many oak seedlings, only happened about once in 35–100 years. Therefore, many of the isolated patches of trees that stood on the savannas were of a similar age because they represented infrequent events that produced widespread surges of growth.

Savannas were not just tallgrass prairies sprinkled with patches of oak, nor were they open oak forests with an understory of tall grasses. They were two communities of plants; oak groves or woodlands and treeless prairies. The spreading branches of the oaks shaded the understory of a grove, and the leaves that dropped each year changed the soil. So forest dwelling plants were more likely to grow under the old oaks than prairie plants.[172] Even the turkeys, doves, squirrels, and other creatures that lived among the trees played a role in distributing the seeds of plants that seldom grew in the open prairies. Hence, forest and woodland plants such as wedge grass, wood reed, blue-stemmed goldenrod, and hazel, lived under the oaks in the savannas of northern Illinois rather than the bluestems and Indian grasses of the prairies.[173]

There were many brushy oak sprouts, or grubs, hidden among the tall grasses of the savannas.[174] Some of them would become trees, but many would not because fires burned their tops too often. Similarly, other areas within savannas contained patches of small trees that were young or too frequently burned to grow very large. Either of these explanations may describe what happened in the ancient savannas that used to exist in what is now Chicago. William H. Keating, a geologist from the University of Pennsylvania, traveled that way in 1823 with Major Stephen H. Long's expedition to explore the headwaters of the Mississippi River. This was a place where the ancient savannas touched the southern shore of Lake Michigan. However, the land did not impress Keating. Where other travelers would have seen an unparalleled panorama, with colorful birds swooping over endless seas of green grass and blue water, Keating only saw "fatiguing monotony." "There is too much uniformity in the scenery," he wrote, "the extensive water prospect is a waste" and "the land ... affords no relief to the sight." Keating used the same somber tone when he described the patches of trees on the savanna. "It consists merely of a plain, in which but few patches of thin and scrubby woods are observed scattered here and there," he said.[175]

Not so, thought Harriet Martineau, an English author, journalist, abolitionist, and leading women's rights advocate. She spent 2 years traveling in America in the early 1830s. Seeing the prairies and oak savannas near Chicago was one of the goals of her trip. So she rode 40 miles by wagon from Chicago just to stand on top of "Mount Joliet" for a panoramic glimpse of this ancient landscape. The bumpy ride did not dampen her enthusiasm.

> As we proceeded, the scenery became more and more like what all travelers compare it to,—a boundless English park. The grass was wilder, the occasional footpath not so trim, and the single trees less majestic; but no park ever displayed anything equal to the grouping of the trees within the windings of the blue, brimming river Aux Plaines [Des Plaines River].[176]

The small size of the mountain, "the mount is only sixty feet high," she said, did not disappoint her because it still provided a spectacular view of this flat expanse of prairie and savanna. "It commands a view which I shall not attempt to describe, either in its vastness, or its soft beauty," wrote Martineau. "I thought I had never seen green levels till now; and only among mountains had I before known the beauty of wandering showers."[177]

Antoine la Mothe de Cadillac was equally enthusiastic when he wrote a letter to the governor of New France in 1701 describing the oak savannas around Detroit. "Across the prairies are scattered large clusters of trees, yet these trees are marvelously tall, and except for the great oaks, have almost no branches but near the top," wrote Cadillac.[178] Likewise, the "Illinois Country" that Father Antoine Silvy saw in 1710 was to him "the most beautiful I have ever seen." "As far as the eye can see there is prairie," he said, "broken only by scattered groves of trees."[179] General Josiah Harmar, commander of the army stationed on the Ohio frontier near Kaskaskia, was more direct when he jotted down his description of the savannas in 1787. He only said that "copse of woods" were "interspersed" over the prairie "here and there." He added that "they are free form brush and underwood."[180]

Groves or patches of trees became smaller and farther apart toward the west until the oak savanna finally vanished in the dry grasslands of the Great Plains. At the western edge of the savanna a traveler might see only one sturdy oak left standing on the prairie, like a true pioneer that leads the way in spite of the hazards. Lewis and Clark wrote a note in their journal about such a noble tree. It was 1804, and they were on their way up the Missouri River:

> About two miles off from the mouth of the river the party on shore saw another of the objects of Ricara [Arikara] superstition; it is a large oak tree, standing alone in the open prairie, and as it alone has withstood the fire which has consumed everything around, the Indians naturally ascribe to it extraordinary powers.[181]

Bottomland and Protected Forests

Meandering rivers and streams fringed with trees dissected the ancient prairies and savannas. The soils along riverbanks were wetter and burned less often than the surrounding countryside. Even so, areas like this burned an average of once every 19 years, although some places went without fire for as long as 40 years.[182] As a result, many different trees grew here, and they were often taller than those on the more frequently burned savannas. Patrick Shirreff, a Scottish farmer, noticed this as he traveled through Illinois in 1833. He said that the "trees on the margins of the prairie are of small size, and chiefly oak," while "those on bottom, or interval, land, on the banks of rivers, are of immense size." However, Shirreff also pointed out that "sometimes prairies descend to the water's edge, on both sides, and no general rule can be laid down for the prevalence or want of timber."[183]

Even on the Great Plains, some pockets of trees still hung on in ravines, along rivers, or on islands, where the annual fires that swept the prairies seldom reached them. Lewis and Clark noted this in their journal while standing on a hill above the Missouri River in 1805:

> On both sides of the Missouri ... one fertile unbroken plain extends as far as the eye can reach, without a solitary tree or shrub, except in moist situations or in the steep declivities of hills, where they are sheltered from the ravages of fire.[184]

Many groves of trees stood tucked away below river bluffs, often going unseen by people crossing the prairies. These were mostly ancient elm-ash-cottonwood forests. They grew on the banks of rivers that snaked through the Great Plains and the oak-hickory forests of the upper Midwest. Sioux guides took Joseph Nicollet, a French scientist, to such a secluded grove in 1838 during his expedition to map the region between the Missouri and Mississippi Rivers. He was in the southwestern corner of Minnesota when he described it in his journal:

> This little river follows a deep ravine, its sides wooded since they are protected against fire. But the tops of these trees are not visible above the level of the prairie, and as wood is so rare, knowledge of it is precious; the Sioux name this *tchan narhambedan*—the hidden wood.[185]

"The hidden wood" of the Sioux still stands near the town of Chandler in Murray County. Nicollet said that "the prairies only rarely present these isles," or "groups of trees ... which are protected from fires by streams ... or other irregularities of the terrain." So, with the aid of Sioux guides, he planned his route to camp in woods like these "to do our cooking." Nicollet noted that one of the groves he camped in was "celebrated among the Sioux as the rendezvous of the different tribes for hunts since time immemorial." The Indians named it "the isle of the fruit that cracks when one eats it" because of the hackberry trees that grew there.[186]

River islands provided more protection from fires than ravines and bluffs. Robert Stuart noticed this on the Platte River in 1813. It was the end of the journey in which he discovered the Oregon Trail. Stuart was traveling eastward, near where the Platte enters the Missouri, when he made the following entry in his journal:

> The opposite shore seems equally destitute of Timber with this, and the Hills of either side ... are also Prairie. Indeed, it is extraordinary though no less true that while the mainland possesses but few trees and sometimes not even a Willow; yet there is scarcely an Island of however diminutive dimensions but is for the most part wholly covered with wood.[187]

Father Marquette noticed the same thing about the upper Mississippi River in 1673. He was with Joliet and five men in two canoes on the "Great River" between Iowa and Illinois when he said that there is "almost no wood" along the river. However, he noted, "the islands are more beautiful and covered with finer trees."[188] Things changed as Father Marquette and Joliet went farther down the Mississippi.

The huge river cut a wide swath as it meandered from side to side through the ancient oak-hickory forest, creating an extensive floodplain laced with wetlands and sloughs. Over a century after Father Marquette had been there, in 1797, Nicolas de Finiels stood on a bluff in Illinois where he could "admire the confluences of the Mississippi, Missouri, and Illinois rivers." "Your gaze is consumed by the vast plain that proceeds up the Mississippi," wrote this French engineer, "prairies, clumps of woods, ponds, and streams dissect it, and the irregular loops of the river seem to want to imprison it."[189] This network of water and marshlands made it more difficult for fires to penetrate the floodplain. Thus, the trees that Father Marquette and Joliet saw along the edge of the Mississippi became denser and grew taller as they paddled south, and the prairies and savannas receded into the background.

This was the first time Europeans had explored the upper Mississippi River. Father Marquette and Joliet began their journey from Mackinac Island in Lake Michigan. They paddled across the lake to Green Bay. "We made our paddles play merrily over the Lake of the Illinois [Lake Michigan]," wrote Father Marquette. From Green Bay they traveled down the Fox River and portaged to the Wisconsin River. "After 100 miles on this river, we reached the mouth of the Great River," he noted. Then Father Marquette recorded what no European had seen before:

> We safely entered the Mississippi on the 17th of June, with a joy I cannot express. We gently followed the Mississippi southward, seeing deer and buffalo, geese and swans. . . . When we reach the latitude of 41 degrees 28 minutes, we found that turkeys took the place of other game, and buffalo took the place of other animals. . . . As we . . . paddled quietly in clear and calm water, we heard the noise of rapids ahead. I have seen nothing more dreadful. A tangle of large trees, branches and floating islands was issuing from the mouth of a river called *Pekistanouis* [Missouri River] with such force that we passed it with great danger. So great was the agitation, that the water became very muddy and did not become clear again. After making about 50 miles due south, and a little less to the southeast, we came to a river called *Ouaboukigou* [Ohio river]. . . . Here we began to see canes growing on the banks of the river. They are a beautiful green with narrow, pointed leaves. They grow very tall, and so thick that the buffalo have trouble passing through them. . . . We began to see less prairie land, because both banks of the river were lined with lofty trees. The cottonwood, elm and basswood were admirably tall. Yet we still heard buffalo bellowing, which made us think the prairies were near.[190]

Father Marquette and Joliet were now leaving the southern end of the ancient elm-ash-cottonwood forest that grew on the banks of the upper Mississippi River. Cottonwood dominated it, especially on islands and low-lying areas subjected to annual flooding, and elm and ash were common. Various mixtures of other trees grew there as well, such as sycamore, hackberry, oak, and maple. The dense forest along the river looked almost exotic in the open prairies and woodlands of the Midwest. This was especially true when the cottonwoods were "enveloped, from the ground to the branches, by a drapery of dark green vines." That is the way Father De Smet saw them near the mouth of the Missouri River in 1840. This was "one of nature's charms," he said, and another was the "*bignonia*" vine because it "lets loose a profusion of great flame-colored trumpet-shaped blossoms."[191]

Some of these bottomland forests could extend "from the bank of the river to the high hills that skirt it," said Father De Smet while traveling in the Missouri River Valley again in 1851.[192] In other areas, however, it was only a narrow strip of forest, especially in the lower Mississippi River Valley. A mosaic of open woodlands, savannas, and prairies replaced it on the floodplain farther back from the waters edge. Extensive corn fields and great Indian cities also spread over these floodplains.[193] However, Father Marquette and Joliet did not see them in 1673. Mississippian people had lived there for 700 years, but their culture collapsed about a century and a half before they came.[194] Regardless, their Indian descendants were still numerous and warlike. Therefore, Father Marquette and Joliet decided to turn their canoes around and head back up the Mississippi when they reached Arkansas.[195]

Prairies covered about 40% of the upper Mississippi River floodplain near St. Louis in Father Marquette's time, and wet prairies covered almost 14%. No doubt many of these prairies were once Mississippian Indian corn fields or settlements. Likewise, some of the open woodlands and savannas that covered another 43% of these floodplains probably developed when trees invaded such clearings. This seems even more likely because the original survey conducted in 1816 considered almost 39% of these bottom-lands "fit for cultivation."[196]

Most of the ancient forests on the floodplain stayed open after the decline of the Mississippian culture because Indian and lightning fires still swept into them from the uplands during dry years. However, fire did not enter the floodplain during wet years. Even so, great floods had a similar effect, although they occurred less often than fires. Like fires, floods ripped holes in a forest where pioneer species could grow. Floods also helped to keep the understory open, either by stripping away trees and shrubs or burying them under sediment.[197]

The ancient oak-hickory forests in the upper Mississippi River Valley were substantially different from those on the surrounding uplands. They both consisted of a mosaic of open woodlands, savannas, and prairies. However, there were twice as many trees per acre on the floodplains (not including the denser bottomland forests) than on the uplands. Likewise, the trees that characterized the forests were not the same. Pin oak and shagbark and shellbark hickory dominated forests on the floodplains. Pin oak is a relatively small, fast growing, but short-lived tree that can withstand periodic flooding, so it does well on floodplains. However, its thin bark makes it sensitive to fire, although it sprouts like other oaks. On the other hand, white and black oak dominated upland forests, along with some hickory. They did so because these oaks are moderately resistant to fire, and they grow well on dry sites. Oak-hickory forests on the uplands were also much thinner than floodplain forests because fires were more frequent. However, both forests were very open.[198]

The savannas of the ancient oak-hickory forest were on the edge of the Great Plains and midwestern prairies because it was dry there, and fires were frequent. Still, forests of fire-sensitive species also managed to survive within these savannas, but they grew on wet bottomlands, in protected ravines, and behind fire barriers such as large lakes and rivers.[199] The "Great Oasis" in southwestern Minnesota was one such place. This forest stood on the plains, but it seldom burned because it was "protected from the spring and fall fires by the lakes which surround it," wrote Nicollet when he camped there in 1838. The forest covered 300 acres and consisted primarily of basswood, birch, oak, ash, and aspen, all of which were relatively small.[200]

Large rivers provided just as much protection as lakes. Indian and lightning fires generally moved from west to east on the prairies, but few of them could jump rivers that crossed their paths. As a result, the west side of such rivers burned nearly every year while the east side burned only occasionally, sometimes going as long as a century without fire. For example, prairie covered about half of the west side of the Pecatonica River in southern Wisconsin, and the other half was oak savanna. Most of the trees were also small on the west side because of frequent fires. In contrast, dense forest covered half of the east side, and prairie covered only 2%. Likewise, the trees were larger on the east side because they suffered fewer injuries from fire. Finally, fire-resistant oaks dominated the west side of the Pecatonica while more fire-sensitive species were prominent on the protected east side, such as northern red oak, sugar maple, elm, and basswood.[201] Such abrupt and dramatic differences in vegetation were common around rivers and lakes in the ancient oak-hickory forest. Amos Parker noted this in Illinois in 1834, but he did not seem to know the reason. "It is somewhat singular and unaccountable," he said, "but we found it universally to be the fact, that the east side of all the streams had much the largest portion of timber."[202]

Oak Woodlands

Oak woodlands are open forests sprinkled with small isolated prairies. They are the opposite of oak savannas. Most of the oak woodlands were in rugged country because hills have a greater affect on the movement of fires than rivers and lakes. Fires burn erratically even on level prairies, rushing here and there and skipping places as the winds shifted, while lingering in other places as the winds subsided. Ridges and ravines block and funnel winds in ways that make fires even more unpredictable.[203] As a result, some spots remained unburned for decades before regular fires burned again.[204] However, it only takes 10–50 years without fire for oak sprouts to grow tall enough to convert a prairie or a savanna into an open woodland.[205] Once established, these woodlands stood for centuries.

A long period without fire allowed self-replacing settler species to displace woodlands. Normally, Indian and lightning fires kept settler trees confined in protected places, but they invaded the surrounding woodlands and forests when the fires stopped, even for a short time.[206] This usually happened during wet periods or after Indians abandoned an area. Northern red oak and shagbark hickory might begin the invasion in the northern half of the oak-hickory forest, but sugar maple, beech, red maple, or possibly even hackberry, took over eventually.[207] Likewise, white oak might be the first to invade in the southern half of the forest.[208] However, sugarberry, eastern redbud, bitternut and pignut hickory, and in some places beech and magnolia were the species most likely to replace woodlands and themselves.[209] There were few settler species in the dry southwestern part of the forest. Therefore, eastern redbud and elm, and even pioneers such as eastern redcedar, are the likely replacements because they are more shade tolerant than the stunted oaks that grew in the original woodlands.[210] Nevertheless, the invaders rarely spread very far or survived very long in the ancient oak-hickory forest before fires pushed them back into their refuges.

The Ozark Plateaus and Ouachita Mountains of Missouri, Arkansas, and Oklahoma, probably held the largest expanse of woodland in the ancient oak-hickory forest.[211] Schoolcraft rode through this region in 1818. Schoolcraft was a geologist, so he wrote about rocks and minerals and, like most early travelers, exciting events and the discomforts he endured along the way. The few journal entries that he made about vegetation on the Ozark Plateaus in Missouri illustrate that an oak-hickory forest did not consist entirely of woodlands, even in hilly country. One day he rode "over barrens [savannas] and prairies, with occasional forests of oak [woodlands] ... with grass ... [and] very little under-brush."[212] On another day he rode over some "gentle sloping hills" that were "well wooded with oak and hickory." However, Schoolcraft noted that even these woodlands contained "some extensive prairies." Later that day he came to a river where he found "a large growth of forest-trees and cane . . . interspersed with prairies."[213] Thus, the ancient oak-hickory forest that Schoolcraft saw in the highlands was a mosaic of woodlands, savannas, and prairies, dissected by dense forests and wet prairies along river banks.

Schoolcraft could not mention wildflowers in his descriptions of the highlands of Missouri because he was traveling late in the year. If he had come earlier, he would have seen a dazzling display of thistlelike purple blazing stars, yellow and purple coneflowers, bush clovers, and countless other flowers blooming among the grasses between the trees.[214] Similarly, he made no comments about the way hills and valleys influenced the forest mosaic. Still, he had plenty of time to see that most woodlands were on the side of a hill that faced north or east because it was cooler and moister there. This also reduced the number of fires. Oak savannas tended to be in valleys, and on ridges and hot south-facing slopes, where Indian and lightning fires were more frequent. However, farther north in the hilly country of southern Illinois, where it was slightly cooler and wetter, savannas were usually in valleys and on lower south-facing slopes. Open woodlands covered the upper slopes and even ridges.[215]

E. James, a U.S. government surveyor, was a keen observer who not only documented the way slopes and valleys affected the ancient oak-hickory forest mosaic, but he also noted which trees were in each part. It was 1844, and he was working on the Ozark Plateaus where Schoolcraft had traveled 26 years earlier. On the "northern sides of the hills" James found "open post oak and black oak woods, undergrowth oak shrubs, grape vines, and prairie grass." The witness trees that James used to mark the corners of his survey lines show that white oak was also prominent, as well as small amounts of black hickory. They also show that the woodlands James saw within this ancient forest were very open, even on wetter and cooler north slopes. There were only about 10 mature trees per acre in the woodlands.[216] A closed oak-hickory forest might have several times that many mature trees on an acre.[217]

South slopes and valleys had even fewer trees than north slopes. "The vallies and southern slopes," James wrote in his field notes, were "prairie or very open barrens [savannas] with scarce any undergrowth but grass." Patches of "oak sprouts" also protruded above the "prairie grass" in places. These oaks just kept sprouting because Indian and lightning fires ran through the grass so often that they had little chance of becoming trees. However, a sprinkling of larger post oak, black oak, and some black hickory, also stood on these savannas. Eastern redcedar grew here as well, but it stayed well away from the fires on dry, rocky slopes and bluffs. James only found "heavy timber" along rivers. Maple, elm, hackberry, sycamore, and other fire-sensitive trees grew on these wet bottomlands with "undergrowth . . . in thickets which are hard to penetrate," noted James.[218]

Most of the woodlands in the ancient oak-hickory forest were in the hilly country toward the east because rainfall was higher and fires burned less often than in the West. However, fire was still an essential part of this forest. Fire favored the oaks and hickories, and it pushed back the settler species that tried to take over the forest. It also kept the woodlands, savannas, and prairies open and luxuriant. Therefore, even in the rugged highlands of the Ozark Plateaus and Ouachita Mountains, the average time between fires may have been only 27 years. It was probably somewhat longer between fires on slopes with cooler northerly exposures and shorter on slopes with hotter southerly exposures.[219] Savannas burned an average of once every 3 years.[220]

Millions of acres of oak woodland also sprawled over a line of rolling hills and sharp ridges to the west of the Ozarks and Ouachitas. This hot, dry, and isolated part of the oak-hickory forest, known as the Cross Timbers, extended from northern Texas through central Oklahoma and into southeastern Kansas. It faded into the Great Plains on the west side and the Blackland Prairies of Texas on the southeast side. This was a forest of stunted trees kept small by a constant barrage of prairie fires. In some places, the trees were "killed almost annually," and the sprouts from these oaks created "scions of undergrowth," according to Josia Gregg who described this forest in 1844. However, in other places Gregg said that the "oaks are of considerable size, and able to withstand the conflagrations."[221]

A U.S. War Department exploration party traveled through the ancient Cross Timbers in 1852 and included a matter-of-fact account of it in their report:

> This extensive belt of woodland, which forms one of the most prominent and anomalous features upon the face of the country, is from five to thirty miles wide, and extends from the Arkansas river in a southwesterly direction to the Brazos, some four hundred miles. . . . the trees, consisting principally

of post-oak and black-jack [oak], standing at such intervals that wagons can without difficulty pass between them in any direction. The soil is thin, sandy, and poorly watered. This forms a boundary-line, dividing the country suited to agriculture from the great prairies.[222]

Count de Pourtales saw the Cross Timbers 20 years earlier while on a tour with Washington Irving. These two men had completely different impressions of this ancient forest even though they saw the same thing. To de Pourtales it was "one of the most beautiful stretches of forest that I have ever seen." "There were magnificent, sparsely scattered trees and twenty varieties of climbing plants," he wrote, and the "entire wood seems to burst with the many colors of autumn."[223] Washington Irving was less exuberant:

> Unfortunately, we entered too late in the season. The herbage was parched [and] the foliage of the scrubby forest was withered. . . . The fires made on the prairies by the Indian hunters, had frequently penetrated these forests, sweeping in light transient flames along the dry grass, scorching . . . the lower twigs and branches of the trees, and leaving them black and hard, so as to tear the flesh of man and horse that had to scramble through them. . . . It was like struggling through forests of cast iron.[224]

The Cross Timbers was different from the rest of the ancient oak-hickory forest. Still, it shared the patchwork of woodlands, savannas, and prairies that was common in other parts of the forest. A smaller scale mosaic of patches containing trees of about the same age also existed within the woodlands, just as it did in most ancient forests. This was a pioneer forest, so light was critical. Periodic droughts, frequent surface fires, and occasional tornadoes were the forces that provided the needed sunny openings where oak sprouts could grow into thickets of young trees. However, oak sprouts remained small under the shade of larger trees, even though the forest was open enough to have a grassy understory.[225]

Very few patches reached the old transitional stage of development with an overstory of old oaks and an understory of settler trees. Likewise, fires confined dense self-replacing forests to ravines and other protected places. Still, early travelers to the Cross Timbers such as Josia Gregg often complained about dense undergrowth. He wrote in 1844 that the "underwood is so matted in many places with grapevines, green-briars, etc., as to form almost impenetrable 'roughs.'"[226] Shrubs and vines grow quickly after a fire, so it is not surprising that they choked the understory of some oaks, although surface fires repeatedly burned them back. This changed when the highly flammable eastern redcedar invaded a patch of oak. Then a fire flared up into the crowns of the old oaks and destroyed them. This killed the redcedars but not the oaks. They just sprouted into vigorous thickets that eventually thinned and became patches of gnarled, flat-topped, fire-scarred, veteran oaks that made the ancient oak-hickory forest look its age.[227]

The Trapper's Forests

The snowy summits of the Stony Mountain like vast heaps of
white clouds appeared in view presenting to our sight (as the
sun shone full upon it) the most grand and romantic views.
James Bird, fur trader
View from the Rocky Mountain House
Alberta, Canada, 1799[1]

Trappers and fur traders were the first Europeans to see most of the ancient forests of the West. Some of them wrote accounts of what they saw, but many did not because they were scouring the mountains for beaver, not trees. Likewise, some gained glory for their efforts and others, probably one in five, died in the mountains unknown and forgotten.[2] These mountain men were a diverse lot, yet they shared an inner strength, independence, and lust for adventure. They were all exceptional; not just in their time, but in any time.

Author and adventurer George Ruxton knew the trapper as well as anyone. He was a British officer, and a gentleman of his day, but he was as much at home alone in the mountains during the 1840s as any trapper or Indian. In his opinion, the trapper's "habits and character" consisted of "simplicity mingled with ferocity." They are "callous to any feeling of danger," he said, and "laws, human or divine, they neither know nor care to know." Their "good qualities" were "those of the animal," because they rivaled "the beasts of prey in discovering the haunts and habits of game, and in their skill and cunning in capturing it." They were also "strong, active, hardy as bears, daring [and] expert in the use of their weapons." Unwilling to say so himself, Ruxton added that "people fond of giving hard names call them revengeful, bloodthirsty, drunkards." Still, he admitted, "there are exceptions, and I have met honest mountain men."[3]

Trappers led a perilous existence. They had to be tough and resourceful. Above all, they had to be knowledgeable and keenly aware of their surroundings. Even a small mistake was often fatal. Trapper Warren Ferris stressed this point in his journals. He survived 6 years in the western mountains during the early 1830s, so he spoke with authority:

> It may seem ... a trifling matter to note the track of footmen, the report of firearms, the appearance of strange horsemen, and the curling vapour of a far off fire, but these are far from trivial incidents in a region of country where the most important events are indicated by such signs only. Every man carries here emphatically his life in his hand, and it is only by the most watchful precaution, grounded upon and guided by the observation of every unnatural appearance however slight, that he can hope to preserve it.[4]

We know what many trappers looked like, but the photographs we see usually show them when they were older. Most of them dressed in their best for the photographer. But Ruxton saw trappers as they were in their prime. Their faces were those of men who endured the rigors and privation of life in the mountains. They were all "browned by ... constant exposure to the sun and wind." Otherwise, some were "fine, hardy-looking young fellows," and others were wrinkled and scarred. Still, they dressed alike because they lived alike. Ruxton used his genius with words to paint a vivid picture of the average trapper's attire:

> On starting for a hunt, the trapper fits himself out with the necessary equipment.... This equipment consists usually of two or three horses or mules ... and six traps, which are carried in a bag of leather called a trap-sack. Ammunition, a few pounds of tobacco, dressed deer-skins for moccasins, &c., are carried in a wallet of dressed buffalo-skin called a possible-sack.... The costume of the trapper is a hunting-shirt of dressed buckskin, ornamented with long fringes; pantaloons of the same material, and decorated with porcupine-quills and long fringes down the outside of the leg, a flexible felt hat and moccasins clothe his extremities. Over his left shoulder and under his right arm hang his powder-horn and bullet-pouch, in which he carries his balls, flint and steel, and odds and ends of all kinds. Round the waist is a belt, in which is stuck a large butcher-knife in a sheath of buffalo-hide, made fast to the belt by a chain or guard of steel; which also supports a little buckskin case containing a whetstone. A tomahawk is also often added; and, of course, a long heavy rifle is part and parcel of his equipment. I had nearly forgotten the pipe-holder, which hangs round his neck, and is generally a *gage d'amour*, and a triumph of squaw workmanship, in shape of a heart, garnished with beads and porcupine-quills. Thus provided, and having determined the locality of his trapping-ground, he starts to the moun-

tains, sometimes alone, sometimes with three or four in company, as soon as the breaking up of the ice allows him to commence operations.[5]

This was the trapper as Ruxton saw him, but he was more than just rough, simple, and independent. Certainly Jim Bridger was the archetype of a simple trapper. David Brown, who knew him, wrote in the early 1800s that Bridger was "thoughtful" and "muscular," with a neck "which rivaled his head in size and thickness." Brown added that his "bravery was unquestioned, his horsemanship equally so, and . . . he had been known to kill twenty buffaloes by the same number of consecutive shots." However, Brown noted that Jim Bridger "was perfectly ignorant of all knowledge contained in books, not even knowing the letters of the alphabet." Still, he had "a complete and absolute understanding of the Indian character."[6]

Some trappers were more sophisticated than Jim Bridger, although few could have survived two arrows in the back as Bridger did during a ruckus with Blackfeet in 1832.[7] However, most were equally brave and resourceful. Jedediah Smith is one example. He hardly complained when a grizzly nearly tore off his head in 1823. "If you have a needle and thread git it out," Smith said calmly, "and sew up my wounds around my head."[8] Smith did not use profanity or boast. He was clean-shaven, serious, and prayerful. Smith was a Methodist and read the Bible constantly. But a fellow trapper commented that "when his party was in danger, Mr. Smith was always among the foremost to meet it, and the last to fly." His bravery led him to confront 20 Comanche warriors rather than flee. They caught him alone while searching for water and he had no chance to escape. "The Indians . . . frightened his horse, so that he would turn from them . . . and then fired on him," wrote his brother Austin in a letter to their father in 1831. The ball struck Smith in the left shoulder. Then Smith "faced about" and calmly aimed his musket at their chief. The chief fell dead just as Comanche lances ended Smith's life at the age of 32.[9]

Wealth, glory, and adventure motivated most fur trappers and traders, but the nations they represented wanted nothing less than to possess the land. This rivalry among nations, a president's vision, and two historic events involving fur traders sparked the opening of the West. On May 11, 1792, Captain Robert Gray, an American fur trader, rediscovered the river that he named after his ship *Columbia*. The Spaniard, Bruno de Hezeta, detected currents from the river in 1775, although he did not see it.[10] Therefore, Captain Gray was first. He had stumbled on the key to America's claim to the Pacific Northwest. Then, in 1793, fur trader Alexander Mackenzie paddled and trudged from Fort Chipewyan in Alberta, over the Canadian Rockies, and down the Bella Coola River to the Pacific Ocean in British Columbia. It took him 74 days to travel the 1200 miles to the coast, and 33 days to return to the fort; an astounding and difficult feat. He was the first European ever to make the crossing.[11] But he was British. So, these events, and the Louisiana Purchase of 1803, reinforced President Thomas Jefferson's determination to explore the West.

Captain Gray and Alexander Mackenzie unknowingly helped to launch Lewis and Clark's Corps of Discovery. They, in turn, helped to open the way for waves of fur trappers and traders. These hardy adventurers scoured the mountains and relearned the routes that Indians had known for millennia. They seldom wrote down what they saw, but in their later years they served as guides and scouts for the U.S. Army Corps of Topographical Engineers. These unsung military explorers filled in the blanks on maps and made it possible for scientists and pioneers to enter the last of America's ancient forests.[12]

The trapper's forests were some of the most magnificent in North America. They included the ponderosa pine forest, the lodgepole pine forest, the Pacific Douglas-fir forest, and the giant sequoia forest. Trappers probably first entered the ancient ponderosa pine forest because it grew on the lower mountain slopes. This forest consisted of tall trees with an open, grassy understory, much like southern pine forests except the trees were much bigger. They struggled through thick lodgepole pine forests next because they grew higher up the mountains. In the Pacific Northwest, shipborne fur traders were the first to see the ancient Pacific Douglas-fir forest. Even so, trappers arrived soon after Lewis and Clark's Corps of Discovery walked into this damp, dark forest of immense trees. The giant sequoia forest stayed hidden from the trappers longer than any other ancient forest because it consisted of small, widely dispersed groves. Each of these forests was impressive in a different way, but they all shared a grand mountain setting, and it is fitting that mountain men were among the first to see them.

PONDEROSA PINE FOREST

An infinite colonnade.
C. E. Dutton, geologist, 1887[13]

The ancient ponderosa pine forest is a symbol of the West (Fig. 12.1). This

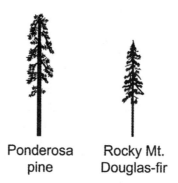

Ponderosa   Rocky Mt.
pine   Douglas-fir

Fig. 12.1   Trees of the ancient ponderosa pine forest.

was a forest of sunny slopes and long summers, just the kind of place where Indians lived and European explorers traveled. Therefore, everyone who journeyed west knew this forest, especially trappers. Spanish conquistadors roaming through the Southwest in the sixteenth century probably were the first Europeans to see the ancient ponderosa pine forest.[14] However, Lewis and Clark were probably the first to write about it. They called the tree "long-leaved pine" at the time because of its 6- to 8-inch-long yellow-green needles. They wrote their brief description of the forest on September 9, 1805, while traveling through the Bitterroot Valley in Montana:

> The valley of the river through which we have been passing is generally a prairie from five to six miles in width. . . . The timber which it possesses is almost exclusively pine, chiefly of the long-leaved kind, with some spruce, and a species of fir. . . . Near the water-courses are also seen a few narrow-leaved cottonwood trees.[15]

The next time Lewis and Clark mentioned seeing the ancient ponderosa pine forest they were traveling through western Idaho after wintering on the Columbia River. It was May 7, 1806, and they had just climbed a hill after traveling "a difficult and stony road." The land was far more agreeable on the other side. From "the heights," they said, "we saw before us a beautiful level country, partially ornamented with long-leaved pine, and supplied with an excellent pasture of thick grass and a variety of herbaceous plants."[16]

Botanist David Douglas traveled through the same region just 20 years later. His description helps to complete our image of the scene.

Douglas said that this ancient ponderosa pine forest consisted of "extensive plains with groups of pine-trees, like English lawn, with rising bluffs or little eminences covered with small brushwood, and rugged rocks covered with ferns, mosses, and lichens."[17] The savannalike ponderosa pine forest that Lewis and Clark, and Douglas, described often grew on rocky or dry plains and hillsides. However, continuous forests of ponderosa pine were also common.

Early travelers found it easy and pleasant to ride through a ponderosa pine forest. Dr. J. D. Cooper, who served as the naturalist for an 1853–1855 railroad survey, noted this on the east slope of the Washington Cascades.

> There is so little underbrush in these forests that a wagon may be drawn through them without difficulty ... The level terraces, covered everywhere with good grass and shaded by fine symmetrical trees of great size, through whose open light foliage the sun's rays penetrate with agreeable mildness, give to these forests the appearance of an immense ornamental park. ... This beautiful forest continued for about twelve miles.[18]

The ancient ponderosa pine forest was rarely dense. However, it was more open on lower slopes than on moist upper slopes, often growing as a pine savanna like the one portrayed by Dr. Cooper. Still, even on upper slopes the trees stood apart and fallen logs seldom blocked the way. Interspersed within this forest were also many fine prairies and meadows, so there was usually plenty of grass for the traveler's horses and mules to graze. Lieutenant Edward Beale, who commanded the Army's experimental Camel Corps, penned a concise description of a southwestern ponderosa pine forest while traveling across central Arizona in 1857:

> It is the most beautiful region I ever remember to have seen in any part of the world. A vast forest of gigantic pines, intersected frequently with open glades, sprinkled all over with mountains, meadows and wide savannahs, and covered with the richest grasses.[19]

The "glades" or "parks," as many early travelers called them, were a common part of ancient ponderosa pine forests. In 1887, C. E. Dutton, a geologist, described these "glades" as "dreamy avenues of grass and flowers winding between sylvan walls, or spreading out in broad open meadows."[20]

Dutton and Lieutenant Beale both wrote their descriptions after seeing one of the largest tracts of ponderosa pine forest in existence, a forest that Lieutenant Beale said "was traversed by our party for many days." It stretched over central Arizona and eastern New Mexico in a strip 25–40 miles wide and almost 300 miles long.[21] Still, it represented only a fraction of the 57 million acres of ancient ponderosa pine forest that spread throughout the West.[22] This was the most widely dispersed pine forest in North America.[23]

Strips and blocks of ancient ponderosa pine forest grew on the lower slopes of many western mountains in a broad horseshoe-shaped band. This band nearly encircled the Sonoran and Mojave Deserts and the Great Basin. It began in the huge unbroken forest that spread across Arizona and New Mexico. From here, stringers of ponderosa pine forest wound northward around lower mountain slopes through Colorado, Wyoming, and western Montana, curving westward through northern Idaho. The forest continued west using the edges of the Blue Mountains of eastern Oregon as stepping stones to cross the dry floor of the upper Great Basin. Then it turned southward and followed the eastern and western slopes of the Cascade Range in Washington and Oregon. Ponderosa pine forests spread out in southern Oregon and northern California and then they narrowed again, continuing south along the slopes of the Sierra Nevada. Finally, the west side of the horseshoe ended in the coastal mountains of southern California. Here it splintered into tiny, scattered islands of ponderosa pine forest trapped high above hot, dry hills covered with chaparral.[24]

The age of this ancient forest depended on where it grew. Ponderosa pine trees took refuge in the mountains of southern Arizona and New Mexico, and in Mexico, during the Ice Age. They moved north when the climate warmed. So southwestern forests were much older than northern forests. Consequently, the ponderosa pine forest that Lieutenant Beale saw in central Arizona was about 10,600 years old.[25] The ponderosa pine forest in western Idaho that delighted Lewis and Clark was probably about 4000 years old.[26] California was different because ponderosa pine trees only had to move uphill from the lowlands. Therefore, they arrived at their modern locations between 12,000 and 11,000 years ago. However, in the Sierra Nevada ponderosa pine had to share the forest with other trees such as lodgepole pine that could not migrate higher because cold air still flowed down from melting glaciers. So nearly pure ponderosa pine forests did not appear in these mountains until 9000–8000 years ago.[27]

This ancient forest consisted mostly of ponderosa pine, but the nearly identical Jeffrey pine dominated it on the east side of the Sierra Nevada in California. Ponderosa pine is an unforgettable tree. Young sapling and pole size ponderosa pines have brown bark like most trees. Early travelers and foresters called them "bull pines."[28] However, when the trees grow older and thicker, the bark splits into long, narrow yellowish plates separated by deep brown fissures. Little red-brown scales shaped like puzzle pieces cover these plates. The scales on Jeffrey pine tend more toward purple than ponderosa pine, otherwise their bark looks the same. The bark of both trees is so distinctive that when they matured early travelers called them "yellow pines." Their colorful trunks were all the more conspicuous in the ancient forest because they were free of branches for half or more the height of the tree.

Most ponderosa pines in the ancient forest were large, but not huge, as Isaac Stevens, governor of the Washington Territory, reported in his 1855 railroad survey:

> Entering the [Rocky] mountains on the eastern side.... The pines, and especially the pines of the valleys, ... and in the lower and easier mountain slopes, are tall and straight, and have a diameter of about three feet, and a height of from one hundred to one hundred and forty feet. The streams intruding into these wooded regions have in them a considerable amount of open and grassed lands.[29]

These are still big trees, and some were even larger than those seen by Governor Stevens, especially in California, reaching heights of 260 feet and diameters of up to 9 feet. Some old Jeffrey pines were nearly as large.[30]

The red, purple, and brown streaks and blotches that decorated the massive yellow trunks of ponderosa and Jeffrey pine trees made them look like marble pillars in a great green cathedral. Some of these forest cathedrals stood for over 700 years before European explorers entered them for the first time.[31] No wonder C. E. Dutton, a geologist, seemed reverential when he described an ancient ponderosa pine forest near the Grand Canyon in 1887:

> The trees are large and noble in aspect and stand widely apart.... Instead of dense thickets where we are shut in by impenetrable foliage, we can look far beyond and see the tree trunks vanishing away like an infinite colonnade....

Fig. 12.2   Ponderosa pine. (Courtesy of the U.S. Forest Service; photograph by Tom Iraci)

All is open, and we may look far into the depths of the forest on either hand.[32]

Ancient ponderosa pine forests were the first forests of large trees that trappers and explorers saw as they worked their way into the western mountains (Fig. 12.2). Captain Lorenzo Sitgreaves noticed this and wrote a brief description in his report on how forests changed as he climbed a mountain in western Arizona. It was October 1851, and trapper Antoine Leroux was guiding his survey expedition across the Southwest. They could not have picked a more able guide even though he had to quit before the expedition ended due to an arrow wound inflicted by angry Yampais.[33] "As we ascended the mountain," Captain Sitgreaves wrote, "the cedar gave place to the nut-bearing pine [piñon pine]; and this, when near the summit, to a pine of larger growth with long leaves [ponderosa pine]." Then, he said, the "tall pines, mingled, after attaining a certain altitude, with aspens of a brilliant yellow."[34] Higher still the pines merged with Rocky Mountain Douglas-fir and other conifers in the mountains of the Southwest.

Mixed-conifer or Rocky Mountain Douglas-fir forests that often included ponderosa pine usually sat on the moister slopes directly above ponderosa pine forests. The mixture of trees differed somewhat from north to south, but the change in vegetation was more dramatic on the lower and drier side of these forests. Shrub-steppes or grasslands blended into the lower edge of ponderosa pine forests in northern mountains, while piñon-juniper woodlands graded into them in the central and southwestern mountains. Captain Sitgreaves also noted "tall pines, oaks, and the low, spreading cedar" on some lower slopes in Arizona that "were mingled so as to produce a park-like effect."[35] Chaparral also took over the foothills below these forests in many parts of southern Arizona and New Mexico. The same was true in southern California, but oak woodlands grew below them elsewhere in California and southern Oregon.[36]

This forest was rich in wildlife. For example, approximately 152 mammals, birds, reptiles, and amphibians lived in the ancient ponderosa pine forest of the Blue Mountains of northeastern Oregon and southeastern Washington. About 110 of them also raised their young in this forest. Several animals also preferred ponderosa pine to other forests, such as the white-headed woodpecker, white-breasted nuthatch, pygmy nuthatch, and the flammulated owl.[37] Likewise, the Abert or tassel-eared squirrel depended almost entirely on ponderosa pine forests in the Southwest.[38]

Deer were abundant in the ancient ponderosa pine forest and black bears were common. Elk also lived in this forest, but they were spotty because the herds moved constantly.[39] In addition, Captain Sitgreaves noted: "The cry of the panther … was occasionally to be heard."[40] Bison, bighorn sheep, and pronghorn antelope wandered into this forest as well.[41] Captain Sitgreaves mentioned seeing pronghorns when he entered a ponderosa pine forest in western Arizona. He wrote in his report that "herds of antelope were seen in all directions, but they kept to the open country, and were shy and difficult to approach."[42] The pronghorns he saw were probably within the lower edge of the forest where it merged with a piñon-juniper woodland. However, Lieutenant Joseph Ives also found pronghorns high in the mountains of Arizona. He was leading an expedition in 1857–1858 that made history when they became the first Europeans to explore the floor of the Grand Canyon.[43] They rode into the forest shortly after their trip into the "Big Cañon," as Lieutenant Ives called it.[44]

The route continued through an open park, dotted with flowery lawns and pretty copses, and then reached the edge of the great forest that surrounds the San Francisco mountain, and entered its sombre precincts. It was delightful to escape from the heat of the sun, and travel through the cool underwood. Across the dark shady glades a glimpse would sometimes be caught of a bright tinted meadow glowing in the sunlight. Antelope and deer were constantly seen bounding by, stopping for a moment to gaze at us, and then darting off into the obscure recesses of the wood.[45]

Deer and pronghorn were abundant because this was an open and airy forest of tall trees standing in a sea of grass. These were features that naturalist C. Hart Merriam vividly recalled in 1889 in an account he wrote of the same area that Lieutenant Ives had explored.

The pine forest is thoroughly mature, nearly all of the trees being of large size, and rarely crowded. It is a noteworthy forest, not alone on account of the size and beauty of the single species of tree of which it is composed (*Pinus ponderosa*), but also because of its openness, freedom from undergrowth, and its grassy carpet.[46]

Lieutenant Beale also emphasized the parklike character of the ponderosa pine forest in a report he wrote over 30 years earlier. He traveled for "ten miles" through this "glorious forest of lofty pines" that, he said, was "perfectly open."[47] Foresters D. M. Lang and S. S. Stewart went even further, saying in their 1910 timber survey for the Kaibab Plateau in Arizona that "practically all of the yellow pine type is open."[48]

No force other than fire could have sustained the openness of ancient ponderosa pine forests. This was a fire forest, but the fires that burned here were mostly gentle. They carved fine details into the forest just as an artist lightly taps a chisel to finish a sculpture. These were surface fires that only flared up occasionally, much like those that burned in southern pine forests. Fires in ponderosa pine forests not only burned more gently but they also burned more frequently than in most other western forests.[49]

Indian and lightning fires burned in ancient ponderosa pine forests an average of about once each 2–14 years.[50] The shortest intervals between fires were in Arizona, New Mexico, and California where Indian populations were high.[51] Longer intervals between fires occurred on the east side of the Cascades in Oregon and in the Black Hills of South Dakota.[52] However, the period between fires was much longer in wet and extremely dry areas. In wet areas, such as along streams, fires burned about 20–50 years apart.[53] They burned about 46 years apart in forests that grew on shallow, rocky soils where the trees were sparse and fuels accumulated slowly.[54] Here as elsewhere in ponderosa pine forests, fires still started more often on drier south slopes than on moister north slopes.[55]

Indians greatly increased the number of fires that would have burned in ancient ponderosa pine forests from lightning alone.[56] S. J. Holsinger, an examiner for the General Land Office in Arizona, wrote in 1902:

> These prehistoric aborigines must have exerted a marked influence upon the vegetation of the country. Their fires, and those of the historic races, unquestionably account for the open condition of the forest.[57]

Even so, more lightning fires started in ponderosa pine forests than in any other forest in the northern Rockies, and probably elsewhere in the West. So the fire frequency would still be high without Indian burning.[58] Regardless, Indian people kept these forests even more open and diverse because they set additional fires.

The Flathead, Pend d'Oreille, and Kootenai set many fires.[59] They more than doubled the fire frequency in ponderosa pine forests that grew near their villages in Montana. They also increased fires along trails and at favored campsites, as H. B. Ayres noted during a U. S. Geological Survey inspection near the Bitterroot Valley in the late 1890s. "The trails most frequented by Indians ... are noticeably burned," he wrote, "especially about the camping places."[60] Consequently, nearby forests burned about once in 9 years, while forests farther away burned about once in 18 years.[61] However, Indians set fires in these more remote forests as well. Thus, lightning probably ignited fires only about once in 23 years, perhaps even less frequently.[62] Lightning fires were rarer still in the ponderosa pine forests above the central Rio Grande Valley in New Mexico, yet the forests burned an average of once every 5–6 years. Therefore, the Pueblo Indians who lived in the valley below the forests must have set most of these fires.[63] Likewise, lightning alone could not account for an average of 11 years between fires in Kings Canyon, in the southern Sierra Nevada. Lightning tended to strike the tops of the surrounding cliffs rather than the valley floor, so Indians must have been responsible for most of the fires that burned in the bottom of the canyon.[64]

Fires that burned less than 20 years apart left little heavy fuel in a ponderosa pine forest for the next fire. Therefore, most fires burned in grass and pine needles, only occasionally did a fire meet a pile of dead trees. So these were creeping surface fires with flames that licked a few feet up the sides of large trees, and then only briefly. Although many fires burned for months and covered thousands of acres, they usually killed few large trees.[65] Foresters Lang and Stewart noted this in their 1910 timber survey for the Kaibab Plateau in Arizona. They said there was evidence of "light ground fires over practically the whole forest," yet "some of the finest stands of yellow pine show only slight charring of the bark."[66] Still, even these ancient forests were not immune to destructive fires. Lang and Stewart also noted "vast denuded areas, charred stubs and fallen trunks and the general, prevalence of blackened poles" in the forest.[67] There is little doubt that bleached or blackened dead trees, standing and fallen, were a part of this forest as they were in most other ancient forests. They were just less common in ponderosa pine forests.

The Black Hills of South Dakota and southeastern Wyoming were a dramatic exception. The fires that burned in the forests that grew there were more severe than those that burned in ponderosa pine forests elsewhere in the West, although some were gentle. These forests and the fires that ravaged them were unusual because the Black Hills were islandlike mountains within the northern Great Plains. Colonel Richard I. Dodge called the hills "a true oasis in a wide and dreary desert" when he saw them in 1875.[68] However, Trapper Warren Ferris had a different perspective. When traveling near the Black Hills in the early 1830s, he noted that the pine forests gave them "a dark forbidding appearance."[69] Ferris also knew that Indians made the Black Hills dangerous.

According to the treaty of 1868, the Black Hills belonged to the Sioux.[70] The Sioux vigorously defended the hills after having taken them from the Cheyenne, Kiowa, and Crow nearly 100 years earlier.[71] However, that became more difficult when rumors spread that the Black Hills were full of gold. Public opinion forced the U.S. Army to send the Seventh Cavalry, under General George A. Custer, into the Black Hills in 1874 to find out whether gold was there or not. General Custer found gold, but his findings were inconclusive, so a second scientific expedition was sent the following year under the command of Colonel Dodge. He also found gold and the Sioux lost the Black Hills.[72]

The reports from these two military expeditions show that extensive hot fires burned through the ancient ponderosa pine forests in the Black Hills. Lightning fires were frequent because of the winds that blew into the hills from the northern Plains. "Throughout the Hills," wrote Colonel Dodge, "the number of trees which bear the marks of the thunderbolt is very remarkable."[73] However, Indians regularly set fires in the surrounding prairies as well, and many of them swept into the ponderosa pine forests on the hills. No doubt they also set some fires within the forests.

Indians spent little time in the Black Hills, including the Sioux, preferring the Plains instead. One old Indian told Colonel Dodge through an interpreter that "the Hills are 'bad medicine,' and the abode of spirits." Nonetheless, some Indians used them for "hunting-grounds and asylum," according to Colonel William Ludlow, Chief Engineer for General Custer's 1874 expedition.[74] Game was abundant even if it was less plentiful than on the plains. Several decades earlier trapper Osborne Russell visited the hills and said "this section of the country . . . has always been celebrated for the game with which it abounds."[75] Indians also visited the Black Hills to cut lodge poles for their tepees.[76]

Indian fires most likely burned less often in the Black Hills than they did in other ponderosa pine forests, at least until the Sioux invaded the area in 1770. Then the average time between fires dropped from 27 to 14 years.[77] This may have led to bigger fires since fuels had time to build up before the Sioux arrived. No doubt future fires would have been smaller and less severe if the Sioux could have continued burning. The Sioux may have increased the frequency of fires because they used the hills for hunting and other purposes more than other Indians in the area did. However, it is likely that some of these fires started during battles the Indians fought with one another for the hills, and later when the Sioux fought the U.S. Cavalry.[78]

No matter what the reason, large fires had ravaged the ponderosa pine forests in the Black Hills by the time that Europeans first saw them. Colonel Dodge recalled the aftermath of these Indian and lightning fires:

> There are many broad belts of country covered with the tall straight trunks of what was only a short time before a splendid forest of trees. . . . The largest of these fires occurred on the head waters of Box Elder Creek. What was evidently a beautiful body of timber fifteen miles long by at least five broad, is now only dead trunks, some standing, . . . the larger portion prostrate . . . making travel through them a trial . . . more especially as the standing trunks, partially decayed, are swayed with every breeze, and seem "just tottering to a fall."[79]

Young pioneer trees moved in to replace those that died in the fires. Aspen took over some areas while ponderosa pine seedlings came back in others. However, since the fires were recent and extensive, many ancient forests in the Black Hills were in the early stages of recover, as Colonel Ludlow pointed out in his report:

> There are few standing sound pines, but there seems to have been here, as in many other parts of the Black Hills region, very extensive fires, that have burned the former forest and left the charred trunks and limbs scattered on the surface. Among these have sprung up a perfect mesh of shrubs and small deciduous trees, mostly trembling aspen. We have passed some pretty good ... pine, but by far the greatest portion is small, ... and low branched.[80]

Colonel Dodge expanded on this description the following year. He wrote in his report that "large, fine saw-logs are to be found, and in very considerable numbers, but rarely in large bodies." Many of these "old and large trees, which by some means escaped the fires," he said, were "scattered through the young forest."[81] Colonel Dodge estimated that 40% of the Black Hills forests was "young," probably with scattered large trees, 10% was "wind-shaken, or injured by lightning or fire," and 10% was "good lumber." Grassy openings made up the remaining 40% of the forest.[82]

The hot fires that burned in the Black Hills during the eighteenth and nineteenth centuries were notable exceptions for ponderosa pine forests. Likewise, the large patches and abundance of young trees that resulted from these fires were unusual. Still, in a 1905 Geological Survey report, T. F. Rixon noted "a phenomenal growth of young pine" that was 30 feet tall along a creek in southwestern New Mexico.[83] They most certainly grew after a hot fire. So severe fires also occurred in ancient ponderosa pine forests outside the Black Hills, although they were rare.[84] Most ponderosa pine forests stood for centuries even as surface fires swept through them every few years. These light fires did little damage to the mature trees because their thick, fire-resistant bark protected the living tissue underneath. Moreover, they burned too often for young trees and debris to become dense enough to support catastrophic fires.

There is no doubt that large trees dominated the scene within most ancient ponderosa pine forests. However, no forest can consist of only old trees. Early travelers who first described these forests tended to ignore small trees much as the Spanish conquistadors had done in longleaf pine forests in the South. That is understandable because they were remarkably similar forests. The openness of even the densest ponderosa pine forests focused the traveler's eye on the thick, straight trunks of tall pines fading into the distance. It must have been an overpowering sight. Most travelers would not notice a few little patches of seedlings, saplings, or even pole-size trees standing here and there among these imposing trunks, much less write about them. Likewise, only the most astute observer would notice that since little trees grew in patches the large trees had to grow in patches as well.

One of the earliest hints of patchiness in ancient ponderosa pine forests came from Clarence King's 1871 account of riding up the west slope of the Sierra Nevada:

> At last, after climbing a long, weary ascent, we rode out of the dazzling light of the foothills into a region of dense woodland, the road winding through avenues of pines so tall that the late evening light only came down to us in scattered rays. . . . Passing from the glare of the open country into the dusky forest, one seems to enter a door, and ride into a vast covered hall. The whole sensation is of being roofed and enclosed. You are never tired of gazing down long vistas, where, in stately groups, stand tall shafts of pine. Columns they are, each with its own characteristic tinting and finish, yet all standing together with the air of relationship and harmony.[85]

The "groups" King saw were small patches of old pioneer forest. Even though King and most other early travelers overlooked groups of young trees, they were also part of this forest. As in most ancient forests, a ponderosa pine forest was a mosaic of different patches, and the trees in a given patch were about the same size and roughly the same age.[86] Most of them consisted of young, middle-aged, and old pioneer forests. Grassy openings were also abundant, but denser patches of old transitional and self-replacing forest were rare because of repeated thinning of understory trees by Indian and lightning fires.

The patchiness of ponderosa pine forests only became generally recognized during the first timber surveys in the early 1900s. Lang and Stewart reported in 1910 that the forest on the Kaibab Plateau was "irregular in density, age classes, and quality of timber represented." They added that old trees ranged "from moderately dense stands to isolated trees."[87] Lone saplings and poles also grew here and there in the ancient forest.[88] But the group structure of ponderosa pine forests was unmistakable. A. C.

Ringland, a forester who worked in New Mexico, made a point of this in his 1905 timber survey. He said "reproduction is ... in small groups," and "small pole bull pines occur in scattered small and compact even aged groups."[89] These groups or patches were indeed small. Scientists now know that most were less than two-tenths of an acre, although some covered three acres or more.[90]

Much like ancient longleaf pine forests in the South, some ponderosa pine forests still looked to early travelers like one continuous stand of old trees. Even foresters such as Lang and Stewart, who often mentioned the patchiness of the forest in their survey, still commented that the Kaibab Plateau was "an unbroken body of mature timber."[91] This visual impression was misleading but fitting given that early travelers could look so far into the open forest. Thus the widely scattered groups of big trees seemed to merge into one uniform forest. Moreover, old trees were larger and more abundant in some parts of the forest than others. In the Southwest, patches dominated by trees over 150 years old normally varied from 3 trees on one-twentieth of an acre to 44 trees on three-quarters of an acre.[92] Some covered even larger areas. Nevertheless, the number of old trees differed considerably from place to place since there was only about one old patch on each acre. No doubt large patches with many big pines made a greater impression on early travelers than sparse areas, so that is what they remembered best about the ancient forest.

Overall, patches of old pioneer forest probably made up less than 40% of ponderosa pine forest mosaics. More commonly they covered only 25% to as little as 17% of the mosaic.[93] The remainder consisted of young and middle-aged forests, and grassy openings, that varied in proportions due to sporadic events such as fires, insects, disease, and droughts. Shrubs also grew in openings within this ancient forest in some regions, such as on the east slope of the Washington Cascades. Here, noted Dr. Cooper during a railroad survey in the early 1850s, "Almost the only shrub is a *Ceanothus*," which grows, he said, "in scattered thickets."[94] Single large trees, or pairs of trees, also stood here and there within the ancient forest.

The trees within a patch looked "even aged" to Ringland and most other early travelers because they were similar in size, but they were not always the same age. Generally, the younger a patch the closer would be the ages of trees within it. Therefore, groups of seedlings, saplings, and even pole-size trees were often about the same age. This was especially true in dry areas where soil moisture was adequate for seedling establishment only in certain years, and sometimes during only a few periods within a century.[95] So the trees usually filled an opening at the same time. In addition, they grew so dense or "compact," as Ringland noted, that they kept younger trees from invading.[96] However, young trees started to gain a foothold when a patch grew older and began to fall apart.

The tiny gap that opened in a patch when a single large tree died was still too small to let in enough sunlight for ponderosa pine. The remaining trees also drained most of the moisture from the upper part of the soil. This made it even more difficult for pine seedlings to survive.[97] Consequently, a ponderosa pine seedling that tried to grow under larger trees or in small openings usually died or became stunted.[98] Lang and Stewart recognized this and noted it in their 1910 report. "It appears," they said, "that seedlings endure the shade for a while but suffer unless the parent trees are removed." They saw first hand that in the ancient forest the "most thrifty reproduction occurs in the openings."[99]

Occasionally a young pine tree grew tall enough to enter the canopy of larger trees. More often they had to wait until surface fires, lightning strikes, and other disturbances thinned the overstory. This created more space for another generation of trees to grow among the survivors. Even so, about a third of the old pioneer patches in the ancient ponderosa pine forest still contained trees of roughly the same age, some for as long as four centuries.[100] The remaining patches included one or more generations even though all but the stunted trees looked like old veterans of about the same age.

Old pioneer forests might stand for centuries and then break apart in only a few decades.[101] The nesting sites for goshawks and other animals that lived in old forests also disappeared as the trees fell.[102] This rapid decline could happen due to old age, drought, or injury from lightning or even light surface fires. Although mature ponderosa pine trees have fire-resistant bark, sometimes a light fire killed them outright just by cooking their roots. Each year over 2 tons of needles and twigs fall to the ground per acre under old pioneer forests. These fuels decompose slowly in the hot dry climates where ponderosa pine grows, so it only takes about 20 years for fuels to stabilize at their maximum weight of about 43 tons per acre. The heavier the fuel the deeper a fire burns into the soil. The result can be soil temperatures as high as 210°F, and temperatures as low as 140°F instantly kills tree roots. Prolonged exposure to slightly lower temperatures also is lethal for roots.[103] George Sudworth saw the consequences of this during his 1899 survey of the forests near Lake Tahoe. He noted that "deep burning at the roots" killed two "older forests." One forest was less than an acre and the other covered several hundred acres. However, Sudworth considered them "exceptional cases."[104]

More often trees became weak and defenseless against insect and disease attack when fires scorched some of their roots. Overcrowded and lightning-struck trees also were vulnerable. Insects such as bark beetles could even multiply enough to spread into the surrounding forest and overwhelm healthy trees. However, infestations were typically small and short lived in the ancient forest. Young vigorous trees or grassy openings surrounded patches of less resistant and older trees and isolated them from the rest of the forest.[105] This led to spotty outbreaks that seldom destroyed whole forests, as Lang and Stewart noted on the Kaibab Plateau:

> Insect infestation has attained enormous proportions over the whole forest. . . . The old fallen trunks, existing in all stages of decay. . . . As a rule, bug-killed trees occur in patches, ranging in extent from a few trees to several acres. The largest single area . . . approximates 80–100 acres.[106]

Diseases such as dwarf mistletoe took a toll on patches of young and old pioneer forests alike. Dwarf mistletoe was widespread in ancient ponderosa pine forests. The only place where mistletoe did not grow was in the Black Hills. Mistletoe provided food for elk and other animals, but this parasitic plant caused tree limbs to swell and generate multiple branches or "witches brooms."[107] Witches brooms were ideal nesting places for many birds, at least until the trees weakened and died.[108]

Witches brooms burned easily, and mistletoe-infected trees retained their lower branches longer than other trees. Some of the branches of infected trees also hung near the ground where light surface fires could reach them. However, fires were flashy and erratic. Sometimes long curving lines of fire crawled slowly against the wind. These backing fires killed pine seedlings as they moved through the forest, but they seldom reached the branches of saplings or pole-size trees because the flames were only a foot or two high. At other times, fires rushed through the forest with the wind, their flames stretching upward and leaning forward like a runner in a foot race. These head fires easily ignited the lower limbs of mistletoe-infected trees, causing them to burn like torches. Thus, mistletoe created the fuel that often led to its destruction. These fires also kept the disease confined within scattered patches and prevented it from devastating whole forests.[109]

Occasionally a surface fire crept into a patch of dense old transitional forest and grew hotter in the deep carpet of needles and twigs that covered the ground under the trees. This boosted it into the branches of small settler trees in the understory, which exploded in fountains of flames. The young settler trees, in turn, acted like ladders, allowing the flames to climb still higher into the canopy where they ultimately reached the tops of the older trees. Then the entire patch of old transitional forest erupted into a small crown fire. Nearby grassy openings, and less flammable patches of young and old pioneer forest, kept the crown fire from spreading.[110] So the flames flickered out after doing their work on the more easily burned branches and needles, and the fire dropped back to the ground and moved on as if searching for another victim.

Old transitional patches were rare in ancient ponderosa pine forests because surface fires were so frequent. Governor Stevens mentioned this in his 1860 railroad survey report. He said that on the east side of the Washington Cascades "the timber (yellow pine) . . . is open, and with the undergrowth only thick in places."[111] These were places that fires had skipped, at least for a time, which usually occurred in valleys or on shady north slopes where conditions were wetter than other parts of the forest. Fires also burned less often during occasional wet periods. This gave settler trees and shrubs a chance to invade old pioneer forests. So there were usually a few old transitional forests available to burn. Patches of self-replacing settler trees were even harder to find in the ancient ponderosa pine forest. Nevertheless, a 1904 U. S. Geological Survey report in Arizona noted that gamble oak, which can behave like a settler species, occurred "as isolated trees, or in small clumps." More often, it grew "as mere brush forming undergrowth" rather than as distinct patches of self-replacing forest.[112]

The settler species that invaded the ancient forest differed from place to place because of the wide distribution of ponderosa pine. Gamble oak was the primary settler tree invading ponderosa pine forests on lower slopes in the Southwest, and Douglas-fir invaded it in many parts of the Rocky Mountains, the Pacific Northwest, and northern California.[113] Rocky Mountain Douglas-fir normally grew on the moist slopes above ponderosa pine, moving down into the forest whenever there was a lull in fires. The same was true with Douglas-fir on the west slope of the Sierra Nevadas in California. Still, it rarely had a chance to gain a foothold because surface fires kept pushing it back.[114] However, gamble oak quickly sprouted after a fire, although the combination of surface fires and browsing by deer and elk tended to keep it in check.[115]

Widely scattered patches of graying or blackened standing dead trees, or snags, and fallen trees were always present in ancient ponderosa pine forests. Individual snags also stood here and there within otherwise healthy old pioneer and old transitional forests. Some groups of dead trees were remnants of fires that failed to consume an entire patch. Others consisted of trees that died from insects or disease but which had not yet burned. These snags provided nesting and perching sites, and plenty of insects and prey, for owls, woodpeckers, hawks, flycatchers, swallows, bats, and many other birds and small animals. Fallen trees also gave animals a welcome place to hide in a forest that was mostly open.[116] However, most dead trees probably only lasted a few decades before fire consumed them and forced the animals to look elsewhere for food and shelter.

Intense flashes of light appeared here and there within the ancient forest as surface fires moved among the trees. Each flash represented the torching of scraggly mistletoe-infected trees, snags and piles of logs, shrubs, or old transitional forests that stood in the way of the flames. These bursts of light and heat also marked the locations of future patches of young ponderosa pine. Pine seedlings popped up throughout the ancient forest after a surface fire, even in the understory.[117] Still, they had a better chance of surviving and grew faster on bare soil in small sunny clearings of about one-tenth to one-half acre. Here shade from nearby trees gave them a little protection from the sun. Charred logs and branches also helped to shield new pine seedlings from the hot sun.[118] They grew even better in these openings after a hot fire.[119] That happened where surface fires burned dead trees or even a single log that crashed into a grassy opening.[120] Such fires were often hot enough to temporarily remove most of the grasses and herbs, which gave the pine seedlings even more room to grow.[121] In addition, the ashes from burned wood enhanced seedling growth by fertilizing the soil with nitrogen and other nutrients.[122] There were exceptions, however. The dry, nitrogen-poor volcanic soils of central Oregon declined in productivity during the first decade or so after a fire.[123]

Even as pine seedlings flourished, the number of deer and elk declined the first year after an extensive surface fire because they no longer had shrubs and grass to eat. But they quickly recovered when fresh green grass and forbs reappeared, and patches of shrubs, oak, and aspen began to produce succulent young sprouts.[124] Thus, after the initial drop, deer populations normally rose almost continuously for up to 20 years. Then they declined as the shrubs aged or diminished due to heavy browsing, and grass became thinner in the shade of the young trees. Elk populations also rose for about 10 years, but they slowly fell to their former levels during the next 10 years.[125] However, even at their peak elk were not as plentiful as deer.[126]

The young pioneer patches of ponderosa pine that came into openings after fires continued to progress from the seedling stage to the sapling and pole stages. However, those seedlings that grew in the shade of old pioneer patches became stunted and unhealthy. Even in openings, seedlings took longer to reach sapling and pole size in the hot, dry climate of the Southwest than elsewhere. There it might take up to 71 years for a patch of seedlings to reach pole size. In Oregon, where the climate is wetter, it took only 24 years for them to reach pole size. However, the best sites in the Sierra Nevada of California could produce pole-size ponderosa pines in as little as 14 years.[127]

Surface fires constantly threatened the young pines. The few seedlings that survived in the shade among piles of needles and branches under mature trees usually died in the flames. Fire also thinned the clusters of young trees that grew in openings, but it seldom destroyed an entire group. The young trees dropped too few needles to carry hot fires. So the trees that survived had more room, and grew even faster. The flames also seared their lower branches, causing them to fall off so that the saplings became even more resistant to fire. This process helped to produce tall branch-free trunks.[128]

Mature trees that stood around patches of young pioneer forest often cast heavy loads of needles and branches part way into them. Therefore, a ring of fuel accumulated inside these young patches to a distance equivalent to about half the height of the mature trees. Here fires often became hot enough to kill the young trees on the edge of the group while leaving those in the center unharmed. Thus patches that spanned about twice the height of the surrounding trees were ideal for young ponderosa pine because they only lost their outer fringe to surface fires. Therefore, frequent surface fires not only regenerated the pines, and kept them thinned and free of undergrowth, they also sharpened the boundaries between patches.[129]

Within 200 years, patches of young pioneer trees became the towering yellow columns that distinguish old pioneer ponderosa pine forests.[130] Goshawks, flammulated owls, and tassel-eared squirrels gradually settled among these aging trees, moving into new homes that replaced the ones they lost when fire destroyed patches of old trees elsewhere in the forest.[131] The clearings where old trees died then became habitat for deer, elk, pronghorn, bear, chipmunks, rabbits, and many other animals that lost their homes where trees grew older.[132] These animals furnished mountain lions, wolves, owls, goshawks, and Indians with food as well. So each part of the forest mosaic simply rotated from youth to old age, and back to youth, as plants grew older and Indian and lightning fires kept burning. Thus nothing was lost in the ancient ponderosa pine forest. Groups of trees of different ages and grassy openings just changed locations within the mosaic. The animals followed and life went on as before.

LODGEPOLE PINE FOREST

Thick pines and fallen timber.
Osborne Russell, trapper, 1835

The ancient lodgepole pine forest was less remarkable for the size of its trees than for the majesty of the scenery of which it was a part (Fig. 12.3). So its beauty was best seen from a distance. Those who saw this forest

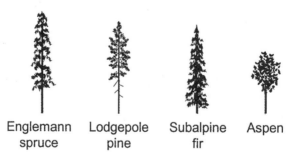

| Englemann | Lodgepole | Subalpine | Aspen |
| spruce | pine | fir | |

Fig. 12.3   Trees of the ancient lodgepole pine forest.

Fig. 12.4 Lodgepole pine, Umatilla National Forest, southeastern Washington and northeastern Oregon. (Courtesy of M.E.I. Productions, Bend, Oregon; photograph by Mike McMurray)

up close usually complained about the difficulty of traveling through its "thick pines and fallen timber," as Trapper Osborne Russell noted while exploring Yellowstone in 1835 (Fig. 12.4).[133] Even so, parts of the lodgepole pine forest were also open or consisted of what Lieutenant Gustavus C. Doane described in 1870 as "undulated prairie dotted with groves of pine and aspen."[134]

This was a forest of jagged mountains, thin soils, and harsh weather. So the trees were small but the forest was vast, spreading over western North America in a great arch that covered high plateaus and wrapped around upper mountain slopes like a blanket. A fringe of spruce-fir and other upper subalpine forests were all that separated it from the icy peaks above. Here the top edge of the forest "presents a ragged and wavy line, apparently advancing and receding from time to time," wrote John Leiberg.[135] Leiberg was a geographer working in the Absaroka Range (named for the Ap-sah-ro-ke or Crow Indians) in Montana in 1904. The ancient lodgepole pine forest occupied about 18 million acres in the Untied States and about 49 million acres in Canada.[136] Scattered areas of lodgepole pine forest also sat within other montane or mountain forests throughout the West from the Yukon Territory to northern Baja California, and even on small areas in the Black Hills.[137]

Trapper Warren Ferris called the lodgepole pine forests he saw in 1832 "Piny Woods" that "are seen stretching darkly along like a belt of twilight" over the mountain sides.[138] Colonel William Ludlow had a similar impression when he gazed at Yellowstone from the top of Mount Washburn in 1875. "We had scarcely time to more than glance at this superb landscape," Colonel Ludlow said, "when a ferocious squall of hail, rain, and snow burst upon us from the northwest, and swept us like dust from the bald summit of the mountain." Even a brief look was enough to create a vivid imprint of this stunning "panorama" in Colonel Ludlow's mind, which he later portrayed in his report.

> The Yellowstone River crooking away . . . was set, as it were, in a vast expanse of green, rising and falling in huge billows, above which here and there jets of steam arose like spray; the encircling peaks, ragged and snow-clad, almost too numerous to count . . . and beyond and above them, ninety miles away, looking almost mysterious from their distance and vast height, the Tetons, of a pale purple hue, with their piercing summits glittering like icebergs.[139]

This is an old forest, but the age of a particular part of the forest depended on when lodgepole pine moved northward, and up and down mountain slopes, at the close of the Ice Age. Lodgepole pine is a rugged pioneer that was one of the first trees to invade land abandoned by glaciers. Therefore, it moved into the cold steppes and tundra on the Columbia Plateau in eastern Washington and Oregon about 12,000 years ago. This was long before the ice sheets finally retreated into Canada.[140] However, Englemann spruce left the lowlands and advanced up the mountains ahead of lodgepole pine. So lodgepole pine followed spruce into the northern Rockies about 10,500 years ago, and subalpine fir came in closely behind it. Lodgepole pine arrived at its modern location on the upper slopes of California's Sierra Nevada about 2000 years later. However, it moved high into the Rockies when mountain glaciers melted during the Great Drought, and then it moved down to its present location about 4000 years ago when the climate cooled again.[141]

Many trappers rode through the ancient lodgepole pine forest in search of beaver. By the time they reached this forest, they had already climbed through the ponderosa pine forest and then the Rocky Mountain Douglas-fir forest.[142] That is, in most areas other than eastern Washington and Oregon where lodgepole pine and ponderosa pine grew side by side. Mount Mazama exploded about 6600 years ago in Oregon and blew volcanic ash over this landscape. Crater Lake is all that remains of this once huge volcano. The eruption created pockets of infertile, dry soil that only lodgepole pine could exploit. As a result, lodgepole pine expanded onto the lower slopes and took over part of the area dominated by ancient ponderosa pine forests. Ponderosa pine trees still grew in places where the volcanic ash was thin enough to allow their roots to reach the nutrients and moisture below. Surgeon and naturalist Dr. Cooper, who traveled with Governor Isaac Stevens's 1853 Pacific Railroad survey expedition, saw this unusual mixture of forests and wrote a brief comment about it in his report:

> On the 9th of August we commenced descending the eastern slope of the [Washington Cascade] mountains, and at once noticed a marked change in the vegetation. Instead of the dense forest of firs [Douglas-fir] covering the western side, the prevailing trees were two species of pines [ponderosa pine and lodgepole pine] and a few oaks; these stood at distances of thirty and fifty feet apart, and the ground underneath was open, smooth, and covered with a good growth of grass.[143]

Dr. Cooper did not mention it, but the boundaries between these two forests were as abrupt as the lines between the soils. Some lodgepole pine forests also grew in low areas where frost settled. Ponderosa pine did not grow in these areas because their seedlings could not withstand the cold air.[144]

Lodgepole pine probably got its name from the Lewis and Clark expedition of 1804–1806 because they called Sioux tepees "lodges" in their journals. Sergeant Patrick Gass, a member of the expedition, also referred to tepees as lodges in his journal, as did Father De Smet and other early explorers.[145] In addition, lodgepole pine is a small tree that often grew in dense patches. So it was abundant, and it made excellent poles since crowded trees stand straight, stay thin, and have few lower branches. As a result, the Sioux used them to construct their tepees or lodges—hence the name lodgepole pine.

"Compared with the giants of the lower zones, this is a small tree," wrote naturalist John Muir. He described lodgepole pine as "a well-proportioned, rather handsome little pine, with grayish-brown bark," and "a sharp, conical top." "The foliage is short and rigid," he said, and the little cones grow "in stiff clusters among the needles."[146] In 1872, Dr. Ferdinand Hayden also remarked on the small size of these trees in his description of the ancient forests that covered Yellowstone's Central Plateau:

> The plateau ridges ... are covered with a dense growth of pines, not large, seldom more than 24 to 30 inches in diameter, averaging not more than 10 inches, but rising as straight as an arrow to the height of 100 to 150 feet, and growing so thickly together that it was with great difficulty we could pass among them with our pack-animals.[147]

Lodgepole pine grew a little larger in the Sierra Nevada than in Yellowstone. Still they only stood about 150 feet tall even on the best sites, although they could grow 7 feet thick and live up to 600 years. Engelmann spruce, subalpine fir, and quaking aspen are small trees that also lived within many ancient lodgepole pine forests. Englemann spruce could become as old as lodgepole pine and they were slightly taller, but subalpine fir was much smaller and lived only half as long. Quaking aspen was smaller still and had a short life span, but it stood out in the ancient lodgepole pine forest.[148] What aspen lacked in size it more than made up in color. John Muir wrote that during the "blessed color-days" of autumn an aspen grove becomes "a gorgeous mass of orange-yellow." "Every leaf," he said, "painted like a butterfly."[149]

Quaking aspen is a pioneer that cannot tolerate shade, so it clustered in openings among the lodgepole pines and in grassy valleys that wound through the ancient forest. In the fall, scattered groups of aspen adorned with yellow and orange leaves blazed like thousands of little fires within the forest. But this was just a grand finale for a tree that filled the ancient forest with splashes of color and light all summer long. Their creamy white trunks and bright green leaves stood out among the stiff and somber pines. Aspen seemed always to be in motion. It only took a gentle breeze to make their leaves flutter causing them to sparkle as their silvery undersides flashed in the sunlight. Aspen groves were unmistakable and alluring even from a distance. Only in the snowy winter landscape did their bare white trunks tend to blend into the background.

Aspen groves were not just decorative highlights within the forest. They were a welcome sight to trappers and Indians alike because they also were oases of life and supplies within relatively sterile high mountain forests. Warren Ferris seemed delighted to see such groves when he arrived at Henry's Fork on the Snake River just west of Yellowstone. Ferris was a member of a trapping expedition for the American Fur Company when he described the scene:

> Noble groves of aspen and cottonwood, and dense thickets of willow border all these streams and channels, and form almost impenetrable barriers around the verdant prairies. . . . Deer and elk in great numbers resort to these fair fields of greenness in the season of growth, and during the inclemency's of winter seek shelter in the thickets.[150]

Such places were common in the mountains of the West. Ferris described many similar valleys in his journals, as did other trappers. Trappers knew they could usually find game around aspen groves because elk, deer, and moose eagerly ate the young sprouts, as well as the shrubs, grasses, and forbs that grew beneath the sparse canopy.[151] "Tall pines," growing nearby also provided "shady retreats for the numerous Elk and Deer during the heat of the day," wrote Osborne Russell in 1836.[152]

Groves that stood along streams served as beacons that announced to trappers where beaver might live. Beaver ate willows and cottonwood, but they preferred to gnaw on aspen not only as food but also to construct their dams. This was especially true in the upper reaches of streams where it was too narrow and steep to support an adequate supply of willows and cottonwood.[153] However, the rolling landscape of Yellowstone provided ideal habitat for beaver. Not only were aspen groves abundant in these valleys, covering as much as 4% of the landscape,[154] but so were willows. Captains John Barlow and David Heap traveled through the area in 1871 and saw "thickets of willows along the river banks." Philetus Norris also commented in 1880 that Yellowstone was "well supplied with rivulets invariably bordered with willows." No wonder Russell trapped beaver in Yellowstone for 3 years starting in 1835 even though Snake Indians had already taken many of the animals for food. As late as 1880, Norris could still say that trappers took "hundreds, if not thousands" of beaver from Yellowstone each year.[155]

The openness and grassy understory of older aspen groves made them excellent campsites, and they usually grew near water. The light soft wood of aspen also made the trees easy to cut and pile into defensive barricades.

Many trappers and Indians raced into aspen groves to build temporary forts so that they could defend themselves against attack, although they preferred fallen lodgepole pines that only needed stacking.[156] Trappers also cut aspen to build lodges in which to pass the winter, and the trees provided fuel and light as well. "At night a good fire of dry aspen wood, which burns clear without smoke, affording a brilliant light, obviates the necessity of using candles," trapper Ferris wrote in his journal.[157] They needed the light to watch their horses at night, not just because of Indian raiders, but also because "lions would make short work with them if an opportunity was affording," wrote Lieutenant Doane.[158]

Patches of aspen were common throughout the high mountain forests of the West. Scattered groves of aspen and cottonwood made up nearly 2% of the forests in the northern Rocky Mountains where they grew mostly in canyons and valleys.[159] However, aspen covered much larger areas in the central and southern Rocky Mountains.[160] Each aspen grove usually occupied about an acre, although a grove could cover more than 100 acres.[161] The presence of aspen meant that fires were common as well. These fires not only stimulated sprouting but they also cleared lodgepole pine and made room for aspen to grow. That is why trappers rarely saw large trees in ancient lodgepole pine forests. Lightning and Indian fires destroyed most of them before they were more than a few centuries old.[162]

Aspen can live up to 200 years, but few trees attained such an old age in the ancient forest.[163] They usually died before they were 100 years old.[164] Settler trees that tolerate shade such as Englemann spruce, subalpine fir, and Douglas-fir often invaded groves, but only occasionally did they escape fire long enough to replace the aspen. Diseases, insects, hungry beavers, and avalanches also killed some aspen. However, fire killed most of the trees because aspen bark is too thin to offer protection from even light fires.[165] Even so, death meant life to aspen. Their roots lived on beneath the soil waiting for the trees above to die so that the bud at the top of the trunk could no longer release a chemical called auxin. Auxin prevented the roots from sprouting, so when the auxin stopped flowing new stems popped up to replace the grove.[166] Scientists think that this sprouting can go on for thousands of years, perhaps tens of thousands of years. That means that many of the aspen patches that adorned ancient lodgepole pine forests were thousands of years old before trappers saw them for the first time. Some of them probably dated back to the Ice Age.[167]

Aspen needs fire, but not just any fire will do. It takes a special kind of fire to ensure that a grove will replace itself. A single light fire usually kills too few trees to stop the flow of auxin into the roots. A grove of aspen is really a clone in which all the trees are not only alike, but they also share the same root system. So most of the trees must die to reduce the flow of auxin enough to generate a large number of sprouts. A light fire will release some sprouts, but they have difficulty growing in the shade of the remaining trees. Repeated light surface fires can kill the sprouts and the trees. Likewise, an especially hot fire kills the trees outright and cooks the roots so that they can no longer sprout.[168] For an aging grove to replace itself with healthy young aspen, it needs a moderately hot fire that kills the overstory trees while leaving the roots undamaged. An aspen grove with an understory of young spruce and fir provides just the right fuel for such a fire. Particularly in the early spring or late fall when it is more likely to be dry. The grove should also be middle-aged so that it is vigorous enough to produce a large number of sprouts. An unhealthy old grove may not be able to replace itself even if the fire is just right.[169]

Even if aspen sprouts vigorously, a large herd of elk might still wander into a grove and eat all the young trees. A recently burned grove is especially attractive to elk because of the lush growth of tender sprouts, shrubs, grasses, and forbs. Aspen needs at least 6–8 years to grow beyond the reach of elk and deer. Thus too many hungry elk and even deer can destroy an aspen grove just as surely as repeated surface fires would destroy it.[170] This seldom happened in the ancient forest, however, because Indians hunted these animals whenever they found them.

Osborne Russell and a man named White witnessed Indians conducting an elk drive in Yellowstone in 1839. When they awoke in pain from arrow wounds inflicted by Blackfeet the day before, Russell said he "hobbled to a small grove of pines" with White. Then "we heard a dog barking and Indians singing and talking." "The sound seemed to be approaching us," Russell recalled. So they hid and watched. Russell's account of what happened next shows that the Indians were making noise to drive a herd of elk into a lake.

> They at length came near to where we were to the number of 60. Then commenced shooting at a large band of elk that was swimming in the lake. Killed 4 of them, dragged them to shore and butchered them, which occupied about 3 hours. They then packed the meat in small bundles on their backs and traveled up along the rocky shore about a mile and encamped.[171]

Such Indian hunts, together with predators such as wolves, probably kept elk and deer populations under control. As a result, game animals rarely if ever exhausted their food supply, so aspen flourished in ancient lodgepole pine forests.[172]

Indian and lightning fires renewed both aspen and lodgepole pine. They grew side by side because neither of these pioneer trees could thrive without fire. "The Two-leaved Pine, more than any other, is subject to destruction by fire," wrote Muir in 1894.[173] "During strong winds" the flames would leap from tree to tree Muir said, "forming one continuous belt of roaring fire that goes surging and racing onward above the bending woods, like the grass-fires of a prairie."[174] Sometimes even a light fire created such an enormous column of heated air that it generated enough wind to turn itself into a crown fire.[175] Likewise, warm air that rose from a valley could push a crown fire up a steep mountain slope.[176]

While sauntering through California's Sierra Nevada, Muir noticed that many trees still died even in less severe fires. Lodgepole pine bark is too thin to insulate the tissue underneath. Those trees that remained alive after a fire often succumbed to bark beetles instead, although a few managed to heal their wounds and survive.[177] Muir's account demonstrated how a hot surface fire could destroy a lodgepole pine forest just as effectively as a crown fire:

> During the calm, dry season of Indian summer, the fire creeps quietly along the ground, feeding on the dry needles and burs; then, arriving at the foot of a tree, the resiny bark is ignited, and the heated air ascends ... dragging the flames swiftly upward; then the leaves catch fire, and an immense column of flame ... rushes aloft thirty or forty feet above the top of the tree.... It lasts, however, only a few seconds ... to be succeeded by others along the fireline at irregular intervals for weeks at a time—tree after tree flashing and darkening, leaving the trunks and branches hardly scarred. The heat, however, is sufficient to kill the trees, and ... in a few years the bark shrivels and falls off. Belts miles in extent are thus killed and left standing with the branches on, peeled and rigid, appearing gray in the distance, like misty clouds. Later the branches drop off, leaving a forest of bleached spars. At length the roots decay, and the forlorn trunks are blown down during some storm, and piled one upon another.[178]

On rare occasions, fires swept through even larger areas of ancient lodgepole pine forest. However, they were so rare that trapper Warren Ferris may have written the only account of the aftermath of such a massive fire. It took place in the forests surrounding a valley in which the town of West Yellowstone, Montana, now sits. Trappers often called valleys "holes," such as Jackson's Hole and Pierre's Hole, so they named this valley the "Burnt Hole." Osborne Russell described it in 1835 as "a prairie Valley about 80 mls in circumference surrounded by low spurs of pine covered mountains which are the source of a great number of streams which by uniting in this valley form the Madison fork."[179] Russell made no comment in his journal about the fire, perhaps because it happened 6–10 years earlier and the burned area was already well forested with young lodgepole pines. The burn was probably more obvious when Ferris saw it in 1832:

> The Burnt Hole is a district on the north side of the Piney Woods, which was observed to be wrapped in flames a few years since. The conflagration that occasioned this name must have been of great extent, and large forests of half-consumed pines still evidence the ravages.[180]

Lightning might have started this fire, although it is just as likely that Indians ignited it. We know, for example, that lightning fires were far less common in ancient lodgepole pine and other subalpine forests than in any other evergreen forest in the northern Rockies. Even so, fires burned through ponderosa pine forests only about three times more often than lodgepole pine, yet they had 22 times more lightning fires.[181] There is no doubt that Indians set many additional fires in ponderosa pine forests. Moreover, ponderosa pine grew on lower slopes within dry grasslands that ignited easily. Lodgepole pine forests were much harder to burn. This means that Indians must have set many fires in ancient lodgepole pine forests. There is no other way to account for the meager difference in the total number of fires that burned in these two ancient forests.

Indians surely started fires in lodgepole pine forests from time to time because they left their campfires burning, and they used fire for war, hunting, and other reasons. Many more Indian fires spread into these forests from mountain valleys where they traveled and lived.[182] They also concentrated campsites near obsidian outcrops in Yellowstone where they made stone tools and arrowheads.[183] As late as 1906, J. E. Stauffer, a forest ranger in the Canadian Rockies, wrote in an annual report that Indians "set out fires in the forests . . . for hunting purposes, in season or out."[184]

Ancient lodgepole pine forests did not all burn the same way or at the same frequency. Fast moving crown fires were most common, but hot and light surface fires also crept through the forests.[185] These cooler fires usually burned in valley bottoms and open forests with little fuel.[186] Some surface fires also spread from the edge of crown fires.[187] In 1904, John Leiberg wrote that in the Absaroka Range, fires in lodgepole pine forests

> are remarkable for their destructive force and intensity. Here and there are uneven aged stands, where extremes in age and a mixed composition prove that occasionally the fires did not consume or kill the entire stand. But as a rule most of the older fires made a clean sweep.[188]

Crown fires and surface fires, or a combination of the two, occurred in the same few hundred acres as often as four times in a century in parts of the northern Rockies.[189] They were somewhat less frequent on the east side of the Cascades in southern Oregon.[190] However, the average interval between fires was over a century in the entire inland Northwest. In the Canadian Rockies it was about 84 years in warm and dry lower subalpine lodgepole pine forests, and about 128 years in cool and moist forests. There was also considerable variation in the area burned from year to year.[191]

On the west slope of the Absaroka Range in Yellowstone, crown fires usually returned to the same patch of lodgepole pine about once every 150–250 years.[192] Forests burned more often in the Canadian and central Rockies.[193] Average intervals between crown fires of three centuries or more were unusual. Such long quiet periods normally occurred in protected places that crown fires often missed, such as moist north slopes, or where surface fires had temporarily reduced fuels by clearing understory trees and logs. This also might happen on relatively level terrain where lodgepole pine grew slowly because of poor soil, such as the Madison Plateau in Yellowstone.[194] More productive areas burned more often. Since conditions varied from place to place, the frequency of crown fires also varied. Consequently, the average interval between crown fires in Yellowstone before 1800 was probably about 200 years.[195]

A crown fire that swept over a single block of trees covering tens of thousands of acres was an extremely rare event in an ancient lodgepole pine forest. Most fires probably burned under 10,000 acres, with the majority of these covering a few hundred to a few thousand acres.[196] During dry years fires sometimes burned in several places at the same time, but they were usually mixtures of small and large fires that remained widely separated. For example, a large fire burned the Cache Creek watershed in Yellowstone about 1756 while only a small fire burned in the Little Firehole River watershed that year. Likewise, a large fire of about 3800 acres burned in the Little Firehole River watershed in 1795, but not in the Cache Creek watershed.[197] Even if a fire did start there, it probably stayed small because the 1756 fire had already cleared so many openings in the forest.[198]

Fires seldom spread over huge areas because there were too many rocky ridges, moist north slopes, canyons, valleys, wet meadows, and lakes that got in their way. What is more important, crown fires need fuel, no matter how hot, dry, and windy the weather. The ancient lodgepole pine forest burned in so many places so often that there was little chance that miles and miles of forest would have enough fuel to carry a crown fire a great distance. Still, strong winds picked up burning embers and blew them over barriers and then dropped them like fiery rain on trees far ahead. These embers often ignited many small fires, but patches of younger forest with little to burn and other obstacles usually kept them confined.[199] Thus, a single crown fire might hop around a forest by flaring up in only the most flammable patches, leaving the less flammable patches unscathed.

A fire driven by strong winds could tower above the forest canopy and then lean over to make a brief but spectacular run through the tree tops before coming to a halt. Such crown fires spanned barriers and helped to enlarge the area burned, although they seldom went far.[200] Regardless, a crown fire of any magnitude still laid waste to the land; turning trees into blackened skeletons, incinerating smaller plants, and clogging streams with silt and debris. It usually took as long for streams and the fish that lived in them to recover as it did the surrounding forest.[201] Bark beetles that bred in fire-injured trees also swarmed into nearby healthy forests and expanded the destruction.[202] Father De Smet described the devastation left by a fire that swept over a lodgepole pine forest in 1845. His Assinboine companions were guiding him through the Canadian Rockies above the Bow River in what is now Banff National Park when he wrote the following note in his journal:

> For the space of six hours we were compelled to trace our route across fragments of broken rocks, through an extensive and parched forest, and

where millions of half-consumed trees lay extended in every direction. Not a trace of vegetation remained, and never had I contemplated so dismal and destructive a conflagration.[203]

Even a fire this powerful often dropped to the ground when it entered an aspen grove with an open understory. Such groves are difficult to burn during most of the fire season. They are almost fireproof because fallen leaves lay flat and form a tightly packed layer that stays moist.[204] So aspen groves helped to restrict the size of crown fires, although they were too small and too few to have more than a local effect. Ironically, patches of young and middle-aged lodgepole pine were also difficult to burn, so they helped to confine crown fires in the ancient forest as well.

A dense carpet of lodgepole pine seedlings normally sprang up a year after a crown fire. These patches of seedlings were nearly immune to fire, especially if there were few dead trees laying among the young pines. They remained resistant to fire as they grew into saplings.[205] This was even true when crown fires roared through forests on the edge of the patch. When they did burn, the young pines usually suffered only light or spotty fires, or a little scorching of foliage.[206] Still, crown fires were likely to destroy a few patches of young trees.[207]

The largest crop of seedlings came up where the lodgepole pine trees had serotinous cones. That is, the cones stayed sealed with sap until a fire melted it and opened the cones so that they could drop their seeds. This resulted in enormous amounts of seed being stored because new cones kept growing, and they stayed on the trees for years. Then fire opened the cones and released the seeds all at once. However, extremely hot crown fires could destroy most of the cones and greatly reduce the number of seedlings. Those lodgepole pine trees that had cones that opened without fire produced even fewer seedlings because they did not store seed from year to year, and the seeds had less protection from the flames.[208]

Sometimes lodgepole pine seedlings in a fresh burn would stay temporarily hidden beneath a mass of pink fireweed made all the more colorful by standing against a background of blackened tree trunks.[209] These patches of fire-killed trees also attracted swarms of roundheaded beetles that bored into the wood under the scorched bark. The beetles quickly multiplied and their larvae became a feast for the little three-toed woodpeckers that concentrated in the burns. Hairy woodpeckers flew in shortly afterward and joined them in feeding on larvae and pecking holes in the dead trees to raise their young. The three-toed woodpeckers gradually left the burn during the next five years as the beetles became scarce, but hairy woodpeckers stayed longer. Now mountain bluebirds moved in to make homes in the nest holes that the woodpeckers abandoned. They lived there until most of the snags fell and the pine seedlings grew into a dense patch of sapling or pole-size trees.[210] Few birds could live in such a crowded forest, so they had to wait until it aged and became more open.[211] By this time another fire had burned somewhere else in the ancient forest and provided new homes for the woodpeckers and mountain bluebirds.

The seedling and sapling stages of development lasted about 20–50 years.[212] During that period snags gradually fell and piled up around the young lodgepole pine trees. It took about 21 years for nearly all the snags to fall, although most of them toppled in 15 years.[213] It might take many more decades for all the logs and other debris lying on the ground to decay.[214] Meanwhile the seedlings grew into nearly impenetrable thickets of saplings. Lieutenant Doane's party struggled through a patch of forest like this in 1870 near Yellowstone Lake. Lieutenant Doane recalled the experience:

> We traveled across a high promontory running into the lake, winding among steep ravines and through fallen timber lying in heaps, with full-grown, living forest above it. This timber must have been deadened by fire, the trunks being bare of limbs and much decayed, but in such masses as to be impassable in many places, causing us to make wide detours to find a trail. The standing forest is very dense; the pack-animals ran between trees, often wedging themselves in so tightly as to require some trouble in extricating them; several of the packs burst, causing numerous delays. Our faces were scratched, clothes torn, and limbs bruised squeezing through between saplings.[215]

It is during this time that a patch of young pioneer forest is most vulnerable to fire, especially where small saplings grow very densely.[216] Consequently, crown fires ravaged both old forests and some of the young forests that started to replace them.[217] However, large downed logs do not burn easily, and there are few highly combustible twigs and branches left after a crown fire, so this was not common in the ancient forest. Moreover, there was only a remote chance that a fire would start during the short period in which a young patch could burn. After that young pioneer forests burned so poorly that they often blocked crown fires and helped to keep them from spreading.[218]

Patches of middle-aged lodgepole pine were just as hard to burn as young forests.[219] These dense forests of small pole-size trees were usually 50–150 years old, depending on the quality of the soils and how fast the trees grew in a given location.[220] Such closely spaced trees blocked strong winds that could make a fire intense, and they cast too much shade over the ground for small flammable plants to grow well. "The lodgepole pines are set so close that there practically can be no additions of any other species," wrote John Leiberg in 1904.[221] By now there were also only a few decaying logs lying around from the previous forest. So fires normally stopped or dropped to the ground when they reached patches of middle-aged pioneer forest, which further limited the spread of crown fires.[222]

A lodgepole pine forest that progressed into the old pioneer stage could also remain nearly fireproof, at least during the first few decades.[223] A patch entered this stage when the trees were 100–200 years old, again depending on their growth rate.[224] Unhealthy trees started dying in dense middle-aged forests, but they died faster during the old pioneer stage. So the canopy became thinner and a little sunlight filtered down to the forest floor. This made it possible for a sparse layer of grasses, forbs, and low shrubs such as grouse whortleberry to grow in the understory. Now there were enough small plants, twigs, and needles covering the ground to sustain a slow-moving surface fire. Crown fires were still unlikely, however, because the flames could not reach the tops of the trees.[225] John Leiberg saw this in the Absaroka Range and made a point of it in his 1904 report:

> The general ground cover consists of moss, . . . a slight sprinkling of pine needles, low shrubs, most species of huckleberry, and more or less of a grassy turf. . . . During the dry season all this material burns readily, but does not make a hot or high flaming fire.[226]

Therefore, crown fires often stopped or went over or around patches of young and middle-aged forests. They also cooled down or even stopped when they entered old pioneer forests that had an open understory. Flaming debris also fell into these forests, but it usually started small fires that only burned holes in the canopy and then went out.[227]

An old pioneer forest became more flammable as it aged, and large dead trees began piling up on the ground. Again, Leiberg said, "the great mass of dry or partly dry wood . . . makes hot and flaming fires, consuming or killing all live timber."[228] Now insects such as the mountain pine beetle started to take a toll. Bark beetles infested and killed many of the weaker trees because they could not plug their holes with sap. The mountain pine beetle also has a secret weapon for stronger trees. It carries the spores of blue-staining fungi in a structure on its head. These fungi reduce a tree's ability to produce sap. So, if a few beetles succeed in penetrating a tree's bark and infect it with fungi, others can follow and kill it.[229] Thus bark beetles also thinned old pioneer forests and added more logs to the piles already accumulating in the understory.[230]

Patches of cluttered old pioneer forest were the places that Osborne Russell, and other trappers and explorers, complained about during their travels. In July 1838, Russell had to ride through what was probably such a forest while "trapping in the YellowStone [*sic*]," as he noted in his journal:

> We came to the junction of two equal forks [Lewis River and Snake River] we took up the left hand, on the west sid thro the thick pines and in many places so much fallen timber that we frequently had to make circles of a quarter of a mile to gain a few rods ahead.[231]

No doubt pine beetles were responsible for some of the forests that Russell found "thickly intermingled with logs and fallen trees."[232] Such outbreaks were more likely on middle slopes than higher slopes because cold weather retards the beetle's growth.[233] Sometimes drought weakened the trees enough for mountain pine beetle numbers to explode and drastically thin or wipe out a patch of old pioneer forest. Outbreaks like this could easily spread to other patches as well and create widespread devastation.[234] Heaps of broken trees and branches produced by such an outbreak could also fuel a massive crown fire.[235] Normally, however, attacks remained small and scattered, acting to thin patches of older trees rather than destroy them.[236]

More and more sunlight reached the understory of old pioneer forests as beetle-infested, wind-blown, or diseased trees fell and opened gaps in the canopy. Settler trees such as Engelmann spruce and subalpine fir took advantage of this light by spreading down from the higher slopes and filling the gaps. Some of them also spread from moist ravines where they found protection from the last fire. On the other hand, Douglas-fir moved into these forests from the dry lower slopes. Lodgepole pine seedlings even filled some of the gaps, particularly where there was adequate sunlight and exposed mineral soil, and where spruce and fir trees were sparse. This was more common where the pines had at least some cones that could drop their seeds without fire.[237] In this way a patch of old pioneer forest gradually became a dense old transitional forest.[238] It might take only 155 years for a patch of lodgepole pine forest to go from seedlings to this stage of development or, on very poor soils, it might take as long as 300 years.[239]

Occasionally, a light surface fire cleared thickets of understory trees from a patch of old transitional forest and returned it to the more open old pioneer stage. However, these fires also increased the number of lodgepole pine seedlings by widening sunny openings and consuming litter and exposing bare soil. So young trees quickly filled the understory. Eventually, however, settler trees took over as it became too shady for pines. Ultimately, a self-replacing forest of settler trees would grow where a pioneer forest of lodgepole pine once stood. The transition from crown fire to young, middle-aged, and old pioneer forest, and then to old transitional and self-replacing forest, normally took more than three centuries to complete.[240] Now and then a windstorm blew down the older lodgepole pines, or mountain pine beetles killed them, and released young spruce and fir growing in the understory, thus greatly shortening the time it took for a patch to reach the self-replacing stage.[241]

Very few self-replacing patches could exist in the ancient lodgepole pine forest. Fires burned too often and in too many places for any but the most well-protected parts of a forest to go unburned more than three centuries.[242] However, a small number of patches went as long as 500 years before fire finally destroyed them.[243] Regardless, not many patches survived the dense old transitional stage because this was the most flammable period in the development of the forest.[244]

The old transitional stage consisted of a mixture of aged and diseased pines, little pockets of young and middle-aged spruce and fir with their low-hanging branches, and loosely stacked piles of dead trees and branches. This was the kind of tangled forest that Lieutenant Ludlow's party had to battle as they descended Mount Washburn in 1875. The storm that forced them off the mountain made the "path slippery and difficult," he said, and it passed "through the densest timber of spruce and pine." "The projecting branches flapped back their freight of rain-drops into our faces and clothing, and many of the broken twigs bore trophies snatched from the packs," wrote Lieutenant Ludlow.[245] A self-replacing forest of spruce and fir looked much the same. A fire that started in this kind of fuel could easily climb the small trees and enter the canopy where the wind blew strongest. Once a fire reached the tops of the old trees and raced forward, it stayed there because the burning saplings and logs underneath fed the flames with superheated air.[246] This was the perfect fuel for a crown fire, so they normally burned in the highly combustible patches of old transitional and self-replacing forest.[247]

The many crown fires that scorched ancient lodgepole pine forests created a mosaic of patches. The age of the trees in each patch date the time the fire occurred. The flames also seared sharp lines into the forest that marked the boundaries between patches.[248] These boundary lines remained visible for centuries, so they crisscrossed the ancient forest, making it look from a distance like a giant quilt. The patches in this quilt or mosaic varied in size and shape because young forests and other barriers blocked or diverted fires, shifting winds pushed them around, and rain and snow put them out. Thus, according to Leiberg, some of the fires that swept over the Absaroka Range in the late 1800s were "very extensive" while others burned "scattered tracts" or "scattered patches." Even so, the largest fires he recorded covered fewer than 15,000 acres, and most were substantially smaller. Some only burned a few hundred acres.[249]

Likewise, the proportions of patches at different stages of development varied, as did their proximity to one another in the mosaic. The magnitude and frequency of insect outbreaks, Indian and lightning fires, and storms and droughts, as well as the condition of the forest at the time and place where they occurred, led to constant change. Yet everything seemed synchronized. A fire might reduce a patch of old transitional forest to ashes while another patch in the mosaic advanced to the same stage and replaced it. Still another patch might then advance to replace that patch, and so forth. However, the next fire might destroy a patch of beetle-riddled old pioneer forest without it being replaced somewhere else. Now there were fewer older forests in the mosaic and more younger forests. The reverse might be true a few decades later. And so it went, century after century.

Even though the proportions of different patches fluctuated over time, they would have varied more if only lightning fires burned these ancient lodgepole pine forests. Since lightning alone would have started fewer fires, there would have been more older forests ready to burn. In addition, many patches of older forests would have been standing side by side instead of being separated by less flammable younger forests. With nothing to stop them, crown fires would have moved from one older forest crammed with fuel to another, running for miles over the landscape.[250] Monstrous fires like these create huge patches. This would have led to wild swings between a landscape dominated by younger forests that filled in the burns and one dominated by older forests a century or two later. However, Indians not only increased the number of fires, they probably made the time and place in which they burned more consistent as well.[251] This higher rate of burning created an ancient forest with smaller crown fires, and a more complex and stable arrangement of large and small patches.

By 1904, Leiberg reported that 40% of the forest mosaic in the Absaroka Range consisted of patches of 120- to 300-year-old trees. These were mostly old pioneer and old transitional forests.[252] Remarkably, the Madison and Central Plateaus in southern Yellowstone contained a similar amount of older forest in 1900, about 44%. However, these older forests had increased from what existed 165 years earlier when they covered only 30% of the plateaus.[253] The same thing probably happened in the Absaroka Range.

Older forests increased because of a decrease in the area burned by crown fires. Crown fires regularly burned from 2 to 12% of the ancient forests that stood on the Yellowstone plateaus. However, fires nearly stopped between 1775 and 1785. Then they increased, although not to the same level as before. Again, fires dropped to almost nothing between 1820 and 1840, only to increase once more and drop again after 1870.[254] Therefore, the area burned by fires cycled up and down consistently until the middle of the eighteenth century; then something happened that made the fires begin a dramatic decline.

Certainly the large amount of less flammable younger forests growing on the plateaus helped to reduce the area burned by crown fires. This gave some of them time to grow and add to the amount of older forest. But that is not the whole story. Weather may also have contributed to these variations, but lightning fires burned Yellowstone's forests half as often as the average for lodgepole pine forests in the northern Rockies.[255] Therefore, the decline in crown fires probably resulted from Indians setting fewer fires, no doubt because there were fewer Indians around who could set fires.

Smallpox and other European diseases decimated the Blackfeet who lived in the region at the same time that fires declined in some of Yellowstone's lodgepole pine forests. The first epidemic occurred about 1781. This was during the period in which fires decreased on the plateaus, even though there should have been more fires since the inland Northwest was in a drought. Fires dropped again in about 1837 when smallpox hit the Blackfeet a second time, wiping out two-thirds of their population. Osborne Russell witnessed their misery. He said that smallpox "made a terrible havoc." Blackfeet were dying in front of him when he rode by a village near Madison Fork. Later that day, Russell said he "passed an Indian lodge standing in the prairie near the river which contained 9 dead bodies."[256] Finally, about 1870, disease ravaged the Blackfeet for the third time, and war took a toll on their numbers as well. Just as before, fires declined on Yellowstone's plateaus even though the area was in the grip of a regional drought.[257]

Indian fires probably decreased on Yellowstone's plateaus because the Blackfeet and other tribes were less likely to visit remote forests after their populations dwindled from disease. Things were different along the heavily traveled Bannock Trail that wound through the river valleys of northern Yellowstone. Blackfeet shared the trail with Bannock, Shoshone, Nez Perce, and Flatheads who followed it to reach the bison that roamed the Great Plains.[258] So, even after an epidemic, the surviving Indians still concentrated on the Bannock Trail in sufficient numbers to keep the fires burning. As a result, fires burned there at the same rate they had since 1703, about 1 every 5 years.[259] However, the Indian presence finally ended there as well, and so did their fires.

The end of fires along the Bannock Trail signaled the passing of the ancient lodgepole pine forest. Things changed quickly after the trappers arrived. Bison, elk, and other game became less abundant because Indians now had guns, so they could kill them with greater efficiency. Trappers and explorers took their share of game as well since they had to live off the land. Guns also increased the carnage from warfare, and Indians died by the thousands because they had no resistance to European diseases. As a result, there were fewer fires and forests began to age. The trappers and explorers of the eighteenth and early nineteenth centuries were the first and perhaps the last Europeans to see the ancient lodgepole pine forest at the peak of its beauty and abundance. By 1900, parts of the forest still looked much as they did in the past. However, even remote forests had to change as the native peoples who lived there gradually disappeared.

PACIFIC DOUGLAS-FIR FOREST

From this place, the noble river can be traced in all its windings ... beyond, the eye sweeps over an interminable forest melting into a blue haze, from which, Mount Hood, capped with its eternal snows, rises in great beauty.
Captain Charles Wilkes
Describing the view from Fort Vancouver overlooking the Columbia River, 1841[260]

Tossing in a heavy sea, with waves and rain pouring over her wooden deck, Captain George Vancouver steered the Royal Navy's ship *Discovery* eastward toward an unseen coast he knew was near. "Being anxious to get sight of the land before night," Captain Vancouver ordered the crew to unfurl "as much sail as we could carry." By late afternoon he saw land "on which the surf broke with great violence." It was April 16, 1792, and he had reached the coast of northern California on the south side of Cape Mendocino. This was close to the southernmost extent of the ancient Pacific Douglas-fir forest (Fig. 12.5).

We stood in for the shore under our topsails for about an hour, and perceived the coast. ... The shore appeared strait and unbroken, of a moderate height, with mountainous land behind, covered with stately forest trees; excepting in some spots, which had the appearance of having been cleared by manual labour; and exhibited a verdant, agreeable aspect.[261]

"Immense numbers of whales were playing about us," wrote Captain Vancouver as he scouted the coastline for a few days. Then he sailed north-

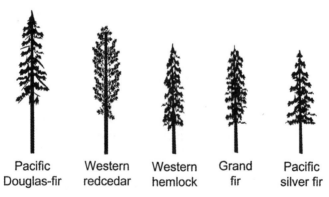

Pacific Douglas-fir   Western redcedar   Western hemlock   Grand fir   Pacific silver fir

Fig. 12.5   Trees of the ancient Pacific Douglas-fir forest.

ward toward Nootka to receive from the Spanish the property and lands they had seized 2 years earlier.[262]

Discovery reached the Destruction Islands off the coast of northern Washington on April 29. Here, "at four o'clock," Captain Vancouver wrote in the ship's log, "a sail was discovered to the westward standing in shore."

> She soon hoisted American colours, and fired a gun to leeward. At six we spoke her. She proved to be the ship *Columbia*, commanded by Mr. Robert Gray. . . . Having little doubt of his being the same person who had formerly commanded the sloop *Washington*, I desired he would bring to, and sent Mr. Puget and Mr. Menzies on board to acquire such information as might be serviceable in our future operations.[263]

Captain Robert Gray was not a famous explorer; he was just an American fur trader. But he and Captain John Kendrick had already seen the coast that Captain Vancouver was exploring. They were trading with Northwest Indians for beaver and sea otter pelts back in 1788. Captain Gray commanded the sloop *Lady Washington* at the time, and Captain Kendrick commanded the ship *Columbia*. During the voyage Robert Haswell, who was a member of *Lady Washington's* crew, wrote one of the earliest descriptions of Oregon's ancient forests:

> This Countrey must be thickly inhabited by the many fiers we saw in the night and collums of smoak we would see in the day time and a delightful countrey thickly inhabited and Cloathed with woods and verdure with maney charming streems of water gushing from the vallies.[264]

After loading *Columbia* with pelts, Captain Gray assumed command and sailed her to China, sold the pelts, went around the Cape of Good Hope, and returned to Boston where the voyage started. In doing so, *Columbia* became the first American ship to sail around the world. Captain Gray added to his accomplishments when he returned to Oregon to meet Captain Kendrick in 1792. He discovered the great river that bears his ship's name less than 2 weeks after helping Captain Vancouver.[265]

It would be almost 13 years before Lewis and Clark would finally reach the mouth of the Columbia. However, in January 1806, Captain Clark left a vivid portrait of the river, the coast, the native people who lived there, and the ancient forests that towered over them. He came upon this picturesque scene while following an Indian guide over a mountain on his way to a beach where he heard a whale had washed ashore.

> Here one of the most delightful views in nature presents itself. Immediately in front is the ocean, which breaks with fury on the coast . . . as far as the eye can see. . . . To this boisterous scene the Columbia, with its tributary waters, widening into bays as it approaches the ocean, and studded on both sides with the Chinnook and Clatsop villages, forms a charming contrast; while immediately beneath our feet are stretched rich prairies, enlivened by three beautiful streams. . . . The mountains are covered with a very thick growth of timber, chiefly pine [Douglas-fir] and fir [grand fir]; some of which . . . perfectly sound and solid, rise to the height of 210 feet, and are from eight to twelve in diameter.[266]

The ancient Pacific Douglas-fir forest occupied a broad belt that extended from southern British Columbia to below Cape Mendocino in northern California, covering about 19 million acres. It spread over much of the Puget Trough and the Olympic Peninsula in Washington. Below this region, one branch of the forest skirted the west side of the Willamette Valley and blanketed the Coast Range in Washington and Oregon. The other branch passed along the eastern edge of the valley and stretched down the west slope of the Cascades. It formed a wide band between the northern oak woodlands and the fir and spruce-fir forests on the upper mountain slopes.[267]

Overton Johnson and William Winter saw this part of the forest in 1843. They were part of the great migration that followed the Oregon Trail to the Northwest. So they traveled into "this fair Valley of the Willamette" and climbed a grassy hill to overlook the farmland they sought. The grand panorama that loomed before them was more than they expected.

> What a prospect! . . . To the East, and extending as far as the eye can reach to the North and South, the Cascades, in one lofty, unbroken range, rise mountain upon mountain, and forest over forest . . . their highest peaks, wrapped in eternal snow. . . . At your feet you can see the Willamette, meandering down the wide fertile Valley . . . further away to the North, St. Helens shows her towering crater of eternal fire. . . . Away to the South, the peering summits of some lofty chain are dimly drawn upon the sky.[268]

The two branches of the forest converged below the Willamette Valley in the Siskiyou and Klamath Mountains that Johnson and Winter saw only "dimly." Then it continued a little farther down the Coast Range of northern California. The Pacific Douglas-fir forest also touched the ocean shoreline in places along the southwestern Oregon and northwestern California coast, but a narrow strip of Sitka spruce-hemlock-redcedar forest, and coast redwood forest, usually kept them apart.[269]

This ancient forest had existed in the Pacific Northwest since the close of the Ice Age, but it remained incomplete for several millennia after it formed. By 10,000 years ago, Douglas-fir, western hemlock, and red alder had already expanded into the Puget Trough.[270] So even then, the forest looked much like the one that fur traders and trappers saw. However, western redcedar was missing. It did not take its place in the forest until 5000 years later.[271] Likewise, northern California was still too dry for Douglas-fir to grow over large areas, that is, until the climate became wetter about 3800 years ago. Then it spread out and completed the forest.[272]

The trees in this forest were "monarchs to whom all worshipful men inevitably lift their hats," wrote Samuel Wilkeson in 1869 while inspecting a railroad route in Washinton Territory.[273] They deserved this approbation. Mild wet winters and dry summers made ideal growing conditions for an imposing forest dominated by conifers such as Douglas-fir, western redcedar, and western hemlock. In 1812, Robert Stuart, a partner in John Jacob Astor's Pacific Fur Company, complained that "from the middle of Octr. to the middle March no man in the woods can possibly keep his arms in order, on account of the unceasing rains." However, to Stuart the dry summers were "the best weather in the world."[274] The few hardwoods that grew among the conifers stayed mostly in the wettest and driest places, particularly in the southernmost part of the forest.[275]

Douglas-fir was the undisputed ruler of this ancient forest, both in size and abundance, making up over 80% of the trees in some areas. Indeed, on the best soils it could stand almost as tall as a coast redwood and reach a diameter of 15 feet. More commonly older Douglas-fir trees stood 180–250 feet tall with trunks 4–6 feet thick. This massive tree could also live more than a thousand years, although few survived that long because it played the role of a pioneer in most of the ancient forest.[276] However, the settler species that replaced it in the wetter part of this forest were nearly as impressive as Douglas-fir.

Western redcedar, a settler species, could also live a thousand years, but it was not quite as tall as Douglas-fir, reaching a maximum height of 250 feet. However, it was slightly thicker, up to 16 feet in diameter. Still, most older trees were under 200 feet tall with diameters less than 6 feet. Western redcedar also required more moisture than Douglas-fir. So it grew primarily in stream bottoms and gulches, on moist flats, and on cool north-facing slopes.[277] Dr. Cooper noticed this in the 1850s. "This cedar is most abundant near the coast," wrote Dr. Cooper, "but common also in damp forests nearly to the top of the Cascade range."[278]

Western hemlock was the most important settler species in this forest. It grew more widely than western redcedar, but it was not as stout nor did it live as long as the other conifers. It could live up to 700 years, although most of them fell in less than 500 years. Likewise, it could grow to a height of nearly 260 feet and a diameter of 9 feet on the best sites. However, it rarely grew larger than 4 feet in diameter and 160–200 feet tall. Western hemlock was also less sturdy and more susceptible to disease and insect attack than Douglas-fir and western redcedar.[279]

Other settler trees played a minor role in this forest such as grand fir, Pacific silver fir, and Pacific yew. Grand fir stayed mostly on the lowlands in moist valleys. Pacific yew also grew in such places. Pacific silver fir grew in the upper part of the forest on mountain slopes, primarily in the northern Cascade Range. Noble fir grew there as well, but it shared the pioneer role with Douglas-fir. However, these trees were more like vassals than royalty. The firs could stand nearly as tall as most of the other conifers in the ancient Pacific Douglas-fir forest, but they were spindly by comparison, rarely exceeding a diameter of 4 feet. Pacific yew was even more lowly. It is a shrubby little tree with scaly bark that lives in moist areas sheltered by large old trees.[280]

To the south, in the Umpqua and Rogue River Valleys, and the Siskiyou and Klamath Mountains, the forest changed. The climate was warmer and drier in southwestern Oregon and northwestern California than elsewhere in this ancient forest. Even so, Douglas-fir still dominated the forest. Pioneers such as ponderosa pine and sugar pine grew among the Douglas-fir trees in interior areas, and they also had a royal stature. Incense-cedar, a settler species, also grew in places. However, the pines and incense-cedar became more important where the Pacific Douglas-fir forest graded into the mixed-conifer forest. Tanoak and Pacific madrone, the principal hardwoods growing in this forest, are much smaller. Tanoak was most abundant in the wetter areas nearer the coast and on moist north slopes farther inland. Pacific madrone grew on many different sites.[281]

California laurel or bay also grew in "the lowest valleys next the streams" of this forest, wrote the Reverend Gustavus Hines in 1840. He was riding "up the valley of the Umpqua river" after preaching to the Indians on the coast when he saw a tree "the like of which I have never seen in any other country." "It appears to be of the laurel family, and is so strongly scented that the air in the groves where it is found, is strongly impregnated with its aromatic odors," said Reverend Hines.[282]

This was not a colorful forest. The canopy cast a heavy shade over the understory, making tree trunks seem even darker than they were. Still, the red-brown bark of western redcedar stood out, and its fernlike foliage was equally distinctive. However, tree leaves were mostly dark yellow-green or a dark shiny-green. Gigantic moss-covered logs that lay piled on the forest floor added more brown, gray, and green, but little else. The feathery fronds of western swordferns that decorated moist places were also just another shade of green. A little relief came in autumn when the leaves of tall bigleaf maple and the shrubby vine maple provided an occasional splash of yellow, orange, or red. The rose pink flowers of Pacific rhododendron also dazzled the eyes here and there, especially in western Oregon.[283] Otherwise, there was little to notice other than the overwhelming size of the forest and the trees.

Most people who saw this ancient forest thought it was spectacular, even though the undergrowth and clutter of fallen trees made it difficult for travelers. Samuel Bowles, editor of the Springfield, Massachusetts, *Republican*, had high praise for the forest he saw as he "bounced" along in an open wagon on a muddy road for two days. It was 1865, and settlers were already carving farms out of the ancient forest. Bowles did not have to climb over logs and hack his way through underbrush, but it was nonetheless an uncomfortable trip. He wrote his account while his wagon rolled north along the west slope of the Washington Cascades to Puget Sound.

But the majestic beauty of the fir and cedar forests ... was compensation for much discomfort. These are the finest forests we have yet met,—the trees larger and taller and standing thicker ... and the undergrowth of shrub and flower and vine and fern, almost tropical in its luxuriance and impenetrable for its closeness. ... We occasionally struck ... perhaps once in ten miles a [farmer's] clearing of an acre or two, rugged and rough in its half-redemption from primitive forest; but for the most part it was a continuous ride through forests, so high and thick that the sun could not reach the road, so unpeopled and untouched, that the very spirit of Solitude reigned supreme.[284]

Others such as Charles Nordhoff, a traveler writer, found the forest "dreary." He came to Oregon in the early 1870s filled with the romantic "dreams of my boyhood," he said, yet even the majesty of the ancient Pacific Douglas-fir forest failed to move him. He left disappointed, and said so when he gave "hints to tourists" in his book. He wrote his views while sitting in a boat on the Columbia River "at anchor abreast of Astoria."

> Wherever you look you see only timber ... On your right is Oregon—its hill-sides a forest so dense that jungle would be as fit a word for it as timber; on the left is Washington Territory, and its hill-sides are as densely covered as those of the nearer shore. This interminable, apparently impenetrable, thicket of firs exercised upon my mind, I confess, a gloomy, depressing influence.[285]

Just like the views of those who saw it, this was a forest of extremes. The trees were huge and so were many of the fires that sustained them. The ancient Pacific Douglas-fir forest was a fire forest. "Extensive tracts of this forest get burnt every year," wrote Dr. Cooper in the early 1850s. He added that, "during our ascent of the western slopes of the Cascade range we passed for days through dead forests." Forests were still burning even as he wrote in his journal. "Large tracts were on fire at the same time," noted Dr. Cooper, "filling the air with smoke, so that we could not see the surrounding country for several days."[286] Lightning seldom caused these fires. Dr. Copper knew this, saying that the "great height to which trees grow may also be due to the rarity of lightning."[287]

Lightning fires were extremely rare in the Coast Range, and they only started occasionally along the west slope of the Cascades and much of southwestern Oregon. They were more frequent in the Siskiyou and Klamath Mountains.[288] Lightning fires were so rare in much of the ancient Pacific Douglas-fir forest that they probably only started in the same area every few thousand years.[289] Moreover, no large lightning fire has burned in the Oregon Coast Range since 1770.[290] Nevertheless, crown fires burned this forest an average of every few hundred years. The only explanation is that Indians ignited most of the fires, not lightning. Even though Indians set some fires in the forest, many of their fires started on the lowlands and in the valleys, such as the Umpqua and Willamette.

Then warm air that rose from the valleys during the heat of the day pushed some Indian fires up the mountain sides and into the forests. Strong easterly winds also carried their fires westward, fanning them into torrents of flame that charged through the Coast Range forests.[291]

Only the white spruce forest in Canada had more extensive fires than those that burned in the Pacific Douglas-fir forest. The Cowlitz Fire in the Cascade Range of Washington was the largest ever recorded in the ancient forest. It may have burned over a million acres about 1800.[292] Another fire burned about 800,000 between the Siuslaw and Siletz Rivers in the Coast Range of western Oregon in September 1849. Other mammoth fires in the Oregon Coast Range that occurred in the 1840s and early 1850s include the Yaquina fire that burned about 480,000 acres and the Nestucca fire that burned about 295,000 acres.[293] Charles Nordhoff saw the aftermath of such a fire as he leaned over the railing of a ship while sailing to Astoria, Oregon.

> The voyage from San Francisco is almost all the way in sight of land; and as you skirt the mountainous coast of Oregon you see long stretches of forest, miles of tall firs killed by forest fires, and rearing their bare heads toward the sky like a vast assemblage of bean-poles.[294]

Huge fires such as this burned the same area of forest about once each 400 years, although they were more frequent in some places and less frequent in a few others.[295] These fires were awesome and dangerous. To be sure, the crown fires that raced over the white spruce and lodgepole pine forests were big, but the trees were small even though some of the fires were extensive. Those fires could not compare to a towering wall of flame roaring through a forest of immense trees. Trees that stood more than 20 stories tall, many thick enough to build a road through, with logs, some the size of a bus, heaped on the ground underneath. Such fuel turned the forest into a colossal furnace. Branches withered in the heat before the trees burst into flames. Dense smoke filled the air, and burning debris rained down in showers of sparks. High winds generated by the fire ripped trees apart, sending limbs flying everywhere, while rocks exploded here and there in the intense heat. Trees crashed into glowing heaps as the ferocious flames rushed on, sometimes racing so fast that they devoured 8000 acres of forest every hour, leaving behind only ghostly snags standing among the smoldering ruins.

Even massive fires spared some of the forest as they twisted and turned in the wind. Long narrow stringers of green trees separated scorched trails where the fire had passed, while other trees huddled in canyons and wet areas out of reach of the flames. When it was over, these remnants of a once grand forest stood scattered over the mountainsides among the swirling clouds of ash and broken trees. They were the future, a source of seeds that would begin the next generation of trees, and a majestic new forest destined to meet a similar fate.

Massive fires created enormous relatively uniform patches of trees that could cover hundreds of thousands of acres. Wind also leveled large areas of trees, but not in huge patches like those that fires normally produced. Lewis and Clark witnessed such a windstorm in November 1805 near Astoria.

> At noon the wind shifted to the northwest, and blew with such tremendous fury that many trees were blown down near us. This gale lasted with short intervals during the whole night.[296]

We do not know how widespread this storm was, but some could cover the entire region west of the Cascades. One such storm hit in 1880. It had sustained winds of 90 miles per hour, with gusts that may have exceeded 135 miles per hour. A windstorm like this could blow down a quarter of a million acres of trees. Unlike a massive fire, however, the trees fell in small groups scattered over a wide area.[297] Likewise, instead of clearing most of the trees, wind only knocked down the larger ones and released the younger ones growing underneath.

The patches that fire created in the ancient Pacific Douglas-fir forest became smaller toward the south.[298] Fires burned more often in the drier areas of the Oregon Coast Range, Oregon Cascades, and southwestern mountains than they did in the wetter or more northerly areas. This kept fuels lower, so the fires were smaller, less dramatic, and highly variable in their behavior. The flames might leap through the tree tops for a short time and then drop to the ground when the wind subsided. Then they might inch forward as a surface fire, flaring up occasionally in pockets of heavy fuel, only to climb into the trees again as the winds stiffened. An erratic fire like this could still cover thousands of acres.[299] However, it burned so unevenly that it created a variety of different patches, each covering a few hundred acres or less.[300] Such mixed fires probably burned the same area of forest about once each 95–145 years in the drier parts of the central region. They burned an average of about 37 years apart in the southwestern region.[301] However, the period between fires was longer than average near the humid coast and shorter in the dry interior areas.[302]

Mixed fires could burn for weeks or months in the southern part of the Pacific Douglas-fir forest.[303] Reverend Hines saw such a fire when he rode over a mountain in southwestern Oregon covered by "the most stately and majestic timber." "We found the fire making sad havoc with the fine timber with which its sides were adorned," wrote Reverend Hines.[304] It was the middle of August 1840 and he was on his way to the coast. Reverend Hines returned along the same route 11 days later "and found that the fire was still raging with increasing violence."

> A vast quantity of the large fir and cedar timber, had been burned down, and in some places the trail was so blockaded with fallen trees, that it was almost impossible to proceed; while now and then we passed a giant cedar, or a mammoth fir, through whose trunk the fire had made a passage, and was still flaming like an oven. Every few moments these majestic spars would come "cracking, crashing, and thundering" to the ground; but while the fire was thus robbing the mountain of its glory, we pushed on over its desolated ridges, and at sun-down arrived on a little prairie at its base, where we made our encampment. Several times during the night we were awakened by the crash of the falling timber, on the mountain, which sometimes produced a noise similar to that of distant thunder.[305]

"Fire, then, is the great governing agent in forest distribution and to a great extent also in the conditions of forest growth," wrote John Muir in 1888 after rambling through Washington's Douglas-fir forests.[306] Like other pioneer species, Douglas-fir seedlings thrive in large openings with bare soil. However, the seedlings need a little shade to protect them from the hot sun when they begin growing, but once they become established they grow best in full sun. These are the conditions that crown fires create. Crown fires consume the litter and expose the soil, and they kill the trees overhead to let in the sun. They also leave behind plenty of standing and fallen dead trees to cast shade over the seedlings. Then, just when they need it, the snags topple and expose the seedlings to more sunlight.[307] Therefore, crown fires provided ideal conditions for growing Douglas-fir.

Unlike most pioneers, Douglas-fir is somewhat tolerant of shade. It can withstand more shade than the most of the other trees that grow in the drier southwestern part of the forest. So it played the role of settler species as well as pioneer there, although it had to share those roles with tanoak, an evergreen hardwood. Tanoak is more tolerant of shade than Douglas-fir, it grows faster when young, and it sprouts vigorously after being burned. On the other hand, Douglas-fir can grow in the understory of other trees, as long as they are not too dense. It also lives much longer and grows much taller than tanoak.[308] Consequently, Douglas-fir and tanoak formed an unusual partnership in the ancient forest because each tree took advantage of slightly different conditions. They could live together in patches of self-replacing forest, and they often joined each other in pioneer forests.[309]

There was ample fuel to carry a fire from the ground to the tops of the largest trees in a forest where Douglas-fir towered over tanoak. Most of the trees died in these crown fires, but the tanoaks quickly came back because they could sprout. Douglas-fir normally returned during the next few years as its seedlings popped up in the bare areas between clumps of tanoak. It would grow slowly at first because tanoak sprouts grew faster than Douglas-fir seedlings, and the sprouts blocked most of the sunlight. Then about 30 years later, the young Douglas-fir trees grew above the tanoaks.[310] Such a patch was well on its way to becoming a middle-aged pioneer forest.

The tanoaks diminished in the deepening shade under the Douglas-fir trees as the patch entered the middle-aged pioneer stage and then the old pioneer stage. Eventually gaps opened here and there in the canopy as the aging trees fell. So tanoak moved back into the patch by filling the gaps, although Douglas-fir seedlings invaded the larger openings. This was the old transitional stage. Thus, Douglas-fir and tanoak often stayed together during the pioneer and old transitional stages of development. They also remained together in the final self-replacing stage because each tree kept filling new openings that formed when older trees died. They continued to grow side by side because neither species was capable of excluding the other.[311]

The southwestern region of the ancient Pacific Douglas-fir forest consisted of a mosaic of small and large patches because of the mixture of crown fires and surface fires that burned there. Each patch was also in a different stage of development, depending on how hot the fire was and how long ago it burned. Crown fires destroyed some patches, while elsewhere in the mosaic surface fires cleared the undergrowth from old pioneer forests of Douglas-fir, leaving the old trees unharmed because of their thick bark. Tanoak grew so dense in other places that there was no room for Douglas-fir seedlings. These tanoak patches could last for a century before enough openings appeared to allow Douglas-fir to invade. Patches of shrubs such as manzanita and deer brush could do the same, and last just as long or longer, especially if they kept burning. Other trees such as Pacific madrone grew in this forest as well, and they became part of additional patches. However, Douglas-fir and tanoak still dominated this part of the ancient forest.[312]

Rainfall was higher in most of the Pacific Northwest than it was in the mountains of the southwestern region of this ancient forest. It was in these rich moist soils to the north that Douglas-fir reached the enormous size and great age that so impressed explorers and early travelers. Even so, the Douglas-fir trees that grew in the drier areas were still big, as Reverend Hines noted in 1840 near the Umpqua River. "We measured one with our lasso as high up as we could reach," he wrote, "and found it to be thirty-six feet in circumference."[313] However, the trees were even bigger in the wetter areas. These were also the places where forests of such impressive trees could stretch to the horizon. Samuel Wilkeson bubbled with enthusiasm when he wrote about them in 1869. "These trees—these forests of trees," he said, "so enchain the sense of the grand and so enchant the sense of the beautiful that I linger on the theme and am loath to depart."[314]

The cathedral-like forests of Douglas-fir that grew in the wetter areas must have seemed immortal to the people who saw them first. However, Douglas-fir could not hold on to the land because it could not grow in the heavy shade that cloaked the forest understory. Even though it might take many centuries, the first generation of Douglas-fir eventually died of old age if a fire did not destroy them first. Western hemlock was the settler species that most often replaced them. Western redcedar replaced them in a few stream bottoms and other wet areas.[315] These settler trees had a princely stature, but Douglas-fir was king. The ancient forest lost much of its majesty when the pioneer forest of Douglas-fir disappeared.

Settler trees seldom had a chance to replace Douglas-fir in the ancient forest. A single generation of Douglas-fir could survive for 750–1000 years before the last tree finally toppled.[316] However, they normally came to a fiery end within less than 400 years. So western hemlock and western redcedar rarely had time to replace Douglas-fir. Even in remote parts of the Washington Cascades, where fires were less extensive than else-where in this forest, only about 12% of the mosaic consisted of self-replacing forests of settler species. They were also small, tucked away in little places where fires had difficulty reaching them. These self-replacing forests most often found refuge high on moist north-facing slopes, in wet valley bottoms, and along steep hillsides at the narrow upper ends of stream channels.[317] Refuges like these were not plentiful where crown fires burned vast areas. Therefore, self-replacing forests certainly would have covered even less of this landscape.

In the late 1800s, John Muir walked through the Pacific Douglas-fir forests of Washington, from Puget Sound to the upper slopes of the Cascades. "We find in passing through them again and again from the shores of the Sound to their upper limits," he said, "that some portions are much older than others."[318] What he saw were forests that fire had fashioned into a mosaic of very large sharply defined patches. Muir did not mention it, but each patch had its complement of wildlife because it was in a different stage of development. Some animals preferred newly burned areas and younger forests, such as Townsend's voles and wrentits, while others lived mostly in older forests, such as marbled murrelets. However, these secretive little birds only nest in older forests along the coast because they seek their prey in the sea.[319] As in ancient forests elsewhere in North America, it was the variety of patches that made wildlife so diverse and plentiful. Few animals lived exclusively in one kind of forest, and many needed more than one kind nearby to thrive.[320]

Muir noted all the major stages of forest development during his trek up the mountain, as well as what created them. For example, he entered a patch of young pioneer forest and said that it "has no trees more than fifty years old, or even fifteen or twenty years old." This forest shows "plainly enough," wrote Muir, that it has "been devastated by fire, as the black, melancholy monuments rising here and there above the young growth bear witness."[321] Muir did not name the trees that grew in this young pioneer forest. Although Douglas-fir was certainly the most abundant, western hemlock saplings may have been growing there as well. The seeds of western hemlock, like eastern hemlock, can germinate and grow either in full sunlight or shade, and on bare soil, rotten wood, or litter. However, western hemlock grows more slowly than Douglas-fir, so it is quickly overtopped. Even so, it persists in the shade and it can enter the middle-aged pioneer stage with Douglas-fir.[322]

Muir also saw a middle-aged pioneer forest on his trip up the mountain. He said it consisted of trees that were "close together with as regular a growth as that of a well-tilled field of grain."[323] This was the least interesting stage of forest development. Few plants could grow in such a dense forest, and wildlife was equally scarce. Plant diversity was greatest in open young forests filled with herbs, shrubs, and young trees. It was lowest in middle-aged forests, and it gradually increased again as forests aged and became less dense.[324]

When Muir saw the old pioneer and old transitional forests, it seemed obvious to him that the dominant trees were "mostly of the same age . . . perhaps from one hundred to two or three hundred years."[325] Muir was a keen observer. So, he rightly assumed that after a fire "a new forest sprang up, nearly every tree starting at the same time or within a few years, thus," he said, "producing the uniformity of size we find in such places."[326] Muir could not know that some burns took longer to fill with trees than others. It could take more than 100 years to replenish the trees in an area where fires burned more than once during a short period or where fires were unusually hot. Such fires left behind few living trees to provide seed for the next forest, so the forest recovered slowly.[327] Even though "the magnificent shafts" of Douglas-fir looked to Muir as if they were the same size and, therefore, the same age, some might have been younger than others.

Immense patches of older forest covered about 60% of the ancient landscape, somewhat less after a massive crown fire destroyed one of them and replaced it with young trees.[328] These were probably mostly dense old transitional forests because that stage of development persisted longer than the more open old pioneer stage. Anyone standing at "the top of some high, commanding mountain," as Muir did, could look in any direction and see nothing but great trees spreading over waves of valleys and ridges. It must have been an incredible sight. Muir thought these forests looked "perfectly solid" when he saw them in the 1870s. Even the rivers, lakes, and prairies that cut through them were to him not "large enough to make a distinct mark in comprehensive views."[329] It would have been difficult for Muir to notice countless gaps scattered about where giants had fallen and young trees were now growing. Likewise, the holes that windstorms had ripped open here and there would not have been obvious from such a distance. Even though they looked "solid," most of these forests were tattered and torn by the ages.

It was the ragged old transitional stage of the ancient forest that early travelers wrote about so eloquently. A forest that was "thick and almost impenetrable from underbrush and fallen trees," as Captain Nathaniel J. Wyeth noted at Fort Vancouver in 1832.[330] Those who ventured into the shady, cluttered interior of an old transitional Douglas-fir forest found an abundance of life. Moss clung to the moist sides of massive trunks and lay draped over those that had toppled, while lichens hung from branches and twigs high overhead. Some of the many fungi that thrived here worked secretly inside the old monarchs, hollowing out their centers, and others spread through the rotting wood strewn beneath them. Some fungi also lived on tree roots such as truffles and chanterelles. Truffles stayed hidden in the soil, but the yellow chanterelle mushrooms decorated the ground after the start of the fall rains. Beetles scurried about, carpenter ants and termites gnawed at the trees, and salamanders hid under the loose bark of fallen veterans. There was also the constant hum of insects or the occasional chattering of a squirrel or the hoot of an owl to break the quite. The sound of a pileated woodpecker rapping on a tree also might echo through the forest now and then. There was much life in the ancient forest, but it was unpretentious.

Anyone who experienced this serenity long enough would find it briefly and dramatically interrupted every so often—not by the sounds of life but by the sounds of age and decay. A massive limb might break high in a tree and come rattling through the branches on its long plunge to earth, thudding heavily to the ground as debris continued to rain over it. Any traveler hearing these sounds, no matter how distant, surely looked up and wondered if another limb might be ready to fall where they stood. Even more unnerving was the cracking of some great tree giving up its life, shattering branches and other trees as it tilted and then fell with a thunderous crash. These too were the sounds of old transitional forests.

The periodic thudding of limbs and crashing of trees filled the forest with piles of branches and, noted Muir, "immense trunks in every stage of decay."[331] Debris was always falling somewhere, and it often landed on shrubby little vine maples. However, instead of crushing and killing the vine maples, it usually just flattened them. Vine maple does not grow well from seed, especially in thick litter and heavy shade, but its stems can take root when they touch the ground, and it sprouts vigorously. So falling debris pinned vine maple to the ground where it could expand and form large clones in the understory of the old Douglas-fir trees.[332]

Falling debris also meant that the forest was changing. Each tree that dropped not only helped to spread vine maple, it also opened a gap where other young trees could grow. However, the openings were normally less than a tenth of an acre, so the height of the surrounding trees made them too shady for Douglas-fir seedlings.[333] Not only that, but less than 30% of the openings contained exposed soil, which Douglas-fir seedlings also needed.[334] As a result, settler species such as western hemlock filled most of the openings, and slowly replaced the old Douglas-fir trees, one at a time. As Douglas-fir diminished and hemlock grew denser, the shade in the understory became heavier until it was too dark in many places for even the hardy vine maple to grow.[335] However, vine maple usually avoided complete destruction. Old transitional forests rarely advanced very far before the next great fire turned them to ashes and snags. Then the little vine maple sprouted once more in the bright sun, intertwining with the young trees that filled the blackened landscape. They remained where they were. These humble plants were a more permanent part of the ancient forest than the mammoth trees that towered over them.[336]

The old transitional forest, 200–500 years old, was the pinnacle of earthly perfection, but it withered like a flower if it bloomed too long. Grand old Douglas-fir trees, many as tall as they would ever become, towered above their fallen comrades that had invaded the area with them after the last great fire. Western hemlock, and perhaps other settler trees, stood in the gaps they left behind. Since gaps opened constantly, the western hemlocks were of all ages, young and old, the eldest striving to reach as high as Douglas-fir, and falling short. Together they formed a magnificent forest of many interlacing layers, but it was the pioneer Douglas-fir not the settler trees that gave the ancient forest grandeur, and it was fire that gave them life.

GIANT SEQUOIA FOREST

Giants among giants.
Joseph Le Conte, professor, 1870.

Jedediah Smith was a trapper, but he also had an insatiable desire to explore. "I wanted to be the first to view a country on which the eyes of a white man had never gazed," Smith admitted.[337] So he and his band of 15

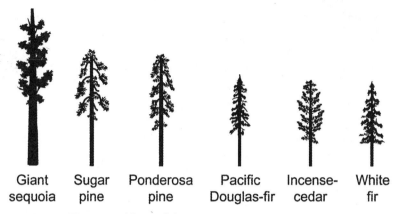

| Giant sequoia | Sugar pine | Ponderosa pine | Pacific Douglas-fir | Incense-cedar | White fir |

Fig. 12.6   Trees of the ancient giant sequoia forest.

trappers rode into the unknown country southwest of Great Salt Lake in 1826 searching for adventure and beaver. Instead they found mostly rocks, sand, and salt flats, and precious few beaver. They finally straggled out of the Mojave Desert, which Smith called "Starvation Country," in November, and showed up at San Gabriel Mission in California, thirsty, hungry, and mostly on foot. Governor Jos Mara Echeanda promptly placed them under house arrest in San Diego and then released them a month later. Governor Echeanda told Smith to return the way he had come. Smith went northward instead, trapping beaver until the next spring. Then he headed up the North Fork of the Stanislaus River and went over the Sierra Nevada near Ebbetts Pass. In so doing, Smith and his companions became the first Europeans to cross the Sierras.[338] They also came close to being the first Europeans to see a giant sequoia (Fig. 12.6).

Unknown to Smith, his route took him very near a giant sequoia forest that is now the Calaveras Grove. It was one of the most remote groves and easy to miss. Ironically, the largest tree in the world grew in one of the smallest forests in North America. This ancient forest only covered 35,607 acres. It also consisted of 75 widely scattered little groves within a narrow belt that stretched 260 miles along the western slope of the Sierra Nevada. Not only that, but most of these groves, and the largest ones, were in the southern Sierra Nevada, far from where Jedediah Smith crossed the mountains.[339]

It would be another trapper, Captain Joseph Walker, who would see the giant trees first. His party spent a grueling month wandering over the Sierras in 1833 from Nevada to California. Thick patches of snow made travel difficult, boulders and cliffs blocked the way, and there was no dried meat left to eat. Cold and starvation forced them to kill many of their horses for food while stumbling through the mountains. However, this epic journey yielded two historic discoveries. After weeks of hardship Captain Walker's party became the first Europeans to peer into the gaping chasm of Yosemite Valley. Clerk of the expedition Zenas Leonard would later recall standing on the edge of a cliff that "appeared to us to be more than a mile high." There he saw "many small streams which would shoot out from under these high snow-banks, and after running a short distance . . . precipitate themselves from one lofty precipice to another, until they are exhausted in rain below." They tried to reach the valley, but Leonard said, after "making several attempts we found it utterly impossible for a man to descend, to say nothing of our horses." So Captain Walker continued westward along a ridge on his way to a second historic discovery.[340]

It took two more weeks of hardship before Captain Walker's party finally reached the middle slopes of the Sierras. There Leonard said they found "the timber much larger and better, game more abundant and the soil more fertile." He added that "the land is well watered by a number of small streams." This was just the kind of place you might expect to find giant trees, and so they did.

> In the last two days travelling we have found some trees of the Red-wood species, incredibly large—some of which would measure from 16 to 18 fathoms round the trunk [31–34 feet in diameter] at the height of a man's head from the ground.[341]

That was it. Leonard said no more about this momentous discovery. No one at the time seemed to understand the significance of this account in Leonard's journal, and if Captain Walker or Leonard mentioned it, few took note. The giant trees were seen and quickly forgotten.

Captain John C. Fremont probably saw the giant sequoia next when he traveled near the Kings River in the winter of 1845. He said in his 1848 memoirs that he found an open forest with "some trees extremely large," but that comment also sparked little interest.[342] So the giant sequoia forest remained unknown until the California Gold Rush. Then, in 1852, Augustus Dowd walked into the Calaveras Grove that Jedediah Smith passed 25 years earlier and instantly made the "bigtrees" famous. James Hutchings, publisher of the *California Magazine*, wrote an account of the event in 1859.

> In the spring of 1852, Mr. A. T. Dowd, one of the "Nimrods" [hunters] of Calaveras county, was employed by the Union Water Company of Murphy's Camp, to supply the workmen ... with fresh meat.... Having wounded a large bear while engaged in this occupation, he industriously followed in pursuit; when to his momentary confusion, and astonishment, his eyes looked for the first time upon one of those magnificent giants.... All thoughts of hunting, or bear pursuing, were forgotten.... Filled with thoughts inspired by what he had seen, he returned to camp, and there related the story of the wonder he had discovered. His companions laughed.... For a day or two he allowed the matter to rest.... On the Sunday morning following he went out as usual, and returned in haste, evidently excited by some event. "Boys," he exclaimed, "I have killed the largest grizzly bear that I ever saw in my life. While I am getting a little something to eat you make preparations to bring him in." ... On, on, they hurried, with Dowd as their guide ... until their leader came to a dead halt at the foot of the tree he had seen, and to them had related the size. Pointing to the immense trunk and lofty top, he cried out, "now, boys, do you believe my big tree story?"[343]

European explorers, settlers, miners, and other assorted travelers began wandering over California's landscape in the middle 1700s, yet it took nearly a century to bring the ancient giant sequoia forest to the attention of the world. The tiny groves stayed hidden within the mixed-conifer forests of the Sierra Nevada because they were tucked away in canyons and other moist places where even giant trees were difficult to see.

There are several theories about why the giant sequoia forest consisted of so many little groves spread over such a large area. Some scientists believe the original forest may have split into groves during the Ice Age or the Great Drought, while others think they formed separately and were unable to join into a single forest. Regardless, glaciers and droughts certainly contributed to their isolation. However, the latter explanation is also consistent with the way giant sequoias probably migrated to the west side of the Sierra Nevada.[344]

Giant sequoias once flourished throughout the Northern Hemisphere, and their ancestors towered above the dinosaurs. Then they disappeared from the Old World about 30 million years ago, and their range continued to shrink in North America as well. The sequoias finally ended up in Nevada where small groups moved up the east side of the Sierras along rivers and streams. They managed to cross to the west side no earlier than 12 million years ago when the mountains were still young and low. Here the groups of sequoias may have remained isolated from one another in the wetter areas because the trees were unable to spread north and south over the steep dry ridges that separated them. As the Sierras rose they blocked the rains that came from the Pacific Ocean and parched the Nevada landscape, which wiped out the few giant sequoias that still survived outside California. Now the sequoias were making their last stand against extinction, trapped between dry ridges, the hot dry grasslands of the Great Central Valley, and the cold rocky peaks of the Sierras. They also had several Ice Ages and droughts left to endure. We will never know how many of these isolated groups of giant sequoias may have perished before Captain Walker saw them.[345]

In spite of the antiquity of the giant sequoia, the ancient forest that Captain Walker's trapping party found was relatively young. The sequoias had only recently recovered from the last Ice Age and the Great Drought that followed it. Cold air from the glaciers had pushed the little groups of sequoias onto the lower slopes of the Sierras and forced them to take refuge in shady ravines. The sequoias moved back as the glaciers retreated, but were again squeezed into small areas around streams and meadows during the Great Drought. The sequoias dispersed once more after the drought ended in the Sierras about 4500 years ago. Thus, the ancient giant sequoia forest took its modern form as the sequoias gradually spread and mixed with other conifers.[346]

The ancient giant sequoia forest occupied only a tiny part of the 12 million acres of mixed-conifer forest that covered California and southern Oregon.[347] To be sure, there were subtle differences in this forest from place to place. Pacific Douglas-fir was an important part of the mixed-conifer forest, but it did not grow in the southern Sierra Nevada. On the other hand, white fir grew throughout the forest, but it became less important when it mixed with Douglas-fir in the north and with red fir on upper mountain slopes.[348] But unlike the other conifers, giant sequoia made a dramatic difference. Its presence transformed the mixed-conifer forest into something otherworldly. As English botanist John Lindley exclaimed when he first saw specimens in 1853, "What a tree is this!"[349] Even the Mokelumne Tribe and other Indian people of the Sierras sensed

Fig. 12.7   Fire-scarred giant sequoia, Sequoia National Park, California. (Photograph by the author)

magic in the giant sequoia long before Europeans gazed in awe at its enormity and grace. They called it *woh-woh-nau*, which is a Miwok word that imitates the hoot of an owl—the guardian spirit of the giant sequoia.[350]

John Muir called the giant sequoia "the noblest of a noble race." "Even the very mightiest of these monarchs" was to him "exquisitely harmonious."[351] Some trees can grow taller, such as the coast redwood and the Australian eucalyptus. One tree, the African baobab, can grow wider, and the bristlecone pine can grow older, but no tree unites height, width, and age like the giant sequoia. It is, Muir said, an "unrivaled display of combined grandeur and beauty."[352]

The giant sequoia is a pioneer that can reach a height of 320 feet, grow to a diameter of 37 feet, and live at least 3200 years (Fig. 12.7). Even an ordinary sequoia is about 250 feet tall and 10–20 feet wide. Just its branch can rival the size of some other trees, reaching a length of 125 feet and a thickness of 7 feet. However, it can be hard to see the true size of such mammoth branches because the lowest limb might be so high that it could hang over a 12-story building.[353] When Captain Walker first gaped at a giant sequoia he probably saw nothing but a colossal cinnamon-red trunk. The bark was fibrous, almost soft to the touch, and etched with parallel grooves that Muir said looked "like the fluting of an architectural column."[354] Captain Walker would have had to tilt his head way back just to see its immense branches and scaly blue-green leaves, or climb a hill to view its dome-shaped crown. Only rarely would he have been able to glimpse an entire tree, not just because of its enormous bulk, but because many of the other trees that grew in this forest were also huge. "They are giants among giants," wrote professor Joseph Le Conte when he saw the ancient giant sequoia forest for the first time in 1870. "The whole forest is filled with magnificent trees."[355]

First among the other giants in this forest was the sugar pine, a pioneer like other pines. It is the largest pine in the world and it has the largest cones.[356] Captain Fremont knew this was a special tree when he first saw it in the spring of 1844. "The forest was imposing to-day in the magnificence of the trees," wrote Captain Fremont in his journal as he descended the Sierras, "some of the pines, bearing large cones [sugar pine], were 10 feet in diameter."[357] He must have seen the biggest specimen of its kind because 10 feet is probably its maximum diameter. Sugar pine can also grow 250 tall and live 600 years, and its cones can be 2 feet long.[358] Muir said that sugar pine "is the king of all the pines and the glory of the Sierra forests."[359]

> The trunk is a smooth, round, delicately tapered shaft, mostly without limbs, and colored rich purplish-brown, usually enlivened with tufts of yellow lichen. At the top of this magnificent bole, long, curving branches sweep gracefully outward and downward.... The needles are about three inches long, finely tempered and arranged in rather close tassels at the ends of slender branchlets that clothe the long, outsweeping limbs. How well they sing in the wind, and how strikingly harmonious an effect is made by the immense cylindrical cones that depend loosely from the ends of the main branches![360]

Ponderosa pine and Jeffrey pine also stood in this ancient forest, and this is where they grew best. They were nearly as large and lived as long as sugar pine.[361] Douglas-fir did not reach the great size in the Sierra Nevada that it did in the wetter areas of the Pacific Northwest, but it was still a big tree. Incense-cedar and white fir also were prominent. Early travelers often mistook incense-cedar for giant sequoia because of its scaly leaves and cinnamon-red bark. However, its leaves are more yellow than giant sequoia, and its bark is lighter and less fibrous, although the greatest difference is size. Incense-cedar, a settler species, normally grew about as big as the largest limbs on a giant sequoia. Even so, it occasionally reached a height of 225 feet and a diameter of 12 feet. White fir was less conspicuous than the other trees. It has silver-green leaves, and deeply furrowed ashy gray bark with large flattened ridges. It could grow 200 feet tall and 8 feet in diameter, but it was usually much smaller and lived only a few centuries because of its vulnerability to insects and disease.[362]

Color and scent filled the ancient giant sequoia forest because it was open and sunny. "The inviting openness of the Sierra woods is one of their most distinguishing characteristics," wrote Muir.[363] The massive cinnamon-red, purple-brown, and yellow tree trunks stood out, even in the whiteness of winter. However, it was in the sunshine of summer that this forest came to life with a dazzling rainbow of color.

Set in a rich brown carpet of old needles and cones, the tints and hues of trunks and leaves provided a colorful backdrop, if only briefly, for a showy display of flowers. Sprays of white flowers appeared here and there in the ancient forest where Pacific dogwood or deer brush bloomed, but little flecks of color were everywhere. Orange leopard lilies, crimson columbines, and pink shooting stars dangled over streams and meadows. Indian pinks, yellow irises, and golden brodiaeas decorated the drier areas. On warm days, a resiny scent told the traveler that *kit-kit-dizze*, as the Indians knew it, grew nearby. It sprawled over the ground in sunlit places creating a thick fernlike mat dotted with little white flowers shaped like roses. Tall fields of meadow lupine also filled moist and partially shaded openings with blue blossoms and bumble bees, and masses of pink and yellow mustang clover carpeted dry gravely openings.[364] The gaily decorated western tanager added flashes of red and yellow to the spectacle, and the tiny wings of migrating lady beetles made the air sparkle. It was "a scene of enchantment," wrote Muir.[365]

This was not a silent forest. Certainly, quite dominated winter, but like color and scent, the sounds of summer resounded throughout the ancient giant sequoia forest. These were not just the rumblings of an afternoon thunderstorm, or the occasional crash of a falling limb or tree, but sounds of life. Sometimes the muffled scratching of a back bear came through the forest as it tore at an old log to feast on carpenter ants. Deer rustled the brush as they wandered about gently nipping at twigs and leaves. Occasionally, the rapping of a sapsucker filled the air as it drilled rows of holes through a sequoia's bark so that it could lick the sap as it oozed from the tree.[366] However, the incessant and loud *shacking* and *weking* sounds of the Steller's jay were unavoidable as they climbed this tree and then that tree, using limbs like a spiral staircase.[367]

Like the jay, but more musical, were frequent bursts of chirps and barks from the little Douglas squirrel or Sierra chickaree. "A small thing," Muir said, "but filling and animating all the woods."[368] Chickarees fell silent in lean years when few were left to scamper about the ancient forest. However, in good years they announced themselves everywhere as they darted among the branches cutting a small green cone, whittling away its scales to reach the seeds, and then moving on to cut another one. However, sugar pine cones could be too big for this little creature to handle while standing on a limb. So it dropped them and scurried down the tree to eat the large seeds while the heavy cones rested on the ground. Chickarees often let the sequoia cones fall as well, particularly when they wanted to store them for future use. On a busy day, one of these little squirrels could cut over 500 sequoia cones in 30 minutes. They rained down like bombs, hitting the soft litter in a series of dull thumps.[369]

Intermingled with this sensory extravaganza were also the unmistakable signs of people. The Monachi, Miwok, and other Sierra Indians had lived here for thousands of years before the trappers arrived. They came in summer and fall to harvest the banquet of game and food plants that such an airy and sunlit forest provided. They gathered manzanita berries, currents, gooseberries, thimbleberries, raspberries, blackberries, elderberries, and strawberries. They dug camas, brodiaea, wild onion, and other roots, and they made a medicinal tea from *kit-kit-dizze* leaves. They also shared hazelnut and pine nut crops with chickarees. However, they were especially fond of black oak acorns.[370] Patches of black oak grew here and there in the forest, particularly around granite outcrops.[371] It was in such places that Indians often made their camps and carved mortar holes in exposed rock where they pounded acorns into meal. The sounds of work, crackling campfires, and voices filled the ancient giant sequoia forest as children played and people chatted while performing the many chores of daily life.

Pioneer species dominated the ancient giant sequoia forest. Like pioneer forests elsewhere, and other Sierra mixed-conifer forests, it was sustained by Indian and lightning fires.[372] They were similar to the gentle fires that burned in ancient ponderosa pine forests, and they kept this forest open and productive in much the same way.[373] This was a fire forest, yet it was unusual among such forests, and not just because it was incredibly diverse for a conifer forest, its matchless beauty, or the bounty it produced. The ancient giant sequoia forest was unique because it was home to the biggest tree on earth. Otherwise, it was identical to the ancient Sierra mixed-conifer forest.[374]

Like the great sugar pines and ponderosa pines that grew in this ancient forest, giant sequoias would vanish without fire. Sequoia cones are serotinous. Their seeds can stay trapped in the cones for more than 20 years waiting for the heat from a fire to open them.[375] Still their seeds fall in prodigious numbers while the trees wait for the next fire. Chickarees eat young green cones and drop the sequoia seeds when they peel away the scales. They do not bother eating the seeds because they are too tiny.[376] Likewise, the larvae of a little *Phymatodes* beetle bores into the cones, causing them to dry and shrink so that the scales lift and drop the seeds tucked underneath.[377] Ice, wind, and snow also break off cones and cause them to dry.[378] However, not all cones can drop their seeds because a thick crust of lichen keeps them tightly closed.[379] Even so, a steady rain of about 2 million sequoia seeds falls on each acre of the forest floor each year.[380] "The Sequoia race is not likely to fail from an insufficiency of seed," wrote C. B. Bradley in 1886.[381]

In spite of the multitude of seeds that fell in the ancient forest, sequoia seedlings only appeared occasionally, and nearly all of them died within a few weeks. Leaves that littered the forest floor dried out quickly, so nearly all the seedlings that grew in such places died before they could send their roots down far enough to reach the moist soil underneath. Wherever the litter was thick, normally the shade was also too heavy for the seedlings. On the other hand, exposure to the sun kills the seeds within 20 days. So even if they landed in an opening, they had to hit soil that was soft enough to bury them, or slip into a crack. This rarely happened because the seeds are so light, and the litter was usually too compact or the soil too hard to penetrate.[382] Therefore, very few young sequoias grew in the shade of the giants, and most of them stayed stunted and unhealthy, with little chance of ever becoming giants themselves.

The lack of young sequoias in the ancient forest concerned Bradley when he saw the Calaveras Grove. "This mighty race of Titans . . . is fading away," he said. "Young Sequoias are not merely inconspicuous in the groves," wrote Bradley, "they are actually few, and the smallest seem to be the fewest."[383] It does not take many young sequoias to sustain a grove when they may grow into trees that live several thousand years. Even so, one grove was most likely heading toward extinction at the time Bradley expressed his concern, and sequoias were dwindling in several others. Of course, the Indians had left long before Bradley arrived, so there were also fewer fires burning in these groves.[384] We have known for a long time where sequoia seedlings grow best. For example, Bradley noticed that the seedlings grew on bare soil, especially around fallen sequoias.

> They would be found standing in groups of threes and fours upon the rim of a huge crater, caused by the uprooting of some old giant; or ... more strangely still, perched high in the air upon a mass of soil adherent to the upturned roots. Elsewhere they were found ... on banks of bare subsoil exposed by slides or by the wash of streams.[385]

Giant sequoia is a pioneer, so it is not surprising that it grew in clearings rather than shade.[386] However, the gaps that most trees create when they tumble to the ground would be too small to let in enough sunlight for giant sequoia. On the other hand, a giant sequoia is so big that it tears a great hole in the canopy when it falls, and it often takes down some of the surrounding trees as well. As the massive sequoia pitches forward, it also yanks its roots out and opens an enormous pit of loose soil where sequoia seeds might become buried. Thus, young sequoias often grew in these openings, as Bradley noted. However, it did not always turn out so well for sequoia seedlings. Sometimes they died because the soil dried out, a camel cricket or a bird ate them, a chickaree dug them up, a fungus infected them, or a deer or fallen branch crushed them. Other pioneer trees or shrubs might also take over the opening and exclude sequoia seedlings.[387] Regardless of the cause, some old sequoias would plunge to earth without being replaced. So a sequoia grove would gradually disappear if the trees only replaced themselves by filling openings left by falling giants or an occasional landslide.

Fire made the difference between extinction and survival for the giant sequoia. "Fire," Muir pointed out, "furnishes bare virgin ground, one of the conditions essential for its growth from the seed."[388] However, a gentle surface fire that removed the litter was too cool to create an opening. Even a hot fire was not enough. Such a fire could harden the soil and keep out rainwater. It also stimulated the growth of shrubs such as deer brush more than it did giant sequoia.[389] The fire had to be very hot to favor the giant sequoia over other plants, the kind of fire that flares up in a dense group of large trees or in logs and piles of debris. Again, Bradley grasped this distinction in 1886. In one place he found young sequoias growing in the "prostrate shell" of an old tree, "where fire and decay had eaten the central substance quite down to the earth, but had left the sides standing like walls."

> But most remarkable of all was an open space of considerable size quite filled with thrifty young Sequoias of uniform age and height. Upon inquiry I learned that some dozen years ago a fire raged long and with unusual fierceness among the fallen trunks at this point, till not only they, but the whole deep layer of vegetable rubbish which covers the actual soil in these forests, had been reduced to heaps of white ashes. . . . it seems, the infant trees have succeeded in establishing themselves upon veritable earth, and not upon the rubbish-layer.[390]

This fire "had evidently given these trees their long-delayed opportunity," wrote Bradley. Even though he was "indebted to a friend" for the "clue" that helped him to understand what he saw, Bradley had hit on the secret that scientists confirmed nearly a century later. Giant sequoia seedlings grow best in places where fires burn the hottest.[391]

A very hot fire starts a chain of effects that benefit giant sequoia. It begins when the flames clear a large sunny opening in the forest. A quarter acre or smaller clearing might provide room for one or a few giant sequoias, while a half acre might be enough for up to 20 sequoias.[392] The column of superheated air from the fire that rises into nearby sequoias dries and opens their cones so that they drop massive amounts of seed into the fresh burn. Meanwhile, the burned wood provides a fluffy layer of ash that buries the little sequoia seeds when they land and protects them from direct sunlight. There is also more moisture available for the seedlings because high temperatures make the soil crumbly so that water flows into it more quickly. Once established, the seedlings have a good chance of survival because the fire's intense heat penetrates deep into the soil and kills root fungi and plant seeds. This reduces losses from disease, and it eliminates faster growing shrubs that might take the moisture and block the sunlight that sequoia seedlings need to grow. By themselves, and with plenty of water, sunlight and nutrients, the seedlings grow at the astounding rate of 8 inches to 2 feet a year. That is double to triple the rate of the fastest growing seedlings in places where fires were not so hot.[393] These are the sequoias destined to become giants.

Surface fires returned to the same place in the ancient giant sequoia forest about once every 5–18 years.[394] Mixed-conifer forests that lacked giant sequoia burned with about the same regularity.[395] The period between fires was shorter on dry ridges and upper slopes and longer on moist lower slopes and in canyons.[396] Upper slopes burned more often because lightning struck them more often.[397] In addition, uphill winds pushed those Indian and lightning fires that started on lower slopes to the ridges. Yet these uphill winds, as well as creeks, rock outcrops and other barriers, often prevented the slow moving fires that started on ridges from reaching the lower slopes. So ridges burned more often, and they were usually more open than the rest of the forest.

Sierra Indians certainly contributed to the high frequency of fires in the ancient giant sequoia forest.[398] They set some of their fires inside the forest to enhance hunting and for other reasons, which Hector Franco, a Yokut, pointed out in an interview about the way his ancestors burned.

> They would only burn once every two or three years in the giant sequoias, but there were some areas that had more plants. The sourberry [squaw bush] grows up high. They would burn every year in the areas where there were the most sourberries, to stimulate growth for the following year, for the berries and for the basketry material.[399]

Some of their fires also entered sequoia groves from foothill woodlands, as Hector Franco noted in his interview. "They would burn down below and the fire would go up," he said. "I know where they burned there, and it would go up."

Fires that start at the base of a hill move faster when they climb a slope with the wind. They also have taller flames, and burn hotter than fires that start at the top of a hill and move down slope against upward flowing winds. That means that most lightning fires were slow moving and light since they started on ridges, and many Indian fires were fast moving and hot because they often started below the ridges. Such hot fires are more likely to ignite thick patches of trees and piles of debris than cooler fires. Therefore, Sierra Indians unintentionally increased the number of very hot fires that burned in the ancient forest, which probably produced more groups of mammoth sequoias than lightning fires alone might have done. Owls may have been the guardians of the sequoias, but Indians helped to sustain them.

Indian and lightning fires burned so frequently that they cleared away debris and young understory trees before they could accumulate. Still, scattered pockets of fuel occurred in the forest where groups of trees died or became thick, or where branches fell.[400] Consequently, surface fires that crept along the forest floor occasionally flared up and became small but fleeting crown fires. However, those patches of trees that were thick enough to burn were few, and less flammable patches usually kept them apart so that a crown fire could not spread. This kept crown fires confined in small, widely separated patches of trees, usually less than half an acre, although on rare occasions a few hundred acres of trees might go up in flames. Large crown fires were almost unknown in this ancient forest, as Muir pointed out.

> In the main forest belt of the Sierra, even when swift winds are blowing, fires seldom or never sweep over the trees in broad all-embracing sheets as they do in the dense Rocky Mountain woods and in those of the Cascade Mountains of Oregon and Washington.[401]

Several decades later, foresters S. B. Show and E. I. Kotok supported this observation. They said:

> "The virgin forest is uneven-aged, or at best even-aged by small groups, and it is patchy and broken; hence it is fairly immune from extensive, devasting crown fires. . . . a continuous crown fire is practically impossible."[402]

Insects and disease played their roles, as they do in all forests, by killing groups of trees and creating new openings.[403] However, fire sculpted the ancient giant sequoia forest and gave it beauty and diversity. The wounds that even a light fire could inflict on mature trees also increased infections and insect attacks.[404] The aftermath of a fire is a sad sight with its blackened and scared trees and logs, scorched foliage, and ashy soil. However, fire sculpting can be fascinating to watch, and the wounds soon heal with color and life. In the fall of 1875, John Muir witnessed "a great fire" shaping an ancient giant sequoia forest on the southeast side of the Kaweah River. "I stopped to watch it and learn what I could of its works and ways with the giants," he said. Like many fires, this one started outside the forest and entered it from below.

> It came racing up the steep chaparral-covered slopes of the East Fork cañon with passionate enthusiasm in a broad cataract of flames. . . . But as soon as the deep forest was reached the ungovernable flood became calm like a tor-

rent entering a lake, creeping and spreading beneath the trees. . . . Only at considerable intervals were fierce bonfires lighted, where heavy branches broken off by snow had accumulated, or around some venerable giant whose head had been stricken off by lightning. . . . the ground-fire advancing in long crooked lines gently grazing and smoking on the close-pressed leaves, . . . and tall spires and flat sheets with jagged flapping edges dancing here and there on grass tufts and bushes, . . . huge-fire-mantled trunks on the hill slopes glowing like bars of hot iron, violet-colored fire running up the tall trees, tracing the furrows of the bark in quick quivering rills, and . . . with a tremendous roar and burst of light, young trees clad in low-descending feathery branches vanishing in one flame two or three hundred feet high. . . . Another grand and interesting sight are the fires on the tops of the largest living trees flaming above the green branches . . . and looking like signal beacons on watch towers. . . . the big lamps burn with varying brightness for days and weeks, throwing off sparks like the spray of a fountain. . . . Nearly all the trees that have been burned down are lying with their heads uphill, because they are burned far more deeply on the upper side, on account of broken limbs rolling down against them to make hot fires, while only leaves and twigs accumulate on the lower side and are quickly consumed without injury to the tree. But green, resinless Sequoia wood burns very slowly, and . . . only a shallow scar is made, which is slowly deepened by recurring fires. . . . Fire attacks the large trees only at the ground, consuming the fallen leaves and humus at their feet, doing them but little harm unless considerable quantities of fallen limbs happen to be piled about them. . . . The saddest thing of all was to see the hopeful seedlings, . . . helplessly perishing, and young trees, perfect spires of verdure . . . suddenly changed to dead masts.[405]

This fire was probably typical of those that burned in the ancient giant sequoia and mixed-conifer forests when the trappers arrived. The regional climate was cool then, as it had been since AD 1300. It was also cool between AD 500 and 1000. However, the climate in the Sierras was unusually warm between AD 1000 and 1300. Consequently, the giant sequoia forest burned nearly twice as often. As a result, these fires were most likely even lighter than they were in cooler periods, and there was probably less debris and undergrowth in the forest as well.[406] Thus, in cool climates or in warm, surface fires burned often enough to keep the ancient giant sequoia and Sierra mixed-conifer forests open and patchy.[407]

The lack of undergrowth and the grouping of trees were such striking features of ancient forests in the Sierra Nevada that they could not go unnoticed by Muir. "The trees of all the species," he said, "stand more or less apart in groves, or in small irregular groups, enabling one to find a way nearly everywhere, along sunny colonnades and through openings that have a smooth, park-like surface."[408] In 1916, a few years before Show and Kotok noted the patchy character of the forest, foresters L. T. Larsen and T. D. Woodbury also noticed that these forests were "made up of groups." In addition, they found that the trees in each group were "approximately even-aged" because they were mostly pioneers that filled fresh clearings at about the same time. However, they said that individual trees "of various ages and species" also stood here and there within the forest.[409]

"Fire is the . . . controller of the distribution of trees," wrote Muir.[410] Each tree, shrub, flower, and animal that lived in the ancient giant sequoia forest had its needs, and fire provided for all of them by creating a variety of conditions. It burned hot or cool in various places, returning often or sporadically, and it skipped some places for long periods. Therefore, the forest consisted of a mosaic of patches in different stages of development, and each patch contained a different mixture of trees, shrubs, or other plants. Because fires were so frequent, these were mostly very small patches, usually about two-tenths of an acre. Some patches also covered half an acre or more, and a few covered tens or hundreds of acres.[411] However, these larger patches gradually disappeared as they dissolved into the smaller patches that characterize this ancient forest.

The species that grew in each patch depended on the availability of seed, which varied from year to year, and their differing abilities to take advantage of the conditions created by the last fire. Giant sequoia germinates best on bare soil in full sunlight and survives best on light litter.[412] This is just what happened in the ancient forest as leaves fell on seedlings while they grew in a fresh clearing. Even so, it was only in places where hot fires burned piles of debris and logs that giant sequoia had an advantage over other pioneers.[413] Otherwise, ponderosa pine or shrubs, such as deer brush or manzanita, might take over the clearing.[414] White fir and sugar pine seedlings can also grow on bare soil in clearings, although they do better in light litter.[415] That means that ponderosa pine, white fir, sugar pine, and shrubs, mixed or separately, filled most new openings, and patches of giant sequoia seedlings were few and widely scattered. Incense-cedar moved into smaller or older openings because it grows well in light litter and full sun or partial shade.[416] Consequently, patches of giant sequoias only dominated about 5–8% of the ancient forest mosaic, and these were mostly old veterans.[417]

Incense-cedar and white fir seedlings can tolerate shade under shrubs, so they could grow through them and eventually take over a brushy opening.[418] As a result, nearly 13% of the ancient forest consisted of patches of shrubs intermixed with young and middle-aged settler trees, some towering over the shrubs and robbing them of sunlight.[419] Incense cedar, white fir, and sugar pine can also move into the understory of trees where the canopy is not too dense because they tolerate light litter and partial shade.[420] However, white fir was the ultimate settler species in the ancient giant sequoia forest because only its seedlings can tolerate both deep litter and heavy shade.[421]

White fir was the most abundant tree in this ancient forest because it could take advantage of so many situations, but the pines were almost as numerous.[422] White fir grew well in openings and shady places where the pioneer sequoias and pines could not grow. Even so, surface fires that crept beneath the trees often killed young white fir in the understory before they could mature. Still, they came back quickly. The thick bark of older trees usually protected them as passing fires cleared the under-growth, and reduced litter and duff to ash. These were ideal places for white fir seedlings because they grow best where overstory trees shade bare soil.[423] So a new generation of seedlings soon replaced the young trees that fire had cleared from the understory. As a result, young white fir trees grew under some patches of older forest, even though fire killed them repeatedly. These patches of old transitional forests covered about 8% of the forest mosaic, and open old pioneer forests covered about 10%.[424] A scattering of single old trees, often the lone survivors of once thriving patches, also stood within the ancient forest.[425]

Indian and lightning fires were so frequent in the ancient giant sequoia forest that patches of settler trees seldom had enough time to develop into self-replacing forests. Settler trees that grew in patches of brush rarely escaped the many fires that burned in this forest. Shrubs burn easily, but fire also stimulates their growth.[426] So fire killed the settler trees, and the shrubs returned within a year. It took many years for the settler trees to grow above the brush again, and then they often met the same fate. Patches of self-replacing forest are also extremely vulnerable to fire. White fir sheds a large number of small branches,[427] and the trees are constantly falling as age and disease thin a patch. These fuels pile up in an understory already choked with young fir trees. A creeping surface fire that enters such fuel often becomes intense. It then climbs the young trees until it reaches the canopy, where it becomes a seething crown fire that envelops the entire patch and kills all the trees.[428] The fire drops back to the ground and moves away when it runs out of fuel. These and other destructive forces such as insects and disease constantly devastated settler trees in the ancient forest. As a result, patches of self-replacing white fir forest only accounted for about 4% of the mosaic.[429]

Clearings also appeared here and there in the ancient forest after an extensive surface fire. Some of them formed where young pioneer forests burned before they could grow tall enough to escape the flames. Others formed where self-replacing or old transitional forests burned, or where insects or disease had killed a group of trees. Regardless, these openings developed often enough to produce a continuous supply of young trees. Consequently, young and middle-aged pioneer forests covered about 42% of the forest mosaic. Most of these younger forests consisted of conifers. Black oak, the principal hardwood, dominated less than 1% of the forest mosaic.[430]

This ancient forest was rich with meadows, flowery or grassy areas, and patches of shrubs. Shrubs filled openings in nearly 21% of the forest, and an additional 14% of the forest consisted of flower or grass-covered openings, and clearings.[431] This abundance of shrubs and small plants supported a variety of game, especially deer and bear. Captain Walker's starving trappers were "greatly cheered" after shooting "two large black tailed deer and a black bear, and all very fat" just before they discovered the giant sequoia in 1833.[432] Pioneer Hale Tharp also found "a great many deer and a few bear in the meadow [Log Meadow]" when he discovered the Giant Forest Grove in 1858.[433]

The large number of combinations of sun and shade, and bare soil and litter, that could exist at a particular time generated an equally large number of different patches in the ancient giant sequoia forest. A single species usually dominated each patch, primarily because the parent trees produced a large crop of seeds when fire or something else prepared a suitable place for them to germinate and grow. Other species often grew among them, but they were less prominent. Thus, the selective action of frequent surface fires, as well as insects, disease, drought, and other disturbances created a mosaic of amazing diversity. These disturbances also kept the forest youthful and vigorous. So most of the mosaic consisted of grassy and shrubby openings, and small patches of young and middle-aged forests, but older forests still dominated the scene because the trees were huge. No doubt, the proportions of patches in older forests also varied from grove to grove, and they varied over time as well. This further increased the diversity of the ancient giant sequoia forest. However, like so many of America's ancient forests, it was the Indian's mastery of fire, as well as lightning fires, that sustained this forest of giants.

Some pine trees were cut down and stripped of their branches, and these being laid on the snow furnished us with a bed, whilst a fire was lighted on a floor of green logs. To sleep thus—under the beautiful canopy of the starry heavens—in the midst of lofty and steep mountains—may appear strange to you, and to all lovers of rooms, rendered comfortable by stoves and feathers; but you may think differently after having come and breathed the pure air of the mountains, where in return coughs and colds are unknown. Come and make the trial, and you will say that it is easy to forget the fatigues of a long march, and find contentment and joy even upon the spread branches of pines, on which, after the Indian fashion, we extended ourselves and slept, wrapped up in buffalo robes.

Father De Smet, Northern Rocky Mountains
Alberta, Canada, 1845

Notes and Citations

PART ONE. *The Making of America's Ancient Forests*

1. Ralegh, Sir Walter 1751.

CHAPTER ONE *The Great Cold*

1. Seno 1985:13–14.
2. Goudie 1992:39, 50.
3. Wicander and Monroe 1993:320–323.
4. Davis 1981; Nilsson 1983:38, 386–387; Porter 1988; Wright 1981.
5. Pielou 1991:16, 33; Porter 1988; Whitlock 1993; Wicander and Monroe 1993:528.
6. Porter 1988.
7. Barry 1983; Pielou 1991:8.
8. Imbrie and Imbrie 1979:104–105.
9. Berger and Loutre 1991; Kerr 1993a, 1994, 1995; Wicander and Monroe 1993:535.
10. Delcourt and Delcourt 1991a.
11. Whitlock 1993.
12. Pielou 1991:12; Graham 1990.
13. Friis-Christensen and Lassen 1991.
14. Kerr 1993b.
15. Pielou 1991:16.
16. Kunz and Reanier 1994.
17. Erickson 1990:136.

18. Erickson 1990:63.

19. Kunz and Reanier 1994; Mayle and Cwynar 1995.

20. Delcourt and Delcourt 1991a:65; Imbrie and Imbrie 1979:182–183.

21. Pielou 1991:308.

22. Friis-Christensen and Lassen 1991; Imbrie and Imbrie 1979:179–180; Pielou 1991:15; Wicander and Monroe 1993:517–518.

23. Creager and McManus 1967; Porter 1988:7–8; Bloom 1983.

24. Hostetler et al. 1994; Imbrie and Imbrie 1979:56; Porter 1988:10–12; Smith and Street-Perrott 1983; Wicander and Monroe 1993:529, 532.

25. Pielou 1991:26; Porter 1988; Whitlock 1993.

26. Illinois State Museum 1995; Nilsson 1983:367–368; Porter 1988; Rube 1983; Wolfe 1972; Wright 1981.

27. Porter 1988; Wright 1984.

28. Bliss 1995; Erickson 1990:48–49.

29. Pielou 1991:23, 109, 175–176.

30. Bloom 1983; Wicander and Monroe 1993:532–534.

31. Baker, V. R. 1983; Pielou 1991:25, 187; Woodford 1965:433.

32. Baker, V. R. 1983; Pielou 1991:186; Porter 1988; Waitt and Thorson 1983; Wicander and Monroe 1993:530–531.

33. Pielou 1991:192–193.

34. Gates et al. 1983; Pielou 1991:194–197.

35. Denton and Huges 1980; McAndrews 1966; Nilsson 1983:387; Pielou 1991:196–197; Wicander and Monroe 1993:532.

36. Balnchon and Shaw 1995; Goudie 1992:52; Nilsson 1983:405; Pielou 1991:20, 211–215.

37. Bloom 1983; Whitmore et al. 1967; Wicander and Monroe 1993:528; *Concise Columbia Encyclopedia* 1995.

CHAPTER TWO *Ice Age Forests*

1. Spaulding et al. 1983.

2. Graham 1990.

3. Guthrie 1984; Martin 1990.

4. Anonymous 1994a, 1995.

5. Dahl 1946; Pielou 1991:35–38, 138.

6. Pielou 1991:84; Porter 1988; Wright 1981.

7. Graham 1990; Mandryk 1990; Solomon and Webb 1985.

8. Whitehead 1973; Graham 1990; Wright 1981.

9. Graham 1990; Pielou 1991:168.

10. Baker, R. G. 1983; Whitlock 1993, 1995; Porter 1988.
11. Barnosky 1984; Worona and Whitlock 1995.
12. Guthrie 1990; Barnosky 1984; Pielou 1991:133. Rhode and Madsen 1995; Worona and Whitlock 1995; Whitlock 1995.
13. Johnson and Packer 1967.
14. Hopkins 1967; Zimov et al. 1995; Porter 1988.
15. Pielou 1991:151.
16. Burns 1990; Muller-Beck 1967.
17. Guthrie 1990.
18. Whitehead 1973; Porter 1988.
19. Wright 1981.
20. Martin 1975; Wells 1970.
21. Watts 1983; Wells 1965, 1970; Wells and Stewart 1987.
22. Hafsten 1961; Hall 1995; Wendorf 1961.
23. McMillan and Wood 1976; Solomon and Webb 1985; Watts 1983; Wells and Stewart 1987.
24. Davis 1981; Delcourt and Delcourt 1991a:142, 1991b; Jackson et al. 1986; Watts 1983; Wells 1965; Wells and Stewart 1987.
25. Pielou 1991:108–109.
26. Guthrie 1990; Illinois State Museum 1995.
27. Pielou 1991:263–264.
28. Whitehead 1967; Wright 1981.
29. Davis 1983; Jacobson and Dieffenbacher-Krall 1995.
30. Davis 1981; Delcourt and Delcourt 1991b, 1994.
31. Delcourt and Delcourt 1979; Porter 1988; Wells 1970; Whitehead 1973; Wright 1981.
32. Baker, R. G. 1983; Peet 1989; Porter 1988; Whitlock 1993.
33. Anderson 1989; Spaulding et al. 1983.
34. Whitlock 1995.
35. Adam 1967; Worona and Whitlock 1995.
36. Barnosky 1984; Spaulding et al. 1983; Worona and Whitlock 1995.
37. Barnosky 1984; Worona and Whitlock 1995.
38. West 1989.
39. Wisner 1998a.
40. Baker, R. G. 1983; Barnosky 1984; Pielou 1991:133. Rhode and Madsen 1995; Worona and Whitlock 1995; Whitlock 1995.
41. Betancourt 1984; Porter 1988; Spaulding et al. 1983.
42. Wright 1993.

43. Allen et al. 1997.
44. Delcourt and Delcourt 1991a:116–117; Spaulding et al. 1983.
45. Miller and Wigand 1994.
46. Bentancourt et al. 1993; Woolfenden 1996.
47. Anderson 1989.
48. Betancourt 1984; Delcourt and Delcourt 1991a:116–117; Spaulding et al. 1983.
49. Cartledge and Propper 1993.
50. Chaney 1965.
51. Barry 1983.
52. Axelrod 1977.
53. Fowells 1965.
54. Baker, R. G. 1983.
55. Woolfenden 1996.
56. Anderson 1989; Barry 1983.
57. Anderson, R.S. 1990; Woolfenden 1996.
58. Fagan 1991:197.
59. Anderson 1989; Baker, R. G. 1983; West 1989.
60. Watts 1983.
61. Watts 1983; Wright 1981.
62. Whitehead 1967, 1973.
63. Delcourt and Delcourt 1991a:207; Whitehead 1973; Wright 1981.
64. Davis 1981.
65. Delcourt and Delcourt 1981; Porter 1988; Webb 1987.
66. Delcourt and Delcourt 1979; Watts 1983.
67. Borremans 1990; Davis 1981; Wright 1981.
68. Wells 1970.
69. Hafsten 1961; Wendorf 1961.
70. Hall 1995.
71. Hall 1995.
72. Brown 1993.
73. Graham 1990.
74. Churcher et al. 1989.
75. Graham 1990; Guthrie 1990.
76. Flerow 1967.
77. Silverberg 1970:175.
78. Nilsson 1983:415, 429; Illinois State Museum 1995.
79. Fagan 1987:112; Guthrie 1990.

80. Pielou 1991:111–112, 254.
81. Martin, P. S. 1990; Nilsson 1983:409; Spaulding et al. 1983.
82. Guthrie 1990; Pielou 1991:118.
83. Nilsson 1983:413–415.
84. Wells and Stewart 1987; Pielou 1991:112, 261.
85. Pielou 1991:111.
86. Barry 1983; Nilsson 1983:384.

CHAPTER THREE *The Birth of Modern Forests*

1. Stark 1948.
2. Tralau 1973 (cited in Peters 1988).
3. Davis 1983.
4. Pielou 1991:19.
5. Pielou 1991:18–19.
6. Watts 1979 (cited in Davis 1983).
7. Johnson and Adkisson 1985, 1986; Johnson and Webb 1989; Webb 1986, 1987.
8. Davis 1981, 1983; Webb et al. 1983.
9. Wright 1976.
10. Cogbill 1996; Davis 1981, 1983; Watts 1979.
11. Davis 1981, 1983.
12. Ager 1983.
13. Burns 1990; Mandryk 1990.
14. Ritchie and MacDonald 1986.
15. Watts 1983; Wright 1981.
16. Hansen et al. 1974
17. Mayle and Cwynar 1995; Nilsson 1983:372; Pielou 1991:18.
18. Nilsson 1983:391.
19. Webb et al. 1983.
20. Watts 1983; Wright 1981.
21. Davis 1983.
22. Bonnichsen et al. 1985; Mayle and Cwynar 1995.
23. Mayle and Cwynar 1995.
24. Watts 1983.
25. Bonnichsen et al. 1985; Watts 1983; Wright 1981.
26. Huntley and Webb 1989; Webb 1987.
27. Delcourt and Delcourt 1991b.

28. Watts 1983; Wright 1976.
29. Weedon and Wolken 1990.
30. Fowells 1965.
31. Pielou 1991:283; Weedon and Wolken 1990.
32. Weedon and Wolken 1990; Wells 1970.
33. Davis 1981, 1983; Delcourt and Delcourt 1991a:41.
34. Davis 1981, 1983; Delcourt and Delcourt 1991a:42.
35. Davis 1981, 1983; Watts 1983.
36. Davis 1981, 1983; Watts 1983.
37. Harlow et al. 1996.
38. Delcourt and Delcourt 1991b.
39. Watts 1983.
40. Davis 1981; Webb et al. 1983; Weedon and Wolken 1990.
41. Delcourt and Delcourt 1991b; Davis 1981; Solomon and Webb 1985; Watts 1983; Wright 1981.
42. Davis 1981.
43. Johnson and Adkisson 1985.
44. Harlow et al. 1996.
45. Watts 1983.
46. Davis 1981, 1983.
47. Heinselman 1973; Watts 1983; Wright 1976.
48. McAndrews 1966.
49. Davis 1981; Watts 1983.
50. Beaudoin et al. 1997; Webb et al. 1983.
51. Jacobson and Dieffenbacher-Krall 1995; Craig 1969.
52. Delcourt and Delcourt 1991a:201; Heinselman 1973; Jacobson and Dieffenbacher-Krall 1995; McAndrews 1966; Swain 1973; Watts 1983; Webb et al. 1983.
53. McAndrews 1966; Wright 1976.
54. Davis 1983.
55. Davis 1981.
56. Davis 1983; Watts 1983; Watts et al. 1996.
57. Davis 1983; Watts 1983; Watts et al. 1996.
58. Davis 1981, 1983; Delcourt and Delcourt 1991a:32–37.
59. Delcourt and Delcourt 1991a:36; Gaudreau 1988.
60. Johnson and Adkisson 1985, 1986; Johnson and Webb 1989; Webb 1986, 1987.
61. Davis 1981.

62. Delcourt and Delcourt 1991a:36–37.
63. Webb 1986, 1987.
64. Delcourt and Delcourt 1991a:36–37.
65. Foster and Zebryk 1993; Paillet 1994.
66. Davis 1981, 1983.
67. Barnosky 1984; Delcourt and Delcourt 1991a:45–47; Peet 1989; Whitlock 1995.
68. Barnosky 1984; Heusser 1983; Whitlock 1995.
69. Worona and Whitlock 1995.
70. Rhode and Madsen 1995.
71. Barnosky 1984; Heusser 1983; Rhode and Madsen 1995; Whitlock 1995.
72. Barnosky 1984; Heusser 1983; Rhode and Madsen 1995; Whitlock 1995.
73. Fowells 1965.
74. Worona and Whitlock 1995.
75. Harlow et al. 1996.
76. Hebda and Mathewes, 1984; Worona and Whitlock, 1995.
77. Barnosky 1984; Heusser 1983.
78. Cwynar 1987.
79. Barnosky 1984; Peet 1989; Whitlock 1995.
80. Baker, R. G. 1983.
81. Barnosky 1984; Wright 1981.
82. Baker, R. G. 1983.
83. Whitlock 1993.
84. Betancourt et al. 1993.
85. Cartledge and Propper 1993.
86. Baker, R. G. 1983.
87. Fowells 1965.
88. Anderson, R. S. 1989; Whitlock 1995.
89. Anderson, R. S. 1989.
90. Wells 1965; Weedon and Wolken 1990.
91. Anderson, R. S. 1989; Axelrod 1977.
92. Anderson, R. S. 1990; Woolfenden 1996.
93. Wells 1965.
94. Webb et al. 1983; Wright 1976.
95. Delcourt and Delcourt 1991a:200–201; Pielou 1991:18, 272, 278, 291.
96. Overpeck 1996.
97. Hoffman and Jones 1970 (cited in Pielou 1991:283).
98. McMillan and Wood 1976; Webb et al. 1983.

99. Delcourt and Delcourt 1981.

100. Whitlock 1995.

101. Anderson, R. S. 1994; Woolfenden 1996.

102. Anderson, R. S. 1990, 1994; Woolfenden 1996.

103. Baker, R. G. 1983; Delcourt and Delcourt 1991a:200–201.

104. Fagan 1991:95, 323; Wright 1976.

105. Woolfenden 1996.

106. Anderson, R. S. 1989; Baker, R. G. 1983.

107. Baker, R. G. 1983; West, G. J. 1989.

108. Barnosky 1984; Cwynar 1987; Heusser 1983; Worona and Whitlock 1995

109. Aikens 1983; Pielou 1991:291; Fagan 1991:261.

110. Jacobson and Dieffenbacher-Krall 1995.

111. Spear et al. 1994.

112. Spear et al. 1994.

113. Davis 1981, 1983.

114. Baker, R. G. 1983.

115. Baker, R. G. 1983; Whitlock 1993.

116. Miller and Wigand 1994.

117. Whitlock 1995.

118. Baker, R. G. 1983.

119. Anderson, R. S. 1989; Baker, R. G. 1983.

120. Baker, R. G. 1983; West 1989.

121. Woolfenden 1996.

122. Anderson, R. S. 1990, 1994.

123. Fagan 1991:174; Goudie 1992:163; Woolfenden 1996.

124. Delcourt and Delcourt 1991a:65; Imbrie and Imbrie 1979:182–183.

125. Delcourt and Delcourt 1991a:201.

126. Davis 1981.

127. Davis 1983.

CHAPTER FOUR   *Ancient People in a New World*

1. Fagan 1991:71; Fagan 1987:133; West 1983.

2. Laughlin 1967.

3. Lysek 1997.

4. Dietrich 1996; Fiedel 1987:45.

5. Fagan 1991:69.

6. Ager 1983; Hall 1997; Hopkins 1967; Zimov et al. 1995; Porter 1988.

7. Harrington 1970, 1996; Pielou 1991:111.

8. Hall 1997; Laughlin 1967.

9. Harrington 1970; Hopkins 1967.

10. Fagan 1991:197.

11. Hopkins 1967; Laughlin 1967.

12. Laughlin 1967.

13. Russell 1980:32.

14. Fagan 1991:64; Fiedel 1987:31–32, 36.

15. Fiedel 1987:32; Laughlin 1967; Wilson 1978.

16. Wilson 1978.

17. Fiedel 1987:31, 36, 38, 44; Frison 1991:56.

18. Fiedel 1987:36, 38.

19. Kunz and Reanier 1994.

20. Fagan 1987:130; Fagan 1991:65.

21. Fagan 1987:131.

22. Fiedel 1987:23.

23. Fiedel 1987:27.

24. Fiedel 1987:65–66; Frison 1991:41.

25. Fagan 1987:180–181.

26. Frison 1991:149, 295.

27. Frison 1991:293; Lahren and Bonnichsen 1974.

28. Fagan 1987:180–181, 201; Fiedel 1987:65–66; Frison 1991:41, 107–108, 177, 293–295; Waldman 1985:3–4.

29. Muller-Beck 1967.

30. Ager 1983; Zimov et al. 1995.

31. Flerow 1967; Hopkins 1967; Zimov et al. 1995.

32. Pielou 1991:19; Porter 1988.

33. Hall 1997; Hopkins 1967.

34. Kunz and Reanier 1994.

35. Burns 1990; Muller-Beck 1967.

36. Pielou 1991:111.

37. Burns 1990; Mandryk 1990.

38. Martin 1990; Ritchie and MacDonald 1986.

39. Nilsson 1983:397; Pielou 1991:192–193.

40. Denton and Huges 1980; Nilsson 1983:387.

41. Watts 1983; Weedon and Wolken 1990; Wright 1976.

42. Pielou 1991:178–180.

43. Porter 1988; Rube 1983; Wright 1981.

44. Watts 1983.

45. Pielou 1991:84.

46. Watts 1983; Wright 1981.

47. Graham 1990; Graham and Lundelius 1996; Graham et al. 1996; Illinois State Museum 1995; Silverberg 1970:170.

48. Fiedel 1987:65.

49. Fiedel 1987:49; Martin 1967, 1973; West 1983.

50. Stoltman and Baerreis 1983.

51. Jaffe 1992.

52. Fiedel 1987:49.

53. Kelly and Todd 1988.

54. Frison 1988.

55. Fiedel 1987:76.

56. Fisher A. C. (cited in McDonald 1995).

57. Whitlock 1993.

58. Whitlock 1993.

59. Baker, V. R. 1983.

60. Barnosky 1984; Wright 1981.

61. Porter 1988; Waitt and Thorson 1983; Wicander and Monroe 1993:530–531.

62. Pielou 1991:186; Porter 1988; Waitt and Thorson 1983; Wicander and Monroe 1993:530–531.

63. Hoblitt et al. 1996; Porter 1988.

64. Fagan 1991:76; Fagan 1987:145.

65. Barnosky 1984; Peet 1989; Whitlock 1995.

66. Meighan and Haynes 1970; West 1983.

67. Wisner 1998a.

68. Stepp 1997.

69. Whitehead 1973; Porter 1988.

70. Graham and Lundelius, Jr. 1996; Graham et al. 1996.

71. Hall 1998.

72. Wicander and Monroe 1993:532–534.

73. Graham and Lundelius, Jr. 1996; Graham et al. 1996.

74. Kelly and Todd 1988.

75. Hall 1998.

76. Wright 1994.

77. Wisner 1998b.

78. Bonnichsen et al. 1985.

79. Fiedel 1987:64.

80. Bonnichsen et al. 1985.
81. Bonnichsen et al. 1985; Nilsson 1983:405.
82. Bonnichsen et al. 1985; Watts 1983.
83. Kelly and Todd 1988.
84. Silverberg 1970:170.
85. Woodford 1965:451; Graham and Lundelius 1996; Graham et al. 1996; Nilsson 1983:415.
86. Blackwelder et al. 1979.
87. Canby 1979.
88. Fagan 1991:112; Fiedel 1987:64.
89. Agenbroad 1988.
90. Graham et al. 1981.
91. Davis 1981; Delcourt and Delcourt 1991a:207; Watts 1983; Webb et al. 1983; Weedon and Wolken 1990.
92. Watts 1983; Davis 1981, 1983.
93. Black 1980; Parker 1910.
94. Parker 1910.
95. Watts 1983.
96. Frison 1991:25, 149–151.
97. Wright 1981.
98. Hall 1995.
99. Watts 1983; Wright 1976.
100. Nilsson 1983: 407.
101. Colinvaux et al. 1996.
102. Fiedel 1987:81
103. Fagan 1987:167.
104. Roosevelt et al. 1996.
105. Fiedel 1996.
106. Dillehay 1996; Fiedel 1996; Meggers 1996.
107. Fiedel 1987:49; Roosevelt et al. 1996.
108. Fiedel 1987:77; Meltzer 1997; Menon 1998; Roosevelt et al. 1996.
109. Fiedel 1987:49.

CHAPTER FIVE *Taming a Wilderness*

1. Quoted in Nash 1967:xiii.
2. Fagan 1987:179.
3. Kunz and Reanier 1994.

4. Fiedel 1987:48.

5. Fiedel 1987:56.

6. Fagan 1987:182; Fagan 1991:79; Frison 1991:41, 44, 373.

7. Frison 1991:291.

8. Frison 1991:41, 355; Lahren and Bonnichsen 1974, (cited in Fiedel 1987:71–72).

9. Martin, P. S. 1967, 1973.

10. Fagan 1991:79; Frison 1991:143–155.

11. Frison 1988.

12. Fagan 1991:124; Frison 1988.

13. Fagan 1991:79; Frison 1988; Frison 1991:143–155,301–302; Kelly and Todd 1988.

14. Fagan 1987:212.

15. Agenbroad 1988; Frison 1991:150.

16. Delcourt and Delcourt 1991b; Watts 1983; Wright 1976.

17. Fiedel 1987:64.

18. Betancourt et al. 1993.

19. Agenbroad 1988; Lundelius 1988.

20. Kelly and Todd 1988.

21. Weedon and Wolken 1990; Wells 1970.

22. Barnosky 1984; Peet 1989; Whitlock 1995; Worona and Whitlock 1995.

23. Guthrie 1990.

24. Guthrie 1990; Martin, P. S. 1990.

25. Graham 1990; Lundelius 1988.

26. Agenbroad 1988; Nilsson 1983:391; Watts 1983; Webb et al. 1983; Wright 1981.

27. Borremans 1990.

28. Webb 1987.

29. Haynes and Stanford 1984; Agenbroad 1988.

30. Janetski 1987:15.

31. Fiedel 1987: 65.

32. Mead and Meltzer 1984; Meltzer and Mead 1983.

33. Barnosky 1989.

34. Agenbroad 1988; Martin, P. S. 1990.

35. Martin, P. S. 1990; Semken 1983.

36. Martin, P. S. 1967, 1973, 1975, 1990; Semken 1983; West 1983.

37. Fisher 1997.

38. Agenbroad 1988; Fagan 1991: 99; West 1983.

39. Webb et al. 1983; Whitlock 1993; Wright 1976.

40. Barnosky 1989.
41. Fagan 1987:199–200; Frison 1988; Frison 1991:155,176; Guthrie 1990; Martin 1990.
42. West 1983.
43. Frison 1988; Frison 1991:38–67.
44. Frison 1991:163–164.
45. Seno 1985:77.
46. Coues 1965:1081.
47. Frison 1988; Frison 1991:38–67.
48. Frison 1991:155.
49. Seno 1985:34.
50. Seno 1985:85.
51. Seno 1985:122.
52. Seno 1985:119.
53. Smith 1624a.
54. Quoted in Seno 1985:31 from Thwaites 1899.
55. Chittenden and Richardson 1969:666.
56. Quoted in Seno 1985:31–34 from Thwaites 1899.
57. Chittenden and Richardson 1969:248.
58. Quoted in Seno 1985:45–46 from Thwaites 1898a.
59. Quoted in Seno 1985:45–46 from Thwaites 1898a.
60. Quoted in Seno 1985:31 from Thwaites 1899.
61. Quoted in Seno 1985:34 from Thwaites 1899.
62. Frison 1991:158–186.
63. Frison 1991:201–204.
64. Frison 1991:241.
65. Frison 1991:170.
66. Fagan 1991:131.
67. Frison 1991:179.
68. Chittenden and Richardson 1969:1025–1032.
69. Quoted in Fagan 1991:130.
70. Frison 1991:179.
71. Coues 1965:334–335.
72. Fagan 1991:130.
73. Kauffman 1977 (cited in Fiedel 1987:67).
74. Fagan 1991:106, 112.
75. Fagan 1991:95, 323.
76. Aikens 1983; Fagan 1991:232–233; Jaeger 1966:254.

77. Aikens 1983.

78. Stoltman and Baerreis 1983.

79. Anderson and Hanson 1988.

80. The northern part of the northeastern cultural area defined by Waldman 1985:32.

81. Jaeger 1966:380–383.

82. Most names interpreted by Phil Konstantin 1997, a member of the Cherokee Nation of Oklahoma, Waldman 1985, and other sources.

83. Catlin 1875:179–180.

84. Newcomb 1993:293.

85. Aikens 1983.

86. Jaeger 1966:374–375.

87. Gustafson 1966:111–113 (cited in Hadley and Sheridan 1995:40–41).

88. Heizer and Elsasser 1980:101–102.

89. Barnosky 1989; Imbrie and Imbrie 1979:178–179; Worona and Whitlock 1995.

90. Semken 1983.

91. Woolfenden 1996.

92. Barry 1983; Bentancourt et al. 1993; Imbrie and Imbrie 1979:56; Porter 1988; Rhode and Madsen 1995; Wicander and Monroe 1993:529, 532.

93. Delcourt and Delcourt 1991a:200–201; Fagan 1991:121–124; Frison 1991:186, 191; Pielou 1991:18, 272, 278, 291.

94. Anderson 1989; Baker, R. G. 1983; West, G. J. 1989; Woolfenden 1996.

95. Cwynar 1987; Worona and Whitlock 1995.

96. Gremillion 1996; Stoltman and Baerreis 1983; Watts 1983.

97. Stoltman and Baerreis 1983.

98. Semken 1983; Wells 1965; Weedon and Wolken 1990; Whitlock 1995.

99. McAndrews 1966.

100. Martin 1975; Morrow 1996; Stoltman and Baerreis 1983; Webb et al. 1983.

101. Fagan 1991:124; Frison 1991:186.

102. Fagan 1991:124; Frison 1991:186.

103. McMillan and Wood 1976.

104. Aikens 1983.

105. Baker, R. G. 1983; Nilsson 1983: 409; Spaulding et al. 1983.

106. Dillehay 1974; Semken 1983.

107. Byars 1993.

108. Frison 1991:333.

109. Frison 1991:258.

110. Frison 1991:246–258.

111. McAndrews 1966.

112. Martin 1975.

113. Fagan 1991:133–134; Fiedel 1987:208.

114. Haines 1965:36.

115. Davis 1983; Delcourt and Delcourt 1991a:200–201; Pielou 1991:18, 272, 278, 291; Stoltman and Baerreis 1983; Watts 1983; Watts et al. 1996.

116. Van Doren 1928:90.

117. Russell 1980:83.

118. Fagan 1991:109, 111–112.

119. Russell 1980:21.

120. Anderson and Hanson 1988.

121. Fagan 1991:108; Johnson 1994; Morrow 1996; Stoltman and Baerreis 1983.

122. Smith 1624a.

123. Anderson and Hanson 1988.

124. Fagan 1991:113–114; Johnson 1994.

125. Anonymous 1994b.

126. Balls 1962:14; Fagan 1991:313, 345; Tantaquidgeon 1942:25, 74.

127. Benson 1964:394.

128. Seno 1985:122.

129. Fagan 1991:221.

130. Hodge and Lewis 1990a:90–91.

131. Aikens 1983; Fagan 1991:221–224,234.

132. Hodge and Lewis 1990a:3–4.

133. Hodge and Lewis 1990a:65 (from the account given to Cabeza de Vaca by Figueroa).

134. Hodge and Lewis 1990a:68 (from the account given to Cabeza de Vaca by Figueroa).

135. Hodge and Lewis 1990a:89; Newcomb 1993:29–57.

136. Hodge and Lewis 1990a:61.

137. Chittenden and Richardson 1969:372.

138. Hodge and Lewis 1990a:66–67 (from the account given to Cabeza de Vaca by Figueroa).

139. Newcomb 1993:29.

140. Hodge and Lewis 1990a:66 (from the account given to Cabeza de Vaca by Figueroa).

141. Cartledge and Propper 1993.

142. Miller and Albert 1993.

143. Betancourt et al. 1993.

144. Allen et al. 1997.

145. Betancourt et al. 1993.

146. Glassow et al. 1987 (cited in Fagan 1991:196–197).

147. Heizer and Elsasser 1980:93.

148. Balls 1962:10–14; Fagan 1991:189; Glassow et al. 1987 (cited in Fagan 1991:196–197); Fiedel 1987:67; Helen McCarthy of the University of California at Davis, quoted by Martin 1996.

149. Heizer and Elsasser 1980:91.

150. Heizer and Elsasser 1980:98–99; Koenig and Knops 1997; McCarthy 1993.

151. Seno 1985:28–30.

152. Coues 1965:424.

153. Chittenden and Richardson 1969:509; Leighton 1985:50.

154. Mackenzie 1904:254–255.

155. Black 1980:126.

156. Balls 1962:29; Herrick 1977:267.

157. Herrick 1977:413.

158. Black 1980:124; Hart 1992:2.

159. Speck 1941.

160. Compton 1993:179; Merriam 1966:310.

161. Hart 1992:52; Herrick 1977:302.

162. Chestnut 1902; Herrick 1977:413; Densmore 1928; Tantaquidgeon 1942:26, 80.

163. Denevan 1992; Waldman 1985:50–52.

164. McCracken 1959:84.

165. Chittenden and Richardson 1969:191.

166. Frison 1991:126; Coues 1965:423–424.

167. Hodge and Lewis 1990a:66 (from the account given to Cabeza de Vaca by Figueroa).

168. Hodge and Lewis 1990c:362–363.

169. Grinnell 1995:2.

170. Basehart 1974:43; Catlin 1875:64; Hellson 1974:116; Hart 1981; Gilmore 1919:63; Mahar 1953:41.

171. Leighton 1985:32, 49.

172. Weaver and Clements 1938:405; Hart 1981; Hellson 1974:116.

173. Frison 1988.

174. Smith 1933:112.
175. Seno 1985:78.
176. Bean and Saubel 1972:85; Chestnut 1902.
177. Catlin 1875:65–66.
178. Seno 1985:20–21.
179. Chittenden and Richardson 1969:467.
180. Chittenden and Richardson 1969:535.
181. Suphan 1974.
182. Quaife 1978:226–227.
183. Chittenden and Richardson 1969:535.
184. Morgan 1993:430.
185. Coues 1965:1197.
186. Seno 1985:95.
187. Chittenden and Richardson 1969:243.
188. Coues 1965:424.
189. Johnson and Winter 1846:53.
190. Coues 1965:664–665.
191. Coues 1965:669; Vastoka 1969.
192. Heizer and Elsasser 1980:30–31.
193. Coues 1965:669.
194. Quoted in Vastoka 1969 from Meares 1790.
195. Vastoka 1969.
196. Hebda and Mathewes 1984.
197. Gunther 1973:40; Coues 1965:664–665, 730; Wissler 1989:215–217.
198. Coues 1965:771.
199. Hebda and Mathewes 1984; Coues 1965:662–663, 770–771.
200. Swan 1857:34–35.
201. Heizer 1975:37.
202. Heizer and Elsasser 1980:31.
203. Swan 1857:35–36.
204. Coues 1965:827; Hebda and Mathewes 1984.
205. Hebda and Mathewes 1984; Worona and Whitlock 1995.
206. Anonymous 1998; Mackenzie 1904:268.
207. Cooper 1860:26.
208. Mackenzie 1904:255, 298–299.
209. Agee 1993:187–225; Franklin and Hemstrom 1981; Franklin 1995.
210. Sauer 1971:253.
211. Cronon 1983:45.

212. Beaudoin and Bizier 1997.

213. Russell 1980:195–198.

214. Burrage, 1932 (cited in Calloway, 1991:39).

215. Seno 1985:81.

216. Seno 1985:83.

217. Fagan 1991:216–217; Heizer and Elsasser 1980:62–63.

218. Van Doren 1928:193.

219. Smith 1624a.

220. Smith 1624a.

221. Hammett 1992.

222. Smith 1624a.

223. Askew School of Public Administration and Policy's Information Server 1997. The Florida State University.

CHAPTER SIX   *Enhancing Nature's Bounty*

1. Lewis 1982a:50.

2. Quoted in Ortiz 1993.

3. Balls 1962:68.

4. Bonnicksen et al. 1999; Schlick 1994:97.

5. Virgil Bishop, North Fork Mono, personal communication, 1991 (cited in Anderson 1993b).

6. Dobyns 1956:18 (cited in Bonnicksen et al. 1999); Newcomb 1993:41.

7. Seno 1985:24.

8. Haines 1965:26–27.

9. Frison 1991:363–365; Wilke 1993.

10. Chittenden and Richardson 1969:488.

11. Harbinger 1964:25 (cited in Bonnicksen et al. 1999).

12. Coues 1965:603.

13. Balls 1962:31.

14. Chittenden and Richardson 1969:488–489.

15. Turner and Kuhnlein 1983:211 (cited in Bonnicksen et al. 1999).

16. Bonnicksen et al. 1999; Harbinger 1964:11; Hart 1992:14; Turner and Kuhnlein 1983.

17. Quoted in Ortiz 1993, from Peri and Patterson 1978:17.

18. Bonnicksen et al. 1999; Harrington 1967:42; Kennedy 1961:81.

19. Coues 1965:603.

20. Haines 1965:34–35.

21. Quoted in Kindscher 1987:186 from Johnston 1970 (cited in Bonnicksen et al. 1999).
22. Waldman 1985:129–130.
23. Anderson 1993a.
24. Russell 1980:128.
25. Russell 1980:85.
26. Peri and Patterson 1979:39 (cited in Bonnicksen et al. 1999).
27. Martin, G. 1996.
28. McCarthy 1993.
29. Anderson 1993b.
30. Seno 1985:28–30.
31. Chittenden and Richardson 1969:358.
32. Coues 1965:995.
33. Richards 1997.
34. Anderson 1993b.
35. Murphy 1959 (cited in Bonnicksen et al. 1999).
36. Peri et al. 1982:120 (cited in Bonnicksen et al. 1999).
37. Chestnut 1902:357 (cited in Bonnicksen et al. 1999).
38. Lydia Beecher, Mono, personal communication, 1991 (cited in Anderson 1993b).
39. Balls 1962:20–22; Bean and Saubel 1972:107.
40. Bonnicksen et al. 1999.
41. Hodge and Lewis 1990a:89.
42. Chittenden and Richardson 1969:535.
43. Smith 1624a.
44. Day 1953; Keener and Kuhns 1997.
45. Spurr and Barnes 1980:568.
46. Bye 1972:91; Nabhan 1985; Wyman and Harris 1951:16 (all cited in Bonnicksen et al. 1999).
47. Nabhan 1985:26; Cornett 1987:17 (all cited in Bonnicksen et al. 1999).
48. Parker 1913 (cited in Bierhorst 1994).
49. Driver and Massey 1957:228 (cited in Bonnicksen et al. 1999); Hurt 1987:37.
50. Anderson 1993b.
51. Fagan 1991:328–329.
52. Smith 1997.
53. Fagan 1991:269–270, 358; Fiedel 1987:109–111, 201–203.
54. Fiedel 1987:206.

55. Fagan 1991:272; Hurt 1987:17.

56. Fagan 1991:270–271.

57. Aikens 1983; Fagan 1991:94, 239, 261.

58. Berry and Berry 1986 (cited in Fagan 1991:261); Fagan 1991:270; Fiedel 1987:203–205.

59. Fiedel 1987:205–207.

60. Fagan 1991:273; Hurt 1987:17.

61. Carter 1967:79–83.

62. Fiedel 1987:208.

63. Spoerl and Ravesloot 1995.

64. Aikens 1983.

65. Cartledge and Propper 1993.

66. Truett 1996.

67. Cartledge and Propper 1993; Kohler and Matthews 1988; Spoerl and Ravesloot 1995.

68. Jennings 1975:65.

69. Hodge and Lewis 1990c:359.

70. Fagan 1991:254; Thomas 1994:112, 116.

71. Thomas 1994:112.

72. Quoted in Doolittle 1992 from Hammond and Rey 1928:160.

73. Quoted in Doolittle 1992 from Hammond and Rey 1966:220.

74. Doolittle 1992.

75. Hurt 1987:20.

76. Spoerl and Ravesloot 1995.

77. Thomas 1994:113.

78. Hurt 1987:20–21.

79. Thomas 1994:113.

80. Doolittle 1992; Fagan 1991:294; Fiedel 1987:203–205.

81. Doolittle 1992; Fagan 1991:294.

82. Quoted in Lawton et al. 1993 from Los Angeles Star 1859:2.

83. Fagan 1991:271.

84. Doolittle 1992; Thomas 1994:98.

85. Hurt 1987:25.

86. Aikens 1983; Fagan 1991:94.

87. Cremony 1868.

88. Waldman 1985:34–35.

89. Fagan 1991:283–291.

90. Fagan 1991:280.

91. Hodge and Lewis 1990c:353.
92. Hodge and Lewis 1990c:300, 314, 315–316.
93. Waldman 1985:18.
94. Cartledge and Propper 1993; Fagan 1991:473.
95. Fiedel 1987:216–217.
96. Fiedel 1987:210.
97. Grinnell 1972:65 (cited in Bonnicksen et al. 1999).
98. Barrett and Gifford 1933 (cited in Bonnicksen et al. 1999).
99. Quoted in Cronon 1983:49 from Morton, T. 1632.
100. Kohler and Matthews 1988.
101. Kohler 1992; Kohler and Matthews 1988.
102. Kohler 1992; Kohler and Matthews 1988.
103. Cartledge and Propper 1993.
104. Kohler and Matthews 1988.
105. Kohler 1992; Kohler and Matthews 1988.
106. Fagan 1991:358; Fiedel 1987:109–110.
107. Gremillion 1996.
108. Beardsley 1940:511 (cited in Bonnicksen et al. 1999).
109. Little 1980:495.
110. Adair 1775:128 (cited in Hammett 1992).
111. Quoted in Hammett 1992 from Bartram 1973.
112. Hammett 1992; Lafferty 1994; Newcomb 1993:79; Van Doren 1928:357–359.
113. Sauer 1971:71.
114. Quoted in Hammett 1992 from Adair 1775:128.
115. Hodge and Lewis 1990a:87–88.
116. Fagan 1991:358–360; Stoltman and Baerreis 1983.
117. Fiedel 1987:110.
118. Fagan 1991:358–360; Fiedel 1987:110; Stoltman and Baerreis 1983.
119. Coues 1965:418.
120. Saunders et al. 1997.
121. Fagan 1991:351–354; Fiedel 1987:111; Gibson 1993.
122. Fagan 1991:351–354; Gibson 1993; Thomas 1994:79–84.
123. Gibson 1993.
124. Gibson 1993; Thomas 1994:83.
125. Gibson 1993.
126. Fagan 1991:351–354; Gibson 1993; U.S. National Park Service 1996a.
127. Pringle 1997.
128. Fiedel 1987:231.

129. Hurt 1987:12.
130. Waldman 1985:19–20.
131. Fagan 1991:361–366; Thomas 1994:128–129
132. Fiedel 1987:232.
133. Fagan 1991:362–363; Thomas 1994:128.
134. Fiedel 1987:231; Kenyon 1986; Thomas 1994:130–133.
135. Clay 1996.
136. Fiedel 1987:231; Kenyon 1986; Thomas 1994:128–129.
137. Fiedel 1987:235; Hurt 1987:13.
138. Fagan 1991:369–370; Fiedel 1987:235.
139. Fiedel 1987:238.
140. Fagan 1991:373–374.
141. Fiedel 1987:240–241.
142. Fagan 1991:373–374, 389; Fiedel 1987:236; Stoltman and Baerreis 1983.
143. Clay 1996; Fiedel 1987:239; Stoltman and Baerreis 1983.
144. Fiedel 1987:236–237.
145. Stoltman and Baerreis 1983; Thomas 1994:139–141.
146. Fiedel 1987:257.
147. Fagan 1991:382–383; Fiedel 1987:243.
148. Fagan 1991:385–388; Fiedel 1987:244–246.
149. Fiedel 1987:244–245.
150. Hodge and Lewis 1990c:330.
151. Fagan 1991:389; Fiedel 1987:247; Hurt 1987:15; Stoltman and Baerreis 1983.
152. Fiedel 1987:110, 246; Hurt 1987:13; Stoltman and Baerreis 1983.
153. Carter 1967:74–75.
154. Rollins 1995:221–222.
155. Grant 1952:62.
156. Hurt 1987:17; Russell 1980:151, 168.
157. Fagan 1991:272; Fiedel 1987:174; Stoltman and Baerreis 1983.
158. Hurt 1987:16.
159. Fiedel 1987:175; Stoltman and Baerreis 1983.
160. Woods 1996.
161. Anonymous 1997a; Stoltman and Baerreis 1983; Woods 1996.
162. Lynn 1996.
163. Russell 1980:154–158.
164. Fiedel 1987:246.
165. Neumann 1995.

166. Fagan 1991:395; Fiedel 1987:230, 257; Newcomb 1993:279–313.

167. Fiedel 1987:247.

168. Fagan 1991:404; Fiedel 1987:245.

169. Fagan 1991:399; Fiedel 1987:230; Thomas 1994:152, 171.

170. Fagan 1991:391, 394–399; Fiedel 1987:249–250; Thomas 1994:152–158; Woods 1996.

171. Fagan 1991:392; Hurt 1987:32; U.S. National Park Service 1996b.

172. Wissler 1989:163.

173. Fiedel 1987:251.

174. Charlevoix 1966, Vol. II:260–261.

175. Fiedel 1987:251–252.

176. Fiedel 1987:247–248; Hurt 1987:16.

177. Dietrich 1996.

178. Frison 1991:433–434.

179. Anonymous 1997a.

180. Woods 1996.

181. Fiedel 1987:255.

182. Woods 1996.

183. Waldman 1985:105–106.

184. Denevan 1992.

185. Waldman 1985:29.

186. Van Doren 1928:296–297.

187. Quoted in Denevan 1992.

188. Denevan 1992; Dobyns 1983:42; Ramenofsky 1987.

189. Quoted in Jennings 1976:18.

190. Jennings 1976:18.

191. Quoted in Rostlund 1957.

192. Hodge and Lewis 1990b:195.

193. Fiedel 1987:257–259.

194. Coues 1965:197; Wissler 1989:197.

195. Quoted in Will and Hyde 1964:63.

196. Quoted in Will and Hyde 1964:59.

197. Quoted in Rostlund 1957.

198. Hodge and Lewis 1990b:215.

199. Hodge and Lewis 1990b:216.

200. Van Doren 1928:68.

201. Gremillion 1996.

202. Day 1953.

203. Van Doren 1928:173.
204. Smith 1624a.
205. Smith 1624c; May 1953.
206. Russell 1980:10–11.
207. Quoted in Russell 1980:153.
208. May 1953; Russell 1980:149, 169.
209. Russell 1980:9.
210. Quoted in Cronon 1983:43.
211. Smith 1624c.
212. Quoted in Bakeless 1961:112–113.
213. Benson 1964:440–443.
214. Day 1953; Grant 1952:314; Van Doren 1928:287–288.
215. Quoted in Doolittle 1992.
216. Doolittle 1992.
217. Russell 1980:12, 144.
218. Hunt 1987:34.
219. Quoted in Jennings 1976:63.
220. Van Doren 1928:287.
221. Jennings 1976:63.
222. Smith 1624a.
223. Hunt 1987:33.
224. Hunt 1987:34.
225. Wright 1994.
226. Grant 1952:313.
227. Grant 1952:314.
228. Heidenreich 1971 (cited in Williams 1989:37 and Clark and Royall 1996).
229. Russell 1980:142.

CHAPTER SEVEN   *Fire Masters*

1. Lewis 1982a:26.
2. McDonald 1995.
3. Balter 1995; Fiedel 1987:25.
4. Eiseley 1954; Goudsblom 1992:12–23; Lewis and Schweger 1973; Silverberg 1970:181.
5. Alexander 1980; Wells and Stewart 1987.
6. Heinselman 1985; Suffling et al. 1982.
7. Coues 1965:185.

8. Quoted in Dary 1975:40.
9. Quoted in Peterson 1994–1995.
10. Wells and Stewart 1987.
11. Taylor 1971.
12. Fuquay 1980.
13. Arno 1985; Aschmann 1977; Barrett and Arno 1982; Kilgore and Taylor 1979; Minnich 1987; Reynolds 1959.
14. Arno and Fishcher 1995 (charcoal and ashy silt data from P. J. Mehringer, Department of Anthropology, Washington State University, Pullman).
15. Miller and Wigand 1994.
16. Ager 1983.
17. Swain 1973; Watts 1983; Wright 1976.
18. Watts 1983.
19. Barnosky 1984; Heusser 1983; Worona and Whitlock 1995.
20. Cwynar 1987.
21. Fagan 1991:76; Fagan 1987:145.
22. Quoted in Bakeless 1961:324.
23. Bakeless 1961:200.
24. Bakeless 1961:241.
25. Quoted in Maxwell 1910.
26. Chittenden and Richardson 1969:213.
27. James 1846 (Chapter 2).
28. Coues 1965:1289.
29. Quoted in Gruell 1983 from Thwaites 1959.
30. Quoted in Gruell 1985.
31. Haines 1965:7.
32. Quoted in Barrett 1980b from Mullan 1861.
33. Young 1899.
34. Marryat 1855:135–136.
35. Quoted in Pyne 1982:499.
36. Boyd 1986; Goetzmann 1993:235–237.
37. Quoted in Boyd 1986 from Wilkes 1849.
38. Davis, J. T. 1966; Lorimer 1993; Russell 1980:200; Wilcox 1970.
39. Hodge and Lewis 1990c:303.
40. Smith 1624a.
41. Haines 1965:26–27.
42. Seno 1985:38 from Blair 1910.
43. Quoted in Day 1953 from Williams 1643; Stewart 1963.

44. Chittenden and Richardson 1969:1047.
45. Henry 1984:205.
46. Seno 1985:4.
47. Stewart 1963.
48. Gayton 1948.
49. Fayel 1886.
50. Chittenden and Richardson 1969:242–243.
51. Morison 1971:675; Russell 1983.
52. Quoted in Maxwell 1910.
53. Coues 1965:75.
54. Quoted in Day 1953 from Williams 1643.
55. Frison 1988.
56. Grant 1952:68.
57. Quoted in Gruell 1985.
58. Fremont 1856:376.
59. Davies 1980:91.
60. Goetzmann 1993:95–99; Gruell 1985.
61. Quoted in Pyne 1982:499.
62. Coues 1965:427.
63. Nevin 1974:150–151.
64. Quoted in Pyne 1982:500 from Bourke 1971:45.
65. Coues 1965:1046.
66. Coues 1965:1053; also quoted in Barrett 1980b.
67. Pyne 1982:72.
68. Lewis 1982a:45.
69. Phillips 1940:215.
70. Doane 1871.
71. Everts 1871.
72. Langford 1873:4.
73. Coues 1965:77.
74. Quoted in Barrett 1980b from Thwaites 1959.
75. Quoted in Gruell 1985 from Davies 1961; Nathniel J. Wyeth recorded a nearly identical observation in his journal entry for August 17, 1834, in Young 1899.
76. Chittenden and Richardson 1969:248.
77. Hodge and Lewis 1990a:64–65 (from the account given to Cabeza de Vaca by Figueroa).
78. Van Doren 1928:183.

79. Quoted in Cooper 1960 from Holsinger 1902.

80. Quoted in Cooper 1960 from Bell 1870.

81. Seklecki et al. 1996.

82. Stevens 1855:265.

83. Quoted in Russell 1983 from Anonymous 1906.

84. Quoted in Boyd 1986 from Wilkes 1849.

85. Quoted in Gruell 1983 from Phillips 1957.

86. Doane 1871.

87. Pyne 1982:72 from Morris 1979:310.

88. Wheeler 1990:286.

89. Quoted in Pyne 1982:73 from Dodge 1959.

90. Gruell 1985.

91. Elliott 1910; Davies 1961.

92. Young 1899.

93. Pyne 1982:72 from James 1822.

94. Kephart 1916 (Chapter IX, Among the Springs).

95. Seno 1985:85 from Hennepin 1880.

96. Haines 1965:30–31.

97. Haines 1965:32.

98. Quoted in Russell 1983 from Van der Donck 1841:20–21.

99. Quoted in Russell 1983 from Van der Donck 1841:20–21.

100. Quoted in Russell 1983 from Van der Donck 1841:20–21.

101. Quoted in Boyd 1986 from Dixon 1907.

102. Barrett and Gifford 1933; Hadley and Sheridan 1995:14.

103. Quoted in Boyd 1986 from Sapir 1907.

104. Hodge and Lewis 1990a:67 (from the account given to Cabeza de Vaca by Figueroa).

105. Joutel 1962:52.

106. Joutel 1962:94, 106.

107. Quoted in Lehmann 1965 from Ehrenberg 1935:29–35.

108. Quoted in Lehmann 1965 from Olmsted 1857:233.

109. Van Doren 1928:139.

110. Smith 1624a.

111. Quoted in Maxwell 1910.

112. Quoted in Hammett 1992 from Lefler 1967:215.

113. Seno 1985:75, from Hennepin 1880.

114. Quoted in Angle 1968:15 from Hennepin 1880.

115. Quoted in Boyd 1986.

116. Chittenden and Richardson 1969:1021–1022.

117. Chittenden and Richardson 1969:1022.

118. Boyd 1986; Lutz 1959:18–20; Pyne 1982:500.

119. Grant 1952:297.

120. Grant 1952:296–299.

121. Grant 1952:298.

122. Bryant 1967 (Chapter 14).

123. Quoted in Bean and Lawton 1993 from Simpson 1938:51.

124. Quoted in Lewis 1993 from Clar 1959:7.

125. Bean and Lawton 1993; Boyd 1986.

126. Boyd 1986; Chittenden and Richardson 1969; Davies 1980; Lewis 1993.

127. Boyd 1986.

128. Marryat 1855:136–137.

129. Heizer and Elsasser 1980:106–107.

130. Beckwith 1854:55.

131. Beckwith 1854:51.

132. Quoted in Boyd 1986 from Armstrong 1857:119.

133. Quoted in Boyd 1986.

134. Davies 1980: 94.

135. Peters and Buntin 1994 from Wislizenus 1912.

136. Peters and Buntin 1994 from Haines 1971.

137. Chittenden and Richardson 1969:1032–1033.

138. Chittenden and Richardson 1969:1033.

139. Hodge and Lewis 1990a:67 (from the account given to Cabeza de Vaca by Figueroa).

140. Coues 1965: 249.

141. Coues 1965:330–331, 1289.

142. Davies 1980: 94.

143. Quoted in Cronon 1983:51 from Dwight 1821.

144. Quoted in Barrett 1980b from Mullan 1861.

145. Quoted in Cooper 1960 from Wheeler 1878.

146. Lewis 1982a:28; Kirsch and Kruse 1972; Komarek 1969.

147. Quoted in Boyd 1986 from Clyman 1960.

148. *Colorado Tribune*, July 12, 1852. V:13. Matagorda, TX.

149. Lewis 1982a:25.

150. Tirmenstein 1989a.

151. Quoted in Pilarski 1994 (from an interview with Hector Franco, a Wukchumni Yokut, conducted by M. Kat Anderson in 1993).

152. Barrett and Gifford 1933.
153. Quoted in Pyne 1982:84 from Fremont 1966.
154. Carvalho 1859:56, 58.
155. Quoted in Gussow 1971:71–73.
156. Higgins 1986:7.
157. Lewis 1982b.
158. Quoted in Wright and Bailey 1980:1 from Fidler 1793.
159. Quoted in Lorimer 1993.
160. Coues 1965:119.
161. Quoted in Hawes 1923 from Wood 1634.
162. Quoted in Russell 1983 from Van der Donck 1841:20–21.
163. Cronon 1983:28.
164. Abrams 1992.
165. Quoted in Day 1953 from Denton 1670.
166. Van Doren 1928:107.
167. Van Doren 1928:159.
168. Quoted in Boyd 1986 from Wilkes 1849.
169. Quoted in Stewart 1955.
170. Quoted in Boyd 1986 from Brown 1878.
171. Lewis 1993.
172. Quoted in Cooper 1960 from Holsinger 1902.
173. Quoted in Gussow 1971:71–73.
174. Quoted in Gussow 1971:71–73.
175. Quoted in Foti and Glenn 1991 from Du Pratz 1774:54.
176. Quoted in Foti and Glenn 1991 from Du Pratz 1774:54.
177. Quoted in Foti and Glenn 1991 from Featherstonhaugh 1844.
178. Quoted in Foti and Glenn 1991 from Gerstacker 1881.
179. Quoted in Day 1953 from Johnson 1654.
180. Russell 1980:10–11.
181. Partly quoted in Thompson and Smith 1970, Day 1953, Russell 1983, and Maxwell 1910.
182. Quoted in Hawes 1923, from Wood 1634.
183. Quoted in Day 1953 from Williams 1643.
184. Quoted by Lorimer 1993.
185. Van Doren 1928:139.
186. Van Doren 1928:192.
187. Russell 1980:51; Smith 1624a.
188. Barrett and Arno 1982.

189. Komarek 1969.

190. Smith 1624a.

191. Lewis 1982a:38–39.

192. Quoted by Ortiz 1993.

193. Quoted in Lewis 1993 from Gayton 1948.

194. Hutchings 1859a.

195. Quoted in Biswell 1967.

196. Quoted in Biswell 1961a.

197. Marryat 1855:136–137.

198. Komarek 1969.

199. Hodge and Lewis 1990a:67 (from the account given to Cabeza de Vaca by Figueroa).

200. Hodge and Lewis 1990a:67 (from the account given to Cabeza de Vaca by Figueroa).

201. Quoted in Day 1953 from Williams 1643.

202. Quoted in Pyne 1982:500 from Lutz 1959.

203. Quoted in Pilarski 1994 (from an interview with Hector Franco, a Wukchumni Yokut, conducted by M. Kat Anderson in 1993).

204. Daniels 1976.

205. Quoted in Bean and Lawton 1993.

206. Daniels 1976; Timbrook et. al. 1993.

207. Quoted in Timbrook et. al. 1993.

208. Quoted in Timbrook et. al. 1993.

209. Quoted in Timbrook et. al. 1993.

210. Quoted in Lewis 1993 from Cook 1960.

211. Quoted in Timbrook et. al. 1993.

212. Quoted in Timbrook et. al. 1993.

213. Timbrook et. al. 1993.

214. Quoted in Bean and Lawton 1993.

215. Quoted in Bean and Lawton 1993 from Simpson 1938:51; Timbrook et. al. 1993.

216. Van Doren 1928:139.

217. Komarek 1969.

218. Hodge and Lewis 1990a:67 (from the account given to Cabeza de Vaca by Figueroa).

219. Quoted in Boyd 1986 from Wilkes 1849.

220. Boyd 1986; Schenk and Gifford 1952.

221. Quoted in Boyd 1986 from Riddle 1920.

222. Quoted in Boyd 1986 from Lyman 1900.
223. Boyd 1986 from Jacobs 1945.
224. Baxley 1865; Bonnicksen et. al. 1999; Boyd 1986; Gibbens and Heady 1964; Lewis 1993; McCarthy 1993; Pilarski 1994; Reynolds 1959; Schenk and Gifford 1952.
225. Quoted in Lewis 1993 from Harrington 1932.
226. Quoted in Lewis 1993 from Schenk and Gifford 1952.
227. Anderson 1993b; Martin 1996; McCarthy 1993; Reynolds 1959.
228. Scott et al. 1977.
229. Marryat 1855:110.
230. Barrett and Gifford 1933; Chestnut 1902; Schenk and Gifford 1952; Zigmond 1981.
231. Quoted in Boyd 1986 from Wilkes 1849.
232. Anderson 1993b; Dayton 1931; Ortiz 1993.
233. Anderson 1993b.
234. Anderson 1993b; Lewis 1993; O'Neale 1932; Pilarski 1994.
235. Lewis and Ferguson 1988.
236. Lewis 1982a:37.
237. Snyder 1991a.
238. Lewis 1982a:35.
239. Mackenzie 1904:27.
240. Beaudoin et al. 1997.
241. Mackenzie 1904:34; also see Lewis 1982a.
242. Quoted in Lewis 1982a from Macoun 1882:125.
243. Lewis 1982a:25.
244. Lewis 1977.
245. Lewis 1982a:26.
246. Lewis 1982a:42.
247. Lewis 1982a:32–33.
248. Lewis 1982a:27.
249. Lewis 1982a:45.
250. Lewis 1982a:31.
251. Lewis 1982a:33.
252. Komarek 1969.
253. Lewis 1982a:42.
254. Lewis 1982a:42.
255. Lewis 1982a.42.
256. Lewis 1982a:25.

257. Lewis 1982a:50.

258. Benson 1964:374.

259. Quoted in Thompson and Smith 1970 from Campbell 1772.

260. Thompson and Smith 1970 from McMartin 1823.

261. Quoted in Day 1953 from Bigelow 1876.

262. Russell 1980:125.

263. Quoted in Day 1953 from Purchas 1906.

264. Parts of quote in Thompson and Smith 1970, Day 1953, Russell 1983, and Maxwell 1910, from Morton 1632.

265. Russell 1980:125.

266. Quoted in Calloway 1991:51 from Purchas 1906.

267. Maxwell 1910.

268. Rostlund 1957.

269. Saleeby 1983:32, 44; Suphan 1974; Taylor 1974.

270. Agee 1993:206–207; Boyd 1986; Deur 1997; Shaw 1997.

271. Agee 1993:206–207; Ripple 1994; Teensma et al. 1991.

272. Deur 1997.

273. Deur 1997; Hunn and French 1981; Turner and Bell 1973.

274. Ripple 1994; Williams 1976:36.

275. Deur 1997; Shaw 1997; Tirmenstein 1989b, 1990a.

276. Nordhoff 1875:209.

277. Coues 1965:940.

278. Shaw 1997.

279. Davies 1980:47.

280. Boyd 1986; Cronon 1983; Harrington 1932; Lewis 1993.

281. Quoted in Gruell 1985 from Thwaites 1966.

282. Boyd 1986; Tirmenstein 1989b, 1990a, 1990b.

283. Coues 1965:753.

284. Ripple 1994; Williams 1976:36.

285. Snyder 1991b, 1991c; Taber and Dasmann 1958; Tirmenstein 1989b, 1990a.

286. Coues 1965:798.

287. Rollins 1995:14.

288. Lewis 1993; Loud 1918.

289. Quoted in Lewis 1993 from Loud 1918.

290. Quoted in Lewis 1993 from Harrington 1932.

291. Lewis 1993.

292. Quoted in Lewis 1993 from Harrington 1932.

293. McMurray 1989a.

294. Quoted in Lewis 1993 from Harrington 1932.

295. Smith 1624a.

296. Cronon 1983:48.

297. Grant 1952:244.

298. Newcomb 1993:293.

299. Quoted in Doolittle 1992.

300. Quoted in Russell 1983 and Hammett 1992 from Percy 1625.

301. Hodge and Lewis 1990a:67–68 (from the account given to Cabeza de Vaca by Figueroa).

302. Rostlund 1960.

303. Joutel 1962:65.

304. Morison 1971.

305. Hodge and Lewis 1990a:43–44.

306. Hodge and Lewis 1990a:50.

307. Stevens 1855:362.

308. Barrett 1980a, 1980b; Cooper 1960; Holsinger 1902; Johnson 1969; Lewis 1982a.

309. Quoted in Barrett 1980b from Johnson 1969.

310. Lewis 1982a:40.

311. Lewis and Ferguson 1988.

312. Lewis 1994.

313. Smith 1624b.

314. Lewis 1994.

315. Barrett 1980a.

316. Quoted in Maxwell 1910.

317. Woolfenden 1996.

PART TWO   Forests at Discovery

1. Quoted in Bakeless 1961:200.

2. Bailey 1961; Goetzmann 1986; Hodge and Lewis 1990c; Morison 1971, 1974; Sauer 1971; Seno 1985.

3. Natural Resources Canada 1998; USDA Forest Service 1989.

CHAPTER EIGHT   Timeless Qualities of Ancient Forests

1. Frelich and Graumlich 1994; Frelich and Lorimer 1991b.
2. Jones 1945.
3. Watt 1947.
4. Johnson et al. 1995; Lorimer 1994; McGarigal 1994; Thomas et al. 1978; Seip and Savard 1995; Temple et al. 1979; Verner and Boss 1980; Welsh and Healy 1993.
5. Quoted in Steele 1985 (quote provided by Randy Robertson, Stephen F. Austin State University).

CHAPTER NINE   The Spanish Explorer's Forests

1. Hodge and Lewis 1990a:13.
2. Morison 1974:507.
3. Bean and Lawton 1993; Daniels 1976; Hodge and Lewis 1990b; Timbrook et. al. 1993.
4. Hodge and Lewis 1990a:29.
5. Hodge and Lewis 1990b:173, 270.
6. Davis 1983; Watts 1983.
7. Daniel et al. 1979:34–35.
8. Frost 1993.
9. Fowells 1965; Frost 1993; Landers et al. 1995; Peet and Allard 1993; Shantz and Zon 1924; Society of American Foresters 1967; White 1984.
10. Van Doren 1928:50–53.
11. Van Doren 1928:71, 146, 180–181.
12. Baker and Langdon 1990; Frost 1993; Harcombe et al. 1993; Lawson 1990; Wahlenberg 1960:22–26.
13. Baker and Langdon 1990; Blaisdell et al. 1973; Cowell 1995; Frost 1993; Harcombe et al. 1993; Lawson 1990; Wahlenberg 1960:22–26; White 1984.
14. Hammett 1992.
15. Van Doren 1928:191.
16. Hodge and Lewis 1990b:130, 270.
17. Fritz 1935 (quote provided by Randy Roberson, Stephen F. Austin State University).
18. Hodge and Lewis 1990b:169.
19. Hodge and Lewis 1990a:27.

20. Hooper and McAdie 1995.

21. Hodge and Lewis 1990b:178.

22. Van Doren 1928:56.

23. Carey 1992c; Harlow et al. 1996.

24. Van Doren 1928:57.

25. Hooper et al. 1996; Scott et al. 1977; Walker and Escano 1992.

26. Engstrom 1993.

27. Landers et al. 1995; Vogl 1972; Walker 1985.

28. Blaisdell et al. 1973; Bridges and Orzell 1989; Cowell 1995; Wahlenberg 1960:22–26.

29. Van Doren 1928:250.

30. Komarek 1968.

31. Chrismer et al. 1995.

32. Van Doren 1928:146.

33. Quoted in Steele 1985 (quote provided by Randy Roberson, Stephen F. Austin State University).

34. Belanger et al. 1993; Carey 1992a, 1992b; Fowells 1965; Boyer 1990; Baker and Langdon 1990; Rebertus et al. 1993.

35. Boyer 1990, 1993; Carey 1992a; Chapman 1932; Fowells 1965; Frost 1993; Landers et al. 1995; Platt and Rathbun 1993; Platt et al. 1988; Platt et al. 1991.

36. Burns and Honkala 1990.

37. Palik et al. 1997.

38. Chapman 1909.

39. Shugart et al. 1981.

40. Harlow et al. 1996; Oliver and Larson 1990:264.

41. Hodge and Lewis 1990a:96.

42. Hodge and Lewis 1990a:96.

43. Hodge and Lewis 1990a:112.

44. Daniel et al. 1979:34–35.

45. Shantz and Zon 1924; Wright et al. 1979.

46. Betancourt et al. 1993.

47. Miller and Wigand 1994.

48. Cartledge and Propper 1993.

49. Gottfried 1992; Leiberg et al. 1904; Short and McCulloch 1977.

50. Calkins 1909; Pieper 1977.

51. Fowells 1965; Gottfried 1992; Pieper 1977.

52. Fowells 1965; Gottfried 1992; Lang and Stewart 1910:7–8; McMurray

1986a, 1986b; Pieper 1977; Tirmenstein 1986; West 1995; Wright et al. 1979.

53. Ringland 1905.

54. Kohler 1992; Kohler and Matthews 1988.

55. Hodge and Lewis 1990a:115.

56. Wright et al. 1979.

57. Agee 1993; Bunting 1994; Calkins 1909; Dealy et al. 1978; Gottfried 1992; Miller 1995; Miller and Rose 1995; Miller and Wigand 1994; Pieper and Wittie 1990; Shaw 1995.

58. Lang and Stewart 1910:7–8.

59. Leiberg et al. 1904:19.

60. Calkins 1909:5.

61. Calkins 1909; Dealy et al. 1978; McMurray 1986a, 1986b; Pieper 1977, 1993; Rumble and Gobeille 1995; Short et al. 1977; Short and McCulloch 1977; Tirmenstein 1986.

62. Quoted in Boyd 1986 from Palmer 1847.

63. Agee 1993; Barbour 1995; Bartolome and Standiford 1992; Burcham 1957; Howard (n.d.)b, 1992b.

64. Daniels 1976.

65. Quoted in Timbrook et. al. 1993.

66. Quoted in Burcham 1957:87–88 from Coues 1900.

67. Fremont 1856:412.

68. Griffin 1977.

69. Anderson, R. S. 1989; Baker, R. G. 1983; Balls 1962:10–14; Barnosky 1984; Cwynar 1987; Fagan 1991:189; Fiedel 1987:67; Glassow et al. 1987 (cited in Fagan 1991:196–197); Helen McCarthy of the University of California at Davis, quoted by Martin, G. 1996; McCarthy 1993; West, G. J. 1989; Woolfenden 1996; Worona and Whitlock 1995.

70. Griffin 1977; Thompson 1961.

71. Quaife 1978:140.

72. Biswell 1956; Burcham 1957:90, 104.

73. Burcham 1957:88–89; Thompson 1961.

74. Quoted in Thompson 1961.

75. Burcham 1957:88–89; Thompson 1961.

76. Quaife 1978:141.

77. Griffin 1977; Sawyer et al. 1977.

78. Quoted in Thompson 1961 and Burcham 1957:88, from Belcher 1843.

79. Quoted in Thompson 1961.

80. Quoted in Griffin 1977, from Vancouver 1798.

81. Howard 1992c; Standiford et al. 1996.

82. Burcham 1957:108–110; Johnson and Winter 1846:95; Thompson 1961.

83. Quaife 1978:137.

84. Bonnicksen 1974.

85. Dufresne 1965:111–114.

86. Bonnicksen 1974; Howard 1992b.

87. Synder and Snyder 1989.

88. Verner and Boss 1980.

89. Quoted in Burcham 1957:105 from Cronise 1868.

90. Barbour 1995; Howard (n.d.)a; Standiford et al. 1996; Vankat and Major 1978.

91. Burcham 1957:86; Barbour 1995; Muir 1961:116–117; Standiford et al. 1996.

92. Burcham 1957:86; McMurray 1990; Tirmenstein 1989c.

93. Howard (n.d.)a.

94. Burcham 1957:86; Barbour 1995; Standiford et al. 1996.

95. Griffin 1977; Sawyer et al. 1977.

96. Barbour 1995.

97. Harlow et al. 1996.

98. Barbour 1995; Burcham 1957:86.

99. Keeley and Keeley 1995; Timbrook et al. 1993.

100. Quoted in Minnich 1987 from Brewer 1949.

101. Quoted in Minnich 1987 from Leiberg 1900.

102. Biswell 1963, 1974, 1976; Bonnicksen 1980; Bonnicksen and Lee 1979; Chou et al. 1993; Lee and Bonnicksen 1978; Minnich 1983; Philpot 1973; Vogl 1967.

103. Timbrook et al. 1993.

104. Biswell 1963; Hanes 1971; Hanes and Jones 1967; Hendricks 1968; Taber and Dasmann 1958.

105. Biswell 1961b, 1963; Gibbens and Schultz 1963.

106. Griffin 1977.

107. Sawyer et al. 1977.

108. Peterson 1961; Whitney 1986.

109. Quoted in Burcham 1957:92 from Gibbs (1860).

110. Quoted in Burcham 1957:92 from Gibbs (1860).

111. Agee 1993:352–370; Franklin and Dyrness 1988; Harlow et al. 1996; Sugihara and Reed 1987; Sugihara et al. 1987.

112. Quoted in Boyd 1986 from Palmer 1847.

113. Quoted in Boyd 1986 from Wilkes 1849.
114. Fowells 1965; Howard (n.d.)b; Sugihara and Reed 1987.
115. Agee 1996.
116. Agee 1993:352–370; Boyd 1986; Hastings et al. 1997; Keter 1987.
117. Bakeless 1961:396–398.
118. Daniels 1976.
119. Brown and Stanger 1969.
120. Daniels 1976.
121. Bakeless 1961:396–398.
122. Farquhar 1969:15–16.
123. Griffith 1992; Olson et al. 1990.
124. Jepson 1967.
125. Daniel et al. 1979:34–35; Griffin and Critchfield 1972; Olson et al. 1990; Shantz and Zon 1924; Stone et al. 1972.
126. Quoted in Burcham 1957:100 from Gibbs (1860).
127. Lewis 1993; Loud 1918.
128. McMurray 1989a; Sugihara et al. 1987.
129. Olson et al. 1990.
130. Veirs 1982.
131. Greenlee and Langenheim 1990; Meighan and Haynes 1970; West 1983.
132. Lewis 1993.
133. Greenlee and Langenheim 1990.
134. Brown and Swetnam 1994; Greenlee and Langenheim 1990.
135. Jacobs et al. 1985.
136. Griffith 1992; Olson et al. 1990.
137. Stone 1968; Stone et al. 1972.
138. Hibbs 1982.
139. Hunter and Parker 1993.
140. Stone et al. 1972.
141. Veirs 1980a, 1980b, 1982.
142. Veirs 1980a, 1980b, 1982.
143. Fowells 1965; Hermann and Lavender 1990; McMurray 1989a; Veirs 1980a, 1980b, 1982; Wilkinson et al. 1997.
144. Fowells 1965; Howard 1992a; Packee 1990; Stone 1968; Stone et al. 1972.
145. Zinke 1977; Stone 1968; Stone et al. 1972.
146. Davies 1980:160.

CHAPTER TEN *Forests of the Colonies*

1. Smith 1624a.
2. Bailey 1961; Goetzmann 1986; Morison 1971, 1974; Sauer 1971; Seno 1985.
3. Day 1953.
4. Cronon 1983:90.
5. Quoted in Spurr and Barnes 1980:560.
6. Quoted in Bakeless 1961:279.
7. Daniel et al. 1979:34–35; Fowells 1965; Shantz and Zon 1924.
8. Quoted in McIntosh 1962.
9. Quoted in Bakeless 1961:279.
10. Waldman 1985:111–112.
11. From the *Maryland Journal and Baltimore Advertiser*, October 19, 1779.
12. Waldman 1985:111–112.
13. Bromley 1935; Fowells 1965; Forthingham 1924; Nowacki and Abrams 1992; Tirmenstein 1991a, 1991b.
14. Benson 1964:597.
15. Abrams 1992; Cook et al. 1998; Frothingham 1924; Greller 1995; Lorimer 1980a, 1993; Stephenson 1986; Weaver and Clements 1938:512–513.
16. Burnham 1988; Keever 1953.
17. Foster and Zebryk 1993; Foster et al. 1992; Griffin 1993; Paillet 1994; Stephenson 1986; American Chestnut Foundation 1998.
18. American Chestnut Foundation 1998.
19. Neumann 1985; Russell 1980.
20. Rothwell 1991:113.
21. Russell 1980:82.
22. Bakeless 1961:132.
23. Smith 1624a.
24. Kidwell 1989.
25. Barnes et al. 1998:438; Burnham 1988; American Chestnut Foundation 1998.
26. Russell 1980:197–198.
27. Burnham 1988.
28. Brockman 1986; Fowells 1965.
29. Rothwell 1991:175.
30. Smith 1631:182.
31. Smith 1624a.
32. Quoted in Maxwell 1910.

33. Quoted in Day 1953 from Johnson 1654.
34. Quoted in Thompson and Smith 1970 from Morton 1632.
35. Bromley 1935.
36. Quoted in Lorimer 1985 from Coyne 1903.
37. Thompson and Smith 1970.
38. Maxwell 1910.
39. Keever 1953.
40. Paillet 1988, 1994, 1996a.
41. Griffin 1993; Paillet 1996a.
42. Day 1953.
43. Paillet 1988; Whitney and Davis 1986.
44. Griffin 1993; Paillet 1994, 1996a.
45. Burnham 1988; Foster and Zebryk 1993; Foster et al. 1992; Fowells 1965; Hodges and Gardiner 1993; Griffin 1993; Hodges and Johnson 1993; Lorimer 1993, 1985; Paillet 1994, 1996a; Regelbrugge and Smith 1994; Van Lear 1993.
46. Regelbrugge and Smith 1994.
47. Quoted in Lorimer 1993 from Bassett 1970.
48. Quoted in Bromley 1935.
49. Barden and Woods 1973; Lorimer 1985; Haines et al. 1975.
50. Abrams 1992; Lorimer 1993; Nowacki and Abrams 1992; Stephenson 1986.
51. Busing 1994; Leak and Filip 1977; Leak and Gottsacker 1985; Paillet 1996a; Rudoph and Lemmien 1976; Sander and Graney 1993; Smith 1980.
52. Burns and Honkala 1990; Daniel et al. 1979:296; Paillet 1996a.
53. Leak and Filip 1977; Leak and Gottsacker 1985; Miller and Kochenderfer 1998; Ward and Parker 1989.
54. Abrams 1992; Foster and Zebryk 1993; Foster et al. 1992; Griffin 1993; Lorimer 1980a, 1985, 1993; Nowacki and Abrams 1992; Paillet 1994, 1996a; Van Lear 1993; Regelbrugge and Smith 1994.
55. Burnham 1988; Forthingham 1924; Paillet 1994.
56. Quoted in Spurr and Barnes 1980:559; also quoted in Bromley 1935.
57. Daniel et al. 1979:34–35; Shantz and Zon 1924.
58. Dean 1994; Natural Resources Canada 1998.
59. Carey 1993a; Fowells 1965; Seymour 1995; Society of American Foresters 1967; Shantz and Zon 1924; Siccama 1971; Southern Appalachian Man and the Biosphere (SAMAB) 1996; Wendel and Smith 1990.
60. Jacobson and Dieffenbacher-Krall 1995; Craig 1969; Wright 1976.
61. Fowells 1965; Seymour 1995; Wendel and Smith 1990.

62. Quoted in Spurr and Barnes 1980:559; also quoted in Bromley 1935.
63. Bromley 1935.
64. Bromley 1935.
65. Bromley 1935; Hawes 1923; Nowacki and Abrams 1992, 1994; Siccama 1971; Wendel and Smith 1990.
66. Quoted in Hawes 1923 from Wood 1634.
67. Quoted in Bromley 1935.
68. Quoted in Bromley 1935.
69. Heidenreich 1971 (referenced in Williams 1989:37 and Clark and Royall 1996); Maissurow 1935.
70. Bromley 1935; Graham 1941; Maissurow 1935.
71. Maissurow 1935; Wein and Moore 1977.
72. Fowells 1965; Maissurow 1935.
73. Maissurow 1935; Quinby 1991.
74. Nowacki and Abrams 1992, 1994.
75. Bromley 1935; Henry and Swan 1974; Lutz 1930a, 1930b.
76. Maissurow 1935.
77. Wendel and Smith 1990; Whitney 1990.
78. Quoted in Whitney 1990.
79. Quoted in Lutz 1930a.
80. Henry and Swan 1974.
81. Waldman 1985:110; Whitney 1990.
82. Lutz 1930a.
83. Henry and Swan 1974.
84. Henry and Swan 1974; Lutz 1930a, 1930b; McIntosh 1962.
85. Henry and Swan 1974; Lutz 1930a, 1930b; McIntosh 1962; Quinby 1991.
86. McIntosh 1962.
87. Carey 1993a; Kudish 1992; Hibbs 1982; Lutz 1930a, 1930b; Maissurow 1935; Stearns 1951.
88. Rothwell 1991:216.
89. Bromley 1935; Lutz 1930a, 1930b; Siccama 1971.
90. Benson 1964:597.
91. Fowells 1965.
92. Bromley 1935.
93. Rothwell 1991:213.
94. Rothwell 1991:215–216.
95. Rothwell 1991:213.
96. Rothwell 1991:213.

97. Henry and Swan 1974.

98. Barnes et al. 1998:623.

99. Henry and Swan 1974; Hibbs 1982; Oliver and Stevens 1977.

100. Henry and Swan 1974.

101. Fowells 1965; Lutz 1930a; Stearns 1951.

102. Rothwell 1991:213.

103. Quoted in Day 1953 from Rosier 1906.

104. Rothwell 1991:39.

105. Buchanan 1824.

106. Eyre 1980; Fowells 1965; Greller 1995; Society of American Foresters 1967; Siccama 1971; Southern Appalachian Man and the Biosphere (SAMAB) 1996; Vankat 1979.

107. Daniel et al. 1979:34–35; Natural Resources Canada 1998.

108. Davis 1981; Gaudreau 1988.

109. Davis 1981, 1983; Delcourt and Delcourt 1991a:32–37.

110. Bourdo 1983; Davis et al. 1996; Delcourt and Delcourt 1991a:36–37.

111. Frelich and Lorimer 1991a.

112. Coladonato 1991; Fowells 1965; Peterson and Pickett 1991; Tirmenstein 1991c; Vankat et al. 1975.

113. Anonymous 1997b.

114. Buchanan 1824.

115. Balls 1962:14; Fagan 1991:313, 345; Tantaquidgeon 1942:25, 74; Hart 1992:52; Herrick 1977:302.

116. Coladonato 1991.

117. Russell 1980:79, 86–87.

118. Quaife 1921:69–70.

119. Coladonato 1991; Fowells 1965; Harlow et al. 1996; American Forestry Association 1978; Tirmenstein 1991c.

120. Barnes et al. 1998:477–479.

121. Fowells 1965; Society of American Foresters 1967.

122. Runkle 1981; Woods and Whittaker 1981

123. Rogers 1978.

124. Eckstein 1980; Godman and Lancaster 1990.

125. Godman and Lancaster 1990.

126. Rogers 1978.

127. Rogers 1978; Woods and Whittaker 1981.

128. Frelich and Graumlich 1994; Rogers 1978.

129. Maissurow 1941.

130. Davis et al. 1996.

131. Rogers 1978; Frelich and Lorimer 1985.

132. Eckstein 1980; Rogers 1978; Stearns 1951.

133. Frelich and Lorimer 1991a.

134. Loucks 1983.

135. Peterson and Pickett 1991.

136. Quoted in Stearns 1949.

137. Frelich and Reich 1996; Maissurow 1941.

138. Fowells 1965; Frelich and Lorimer 1991a; Frelich and Reich 1996; Sullivan 1994; Stearns 1949, 1951; Uchytil 1991d.

139. Davis et al. 1996; Frelich and Reich 1996.

140. Frelich and Lorimer 1991a.

141. Bourdo 1983; Stearns 1951.

142. Frelich and Lorimer 1991b.

143. Frelich and Lorimer 1991a.

144. Morison 1971:277–318.

145. Quoted in Bakeless 1961:206.

146. Davis, R. B. 1966.

147. Eyre 1980; Fowells 1965; Society of American Foresters 1967; Shantz and Zon 1924; Southern Appalachian Man and the Biosphere (SAMAB) 1996:109.

148. Daniel et al. 1979:34–35; Natural Resources Canada 1998.

149. Cogbill 1996; Davis 1981, 1983; Pielou 1991:291; Watts 1979.

150. Bromley 1935; Cogbill 1996; Eyre 1980; Hawes 1923; Reiners and Lang 1979; Society of American Foresters 1967; Siccama 1971; Southern Appalachian Man and the Biosphere (SAMAB) 1996; Westveld 1956.

151. Cogbill 1996.

152. Quoted in Cogbill 1996.

153. Quoted in Day 1953.

154. Beck 1990.

155. Frank 1990.

156. Blum 1990.

157. Cogbill 1996; Davis 1966; Lorimer 1977, 1980b.

158. White et al. 1985.

159. Reiners and Lang 1979; Sprugel 1976.

160. White et al. 1985.

161. Barnes et al. 1998:399.

162. White et al. 1985.

163. Computed from data in White et al. 1985, Table 4.
164. Quoted in Cogbill 1996.
165. Grant 1952:81.
166. Cogbill 1996; Lorimer 1977; Oliver and Larson 1990:379.
167. Benson 1964:589.
168. Calloway 1991:54–55; Day 1953; Thompson and Smith 1970.
169. Lorimer 1980b.
170. Based on Coolidge's 1963 map of forest fires in Lorimer 1980b, and vegetation maps by Westveld 1956; Lorimer 1977.
171. Quotes in Lorimer 1977.
172. Lorimer 1977.
173. Based on Coolidge's 1963 map of forest fires in Lorimer 1980b.
174. Bromley 1935; Davis 1966; Hughes and Bechtel 1997; Leak 1991; Lorimer 1977, 1980b.
175. Cogbill 1996; Davis 1966; Leak 1991; Lorimer 1977, 1980b.
176. Burt and Grossenheider 1980; Ministry of Forests 1995.
177. Lorimer 1977, 1980b.
178. Cogbill 1996.
179. Reiners and Lang 1979; Sprugel 1976.
180. Davis 1981, 1983; Delcourt and Delcourt 1991a:42; Watts 1983.
181. Reiners and Lang 1979; Sprugel 1976.
182. Sprugel 1976.
183. Sprugel 1976.

CHAPTER ELEVEN *Forests of the Fathers*

1. Grant 1952:276.
2. Belton 1997; Goetzmann 1986; Morison 1971.
3. Morison 1971:354.
4. Bakeless 1961:108.
5. Morison 1971:346.
6. Belton 1997.
7. Grant 1952:272–277.
8. Bakeless 1961; Bailey 1961; Belton 1997; Goetzmann 1986; Morison 1971; Muhlstein 1994; O'Hagan 1896; Rodesch 1984; Seno 1985; Smith 1854a, 1854b.
9. Charlevoix 1966,Vol I:245.
10. Foote 1983; Natural Resources Canada 1998.

11. Davis 1981, 1983; Elliott-Fisk 1989; Huntley and Webb 1989; Nilsson 1983:391; Pielou 1991:18–19; Webb 1987; Wright 1976.

12. Van Tramp 1870:16.

13. Mackenzie 1904:20.

14. Wallace 1932:194.

15. Beaudoin et al. 1997; Delcourt and Delcourt 1991a:65; Imbrie and Imbrie 1979:182–183.

16. Grant 1952:279.

17. Dean 1994; Elliott-Fisk 1989; Foote 1983; Uchytil 1991a; Vankat 1979: 96–107; Weaver and Clements 1938:487–489.

18. Lanman 1856:64.

19. Davies 1980:142.

20. Quoted in Twining 1983.

21. Dean 1994.

22. Thwaites 1898b.

23. Dean 1994; Foote 1983; Fowells 1965; Johnston 1990; Neiland and Viereck 1977; Nienstaedt and Zasada 1990; Rowe 1970; Rowe and Scotter 1973; Rudolph and Laidly 1990; Van Cleve et al. 1991; Viereck and Johnston 1990.

24. Davis 1981, 1983; Fowells 1965; Frank 1990; Neiland and Viereck 1977; Rudolph and Laidly 1990; Watts 1983.

25. Wallace 1932:211.

26. Mackenzie 1904:33.

27. Mackenzie 1904:33.

28. Alexander 1980; Lewis 1977, 1982a.

29. Cogbill 1985.

30. Mackenzie 1904:34.

31. Heinselman 1985.

32. Alexander 1980.

33. Cogbill 1985.

34. Belyea 1994:117–126.

35. Rowe 1970; Uchytil 1991c.

36. Lewis 1977, 1982a; Rowe and Scotter 1973.

37. Alexander 1980; Cogbill 1985; Heinselman 1981, 1985; Payandeh 1974; Rowe and Scotter 1973; Suffling 1983; Suffling et al. 1982.

38. Heinselman 1981; Rowe and Scotter 1973; Quirk and Sykes 1971.

39. Rowe 1970.

40. Dean 1994; Foote 1983; Heinselman 1981; Rowe and Scotter 1973; Society of American Foresters 1967; Van Cleve et al. 1991; Viereck 1975.

41. Foote 1983; Heinselman 1974, 1985; Koehler and Brittell 1990; Pavek 1992; Rowe and Scotter 1973; Scotter 1970; Snyder 1991d; Tesky 1992a; Van Cleve et al. 1991; Wolff 1980.

42. Lewis 1982a.

43. Gates et al. 1983.

44. Wallace 1932:340.

45. Wallace 1932:345.

46. Foote 1983; Heinselman 1974, 1985; Rowe and Scotter 1973; Scotter 1970; Snyder 1991a; Tesky 1993; Tirmenstein 1991d; Van Cleve et al. 1991.

47. Foote 1983; Heinselman 1974, 1985; Rowe and Scotter 1973; Scotter 1970; Snyder 1991a; Tesky 1993; Tirmenstein 1991d; Van Cleve et al. 1991.

48. Foote 1983; Heinselman 1974, 1985; Rowe and Scotter 1973; Scotter 1970; Snyder 1991d; Van Cleve et al. 1991.

49. Cogbill 1985.

50. Heinselman 1981; Uchytil 1991a, 1991b; Viereck 1975.

51. Foote 1983; Heinselman 1981; Van Cleve et al. 1991.

52. Interpreted from Suffling et al. 1982.

53. Dean 1994; Heinselman 1974, 1985; Koehler and Brittell 1990; Rowe and Scotter 1973; Suffling 1983; Wolf 1980.

54. Foster 1869:80.

55. Daniel et al. 1979:34–35; Fowells 1965; Shantz and Zon 1924.

56. Eyre 1980; Fowells 1965; Hansen et. al. 1974; McAndrews 1966; Society of American Foresters 1967.

57. Palik and Pregitzer 1992.

58. Pielou 1991:18–19.

59. Pielou 1991:192–193.

60. McAndrews 1966.

61. Davis 1981, 1983; Watts 1983.

62. Heinselman 1973.

63. Davis 1981, 1983; Delcourt and Delcourt 1991a:32–37, 201; Watts 1983; Webb et al. 1983.

64. Wright 1976.

65. Foster 1869:80.

66. Ahlgren and Ahlgren 1983; Bromley 1935; Carey 1993a; Fowells 1965.

67. Carey 1993b; Fowells 1965.

68. Carey 1993c, 1993d; Fowells 1965.

69. Carey 1993a, 1993b, 1993c, 1993d; Eyre 1980; Fowells 1965.

70. Quoted in Seno 1985:206.
71. Quaife 1921:230.
72. Tanner 1977:32.
73. Quaife 1921:231–232.
74. Quoted in Twining 1983.
75. Ahlgren and Ahlgren 1984.
76. Quaife 1921:232.
77. Quaife 1921:233.
78. Peterson 1947:1.
79. Schoolcraft 1966:207.
80. Schoolcraft 1966:206.
81. Schoolcraft 1966:210.
82. Schoolcraft 1966:221–222.
83. Schoolcraft 1966:213–214.
84. Schoolcraft 1966:215–216.
85. Schoolcraft 1966:223.
86. Schoolcraft 1966:250–252.
87. Schoolcraft 1966:252.
88. Schoolcraft 1966:250–252.
89. Heinselman 1973.
90. Frissell 1973.
91. Heinselman 1973.
92. Quoted in Alexander 1980.
93. Ahlgren and Ahlgren 1984:23, 44; Frissell 1973; Hansen et al. 1974; Heinselman 1973.
94. Frissell 1973.
95. Swain 1973.
96. Frissell 1973.
97. Heinselman 1973.
98. Heinselman 1973.
99. Heinselman 1973.
100. Frissell 1973.
101. Ahlgren and Ahlgren 1984:117.
102. Carey 1993c; Heinselman 1973; Lynham and Stocks 1989; Rowe 1970.
103. Estimate based on maps and acreage burned data in Frissell 1973.
104. Quoted in Barnes et al. 1998:244.
105. Bourdo 1983; Carey 1993c; Fowells 1965; Hansen et al. 1974.
106. Schoolcraft 1966:210.

107. Vogl 1970.

108. Ahlgren 1974; Ahlgren and Ahlgren 1983; Heinselman 1973; Fowells 1965.

109. Ahlgren 1974; Heinselman 1973, 1981; Frelich and Reich 1995.

110. Heinselman 1981.

111. Gullion 1977.

112. Heinselman 1973, 1981; Frelich and Reich 1995.

113. Frelich and Reich 1995; Heinselman 1973, 1981.

114. Frelich and Reich 1995.

115. Heinselman 1973, 1981.

116. Ahlgren and Ahlgren 1983, 1984; Cogbill 1985; Frelich and Reich 1995; Hansen et al. 1974; Heinselman 1973, 1981; Uchytil 1991b.

117. Ahlgren 1974, 1976; Heinselman 1973, 1981; Frelich and Reich 1995.

118. Ahlgren 1974; Heinselman 1973, 1981; Frelich and Reich 1995.

119. Heinselman 1973.

120. Fowells 1965.

121. Ahlgren and Ahlgren 1983.

122. Andrews 1857:85.

123. Heinselman 1973.

124. Carey 1993a, 1993b; Fowells 1965; Heinselman 1973, 1981, 1985.

125. Heinselman 1973, 1981, 1985.

126. Heinselman 1973.

127. Van Wagner 1977.

128. Heinselman 1973.

129. Van Wagner 1977.

130. Frissell 1973; Heinselman 1973.

131. Ahlgren 1976; Carey 1993a, 1993b; Fowells 1965; Frissell 1973; Maissurow 1935; Van Wagner 1970.

132. Ahlgren 1976; Carey 1993b; Fowells 1965; Frissell 1973; Van Wagner 1970.

133. Ahlgren 1976; Carey 1993a; Fowells 1965; Frissell 1973; Maissurow 1935, 1941; Peterson and Squiers 1995.

134. Fowells 1965; Howard and Tirmenstein 1996.

135. Graham 1941; Peterson and Squiers 1995.

136. Estimate based on maps and acreage burned data in Frissell 1973.

137. Frelich and Reich 1995; Heinselman 1973, 1981.

138. Graham 1941; Maissurow 1941; Stearns 1950.

139. Frelich and Reich 1996.

140. Frelich and Reich 1995.
141. Frelich and Lorimer 1991a.
142. Heinselman 1973.
143. Daniel et al. 1979:34–35; Fowells 1965; Shantz and Zon 1924.
144. McMillan 1976; Neumann 1985; Smith and Neal 1991.
145. Quoted in Seno 1985:43–44 from Thwaites 1898a.
146. Quoted in Seno 1985:44–45 from Thwaites 1898a.
147. Quoted in Bakeless 1961:324–325.
148. Sullivan 1995a, 1995b; Daniel et al. 1979:34–35; Fowells 1965; Shantz and Zon 1924.
149. Davis 1981; Johnson and Adkisson 1985.
150. Delcourt and Delcourt 1991b; Hall 1995; Watts 1983; Wright 1976.
151. McMillan and Wood 1976.
152. Graham et al. 1981.
153. Fowells 1965;.
154. Seno 1985:211.
155. Quoted in Angle 1968:74 from Baker 1903.
156. Quoted in Lorimer 1993.
157. Quoted in White 1995 from Baldwin 1877.
158. Carey 1992d, 1992e; Fowells 1965; McMillan 1976; Sullivan 1995a, 1995b; White and White 1996.
159. Adams and Anderson 1980; Barnes et al. 1998:366–368; Carey 1992c, 1992d; Dunevitz 1993; Fowells 1965; McMillan 1976; Sullivan 1995a, 1995b; Tirmenstein 1991b, 1988.
160. Coladonato 1992a, 1992b; Fowells 1965; McMillan 1976; Tirmenstein 1991e.
161. Fowells 1965; Harlow et al. 1996; Tirmenstein 1991b, 1991e, 1988.
162. Quoted in Angle 1968:49 from Hennepin 1880.
163. Muhlstein 1994:156–157.
164. Nuzzo 1986.
165. Lorimer n.d. from Curtis 1959.
166. Sullivan 1995a, 1995b.
167. Schoolcraft 1966:56.
168. Brown 1993; Samson and Knopf 1994.
169. Schoolcraft 1966:56.
170. Edwards 1847:24.
171. Seno 1985:224.
172. Bray 1960.

173. Packard 1988.

174. Lorimer n.d.

175. Quoted in Angle 1968:86.

176. Quoted in Angle 1968:175.

177. Quoted in Angle 1968:177.

178. Quoted in Seno 1985:124–125 from Thwaites 1898a.

179. Quoted in Seno 1985:121 from Dickson 1980.

180. Quoted in Angle 1968:48.

181. Coues 1965:175.

182. Sullivan 1995b.

183. Quoted in Angle 1968:135 from Shirreff 1835:238–250.

184. Coues 1965:264.

185. Bray and Bray 1993:69.

186. Bray and Bray 1993:56.

187. Rollins 1995:220.

188. Quoted in Bakeless 1961:327.

189. Quoted in Nelson et al. 1996.

190. Seno 1985:48–58.

191. Chittenden and Richardson 1969:199.

192. Chittenden and Richardson 1969:644.

193. Fiedel 1987:255; Woods 1996.

194. Nelson et al. 1996.

195. Seno 1985:57.

196. Nelson et al. 1996.

197. Nelson et al. 1996.

198. Carey 1992c, 1992f; Fowells 1965; Nelson et al. 1996; Tirmenstein 1991b.

199. Anderson, R. C. 1990; McAndrews 1966; Sullivan 1995a.

200. Bray and Bray 1993:66.

201. Leitner et al. 1991.

202. Quoted in White 1995 from Parker 1836.

203. Anderson, R. C. 1990.

204. Patterson 1992.

205. Apfelbaum and Haney 1990; McAndrews 1966; Sullivan 1995a.

206. Anderson, R. C. 1990.

207. Abrams 1992; Lorimer 1983, 1993; McCune and Cottam 1985; Nowacki and Abrams 1992; Shotola et al. 1992; Tirmenstein 1988, 1991e.

208. McMillan 1976.

209. Abrams 1992.

210. Abrams 1992; Bragg and Hulbert 1976; Burns and Honkala 1990.
211. Sullivan 1995a.
212. Schoolcraft 1966:7.
213. Schoolcraft 1966:54.
214. Stritch 1990.
215. Barnes et al. 1998:366–368; Foti and Glenn 1991; McMillan 1976; Sullivan 1995a, 1995b.
216. McMillan 1976, with quote from James 1844:57.
217. Leitner et al. 1991; Shotola et al. 1992.
218. Carey 1992e; McMillan 1976, with quotes from James 1844:57, 87.
219. Foti and Glenn 1991.
220. Cutter and Guyette 1994.
221. Gregg 1962:283 (quote from the Ancient Cross Timbers Project team, Tree-Ring Laboratory, University of Arkansas, Fayetteville).
222. U. S. War Department 1854:92.
223. de Pourtales 1832:51–53.
224. Irving 1886:78–79 (quote from the Ancient Cross Timbers Project team, Tree-Ring Laboratory, University of Arkansas, Fayetteville).
225. Burns and Honkala 1990; Carey 1992e; Rice and Penfound 1959; Clinton et al. 1993; Fowells 1965; Johnson 1993a, 1993b; Runkle 1991; Tirmenstein 1991b.
226. Gregg 1962:283 (quote from the Ancient Cross Timbers Project team, Tree-Ring Laboratory, University of Arkansas, Fayetteville).
227. Carey 1992d, 1992e; Rice and Penfound 1959.

CHAPTER TWELVE   *The Trapper's Forests*

1. Belyea 1994:184.
2. Gilbert 1973:80.
3. Kephart 1916 (Chapter VIII, The Beaver and His Trapper).
4. Phillips 1940:103–104.
5. Kephart 1916 (Chapter VIII, The Beaver and His Trapper).
6. Morgan 1969:304.
7. Utley 1997:141.
8. Utley 1997:56.
9. Morgan 1969:312, 364–365; Utley 1997:42, 50, 99–100.
10. Goetzmann 1986:108.
11. Mackenzie 1904; Tanner 1977:72–73.

12. Bailey 1961; Goetzmann 1986, 1991.

13. Quoted in Biswell et al. 1973 from Dutton 1887.

14. Scurlock and Finch 1997.

15. Coues 1965:590.

16. Coues 1965:993.

17. Davies 1980: 61.

18. Cooper 1860:14.

19. Quoted in Cooper 1960.

20. Quoted in Biswell et al. 1973 from Dutton 1887.

21. Cooper 1960.

22. Daniel et al. 1979:34–35; U.S. Department of Agriculture 1973.

23. Eyre 1980.

24. Harlow et al. 1996; USDA Forest Service 1993.

25. Anderson, R. S. 1989.

26. Anderson, R. S. 1989; Whitlock 1995.

27. Anderson, R. S. 1990; Woolfenden 1996.

28. Ringland 1905.

29. Stevens 1855:102.

30. Fowells 1965; Habeck 1992a, 1992b; Harlow et al. 1996.

31. Fowells 1965; Habeck 1992a, 1992b; Harlow et al. 1996.

32. Quoted in Biswell et al. 1973 from Dutton 1887.

33. Goetzmann 1993:275.

34. Sitgreaves 1853:10.

35. Sitgreaves 1853:12.

36. Atzet 1996; Barbour 1995; Covington and Sackett 1988; Eyre 1980; Fowells 1965; Franklin and Dyrness 1988: Rundel et al. 1977; Show and Kotok 1929; Storer and Usinger 1968; White and Vankat 1993.

37. Illg and Illg 1994; Reynolds and Linkhart 1992; Scott et al. 1977; Thomas et al. 1979a; Verner and Boss 1980.

38. Burt and Grossenheider 1980; Patton 1977.

39. Burt and Grossenheider 1980; Severson and Medina 1983; Thomas et al. 1979a.

40. Sitgreaves 1853:37.

41. Burt and Grossenheider 1980; Thomas et al. 1979a.

42. Sitgreaves 1853:10.

43. Goetzmann 1993:306–307.

44. Ives 1861:104.

45. Ives 1861:114.

46. Quoted in Scurlock and Finch 1997.
47. Quoted in Cooper 1960.
48. Lang and Stewart 1910:25.
49. Agee 1994; Arno 1980; Arno and Petersen 1983; Bevins and Barney 1980; Martin 1982.
50. Arno 1976, 1980; Brown et al. 1995; Martin 1982; McBride and Jacobs 1980; Show and Kotok 1924; Swetnam 1988; Warner 1980; Weaver 1951.
51. Baisan and Swetnam 1997; Caprio and Swetnam 1995; Danzer et al. 1996; Dieterich 1980; Kaib et al. 1996; Seklecki et al. 1996; Show and Kotok 1924; Swetnam 1988; Swetnam and Baisan 1996; Touchan et al. 1996; Warner 1980; Weaver 1951.
52. Agee 1994; Fisher et al. 1986.
53. Arno 1996.
54. Laven et al. 1980.
55. Kilgore and Taylor 1979; Laven et al. 1980.
56. Baisan and Swetnam 1997; Barrett and Arno 1982; Kilgore and Taylor 1979; Seklecki et al. 1996; Warner 1980.
57. Quoted in Cooper 1960 from Holsinger 1902.
58. Bevins and Barney 1980.
59. Barrett and Arno 1982; Belyea 1994:221.
60. Quoted in Arno et al. 1997.
61. Barrett and Arno 1982.
62. Computed from data in Barrett and Arno 1982.
63. Baisan and Swetnam 1997.
64. Warner 1980.
65. Cooper 1960; Covington and Moore 1994; Swetnam 1988.
66. Lang and Stewart 1910:19.
67. Lang and Stewart 1910:18–19.
68. Dodge 1876:149.
69. Phillips 1940 (Part I, Chapter VII).
70. Jackson 1966.
71. Fisher et al. 1986; Grinnell 1995:34.
72. Jackson 1966; Progulske 1974.
73. Dodge 1876:62.
74. Ludlow 1875:18.
75. Haines 1965:80.
76. Dodge 1876:138.
77. Fisher et al. 1986.

78. Fisher et al. 1986.

79. Dodge 1876:62–63.

80. Ludlow 1875:37–38.

81. Dodge 1876:100–101.

82. Dodge 1876:102.

83. Quoted in Cooper 1960.

84. Arno et al. 1997.

85. King 1874:28–29.

86. Agee 1993:332–333; Arno et al. 1995, 1997; Biswell et al. 1973; Bonnicksen and Stone 1981, 1982a; Cooper 1960, 1961; Franklin and Dyrness 1988; Hallin 1959; Schubert 1974.

87. Lang and Stewart 1910:17.

88. Gila Reconnaissance Party 1912:3.

89. Ringland 1905:16.

90. Arno et al. 1995, 1997; Biswell et al. 1973; Bonnicksen and Stone 1981, 1982a; Cooper 1960, 1961; Covington and Moore 1992; White, A. S. 1985.

91. Lang and Stewart 1910:6.

92. White, A. S. 1985.

93. Cooper 1960; Covington and Moore 1994; White, A. S. 1985.

94. Cooper 1860:14.

95. Cooper 1960; Covington and Moore 1994; Fowells 1965; Heidmann et al. 1982; White, A. S. 1985.

96. Ringland 1905:16.

97. Cooper 1960.

98. Cooper 1960; Fowells 1965; Habeck 1992a; Oliver and Ryker 1990; White and Vankat 1993.

99. Lang and Stewart 1910:28, 30.

100. Based on data in White, A. S. 1985.

101. Cooper 1960.

102. Burt and Grossenheider 1980; Illg and Illg 1994; Patton 1977; Reynolds and Linkhart 1992; Reynolds et al. 1992; Thomas et al. 1979a; Verner and Boss 1980.

103. Harrington and Sackett 1992.

104. Sudworth 1900.

105. Dahms and Geils 1997; Hessburg et al. 1994; Oliver 1995; Oliver and Ryker 1990.

106. Lang and Stewart 1910:19–20.

107. Severson and Medina 1983.

108. Bennetts et al. 1996.

109. Harrington and Hawksworth 1988; Hessburg et al. 1994.

110. Habeck 1976.

111. Stevens 1860:140.

112. Camp et al. 1997; Leiberg et al. 1904:15.

113. Biswell et al. 1973; Bradley et al. 1992; Burns and Honkala 1990; Davis et al. 1980; Fowells 1965; Franklin and Dyrness 1988; Habeck and Mutch 1973; Habeck 1976; Helms 1995; Smith and Fischer 1997; Tesch 1995.

114. Arno 1976.

115. Biswell et al. 1973.

116. Cunningham et al. 1980; Maser et al. 1979; Reynolds and Linkhart 1992; Scott 1978; Thomas et al. 1979b.

117. Lang and Stewart 1910:30.

118. Biswell et al. 1973.

119. Biswell et al. 1973; Harrington and Kelsey 1979; McDonald et al. 1997; Sackett 1984; Weaver 1951.

120. Harrington and Sackett 1988; Sackett and Hasse 1996; Weaver 1951.

121. Sackett 1984.

122. Covington and Sackett 1988.

123. Monleon et al. 1997.

124. Patton 1974; Severson and Medina 1983.

125. Lowe et al. 1978.

126. Irwin et al. 1994; Severson and Medina 1983.

127. Bradley et al. 1992; Cooper 1960; Oliver and Ryker 1990.

128. Biswell 1972; Biswell et al. 1973; Cooper 1960; Harrington and Kelsey 1979; Weaver 1951;

129. Biswell 1970, 1972; Biswell et al. 1973; Cooper 1960; Weaver 1951;

130. Bradley et al. 1992.

131. Burt and Grossenheider 1980; Illg and Illg 1994; Patton 1977; Reynolds and Linkhart 1992; Reynolds et al. 1992; Thomas et al. 1979a; Verner and Boss 1980.

132. Bradley et al. 1992; Irwin et al. 1994; Lowe et al. 1978; Patton 1974; Severson and Medina 1983.

133. Doane 1871:5, 14, 20, 22, 23, 27; Haines 1965:101; Hayden 1872:100–101.

134. Doane 1871:5.

135. Leiberg 1904:29.

136. Daniel et al. 1979:34–35; Koch 1996.

137. Eyre 1980; Fowells 1965; Lotan and Critchfield 1990; Natural Resources Canada 1998; Peet 1989.

138. Phillips 1940:86.

139. Ludlow 1876:20–21.

140. Barnosky 1984; Cwynar 1987; Delcourt and Delcourt 1991a:45–47; Whitlock 1995.

141. Anderson, R. S. 1990; Baker, R. G. 1983; Delcourt and Delcourt 1991a:200–201; Pielou 1991:18, 272, 278, 291; Whitlock 1993; Woolfenden 1996.

142. Lotan and Critchfield 1990; Peet 1989.

143. Stevens 1855:220.

144. Franklin and Dyrness 1988.

145. Chittenden and Richardson 1969:242; Coues 1965:91, 137.

146. Muir 1961:154–155.

147. Hayden 1872:102.

148. Fowells 1965; Harlow et al. 1996; Loope and Gruell 1973; Lotan and Critchfield 1990.

149. Muir 1961:91.

150. Phillips 1940 (Part III, Chapter XV).

151. Canon et al. 1987; DeByle 1984; Howard and Tirmenstein 1996; Mueggler 1985; Snyder 1991a, 1991b, 1991c.

152. Haines 1965:44.

153. DeByle 1985a; Tesky 1993.

154. Houston 1973.

155. Chadde and Kay 1991; Haines 1965:26–27.

156. Quaife 1978:72.

157. Phillips 1940 (Part VII, Chapter LVI).

158. Doane 1871:26.

159. Leiberg 1904:20–21.

160. Jones 1985.

161. Grant 1993; Kay 1997.

162. Fowells 1965; Harlow et al. 1996; Lotan and Critchfield 1990.

163. Loope and Gruell 1973.

164. DeByle 1985b; Fowells 1965; Hinds 1985; Jones et al. 1985; Mueggler 1985.

165. Fowells 1965; Hinds 1985; Jones and DeByle 1985a; Jones et al. 1985.

166. Schier et al. 1985.

167. Grant 1993.

168. Jones and DeByle 1985a.

169. Jones and DeByle 1985a, 1985b; Kay 1997; Schier et al. 1985.

170. Bartos and Mueggler 1979; Basile 1979; DeByle 1979, 1984, 1985b.

171. Haines 1965:104 [punctuation added].

172. Kay 1997; Kay et al. 1994.

173. Muir 1961:155.

174. Muir 1961:156.

175. Rothermel 1991.

176. Arno 1976; Barrettt and Arno 1990, 1995.

177. Amman and Ryan 1991; Arno 1976; Fowells 1965; Peterson 1984.

178. Muir 1961:156.

179. Haines 1965:29 (also see p. 110, Plate IX).

180. Phillips 1940:86.

181. Arno 1980, 1985; Bevins and Barney 1980.

182. Barrett 1980a, 1980b; Barrett and Arno 1982; Haines 1965:30–31; Houston 1973; Leiberg 1904:27; White, C. A. 1985a:128, 1985b:22–24.

183. Houston 1973; Janetski 1987:6–7, 57–61.

184. Quoted in White, A. S. 1985:24.

185. Arno 1976, 1980; Arno and Petersen 1983; Arno et al. 1993; Wadleigh and Jenkins 1996.

186. Davis 1980; Despain 1990:109–110; White, C. A. 1985a:144.

187. Franklin and Laven 1991.

188. Leiberg 1904:27–28.

189. Arno 1976, 1980; Arno and Petersen 1983; Arno et al. 1993.

190. Agee 1994.

191. Barrett et al. 1997; White, C. A. 1985a:129, 131.

192. Barrettt and Arno 1990, 1995.

193. Heinselman 1985; Romme 1982; White, C. A. 1985b:22.

194. Barrettt and Arno 1990, 1995; Romme 1980.

195. Computed using fire cycle equations in Van Wagner 1978 based on cover data from graphs in Romme and Despain 1989, Despain and Romme 1991, and Romme 1980.

196. Leiberg 1904:36; White, C. A. 1985b:79; Romme 1982.

197. Barrettt and Arno 1990, 1995; Romme 1982.

198. Barrettt and Arno 1990, 1995.

199. Rothermel 1991.

200. Van Wagner 1977.

201. Minshall et al. 1989, 1995.

202. Amman and Ryan 1991; Berryman 1982.

203. Chittenden and Richardson 1969:505–506.

204. Fechner and Barrows 1976; Jones and DeByle 1985a.

205. Loope and Gruell 1973; Despain 1990:108–109; Romme and Despain 1989.

206. Omi and Kalabokidis 1990.

207. Arno 1976, 1980; Leiberg 1904:30; Omi and Kalabokidis 1990.

208. Lotan 1976; Tinker et al. 1994; Uchytil 1992.

209. Pavek 1992.

210. Taylor 1979.

211. Austin and Perry 1979.

212. Heinselman 1981; Despain 1990:108–109; White, C. A. 1985a:16.

213. Fischer and Bradley 1987.

214. Brown et al. 1998.

215. Doane 1871:22.

216. Uchytil 1992; Wadleigh and Jenkins 1996; White, C. A. 1985a:145.

217. Arno 1976, 1980; Fischer and Bradley 1987; Wadleigh and Jenkins 1996.

218. Despain 1990:108–109, 173–176; Romme 1982.

219. Despain 1990:109, 173–176; Romme and Despain 1989; White, C. A. 1985a:145.

220. Heinselman 1981; Loope and Gruell 1973; Despain 1990:109; White, C. A. 1985a:16.

221. Leiberg 1904:29.

222. Despain 1990:176.

223. Despain 1990:109–110; Romme and Despain 1989.

224. Davis et al. 1980; Despain 1990:109–110; Heinselman 1981; Loope and Gruell 1973; White, C. A. 1985a:16.

225. Despain 1990:109–110, 173–176; White, C. A. 1985a:145.

226. Leiberg 1904:28.

227. Despain 1990:109–111, 173–176; Romme 1982.

228. Leiberg 1904:28.

229. Amman et al. 1989.

230. Fischer and Clayton 1983; Loope and Gruell 1973; Despain 1990:109–110.

231. Haines 1965:97.

232. Haines 1965:101.

233. Amman 1977.

234. Berryman 1982.

235. Fischer and Bradley 1987.

236. Amman 1977.

237. Cole 1993; Parker 1988.

238. Amman 1977; Davis et al. 1980; Despain 1990:110–111; Fischer and Bradley 1987; Fischer and Clayton 1983; Fowells 1965; Heinselman 1981; Leiberg 1904:22; Loope and Gruell 1973.

239. White, C. A. 1985a:16; Romme 1982.

240. Amman 1977; Arno 1976, 1980; Despain 1990:110–111; Franklin and Laven 1991; Heinselman 1981; Leiberg 1904:22; Loope and Gruell 1973; Uchytil 1992; White, C. A. 1985a:16.

241. Amman 1977; Despain 1990:201–203.

242. Arno et al. 1993.

243. Romme 1982.

244. Despain 1990:109–110, 173–176; ; Heinselman 1981; Loope and Gruell 1973; Romme 1982; Uchytil 1992; White, C. A. 1985a:145.

245. Ludlow 1876:21.

246. Rothermel 1991; Van Wagner 1977.

247. Despain 1990:110–111, 173–176; Despain and Romme 1991; Romme 1980, 1982; Romme and Despain 1989.

248. Romme 1982.

249. Leiberg 1904:31, 36, 38, 46, 49, 56, 59, 61, 63, 90.

250. Arno 1976; Arno et al. 1993; Habeck 1976.

251. Barrett 1980a, 1980b; Barrett and Arno 1982; Haines 1965:30–31; Houston 1973; Leiberg 1904:27; White, C. A. 1985a:128, 1985b:22–24.

252. Leiberg 1904:22.

253. Computed using cover data from graphs in Despain 1989, Despain and Romme 1991, and Romme 1980.

254. Despain 1989; Despain and Romme 1991; Romme 1980.

255. Bevins and Barney 1980; Despain 1990:167.

256. Haines 1965:86.

257. Barrett et al. 1997; Despain 1989; Despain and Romme 1991; Romme 1980; Janetski 1987:32; Wissler 1989:106.

258. Janetski 1987:57–61.

259. Houston 1973.

260. Quoted in Anonymous 1845.

261. Quoted in Gates 1941:14.

262. Gates 1941:13, 15.

263. Quoted in Gates 1941:16.

264. Quoted in Peterson 1994:14 from the *Oregon Historical Quarterly.*

265. Goetzmann 1986:107–108, 237–238.

266. Coues 1965:745–749.

267. Daniel et al. 1979:34–35; Eyre 1980; Fowells 1965; Franklin 1995; Franklin and Dyrness 1988; Sawyer and Thornburgh 1977; Sawyer et al. 1977; U.S. Department of Agriculture 1973.

268. Johnson and Winter 1846:67.

269. Daniel et al. 1979:34–35; Eyre 1980; Fowells 1965; Franklin 1995; Franklin and Dyrness 1988; Sawyer and Thornburgh 1977; Sawyer et al. 1977; U.S. Department of Agriculture 1973.

270. Barnosky 1984; Cwynar 1987; Heusser 1983.

271. Worona and Whitlock 1995.

272. Baker, R. G. 1983; West, G. J. 1989.

273. Quoted in Williams 1976:19.

274. Rollins 1995:5.

275. Franklin and Dyrness 1988.

276. Barnes et al. 1998:399; Burns and Honkala 1990:646–649; Fowells 1965; Harlow et al. 1996; Hermann and Lavender 1990.

277. Barnes et al. 1998:399; Burns and Honkala 1990:646–649; Fowells 1965; Harlow et al. 1996; Tesky 1992b.

278. Cooper 1860:26.

279. Barnes et al. 1998:399; Burns and Honkala 1990:646–649; Dale et al. 1985; Fowells 1965; Harlow et al. 1996; Hennon 1995; Tesky 1992c.

280. Agee 1990; Barnes et al. 1998:399; Burns and Honkala 1990:646–649; Fowells 1965; Franklin, and Hemstrom 1981; Harlow et al. 1996; Whitney 1986.

281. Agee 1993:283–290; Eyre 1980; Fowells 1965; Franklin 1995; Franklin and Dyrness 1988; McMurray 1989a, 1989b; Thornburgh 1982.

282. Hines 1850:112.

283. Crane 1989, 1990; Franklin and Dyrness 1988; Whitney 1986:392, 394–395.

284. Quoted in Gates 1941:177.

285. Nordhoff 1875:210–211.

286. Cooper 1860:24.

287. Cooper 1860:24.

288. Agee 1993:29; Komarek 1967; Ripple 1994.

289. Agee 1993:209.

290. Zybach 1993.

291. Agee 1993:210; Ripple 1994; Teensma et al. 1991.

292. Martin et al. 1974.

293. Martin et al. 1974; Teensma et al. 1991; Walstad et al. 1990.

294. Nordhoff 1875:209.

295. Computed using fire cycle equations in Van Wagner 1978 from data in Booth 1991, Fahnestock and Agee 1983, and Teensma et al. 1991.

296. Coues 1965:723.

297. Teensma et al. 1991.

298. Means 1982; Ripple 1994; Teensma et al. 1991.

299. Agee 1990, 1993:211–112, 285; Means 1982; Means et al. 1996; Ripple 1994; Teensma et al. 1991.

300. Hansen et al. 1991; McComb et al. 1993; Ripple 1994.

301. Agee 1993:210, 284; Means 1982.

302. Agee 1990.

303. Agee 1993:284

304. Hines 1850:96.

305. Hines 1850:118.

306. Muir 1918 (Chapter 18, The Forests of Washington).

307. Agee 1990; Burns and Honkala 1990; Fowells 1965; Hermann and Lavender 1990; Uchytil 1991e.

308. Burns and Honkala 1990; Fowells 1965; McMurray 1989a, 1989b.

309. Thornburgh 1982.

310. Thornburgh 1982.

311. Thornburgh 1982.

312. Agee 1993:283–290; Fowells 1965; Thornburgh 1982.

313. Hines 1850:112.

314. Quoted in Williams 1976:19.

315. Franklin and Dyrness 1988.

316. Dale et al. 1985.

317. Camp et al. 1997.

318. Muir 1918 (Chapter 18, The Forests of Washington).

319. Bunnell et al. 1997:44, Appendix II; Peterson 1961.

320. Bunnell et al. 1997:Appendix II; McGarigal and McComb 1995.

321. Muir 1918 (Chapter 18, The Forests of Washington).

322. Agee 1993:214; Fowells 1965; Franklin and Dyrness 1988; Hermann and Lavender 1990; Packee 1990.

323. Muir 1918 (Chapter 18, The Forests of Washington).

324. Schoonmaker and McKee 1988.

325. Muir 1918 (Chapter 18, The Forests of Washington).

326. Muir 1918 (Chapter 18, The Forests of Washington).

327. Franklin and Waring 1980; Franklin and Hemstrom 1981.

328. Computed for trees 200 or more years old using fire cycle equations in Van

Wagner 1978 from data in Booth 1991, Fahnestock and Agee 1983, and Teensma et al. 1991.

329. Muir 1918 (Chapter 18, The Forests of Washington).
330. Young 1899:177.
331. Muir 1918 (Chapter 18, The Forests of Washington).
332. Barnes et al. 1998:119–120; Uchytil 1989.
333. Barnes et al. 1998:474–475; Fowells 1965; Hermann and Lavender 1990.
334. McComb et al. 1993.
335. Dale et al. 1985; Uchytil 1989.
336. Halpern and Franklin 1990; Henderson 1982.
337. Quoted in Utley 1997:90.
338. Daniels 1976:176–177; Farquhar 1969:26; Utley 1997:90
339. Rundel 1972; Weatherspoon 1990.
340. Quaife 1978:129–130, 134.
341. Quaife 1978:135–136.
342. Farquhar 1969:59.
343. Hutchings 1859b.
344. Hartesveldt et al. 1975:66–77.
345. Axelrod 1977; Engbeck 1973:60–65; Hartesveldt et al. 1975:66–77; Harvey 1986; Ornduff 1994; Rundel 1971, 1972.
346. Anderson, R. S. 1994; Baker, R. G. 1983; Woolfenden 1996.
347. Daniel et al. 1979:34–35.
348. Rundel 1971; Rundel et al. 1977.
349. Quoted in Ornduff 1994.
350. Harvey 1986, Hartesveldt et al. 1975:19.
351. Muir 1961:140.
352. Muir 1961:140.
353. Arno 1973; Harlow et al. 1996; Harvey 1980b; Weatherspoon 1990.
354. Muir 1961:140.
355. Le Conte 1971:31.
356. Arno 1973; Fowells 1965; Habeck 1992c; Harlow et al. 1996.
357. Fremont 1856:407.
358. Fowells 1965; Habeck 1992c; Harlow et al. 1996.
359. Muir 1918 (Chapter 22, The Forests of Oregon and their Inhabitants).
360. Muir 1961:119, 122–123.
361. Fowells 1965; Habeck 1992a, 1992b; Harlow et al. 1996.
362. Fowells 1965; Harlow et al. 1996; Laacke 1990.
363. Muir 1961:113–114.

364. Harvey 1980a; Stocking and Rockwell 1969; Storer and Usinger 1968.

365. Muir 1961:126.

366. Shellhammer 1980a.

367. Peterson 1961; Storer and Usinger 1968.

368. Muir 1918 (Chapter 22, The Forests of Oregon and Their Inhabitants).

369. Shellhammer 1980b.

370. Storer and Usinger 1968.

371. Rundel 1971.

372. Barrett and Gifford 1933; Bonnicksen et al. 1999; Biswell 1967; Kilgore and Taylor 1979; Pilarski 1994; Reynolds 1959; Vankat 1977, 1985; Vankat and Major 1978.

373. Bonnicksen 1975; Bonnicksen and Stone 1978, 1981, 1982a.

374. Rundel 1971.

374. Stecker 1980.

376. Shellhammer 1980b.

377. Stecker 1980.

378. Stecker 1980.

379. Stecker 1980.

380. Harvey 1980b.

381. Bradley 1886.

382. Hartesveldt and Harvey 1967; Harvey 1980b.

383. Bradley 1886.

384. Anderson, R. S. 1994; Kilgore and Taylor 1979; Piirto 1996 (from Stohlgren 1991); Rundel 1971; Stephenson 1994.

385. Bradley 1886.

386. Beetham 1961; Fowells 1965; Hartesveldt and Harvey 1967; Harvey 1980b; Weatherspoon 1990.

387. Hartesveldt et al. 1975; Harvey 1980a, 1980b; Shellhammer 1980a, 1980b; Stecker 1980.

388. Muir 1961:147.

389. Hartesveldt and Harvey 1967; Harvey 1980b.

390. Bradley 1886.

391. Hartesveldt and Harvey 1967; Harvey 1980b; Harvey and Shellhammer 1991.

392. Bonnicksen 1975; Bonnicksen and Stone 1978, 1981, 1982a; Stephenson et al. 1991.

393. Beetham 1961; Donaghey 1969; Hartesveldt and Harvey 1967; Harvey 1980b; Kilgore and Biswell 1971; Stephenson 1994.

394. Kilgore and Taylor 1979.

395. Show and Kotok 1924.

396. Kilgore and Taylor 1979.

397. Komarek 1967; Vankat 1985.

398. Barrett and Gifford 1933; Biswell 1967; Bonnicksen 1975; Kilgore and Taylor 1979; Pilarski 1994; Reynolds 1959; Vankat 1977, 1985; Vankat and Major 1978.

399. Quoted in Pilarski 1994 (from an interview with Hector Franco, a Wukchumni Yokut, conducted by M. Kat Anderson in 1993).

400. Biswell 1967; Kilgore and Sando 1975; Kilgore and Taylor 1979; Vankat 1977; Vankat and Major 1978.

401. Muir 1901 (Chapter IX, The Sequoia and General Grant National Parks).

402. Show and Kotok 1924:31.

403. Eaton 1956; Fowells 1965; Graham and Knight 1965; Keen 1952; Leaphart 1963; Startwell et al. 1971.

404. Fowells 1965; Kimmey 1950; Wagener and Cave 1946.

405. Muir 1901 (Chapter IX, The Sequoia and General Grant National Parks); *see also* Bonnicksen 1975; Caprio and Swetnam 1995; and Kilgore and Taylor 1979.

406. Swetnam 1993.

407. Biswell 1961a, 1967; Bonnicksen 1975; Bonnicksen and Stone 1978, 1981, 1982a, 1982b; Winchell 1933:161–162.

408. Muir 1961:113–114.

409. Larsen and Woodbury 1916; Show and Kotok 1924:31.

410. Muir 1901 (Chapter IX, The Sequoia and General Grant National Parks).

411. Bonnicksen 1975; Bonnicksen and Stone 1978, 1981, 1982a; Stephenson et al. 1991.

412. Agee and Biswell 1969; Beetham 1961.

413. Hartesveldt and Harvey 1967; Harvey 1980b; Harvey and Shellhammer 1991.

414. Biswell et al. 1966a; Fowells 1965; Gordon 1970; Harvey 1980b; Kilgore and Biswell 1971; Oliver and Ryker 1990; Quick 1956; Quick and Quick 1961; Tackle and Roy 1953.

415. Fowells 1965; Gordon 1970; Habeck 1992c; Laacke 1990; Stark 1965.

416. Fowells 1965.

417. Bonnicksen and Stone 1978, 1982a, and unpublished data from Kings Canyon National Park.

418. Fowells and Schubert 1951; Laacke 1990; Sellers 1970.

419. Bonnicksen and Stone 1978, 1982a.

420. Fowells 1965; Sellers 1970; Stark 1965.

421. Gordon 1970; Schubert 1956; Stark 1965. Burns and Honkala 1990.

422. Rundel et al. 1977; also computed from Sudworth's 1900 data in McKelvey and Johnston 1992.

423. Stark 1965.

424. Bonnicksen and Stone 1978, 1982a, and unpublished data from Kings Canyon National Park.

425. Larsen and Woodbury 1916.

426. Biswell et al. 1966a; Hartesveldt and Harvey 1967; Harvey 1980b; Quick 1956; Quick and Quick 1961; Tirmenstein 1989a.

427. Biswell et al. 1966b.

428. Kilgore and Sando 1975.

429. Bonnicksen and Stone 1978, 1982a, and unpublished data from Kings Canyon National Park.

430. Bonnicksen and Stone 1978, 1982a, and unpublished data from Kings Canyon National Park; also computed from Sudworth's 1900 data in McKelvey and Johnston 1992.

431. Bonnicksen and Stone 1978, 1982a, and unpublished data from Kings Canyon National Park; Larsen and Woodbury 1916; Muir 1901 (Chapter IX, The Sequoia and General Grant National Parks).

432. Quaife 1978:135.

433. Fry and White 1966:11.

Bibliography

Abrams, M. D. 1992. Fire and the development of oak forests. *BioScience* 42(5):346–353.

Adair, J. 1775. *The history of the American Indians.* No imprint. London.

Adam, D. P. 1967. Late-Pleistocene and recent palynology in the central Sierra Nevada, California. In Cushing, E. J. and H. E. Wright, Jr. (eds.), *Quaternary paleoecology*, pp. 275–301. Volume 7 of the Proceedings of the VII Congress of the International Association for Quaternary Research. New Haven, CT: Yale University Press.

Adams, D. E. and R. C. Anderson. 1980. Species response to a moisture gradient in central Illinois forests. *American Journal of Botany* 67(3):381–392.

Agee, J. K. 1990. The historical role of fire in Pacific Northwest forests. In Walstad, J. D., S. R. Radosevich, and D. V. Sandberg (eds.), *Natural and prescribed fire in Pacific Northwest forests*, pp. 25–38. Corvallis, OR: Oregon State University Press.

Agee, J. K. 1993. *Fire ecology of Pacific Northwest forests.* Washington, D.C.: Island Press.

Agee, J. K. 1994. Fire and weather disturbances in terrestrial ecosystems of the eastern Cascades. USDA Forest Service General Technical Report PNW-GTR-320. Pacific Northwest Research Station, Portland, OR.

Agee, J. K. 1996. Fire in restoration of oregon white oak woodlands. In Hardy, C. C. and S. F. Arno (eds.), *The use of fire in forest restoration*, pp. 72–73. USDA Forest Service General Technical Report INT-GTR-341. Intermountain Research Station, Ogden, UT.

Agee, J. K. and H. H. Biswell. 1969. Seedling survival in a giant sequoia forest. *California Agriculture* 23:18–19.

Agenbroad, L. D. 1988. Clovis people: The human factor in the Pleistocene megafauna extinction equation. In R. C. Carlisle (ed.), *Americans before Columbus: Ice-age origins*, pp. 63–74. Ethnology Monographs 12. Department of Anthropology, University of Pittsburgh, PA.

Ager, T. A. 1983. Holocene vegetational history of Alaska. In Wright, H. E., Jr. (ed.), *Late-Quaternary environments of the United States, Vol. 2, The Holocene*, pp. 128–141. Minneapolis: University of Minnesota Press.

Ahlgren, C. E. 1974. Effects of fires on temperate forests: North central United States. In Kozlowski, T. T. and C. E. Ahlgren (eds.), *Fire and ecosystems*, pp. 195–223. New York: Academic Press.

Ahlgren, C. E. 1976. Regeneration of red pine and white pine following wild-fire and logging in northeastern Minnesota. *Journal of Forestry* 74(3):135–140.

Ahlgren, C. E. and I. F. Ahlgren. 1983. The human impact on northern forest ecosystems. In Flader, S. L. (ed.), *The Great Lakes forest: An environmental and social history*, pp. 33–51. Minneapolis: University of Minnesota Press.

Ahlgren, C. E. and I. F. Ahlgren. 1984. *Lob trees in the wilderness*. Minneapolis: University of Minnesota Press.

Aikens, C. M. 1983. Environmental archaeology in the western United States. In Wright, H. E., Jr. (ed.), *Late-Quaternary environments of the United States, Vol. 2, The Holocene*, pp. 239–251. Minneapolis: University of Minnesota Press.

Alexander, M. E. 1980. Forest fire history research in Ontario: A problem analysis. Proceedings, Fire History Workshop. USDA Forest Service General Technical Report RM-81. Rocky Mountain Forest and Range Experiment Station, Fort Collins, CO, pp. 96–109.

Allen, C. D., J. L. Betancourt, and T. W. Swetnam. 1997. LUHNA pilot project—southwestern United States. Land Use History of North America (LUHNA). Biological Resources Division, U.S. Geological Survey.

American Chestnut Foundation. 1998. *The history of the American chestnut tree.* Bennington, VT.

American Forestry Association. 1978. National register of big trees. *American Forests* 84(4):18–47.

Amman, G. D. 1977. The role of the mountain pine beetle in lodgepole pine ecosystems: impact and succession. In Mattson, W. J. (ed.), *The role of arthropods in forest ecosystems*, pp. 3–18. New York: Springer.

Amman, G. D. and K. C. Ryan. 1991. Insect infestation of fire-injured trees in the Greater Yellowstone Area. USDA Forest Service Research Note INT-398. Intermountain Research Station, Ogden, UT.

Amman, G. D., M. D. McGregor, and R. E. Dolph, Jr. 1989. Mountain pine beetle. USDA Forest Service Forest Insect and Disease Leaflet 2. Washington, D.C.

Anderson, D. G. and G. T. Hanson. 1988. Early archaic settlement in the Southeastern United States: A case study from the Savannah River Valley. *American Antiquity* 53(2):262–286.

Anderson, M. K. 1993a. The experimental approach to assessment of the potential ecological effects of horticultural practices by indigenous peoples on California wildlands. Ph.D. Dissertation, Department of Environmental Science, Policy, and Management, University of California, Berkeley.

Anderson, M. K. 1993b. Native Californians as ancient and contemporary cultivators. In Blackburn, T. C. and M. K. Anderson (eds.), *Before the wilderness: Environmental management by native Californians*, pp. 151–174. Menlo Park, CA: Ballena Press.

Anderson, R. C. 1990. The historic role of fire in the North American grassland. In Collins, S. L. and L. L. Wallace (eds.), *Fire in North American tallgrass prairies*, pp. 8–18. Norman, OK: University of Oklahoma Press.

Anderson, R. S. 1989. Development of southwestern ponderosa pine forests: What do we really know? In Tecle, A., W. W. Covington, and R. H. Hamre (eds.), *Multiresource management of ponderosa pine forests*. Proceedings of a symposium, pp. 15–22. Northern Arizona University, Flagstaff, November 14–16. USDA Forest Service General Techical Report RM-185.

Anderson, R. S. 1990. Holocene forest development and paleoclimates within the central Sierra Nevada, California. *Journal of Ecology* 78:470–489.

Anderson, R. S. 1994. Paleohistory of a giant sequoia grove: The record from Log Meadow, Sequoia National Park. In Aune, P. S. (tech. coord.), *Proceedings of the Symposium on Giant Sequoias: Their Place in the Ecosystem and Society*, pp. 49–55. USDA Forest Service General Technical Report PSW-GTR-151. Pacific Southwest Research Station, Albany, CA.

Andrews, C. C. 1857. *Minnesota and Dacotah: In letter descriptive of a tour through the Northwest, in the autumn of 1856. With information relative to public lands, and a table of statistics.* Washington, DC: R. Farnham.

Angle, P. M. 1968. *Prairie state: Impressions of Illinois, 1673–1967, by travelers and other observers.* Chicago: University of Chicago Press.

Anonymous. 1845. The exploring expeditions, Part II, by Charles Wilkes, U. S. N. *Southern Quarterly Review* XVI:265–298.

Anonymous. 1906. The Voyage of Martin Pring, 1603. In Jameson, J. Franklin (ed.), *Original narratives of early American history*, pp. 341–352. New York: C. Scribner's Sons.

Anonymous. 1994a. Dwarf mammoths. *Discover* 15(1):54.

Anonymous. 1994b. Mining the Rockies in 6000 B.C. *Discover* 15(3):18.

Anonymous. 1995. Breakthroughs: Paleontology. *Discover* 16(4):56.

Anonymous. 1997a. Cahokia Mounds. Illinois State Preservation Agency. Springfield, IL.

Anonymous. 1997b. Bennington Battlefield. Office of Parks, Recreation, and Historic Preservation—Saratoga-Capital District Region, Hoosick Falls, NY.

Anonymous. 1998. Cedar trees. U'mista Cultural Centre. Alert Bay, BC, Canada.

Apfelbaum, S. I. and A. W. Haney. 1990. Management of degraded oak savanna remnants in the Upper Midwest: Preliminary results from three years of study. In Hughes, H. G. and T. M. Bonnicksen (eds.), *Proceedings of the 1st Annual Conference of the Society for Ecological Restoration*, pp. 280–291. Madison, WI: Society for Ecological Restoration.

Armstrong, A. N. 1857. Oregon. Chicago: Charles Scott (republished by Ye Galleon Press, Fairfield, WA, (1969).

Arno, M. K. 1996. Reestablishing fire-adapted communities to riparian forests in the pondersoa pine zone. In Hardy, C. C. and S. F. Arno (eds.), *The use of fire in forest restoration*, pp. 42–43. USDA Forest Service General Technical Report INT-GTR-341. Intermountain Research Station, Ogden, UT.

Arno, S. F. 1973. Discovering Sierra trees. Yosemite Association and Sequoia Natural History Association. Yosemite National Park, CA.

Arno, S. F. 1976. The historical role of fire on the Bitterroot National Forest. USDA Forest Service Research Paper INT-187. Intermountain Forest and Range Experiment Station, Ogden, UT.

Arno, S. F. 1980. Forest fire history in the northern Rockies. *Journal of Forestry* 78(8):460–465.

Arno, S. F. 1985. Ecological effects and management implications of Indian fires. In Lotan, J. E., B. M Kilgore, W. C. Fischer, and R. W. Mutch (tech. coord.), *Proceedings of the Symposium and Workshop on Wilderness Fire*, pp. 81–86. USDA Forest Service General Technical Repport INT-182. Intermountain Forest and Range Experiment Station, Odgen, UT.

Arno, S. F. and T. D. Petersen. 1983. Variation in estimates of fire intervals: A closer look at fire history on the Bitterroot National Forest. USDA Forest Service Research Paper INT-301. Intermountain Forest and Range Experiment Station, Odgen, UT.

Arno, S. F. and W. C. Fishcher. 1995. *Larix occidentalis*—fire ecology and fire managment. In Schmidt, W. C. and K. J. McDonald (comps.), *Ecology and management of Larix forests: A look ahead*, pp. 130–135. Proceedings of an International Symposium, October 5–9, 1992. Whitefish, Montana. USDA Forest Service General Technical Report INT-GTR-319. Intermountain Research Station, Odgen, UT.

Arno, S. F., E. D. Reinhardt, and J. H. Scott. 1993. Forest structure and landscape patterns in the subalpine lodgepole pine type: A procedure for quantifying past and present conditions. USDA Forest Service General Technical Report INT-294. Intermountain Research Station, Odgen, UT.

Arno, S. F., J. H. Scott, and M. G. Hartwell. 1995. Age-class structure of old growth ponderosa pine/Douglas-fir stands and its relationship to fire history. USDA Forest Service Research Paper INT-RP-481. Intermountain Research Station, Ogden, UT.

Arno, S. F., H. Y. Smith, and M. A. Krebs. 1997. Old growth ponderosa pine and western larch stand structures: Influences of pre-1900 fires and fire exclusion.

USDA Forest Service Research Paper INT-RP-495. Intermountain Research Station, Odgen, UT.

Aschmann, H. 1977. Aboriginal use of fire. In Mooney, H., T. M. Bonnicksen, N. L. Christensen, J. E. Lotan, and W. A. Reiners (eds.) *Fire regimes and ecosystem properties*, pp. 132–141. USDA Forest Service, General Technical Report WO-26.

Atzet, T. 1996. Fire regimes and restoration needs in southwestern Oregon. In Hardy, C. C. and S. F. Arno (eds.), *The use of fire in forest restoration*, pp. 74–76. USDA Forest Service General Technical Report INT-GTR-341. Intermountain Research Station, Odgen, UT.

Austin, D. D. and M. L. Perry. 1979. Birds in six communities within a lodgepole pine forest. *Journal of Forestry* 77(9):584–586.

Axelrod, D. I. 1977. Outline history of California vegetation. In Barbour, M. G. and J. Major (eds.), *Terrestrial vegetation of California*, pp. 139–193. New York: Wiley.

Bailey, T. A. 1961. *The American pageant*. Boston: D. C. Heath.

Baisan, C. H. and T. W. Swetnam. 1997. Interactions of fire regimes and land use in the central Rio Grande Valley. USDA Forest Service General Technical Report RM-RP-330. Rocky Mountain Forest and Range Experiment Station. Fort Collins, CO.

Bakeless, J. 1961. *The eyes of discovery: America as seen by the first explorers*. New York: Dover.

Baker, E. P. 1903. Transactions of the Illinois State Historical Society, Springfield, IL. pp. 155–159.

Baker, J. B. and O. G. Langdon. 1990. *Pinus taeda* L. Loblolly pine. In Burns, R. M. and B. H. Honkala (tech. coord.), *Silvics of North America, Vol. 1, Conifers*, pp. 497–512. Agricultural Handbook No. 654. Washington, DC: USDA Forest Service.

Baker, R. G. 1983. Holocene vegetational history of the Western United States. In Wright, H. E., Jr. (ed.), *Late-Quaternary environments of the United States. Vol. 2, The Holocene*, pp. 109–127. Minneapolis: University of Minnesota Press.

Baker, V. R. 1983. Late-Plestocene fluvial systems. In Porter, S. C. (ed.), *Late-Quaternary environments of the United States. Vol. 1, The Pleistocene*, pp. 115–129. Minneapolis: University of Minnesota Press.

Baldwin, E. 1877. *History of La Salle Country, Illinois*. Chicago: Rand, McNally and Company.

Balls, E. K. 1962. *Early uses of California plants*. Berkeley, CA: University of California Press.

Balnchon, P. and J. Shaw. 1995. Reef drowning during the last degalciation: Evidence for catastrophic sea-level rise and ice-sheet collapse. *Geology* 23:4–8.

Balter, M. 1995. Did *Homo erectus* tame fire first? *Science* 268:1570.

Barbour, M. G. 1995. California upland forests and woodlands. In Barbour, M. G. and W. D. Billings (eds.), *North American terrestrial vegetation*, pp. 131–164. New York, Cambridge University Press.

Barden L. S. and F. W. Woods. 1973. Characteristics of lightning fires in southern Appalachian forests. Proceedings, Annual Tall Timbers Fire Ecology Conference, Number 13, pp. 345–361. Tall Timbers Research Station, Tallahassee, FL.

Barnes, B. V., D. R. Zak, S. R. Denton, and S. H. Spurr. 1998. *Forest ecology.* New York: Wiley.

Barnosky, A. D. 1989. The late Pleistocene event as a paradigm for widespread mammal extinction. In Donovan, S. K. (ed.), *Mass extinctions: Processes and evidence*, pp. 235–254. New York: Columbia University Press.

Barnosky, C. W. 1984. Late Pleistocene and early Holocene environmental history of southwestern Washington State, U.S.A. *Canadian Journal of Earth Sciences* 21:619–629.

Barrett, S. A. and E. W. Gifford. 1933. Miwok material culture. *Bulletin of the Public Museum of the City of Milwaukee* 2(4):117–376.

Barrett, S. W. 1980a. Indian fires in the pre-settlement forests of western Montana. Proceedings, Fire History Workshop, pp. 34–41. USDA Forest Service General Technical Report RM-81. Rocky Mountain Forest and Range Experiment Station, Fort Collins, CO.

Barrett, S. W. 1980b. Indians & fire. *Western Wildlands* 6(3):17–21.

Barrett, S. W., and S. F. Arno. 1982. Indian fires as an ecological influence in the northern Rockies. *Journal of Forestry* 80:647–651.

Barrett, S. W. and S. F. Arno. 1990. Fire history of the Lamar River drainage, Yellowstone National Park. In Boyce, M. S. and G. E. Plumb (eds.), *14th Annual Report 1990*, pp. 131–133. University of Wyoming National Park Service Research Center, Laramie, WY.

Barrett, S. W. and S. F. Arno. 1995. Three contrasting fire regimes in Yellowstone National Park. In Brown, J. K., R. W. Mutch, C. W. Spoon and R. H. Wakimoto (tech. coord.), *Proceedings: Symposium on Fire in Wilderness and Park Management*, pp. 157–158. USDA Forest Service General Technical Report INT-GTR-320. Intermountain Research Station, Odgen, UT.

Barrett, S. W., S. F. Arno, and J. P. Menakis. 1997. Fire exposides in the Inland Northwest (1540–1940) based on fire history data. USDA Forest Service General Technical Report INT-GTR-370. Intermountain Forest and Range Experiment Station, Odgen, UT.

Barry, R. G. 1983. Late-Pleistocene climatology. In Porter, S. C. (ed.), *Late-Quaternary environments of the United States. Vol. 1, The Pleistocene*, pp. 390–407. Minneapolis: University of Minnesota Press.

Bartolome, J. W. and R. B. Standiford. 1992. Ecology and management of Californian oak woodlands. In Ffolliott, P. F., G. J. Gottfried, D. A. Bennett, V.

M. Hernandez C., A. Ortega-Rubio, and R. H. Hamre (tech. coord.), *Ecology and management of oak and associated woodlands.* pp. 115–118. USDA Forest Service General Technical Report RM-218. Rocky Mountain Forest and Range Experiment Station, Fort Collins, CO.

Bartos, D. L. and W. F. Mueggler. 1979. Influence of fire on vegetation production in the aspen ecosystem in western Wyoming. In Boyce, M. S. and L. D. Hayden-Wing (eds.), *North American elk, ecology, behavior and management,* pp. 75–78. Laramie: University of Wyoming Press.

Bartram, W. 1973. *Travel through North and South Carolina, Georgia, and east and west Florida.* Savannah, GA: Bee Hive Press.

Basehart, H. W. 1974. Apache Indians XII. Mescalero Apache subsistence patterns and sociopolitical organization. New York: Garland.

Basile, J. V. 1979. Elk-aspen relationships on a prescribed burn. USDA Forest Service Research Note INT-271. Intermountain Forest and Range Experiment Station, Ogden, UT.

Bassett, J. S. (ed.). 1970. *The writings of Colonel William Byrd (1728).* New York: Burt Franklin.

Baxley, H. W. 1865. *What I saw on the west coast of South and North America.* New York: Appleton.

Bean, L. J. and H. W. Lawton. 1993. Some explanations for the rise of cultural complexity in native California with comments on proto-agriculture and agriculture. In Blackburn, T. C. and M. K. Anderson (ed.), *Before wilderness: Environmental management by native californians,* pp. 27–54. Menlo Park, CA: Ballena Press.

Bean, L. J. and K. S. Saubel. 1972. *Temalpakh (from the earth); Cahuilla Indian knowledge and usage of plants.* Banning, CA: Malki Museum Press.

Beaudoin, A. B., D. S. Lemmen, and R. E. Vance. 1997. Paleoenvironmental records of postglacial climate change in the prairie ecozone. Plenary presentation at the 3rd National Ecological Monitoring and Assessment Network Meeting, Environment Canada. Saskatoon, Saskatchewan. January 22.

Beaudoin, C. and H. Bizier. 1997. An adventure in New France: Pierre Boucher, Interpreter and Explorer. Canadian Museum of Civilization, Hull, Quebec.

Beck, D. E. 1990. *Abies fraseri* (Pursh) Poir., Fraser fir. In Burns, R. M. and B. H. Honkala (tech. coord.), *Silvics of North America, Vol. 1. Conifers,* pp. 47–51. Agricultural Handbook No. 654. Washington, DC: USDA Forest Service.

Beckwith, Lieutenant E. G. 1854. *Report of explorations for a route for the Pacific Railroad, on the line of the forty-first parallel of north latitude.* United States War Department. Washington, DC: Nicholson.

Beetham, N. M. 1961. The ecology of the seedling stage of *Sequoia gigantea.* Ph.D. Dissertation, Duke University, Durham, North Carolina.

Belanger, R. P., R. L. Hedden, and P. L. Lorio, Jr. 1993. Management strategies to reduce losses from the southern pine beetle. *Southern Journal of Applied Forestry* 17(3):150–154.

Belcher, Sir Edward. 1843. *Narrative of a voyage round the world, performed in her Majesty's ship Sulphur, during the years 1836–1842,* Vol. 1. London: Henry Colburn.

Bell, W. A. 1870. *New tracks in North America,* 2nd ed., Vols. 1 and 2. London: Chapman & Hall.

Belton, R. J. 1997. *Important moments in Canadian history.* Kelowna, BC, Canada: Department of Fine Arts, Okanagan University College.

Belyea, B. 1994. *Columbia Journals, David Thompson.* Montreal and Kingston: McGill-Queen's University Press.

Bennetts, R. E., G. C. White, F. G. Hawksworth, and S. E. Severs. 1996. The influence of dwarf mistletoe on bird communities in Colorado ponderosa pine forests. *Ecological Applications* 6(3):899–909.

Benson, A. B. 1964. *The America of 1750, Peter Kalm's travels in North America,* Vols. 1 and 2. Mineola, New York: Dover.

Berger A. and M. F. Loutre. 1991. Insolation values for the climate of the last 10 million years. *Quaternary Science Reviews* 10:297–317.

Berry, C. F. and M. S. Berry. 1986. Chronological and conceptual models of the Southwest Archaic. In Condie, C. J and D. D. Fowler (eds.), *Anthropology of the desert west: Essays in honor of Jesse D. Jennings,* pp. 253–327. Salt Lake City: University of Utah Press.

Berryman, A. A. 1982. Mountain pine beetle outbreaks in Rocky Mountain lodge-pole pine forests. *Journal of Forestry* 80(7):410–413, 419.

Betancourt, J. L. 1984. Late Quaternary plant zonation and climate in southeastern Utah. *Great Basin Naturalist* 44(1):1–35.

Betancourt, J. L., E. A. Pierson, K. A. Rylander, J. A. Fairchild-Parks, and J. S. Dean. 1993. Influence of history and climate on New Mexico piñon-juniper woodlands. In Aldon, E. F. and D. W. Shaw (tech. coord.), *Managing piñon-juniper ecosystems for sustainability and social needs,* pp. 42–62. USDA Forest Service General Technical Report RM-236. Rocky Mountain Forest and Range Experiment Station, Fort Collins, CO.

Bevins, C. D. and R. J. Barney. 1980. Lightning fire densities and their management implications on northern region national forests. In *Proceedings, Sixth Conference on Fire and Forest Meteorology,* pp. 127–131. Seattle, Washington, April 22–24. Washington, DC: Society of American Foresters.

Bierhorst, J. 1994. *The way of the Earth: Native America and the environment.* New York: William Morrow.

Bigelow, T. 1876. *Journal of a tour to Niagara Falls in the year 1805.* Boston: J. Wilson.

Biswell, H. H. 1956. Ecology of California grasslands. *Journal of Range Management* 9(1):19–24.

Biswell, H. H. 1961a. The big trees and fire. *National Parks Magazine* 35:11–14.

Biswell, H. H. 1961b. Manipulation of chamise brush for deer range improvement. *California Fish and Game* 47(2):125–144.

Biswell, H. H. 1963. Research in wildland fire ecology in California. Proceedings, Second Annual Tall Timbers Fire Ecology Conference, pp. 63–97. Tall Timbers Research Station. Tallahassee, FL.

Biswell, H. H. 1967. The use of fire in wildland management in California. In Ciriacy-Wantrup, S. V. and J. J. Parsons (eds.), *Natural resources: Quality and quantity*, pp. 71–86. Berkeley: University of California Press.

Biswell, H. H. 1970. Ponderosa pine management with fire as a tool. Proceedings, The Role of Fire in the Intermountain West, October 27–29. pp. 130–136. Intermountain Fire Research Council. School of Forestry, University of Montana, Missoula.

Biswell, H. H. 1972. Fire ecology in ponderosa pine-grassland, Proceedings, Annual Tall Timbers Fire Ecology Conference, Number 12, pp. 69–96. Tall Timbers Research Station. Tallahassee, FL.

Biswell, H. H. 1974. Effects of fire on chaparral. In Kozlowski, T. T. and C. E. Ahlgren (eds.), *Fire and ecosystems*, pp. 321–364. New York: Academic Press.

Biswell, H. H. 1976. A management plan for restoring fire in chaparral at the Pinnacles National Monument. Pinnacles National Monument Order No. PX845060378.

Biswell, H. H., H. Buchanan, and R. P. Gibbens. 1966a. Ecology of the vegetation of a second-growth *Sequoia* forest. *Ecology* 47(4):630–634.

Biswell, H. H., R. P. Gibbens, and H. Buchanan. 1966b. Litter production by bigtrees and associated species. *California Agriculture* 20(9):5–7.

Biswell, H. H., H. R. Kallander, R. Komarek, R. J. Vogl, and H. Weaver. 1973. Ponderosa fire management. Miscellaneous Publication No. 2. Tall Timbers Research Station. Tallahassee, FL.

Black, M. J. 1980. Algonquin ethnobotany: An interpretation of aboriginal adaptation in south western Quebec. National Museums of Canada. Ottawa. Mercury Series Number 65.

Blackwelder, B. W., O. H. Pilkey, and J. D. Howard. 1979. Late Wisconsin sea levels on the Southeast U. S. Atlantic shelf based on inplace shoreline indicators. *Science* 204:618–620.

Blair, E. (ed.). 1910. *Indian tribes of the Upper Mississippi Valley*, Vol. 1, pp. 308–347. Cleveland: A. H. Clark.

Blaisdell, R. S., J. Wooten, and R. K. Godfrey. 1973. The role of magnolia and beech in forest processes in the Tallahassee, Florida, Thomasville, Georgia area. Proceedings, Annual Tall Timbers Fire Ecology Conference, Number 13, pp. 363–397. Tall Timbers Research Station. Tallahassee, FL.

Bliss, L. C. 1995. Arctic tundra and polar desert biome. In Barbour, M. G. and

W. D. Billings (eds.), *North American terrestrial vegetation*. New York: Cambridge University Press.

Bloom, A. L. 1983. Sea level and coastal morphology of the United States through the late Wisconsin glacial maximum. In Porter, S. C. (ed.), *Late-Quaternary Environments of the United States, Vol. 1, The Pleistocene*, pp. 215–229. Minneapolis: University of Minnesota Press.

Blum, B. M. 1990. *Picea rubens* Sarg., Red Spruce. In Burns, R. M. and B. H. Honkala (tech. coord.), *Silvics of North America, Vol. 1, Conifers*, pp. 250–259. Agricultural Handbook No. 654. Washington, DC: USDA Forest Service.

Bonnichsen, R., G. L. Jacobson, Jr., R. B. Davis, and H. W. Borns, Jr. 1985. The environmental setting for human colonization of northern New England and adjacent Canada in late Pleistocene time. In Borns, H. W., Jr., P. LaSalle, and W. B. Thompson (eds.), *Late Pleistocene history of northeastern New England and adjacent Quebec*, pp. 151–159. Geological Society of America Special Paper 197.

Bonnicksen, T. M. 1974. The last patch of Eden. *Sierra Club Bulletin* 59(3):26–29.

Bonnicksen, T. M. 1975. Spatial pattern and succession within a mixed conifer–giant sequoia forest ecosystem. M.S. Thesis, University of California, Berkeley.

Bonnicksen, T. M. 1980. Computer simulation of the cumulative effects of brushland fire-management policies. *Environmental Management* 4(1):35–47.

Bonnicksen, T. M. and R. G. Lee. 1979. Persistence of the fire exclusion policy in southern California: A biosocial interpretation. *Journal of Environmental Management* 8(3):277–293.

Bonnicksen, T. M. and E. C. Stone. 1978. An analysis of vegetation management to restore the structure and function of presettlement giant sequoia-mixed conifer forest mosaics. Report to the U. S. National Park Service Western Regional Office. San Francisco.

Bonnicksen, T. M. and E. C. Stone. 1981. The giant sequoia–mixed conifer forest community characterized through pattern analysis as a mosaic of aggregations. *Forest Ecology and Management* 3(4):307–328.

Bonnicksen, T. M. and E. C. Stone. 1982a. Reconstruction of a presettlement giant sequoia–mixed conifer forest community using the aggregation approach. *Ecology* 63(4):1134–1148.

Bonnicksen, T. M. and E. C. Stone. 1982b. Managing vegetation within U. S. National Parks: A policy analysis. *Environmental Management* 6(2):101–102, 109–122.

Bonnicksen, T. M., M. K. Anderson, H. T. Lewis, C. E. Kay, and R. Knudson. 1999. American Indian influences on the development of forest ecosystems. In Johnson, N. C., A. J. Malk, W. T. Sexton, and R. Szaro (eds.), *Ecological stewardship: A common reference for ecosystem management.* Oxford: Elsevier Science.

Booth, D. E. 1991. Estimating prelogging old-growth in the Pacific Northwest. *Journal of Forestry* 89(10):25–29.

Borremans, N. 1990. The Paleoindian historical context. Florida Division of Historical Resources, Tallahassee, FL.

Bourdo, E. A., Jr. 1983. The forest the settlers saw. In Flader, S. L. (ed.), *The Great Lakes forest: An environmental and social history*, pp. 3–16. Minneapolis: University of Minnesota Press.

Bourke, Capt. J. G. 1971. *On the border with Crook.* Lincoln: University of Nebraska Press.

Boyd, R. 1986. Strategies of Indian burning in the Willamette Valley. *Canadian Journal of Anthropology* 5:65–86.

Boyer, W. D. 1990. *Pinus palustris* Mill. Longleaf pine. In Burns, R. M. and B. H. Honkala (tech. coord.), *Silvics of North America, Vol. 1, Conifers*, pp. 405–412. Agricultural Handbook No. 654. Washington, DC: USDA Forest Service.

Boyer, W. D. 1993. Regenerating longleaf pine with natural seeding. In Hermann, S. M. (ed.), The Longleaf Pine Ecosystem: Ecology, Restortion and Management. Proceedings of the Tall Timbers Fire Ecology Conference No. 18, pp. 299–309. Tall Timbers Research Station. Tallahassee, FL.

Bradley, A. F., N. V. Noste, and W. C. Fischer. 1992. Fire ecology of the forests and woodlands in Utah. USDA Forest Service General Technical Report INT-287. Intermountain Research Station, Odgen, UT.

Bradley, C. B. 1886. A new study of some problems relating to the giant trees. *Overland Monthly and Out West Magazine* 7(Issue 39):305–316.

Bragg, T. B. and L. C. Hulbert. 1976. Woody plant invasion of unburned Kansas bluestem prairie. *Journal of Range Management* 29(1):19–24.

Bray, E. C. and M. C. Bray (eds.). 1993. *Joseph N. Nicollet on the plains and prairies.* St Paul: Minnesota Historical Society Press.

Bray, J. R. 1960. The composition of savanna vegetation in Wisconsin. *Ecology* 41(4):721–732.

Brewer, W. H. 1949. *Up and down California in 1860–64.* New Haven, CT: Yale University Press.

Bridges, E. L. and S. L. Orzell. 1989. Longleaf pine communities of the west Gulf Coastal Plain. *Natural Areas Journal* 9(4):246–263.

Brockman, C. F. 1986. *Trees of North America.* New York: Golden Press.

Bromley, S. W. 1935. The original forest types of southern New England. *Ecological Monographs* 5(1):61–89.

Brown, A. K. and F. M. Stanger. 1969. Discovery of the redwoods. *Forest History* 13(3):6–11.

Brown, D. A. 1993. Early nineteenth-century grasslands of the midcontinent plains. *Annals of the Association of American Geographers* 83(4):589–612.

Brown, J. K., S. F. Arno, L. S. Bradshaw, and J. P. Menakis. 1995. Comparing the Selway-Bitterroot fire program with presettlement fires. In Brown, J. K., R. W. Mutch, C. W. Spoon, and R. H. Wakimoto (tech. coord.), *Proceedings: Symposium on Fire in Wilderness and Park Management*, pp. 48–54. USDA Forest Service General Technical Report INT-GTR-320. Intermountain Research Station, Odgen, UT.

Brown, Joseph H. 1878. Statement to Hubert Howe Bancroft, 1878. Suzzallo Library, University of Washington, Seattle.

Brown, P. M. and T. W. Swetnam. 1994. A cross-dated history from coast redwood near Redwood National Park, California. *Canadian Journal of Forest Research* 24:21–31.

Brown, P. M., W. D. Shepperd, S. A. Mata, and D. L. McClain. 1998. Longevity of windthrown logs in a subalpine forest of central Colorado. *Canadian Journal of Forest Research* 28(6):932–936.

Bryant, Edwin. 1967. *What I saw in California*. Minneapolis: Ross and Haines.

Buchanan, J. 1824. Sketches of the history, manners and customs of North American Indians. *London Quarterly Review* 61(Dec.):102–03.

Bunnell, F. L., L. L. Kremsater, and R. W. Wells. 1997. Likely consequences of forest management on terrestrial, forest-dwelling vertebrates in Oregon. Report M–7. Centre for Applied Conservation Biology, University of British Columbia, Vancouver.

Bunting, S. C. 1994. Effects of fire on-juniper woodland ecosystems in the Great Basin. In Monsen, S. B. and S. G. Kitchen, *Proceedings—Ecology and Management of Annual Rangelands*, May 18–21, 1992. Boise, ID, pp. 53–55. USDA Forest Service General Technical Report INT-GTR-313. Intermountain Research Station, Odgen, UT.

Burcham, L. T. 1957. *California range land*. Sacramento: California Division of Forestry.

Burnham, C. R. 1988. The restoration of the American chestnut. *American Scientist* 76(5)478–487.

Burns, J. A. 1990. Paleontological perspectives on the ice-free corridor. In Agenbroad, L. D., J. I. Mead, and L. W. Nelson (eds.), *Megafauna and man: Discovery of America's heartland*, pp. 61–66. Scientific Papers, Vol. 1. Hot Springs, SD: The Mammoth Site of Hot Springs.

Burns, R. M. and B. H. Honkala (tech. coord.). 1990. *Silvics of North America, Vol. 1, Conifers*. Agricultural Handbook No. 654. Washington, DC: USDA Forest Service.

Burrage, H. S. (ed.). 1932. *Early English and French voyages. James Rosier, "A True Relation of the Voyage of Captaine George Waymouth."* New York: Charles Scribner's Sons.

Burt, W. H. and R. P. Grossenheider. 1980. *A field guide to the mammals, North America north of Mexico.* Boston, MA: Houghton Mifflin.

Busing, R. T. 1994. Canopy cover and tree regeneration in old-growth cove forests of the Appalachian Mountains. *Vegetatio* 115:19–27.

Byars, C. 1993. Ancient human remains found in Colorado cave. *Houston Chronicle* Sunday, Oct. 3, p. 12A.

Bye, R. A. 1972. Ethnobotany of the Southern Paiute Indians in the 1870's: With a note on the early ethnobtanical contributions of Dr. Edward Palmer. In Fowler, D.D. (ed.), *Great Basin Cultural Ecology: A Symposium.* Desert Research Institute. Social Science Publication No. 8.

Calkins, H. G. 1909. Lincoln National Forest, Gallinas Division, Reconnaissance Report. Alamogordo, NM: USDA Forest Service.

Calloway, C. G. 1991. *Dawnland encounters.* Hanover and London: University Press of New England.

Camp, A., C. Oliver, P. Hessburg, and R. Everett. 1997. Predicting late-succes-sional fire refugia pre-dating European settlement in the Wenatchee Mountains. *Forest Ecology and Management* 95:63–77.

Campbell, A. 1772. *Surveys of the Totton and Crossfield Tract.* New York Department of Public Works Field Book, Albany.

Canby, T. Y. 1979. The search for the first American. *National Geographic* 156(3):330–363.

Canon, S. K., P. J. Urness, and N. V. DeByle. 1987. Habitat selection, foraging behavior, and dietary nutrition of elk in burned aspen forest. *Journal of Range Management* 40(5):433–438.

Caprio, A. C. and T. W. Swetnam. 1995. Historic fire regimes along an elevational gradient on the west slope of the Sierra Nevada, California. In Brown, J. K., R. W. Mutch, C. W. Spoon, and R. H. Wakimoto (tech. coord.), *Proceedings: Symposium on Fire in Wilderness and Park Management*, pp. 173–179. USDA Forest Service General Technical Report INT-GTR-320. Intermountain Research Station, Odgen, UT.

Carey, J. H. 1992a. *Pinus palustris* [longleaf pine]. In Fischer, W. C. (compiler), *The Fire Effects Information System* [data base]. USDA Forest Service, Intermountain Research Station, Intermountain Fire Sciences Laboratory, Missoula, MT.

Carey, J. H. 1992b. *Pinus taeda* [loblolly pine]. In Fischer, W. C. (compiler), *The Fire Effects Information System* [data base]. USDA Forest Service, Intermountain Research Station, Intermountain Fire Sciences Laboratory, Missoula, MT.

Carey, J. H. 1992c. *Quercus velutina* [black oak]. In Fischer, W. C. (compiler), *The Fire Effects Information System* [data base]. USDA Forest Service, Intermountain Research Station, Intermountain Fire Sciences Laboratory, Missoula, MT.

Carey, J. H. 1992d. *Quercus marilandica* [blackjack oak]. In Fischer, W. C. (compiler), *The Fire Effects Information System* [data base]. USDA Forest Service, Intermountain Research Station, Intermountain Fire Sciences Laboratory, Missoula, MT.

Carey, J. H. 1992e. *Quercus stellata* [post oak]. In Fischer, W. C. (compiler), *The Fire Effects Information System* [data base]. USDA Forest Service, Intermountain Research Station, Intermountain Fire Sciences Laboratory, Missoula, MT.

Carey, J. H. 1992f. *Quercus palustris* [pin oak]. In Fischer, W. C. (compiler), *The Fire Effects Information System* [data base]. USDA Forest Service, Intermountain Research Station, Intermountain Fire Sciences Laboratory, Missoula, MT.

Carey, J. H. 1993a. *Pinus strobus* [eastern white pine]. In Fischer, W. C. (compiler), *The Fire Effects Information System* [data base]. USDA Forest Service, Intermountain Research Station, Intermountain Fire Sciences Laboratory, Missoula, MT.

Carey, J. H. 1993b. *Pinus resinosa.* [red pine]. In Fischer, W. C. (compiler), *The Fire Effects Information System* [data base]. USDA Forest Service, Intermountain Research Station, Intermountain Fire Sciences Laboratory, Missoula, MT.

Carey, J. H. 1993c. *Pinus banksiana.* [jack pine]. In Fischer, W. C. (compiler), *The Fire Effects Information System* [data base]. USDA Forest Service, Intermountain Research Station, Intermountain Fire Sciences Laboratory, Missoula, MT.

Carey, J. H. 1993d. *Tsuga canadensis* [eastern hemlock]. In Fischer, W. C. (compiler), *The Fire Effects Information System* [data base]. USDA Forest Service, Intermountain Research Station, Intermountain Fire Sciences Laboratory, Missoula, MT.

Carter, G. F. 1967. *Plant geography and culture history in the American Southwest.* Viking Fund Publications in Anthropology, No. 5. New York: Johnson Reprint Corp.

Cartledge, T. R. and J. G. Propper. 1993. Piñon-juniper ecosystems through time: Information and insights from the past. In Aldon, E. F. and D. W. Shaw (tech. coord.), *Managing piñon-juniper ecosystems for sustainability and social needs*, pp. 63–71. USDA Forest Service General Technical Report RM-236. Rocky Mountain Forest and Range Experiment Station, Fort Collins, CO.

Carvalho, S. N. 1859. *Incidents of travel and adventure in the far West; with Col. Fremont's last expedition across the Rocky Mountains: Including three months' residence in Utah, and a perilous trip across the great American desert to the Pacific.* New York: Derby & Jackson.

Catlin, G. 1875. *Life among the Indians*, London: Bracken Books (1996 edition).

Chadde, S. W. and C. E. Kay. 1991. Tall-willow communities on Yellowstone's northern range: A test of the "natural-regulation" paradigm. In Keiter, R. B.

and M. S. Boyce (eds.), *The greater yellowstone ecosystem: Redefining America's wilderness heritage*, pp. 231–262. New Haven, CT: Yale University Press.

Chaney, R. W. 1965. *Redwoods of the past.* San Francisco: Save-the-Redwoods League.

Chapman, H. H. 1909. A method of studying growth and yield of longleaf pine applied in Tyler County, Texas. *Society of American Foresters Proceedings* 4:207–220.

Chapman, H. H. 1932. Is the longleaf pine a climax? *Ecology* 13:331.

Charlevoix, Pierre de. 1966. *Journal of a voyage to North-America*, Vols. 1 and 2. Ann Arbor, MI: University Microfilms.

Chestnut, V. K. 1902. Plants used by the Indians of Mendocino County, California. *Contributions from the U.S. National Herbarium* 7:295–408.

Chittenden, H. M. and A. T. Richardson (eds). 1969. *Life, letters and travels of Father DeSmet*, Vols. 1–4. New York: Arno Press.

Chou, Y. H., R. A. Minnich, and R. J. Dezzani. 1993. Do fire sizes differ between southern California and Baja California? *Forest Science* 39(4):835–844.

Chrismer, G. M., W. G. Ross, and D. L. Kulhavy. 1995. Forest canopy gap size in red-cockaded woodpecker cavity tree clusters: Implications for management. In Kulhavy, D. L., R. G. Hooper, and R. Costa (eds.), *Red-cockaded woodpecker: Recovery, ecology and management*, pp. 214–218. Nacogdoches, TX: Center for Applied Studies in Forestry. College of Forestry, Stephen F. Austin State University.

Churcher, C. S., P. W. Parmalee, G. L. Bell, and J. P. Lamb. 1989. Caribou from the late Pleistocene of northwestern Alabama. *Canadian Journal of Zoology* 67:1210–1216.

Clar, C. R. 1959. California government and forestry from Spanish days until the creation of the Department of Natural Resources in 1927. California Division of Forestry, Sacramento.

Clark, J. S. and P. D. Royall. 1996. Local and regional sediment charcoal evidence for fire regimes in presettlement north-eastern North America. *Journal of Ecology* 84:365–382.

Clay, R. B. 1996. Essential features of Adena ritual. Glenn A. Black Laboratory of Archaeology, University of Kentucky, Lexington.

Clinton, B. D., L. R. Boring, and W. T. Swank. 1993. Canopy gap characteristics and drought influences in oak forests of the Coweeta Basin. *Ecology* 74:1551–1558.

Clyman, J. 1960. *James Clyman, frontiersman.* Camp, Charles (ed.). Portland, OR: Champoeg Press.

Cogbill, C. V. 1985. Dynamics of the boreal forests of the Laurentian Highlands, Canada. *Canadian Journal of Forest Research* 15:252–261.

Cogbill, C. V. 1996. Black growth and fiddlebutts: The nature of old-growth red spruce. In Davis, M. B. (ed.), *Eastern old-growth forests*, pp. 113–125. Washington, DC: Island Press.

Coladonato, M. 1991. *Fagus grandifolia* [beech]. In Fischer, W. C. (compiler), *The Fire Effects Information System* [data base]. USDA Forest Service, Intermountain Research Station, Intermountain Fire Sciences Laboratory, Missoula, MT.

Coladonato, M. 1992a. *Carya cordiformis* [bitternut hickory]. In Fischer, W. C. (compiler), *The Fire Effects Information System* [data base]. USDA Forest Service, Intermountain Research Station, Intermountain Fire Sciences Laboratory, Missoula, MT.

Coladonato, M. 1992b. *Carya tomentosa* [mockernut hickory]. In Fischer, W. C. (compiler), *The Fire Effects Information System* [data base]. USDA Forest Service, Intermountain Research Station, Intermountain Fire Sciences Laboratory, Missoula, MT.

Cole, D. M. 1993. Problems in lodgepole pine thinnings: Basal live limbs and fill-in regeneration. USDA Forest Service Research Paper INT-466. Intermountain Research Station, Odgen, UT.

Colinvaux, P. A., P. E. De Oliveira, J. E. Moreno, M. C. Miller, and M. B. Bush. 1996. A long pollen record from lowland Amazonia: Forest and cooling in glacial times. *Science* 274(5284):85–88.

Compton, B. D. 1993. Upper north Wakashan and southern Tsimshian ethnobotany: The knowledge and usage of plants. Ph.D. Dissertation, University of British Columbia, Vancouver.

Concise Columbia Encyclopedia. 1995. New York: Columbia University Press.

Cook, J. E., T. L. Sharik, and D. W. Smith. 1998. Oak regeneration in the southern Appalachians: Potential, problems, and possible solutions. *Southern Journal of Applied Forestry* 22(1):11–18.

Cook, S. F. 1960. Colonial expeditions to the interior of California: Central Valley, 1800–1826. *Anthropological Records* 16(6):239–292.

Coolidge, P. T. 1963. *History of the Miane woods*. Bangor, ME: Furbush-Roberts.

Cooper, C. F. 1960. Changes in vegetation, structure and growth of southwestern pine forests since white settlement. *Ecological Monographs* 30:129–164.

Cooper, C. F. 1961. Pattern in ponderosa pine forests. *Ecology* 42:493–499.

Cooper, J. G. 1860. Report upon the botany of the route. In Stevens, Isaac I., *Reports on explorations and surveys to ascertain the most practicable and economical route for a railroad from the Mississippi River to the Pacific Ocean, 1853–1855. Botanical report*, Vol. XII, Book II. House of Representatives, 36th Congress, 1st Session. Executive Document No. 56. Washington, DC: Thomas H. Ford.

Cornett, J. W. 1987. Indians and the desert fan palm. Masterkey. Southwest Museum, Los Angeles.

Coues, E. 1900. *On the trail of a Spanish pioneer: The diary and itinerary of Fran-*

cico Garcés in his travels through Sonora, Arizona, and California, 1775–1776, 2 Vols. New York: Francis P. Harper.

Coues, E. (ed.). 1965. *History of the expedition under the command of Lewis and Clark, in three volumes.* New York: Dover. Reprint of Francis P. Harper, New York, edition of 1893.

Covington, W. W. and M. M. Moore. 1992. Postsettlement changes in natural fire regimes: Implications for restoration of old-growth ponderosa pine forests. In Kaufmann, M. R., W. H. Moir, and R. L. Bassett (tech. coord.), *Old-Growth Forests in the Southwest and Rocky Mountain Regions, Proceedings of a Workshop*, pp. 81–99. USDA Forest Service General Technical Report RM-213. Rocky Mountain Forest and Range Experiment Station, Fort Collins, CO.

Covington, W. W. and M. M. Moore. 1994. Southwestern ponderosa forest structure: Changes since Euro-American settlement. *Journal of Forestry* 92(1):39–47.

Covington, W. W. and S. S. Sackett. 1988. Fire effects on ponderosa pine soils and their management implications. In Krammes, J. S. (tech. coord.), *Effects of fire management of southwestern natural resources*, pp. 105–111. USDA Forest Service General Technical Report RM-191. Rocky Mountain Forest and Range Experiment Station, Fort Collins, CO.

Cowell, C. M. 1995. Presettlement Piedmont forests: Patterns of composition and disturbance in central Georgia. *Annals of the Association of American Geographers* 85(1):65–83.

Coyne, J. H. (ed. and trans.). 1903. Exploration of the Great Lakes 1669–1670 by Dollier de Casson and de Brehant de Galinee. Ontario Historical Society Papers and Records, Vol. 4.

Craig, A. J. 1969. Vegetational history of the Shenandoah Valley, Virginia. In Schumm, S. A. and W. C. Bradley (eds.), *United States Contributions to Quaternary Research*, pp. 283–296. Boulder, CO: Geological Society of America, Special Paper Number 123.

Crane, M. F. 1989. *Polystichum munitum* [swordfern]. In Fischer, W. C. (compiler), *The Fire Effects Information System* [data base]. USDA Forest Service, Intermountain Research Station, Intermountain Fire Sciences Laboratory, Missoula, MT.

Crane, M. F. 1990. *Rhododendron macrophyllum* [Pacific rhododendron]. In Fischer, W. C. (compiler), *The Fire Effects Information System* [data base]. USDA Forest Service, Intermountain Research Station, Intermountain Fire Sciences Laboratory, Missoula, MT.

Creager, J. S. and D. A. McManus. 1967. Geology of the floor of Bering and Chukchi Seas—American studies. In Hopkins, D. M. (ed.), *The Bering Land Bridge*, pp. 7–31. Stanford, CA: Stanford University Press.

Cremony, J. C. 1868. *Life among the Apaches.* San Francisco: A. Roman (reprinted 1981 by Time-Life Books, Inc.).

Cronise, T. Fey. 1868. *The natural wealth of California.* San Francisco: H. H. Hancroft.

Cronon, W. 1983. *Changes in the land: Indians, colonists, and the ecology of New England.* New York: Hill and Wang.

Cunningham, J. B., R. P. Balda, and W. S. Gaud. 1980. Selection and use of snags by secondary cavity-nesting birds of the ponderosa pine forest. USDA Forest Service Research Paper RM-222. Rocky Mountain Forest and Range Experiment Station, Fort Collins, CO.

Curtis, J. T. 1959. *The vegetation of Wisconsin.* Madison: University of Wisconsin Press.

Cutter, B. E. and R. P. Guyette. 1994. Fire frequency on an oak-hickory ridgetop in the Missouri Ozarks. *American Midland Naturalist* 132:393–398.

Cwynar, L. C. 1987. Fire and the forest history of the North Cascade Range. *Ecology* 68(4):791–802.

Dahl, E. 1946. On different types of unglaciated areas during the ice ages and their significance to phytogeography. *New Phytologist* 45:225–242.

Dahms, C. W. and B. W. Geils (tech. eds.). 1997. An assessment of forest ecosystem health in the Southwest. USDA Forest Service General Technical Report RM-GTR–295. Rocky Mountain Forest and Range Experiment Station, Fort Collins, CO.

Dale, V. H., M. Hemstrom, and J. Franklin. 1985. Modeling the long-term effects of disturbances on forest succession, Olympic Peninsula, Washington. *Canadian Journal of Forest Research* 16:56–67.

Daniel, T. W., J. A. Helms, and F. S. Baker. 1979. *Principles of silviculture.* New York: McGraw-Hill.

Daniels, G. G. (ed.). 1976. *The Spanish West.* New York: Time-Life Books.

Danzer, S. R., C. H. Baisan, and T. W. Swetnam. 1996. The influence of fire and land-use history on stand dynamics in the Huachuca Mountains of southeastern Arizona. In Ffolliott, P. F., L. F. DeBano, M. B. Baker, Jr., G. J. Gottfried, G. Solis-Garza, C. B. Edminster, D. G. Neary, L. S. Allen, and R. H. Hamre (tech. coord.), *Effects of fire on Madrean Province ecosystems,* pp. 265–270. USDA Forest Service General Technical Report RM-GTR–289. Rocky Mountain Forest and Range Experiment Station, Fort Collins, CO.

Dary, D. A. 1975. *The buffalo book.* New York: Avon Books.

Davies, J. (ed.). 1980. *Douglas of the forests: The North American Journals of David Douglas.* Edinburgh: Harris Publishing.

Davies, K. G. (ed.). 1961. Peter Skene Ogden's Snake Country Journals 1826–27. The Hudson's Bay Rec. Soc., London.

Davis, J. T. 1966. Trade routes and economic exchange among the Indians of California. Reports of the University of California Archaeological Survey, No. 54. Berkeley.

Davis, K. M. 1980. Fire history of a western larch/Douglas-fir forest type in north-western Montana. Proceedings, Fire History Workshop. USDA Forest Service General Technical Report RM-81, pp. 69–74. Rocky Mountain Forest and Range Experiment Station, Fort Collins, CO.

Davis, K. M., B. D. Clayton, and W. C. Fischer. 1980. Fire ecology of Lolo National Forest habitat types. USDA Forest Service General Technical Report INT-79. Intermountain Forest and Range Experiment Station, Odgen, UT.

Davis, M. B. 1981. Quaternary history and the stability of forest communities. In West, D. C., H. H. Shugart, and D. B. Botkin (eds.), *Forest succession: Concepts and application*, pp. 132–153. New York: Springer.

Davis, M. B. 1983. Holocene vegetational history of the eastern United States. In Wright, H. E., Jr. (ed.), *Late-Quaternary environments of the United States, Vol. 2, The Holocene*, pp. 166–181. Minneapolis: University of Minnesota Press.

Davis, M. B., T. E. Purshall, and J. B. Ferrari. 1996. Landscape heterogeneity of hemlock-hardwood forest in northern Michigan. In Davis, M. B. (ed.), *Eastern old-growth forests*, pp. 291–304. Washington, DC: Island Press.

Davis, R. B. 1966. Spruce-fir forests of the coast of Maine. *Ecological Monographs* 36(2):79–94.

Day, G. M. 1953. The Indian as an ecological factor in the northeastern forest. *Ecology* 34(2):329–346.

Dayton, W. A. 1931. Important western browse plants. USDA Misc. Paper No. 101. Washington, DC.

Dealy, J. E., J. M. Geist, and R. S. Dirscoll. 1978. Communities of western juniper in the Intermountain Northwest. USDA Forest Service General Technical Report PNW-74, pp. 11–29. Pacific Northwest Research Station. Portland, OR.

Dean, W. G. 1994. The Ontario landscape, circa A.D. 1600. In Rogers, E. S. and D. B. Smith (eds.), *Aboriginal Ontario, Historical Perspectives on the First Nations*, pp. 2–20. Ontario Historical Studies Series for the Government of Ontario. Toronto: Dundurn Press.

DeByle, N. V. 1979. Potential effects of stable versus fluctuating elk populations in the aspen ecosystem. In Boyce, M. S. and L. D. Hayden-Wind (eds.), *North American elk, ecology, behavior and management*, pp. 13–19. Laramie: University of Wyoming.

DeByle, N. V. 1984. Game species. In *Proceedings, Aspen Symposium*, pp. 39–52. May 22–24, Colorado Springs, CO.

DeByle, N. V. 1985a. Wildlife. In DeByle, N. V. and R. P. Winokur (eds.), *Aspen: Ecology and management in the western United States*, pp. 135–152. USDA Forest Service General Technical Report RM-119. Rocky Mountain Forest and Range Experiment Station, Fort Collins, CO.

DeByle, N. V. 1985b. Animal impacts. In DeByle, N. V. and R. P. Winokur (eds.), *Aspen: Ecology and management in the western United States*, pp. 115–123.

USDA Forest Service General Technical Report RM-119. Rocky Mountain Forest and Range Experiment Station, Fort Collins, CO.

Delcourt, H. R. and P. A. Delcourt. 1991a. *Quarternary ecology: A paleoecological perspective*. New York: Chapman & Hall.

Delcourt, H. R. and P. A. Delcourt. 1991b. Late-quaternary history of the interior highlands of Missouri, Arkansas, and Oklahoma. In Henderson, D. and L. D. Hedrick (eds.), *Restoration of old growth forests in the interior highlands of Arkansas and Oklahoma*, pp. 15–30. Proceedings of the Conference. Morrilton, AR: Winrock International.

Delcourt, P. A. and H. R. Delcourt. 1979. Late Pleistocene and Holocene distributional history of the deciduous forest in the southeastern United States. In Lieth, H. and E. Landolt (eds.), *Contributions to the knowledge of flora and vegetation in the Carolinas, Vol. I*, pp. 79–107. Zurich: Stiftung Rubel.

Delcourt, P. A. and H. R. Delcourt. 1981. Vegetation maps for eastern North America: 40,000 Yr B.P. to the present. In Romans, R. C. (ed.), *Geobotany II*, pp. 123–165. New York: Plenum Press.

Denevan, W. M. 1992. The pristine myth: The landscape of the Americas in 1492. *Annals of the Association of American Geographers* 82(3):369–385.

Densmore, F. 1928. Uses of plants by the Chippewa Indians. *SI-BAE Annual Report* 44:273–379.

Denton, D. 1670. *A brief description of New York* (ed. by W. Gowans, 1845) New York: William Gowans.

Denton, G. H. and T. J. Huges (eds.). 1980. *The last great ice sheets*. New York: Wiley.

de Pourtales. 1832. On the western tour with Washington Irving; the journal and letters of Count de Pourtales, Spaulding, G. F. (ed.). Norman, OK: University of Oklahoma Press.

Despain, D. G. 1990. *Yellowstone vegetation*. Niwot, CO: Roberts Rinehart Publishers.

Despain, D. G. and W. H. Romme. 1991. Ecology and management of high-intensity fires in Yellowstone National Park. Proceedings, 17th Tall Timbers Fire Ecology Conference, *High intensity fire in wildlands: Management challenges and options*, pp. 43–57. Tall Timbers Research Station. Tallahassee, FL.

Deur, D. 1997. Native American gardening on the Oregon coast: the uses of fire. Commentary from the Oregon North Coast, Cannon Beach, OR.

Dickson, I. A. (trans.). 1980. *Father Antoine Silvy, letters from North America*. Belleville, Ontario: Mika Publishing

Dieterich, J. H. 1980. Chimney Spring forest fire history. USDA Forest Service Research Paper RM-220. Rocky Mountain Forest and Range Experiment Station, Fort Collins, CO.

Dietrich, B. 1996. Found skeleton speaks volumes, but some don't like what is says. *Houston Chronicle*, Sunday, September 1, 1996, page 15A.

Dillehay, T. D. 1974. Late Quaternary bison population changes on the Southern Plains. *Plains Anthropologist* 19(65):180–196.

Dillehay, T. D. 1996. Letter to the editor. *Science* 274(5294):1820–1825.

Dixon, R. 1907. The Shasta. *Bulletin of the American Museum of Natural History* 17(5):381–498.

Doane, G. C. 1871. The report of Lieutenant Gustavus C. Doane upon the so-called Yellowstone Expedition of 1870. 41st Congress, 3rd Session, U.S. Senate Executive Document No. 51, Washington, DC.

Dobyns, H. F. 1956. Preconquest Hualapai Plant Food Gathering. Unpublished report to Marks & Marks, Phoenix and Strasser, Spiegelberg, Fried and Frank, Washington, DC.

Dobyns, H. F. 1983. *Their numbers become thinned: Native American population dynamics in eastern North America.* Knoxville, TN: University of Tennessee Press.

Dodge, Colonel Richard I. 1876. *The Black Hills, a minute description of the routes, scenery, soil, climate, timber, gold, geology, zoology, etc.* New York: James Miller.

Dodge, Col. Richard. 1959. *The plains of the great West and their inhabitants.* New York: G. P. Putnam's.

Donaghey, J. L. 1969. The properties of heated soils and their relationship to giant sequoia (*Sequoiadendron giganteum*) germination and seedling growth. M. A. Thesis, San Jose State College, San Jose, CA.

Doolittle, W. E. 1992. Agriculture in North America on the eve of contact: A reassessment. *Annals of the Association of American Geographers* 82(3):386–401.

Driver, H. E. and W. C. Massey. 1957. *Comparative studies of North American Indians.* Philadelphia: Transactions of the American Philosophical Society, Vol. 47, No. 2.

Du Pratz, L. P. 1774. *The history of Louisiana.* English edition, London: T. Becket. Facsimile reproduction 1975 by LSU Press, Baton Rouge, Louisiana.

Dufresne, F. 1965. *No room for bears.* New York: Holt, Rinehart, and Winston, Inc.

Dunevitz, H. 1993. Classification and GIS mapping of oak savanna natural communities in Minnesota. Proceedings, Midwest Oak Savanna Conference, Chicago.

Dutton, C. E. 1887. *Physical geology of the Grand Canyon District*, pp. 49–166. Washington, DC: United States Geological Survey, 2nd Annual Report.

Dwight, T. 1821. *Travels in new England and New York*, 4 vols. Barbara Miller Solomon (ed.). Cambridge, 1969.

Eaton, C. B. 1956. Jeffrey pine beetle. USDA Forest Pest Leaflet No. 11. Washington, DC.

Eckstein, R. G. 1980. Eastern hemlock (*Tsuga canadensis*). Research Report 104. Wisconsin Department of Natural Resources.

Edwards, F. S. 1847. *A campaign in New Mexico with Colonel Doniphan*. Philadelphia: Care and Hart.

Ehrenberg, H. 1935. *With Milam and Fannin*. Dallas: Tardy Publishing.

Eiseley, L. C. 1954. Man the fire-maker. *Scientific American* 191(3):52–57.

Elliott, T. C. 1910. Peter Skene Ogden, Fur Trader. *Oregon Historical Quarterly* 11:229–278.

Elliott-Fisk, D. L. 1989. The boreal forest. In Barbour, M. G. and W. D. Billings (eds.), *North American terrestrial vegetation*, pp. 33–62. New York: Cambridge University Press.

Engbeck, J. H., Jr. 1973. *The enduring giants*. Berkeley: University of California.

Engstrom, R. T. 1993. Characteristic mammals and birds of longleaf pine forests. In Hermann, S. M. (ed.), *The longleaf pine ecosystem: Ecology, restortion and management*, pp. 127–138. Proceedings of the Tall Timbers Fire Ecology Conference No. 18. Tall Timbers Research Station. Tallahassee, FL.

Erickson, J. 1990. *Ice ages: Past and future*. Blue Ridge Summit, PA: Tab Books.

Everts, T. C. 1871. Thiry-seven days of peril. *Scribner's Monthly* III(1):1–17.

Eyre, F. H. (ed.). 1980. *Forest cover types of the United States and Canada*. Washington, DC: Society of American Foresters.

Fagan, B. M. 1987. *The great journey: The peopling of ancient America*. London: Thames and Hudson.

Fagan, B. M. 1991. *Ancient North America: The archaeology of a continent*. London: Thames and Hudson.

Fahnestock, G. R. and J. K. Agee. 1983. Biomass consumption and smoke production by prehistoric and modern forest fires in western Washington. *Journal of Forestry* 81(10):653–657.

Farquhar, F. P. 1969. *History of the Sierra Nevada*. Berkeley: University of California Press.

Fayel, W. (ed.). 1886. *A narrative of Colonel Robert Campbell's experiences in the Rocky Mountain fur trade from 1825 to 1835*. Typed transcription. Missouri Historical Society, St. Louis. Provided by American Mountain Men.

Fechner, G. H. and J. S. Barrows. 1976. Aspen stands as wildfire fuel breaks. Eisenhower Consortium Bulletin 4. Rocky Mountain Forest and Range Experiment Station, Fort Collins, CO.

Fidler, P. 1793. *Diary of Peter Fidler for the period 1792–1793*. Calgary: Glenbox-Alberta Inst.

Fiedel, S. J. 1987. *Prehistory of the Americas*. New York: Cambridge University Press.

Fiedel, S. J. 1996. Letter to the editor. *Science* 274(5294):1820–1825.

Fischer, W. C. and A. F. Bradley. 1987. Fire ecology of western Montana forest habitat types. USDA Forest Service General Technical Report INT-223. Intermountain Research Station, Odgen, UT.

Fischer, W. C. and B. D. Clayton. 1983. Fire ecology of Montana forest habitat types east of the Continental Divide. USDA Forest Service General Technical Report INT-141. Intermountain Forest and Range Experiment Station, Odgen, UT.

Fisher, D. 1997. Tusk tales. *Discover* 18(2):22.

Fisher, R. F., M. J. Jenkins, and W. F. Fisher. 1986. Fire and the prairie-forest mosaic of Devils Tower National Monument. *American Midland Naturalist* 117(2):250–257.

Flerow, C. C. 1967. On the origin of the mammalian fauna of Canada. In Hopkins, D. M. (ed.), *The Bering land bridge*, pp. 271–280. Stanford, CA: Stanford University Press.

Foote, M. J. 1983. Classification, description, and dynamics of plant communities after fire in the taiga of interior Alaska. USDA Forest Service Research Paper PNW-307. Pacific Northwest Forest and Range Experiment Station, Portland, OR.

Forthingham, E. H. 1924. Some silvicultural aspects of the chestnut blight situation. *Journal of Forestry* XXII(8):861–872.

Foster, D. R. and T. Zebryk. 1993. Long-term vegetation dynamics and disturbance history of a Tsuga-dominated forest in central New England. *Ecology* 74:982–998.

Foster, D. R., T. Zebryk, P. Schoonmaker, and A. Lezberg. 1992. Post-settlement history of human land-use and vegetation dynamics of a hemlock woodlot in central New England. *Journal of Ecology* 80:773–786.

Foster, J. W. 1869. *The Mississippi Valley: Its physcial geography.* Chicago: S. C. Griggs.

Foti, T. L. and S. M. Glenn. 1991. The Ouachita Mountain landscape at the time of settlement. In Henderson D. and L. D. Hedrick (eds.), *Restoration of old growth forests in the interior highlands of Arkansas and Oklahoma—Proceedings of the Conference*, pp. 49–65. Ouachita National Forest and Winrock International Institute for Agricultural Development, Hot Springs, AR.

Fowells, H. A. 1965. *Silvics of forest trees of the United States.* USDA Agriculture Handbook No. 271. Washington, DC: USDA Forest Service.

Fowells, H. A. and G. H. Schubert. 1951. Natural reproduction in certain cutover pine-fir stands of California. *Journal of Forestry* 49(3):192–196.

Frank, R. M. 1990. *Abies balsamea* (L.) Mill., balsam fir. In Burns, R. M. and B. H. Honkala (tech. coord.), *Silvics of North America, Vol. 1, Conifers*, pp. 26–35. Agricultural Handbook No. 654. Washington, DC: USDA Forest Service.

Franklin, J. F. 1995. Pacific Northwest forests. In Barbour, M. G. and W. D. Billings (eds.), *North American terrestrial vegetation*, pp. 103–130. New York: Cambridge University Press.

Franklin, J. F. and C. T. Dyrness. 1988. *Natural vegetation of Oregon and Washington.* Corvallis: Oregon State University Press.

Franklin, J. F. and M. A. Hemstrom. 1981. Aspects of succession in the coniferous forests of the Pacific Northwest. In West, D. C., H. H. Shugart and D. B. Botkin (eds.). *Forest succession,* pp. 212–229. New York: Springer.

Franklin, J. F. and R. H. Waring. 1980. Distinctive features of the northwestern coniferous forest. In Waring, R. H. (ed.). *Forests: Fresh perspectives from ecosystem analysis,* pp. 59–86. Proceedings, 40th Biological Colloquim. Corvallis, OR: Oregon State University Press.

Franklin, T. L. and R. D. Laven. 1991. Fire influences on central Rocky Mountain lodgepole pine stand structure and composition. Proceedings, 17th Tall Timbers Fire Ecology Conference, *High intensity fire in wildlands: Management challenges and options,* pp. 183–196. Tall Timbers Research Station. Tallahassee, FL.

Frelich, L. E. and L. J. Graumlich. 1994. Age-class distribution and spatial patterns in an old-growth hemlock-hardwood forest. *Canadian Journal of Forest Research* 24:1939–1947.

Frelich, L. E. and C. G. Lorimer. 1985. Current and predicted long-term effects of deer browsing in hemlock forests in Michigan, USA. *Biological Conservation* 34:99–120.

Frelich, L. E. and C. G. Lorimer. 1991a. Natural disturbance regimes in hemlock-hardwood forests of the upper Great Lakes region. *Ecological Monographs* 61(2):145–164.

Frelich, L. E. and C. G. Lorimer. 1991b. A simulation of landscape-level stand dynamics in the northern hardwood region. *Journal of Ecology* 79:223–233.

Frelich, L. E. and P. B. Reich. 1995. Spatial patterns and succession in a Minnesota southern-boreal forest. *Ecological Monographs* 65(3):325–346.

Frelich, L. E. and P. B. Reich. 1996. Old growth in the Great Lakes region. In Davis, M. B. (ed.), *Eastern old-growth forests,* pp. 144–160. Washington, DC: Island Press.

Fremont, J. C. 1856. *The life of Colonel John Charles Fremont, and his narrative of explorations and adventures, in Kansas, Nebraska, Oregon and California.* New York: Miller, Orton & Mulligan.

Fremont, J. C. 1966. Report of the exploring expedition to the Rocky Mountains. University Microfilm, University of Michigan, Ann Arbor, MI.

Friis-Christensen, E. and K. Lassen. 1991. Length of the solar cycle: An indicator of solar activity closely associated with climate. *Science* 254(5032):698–700.

Frison, G. C. 1988. Paleoindian subsistence and settlement during post-Clovis times on the Northwestern plains, the adjacent mountain ranges, and intermontane basins. In Carlisle, R. C. (ed.), *Americans before Columbus: Ice-age origins,* pp. 83–106. Selected Papers from the First Columbian Quincentenary Symposium, Smithsonian Institution, September 26, 1987. Ethnology Monographs Number Twelve. Department of Anthropology, University of Pittsburgh.

Frison, G. C. 1991. *Prehistoric hunters of the High Plains.* New York: Academic Press.

Frissell, S. S., Jr. 1973. The importance of fire as a natural ecological factor in Itasca State Park, Minnesota. *Quaternary Research* 3:397–407.

Fritz, L. H. 1935. Diary of the Alarcon Expedition into Texas, 1718–1719. Los Angeles, CA: Quivira Society.

Frost, C. C. 1993. Four centuries of changing landscape patterns in the longleaf pine ecosystem. In Hermann, S. M. (ed.), *The longleaf pine ecosystem: Ecology, restortion and management,* pp. 17–43. Proceedings of the Tall Timbers Fire Ecology Conference No. 18. Tall Timbers Research Station, Tallahassee, FL.

Fry, W. and J. R. White. 1966. *Big trees.* Stanford, CA: Stanford University Press.

Fuquay, D. M. 1980. Lightning that ignites forest fires. *Proceedings, Sixth Conference on Fire and Forest Meteorology,* pp. 109–112. Seattle, Washington. April 22–24. Washington, DC: Society of American Foresters.

Gates, C. M. (ed.). 1941. *Readings in Pacific Northwest history, Washington, 1790–1895.* Seattle, WA: University Bookstore.

Gates, D. M., C. H. D. Clarke, and J. T. Harris. 1983. Wildlife in a changing environment. In Flader, S. L. (ed.), *The Great Lakes forest: An environmental and social history,* pp. 52–80. Minneapolis: University of Minnesota Press.

Gaudreau, D. C. 1988. The distribution of late Quaternary forest regions in the Northeast. In Nicholas, G. P. (ed.), *Holocene human ecology in northeastern North America,* pp. 215–256. New York: Plenum Press.

Gayton, A. H. 1948. Yokuts and Western Mono ethnography. *Anthropological Records* 10(1/2).

Gibbens R P. and H. F. Heady. 1964. The influence of modern man on the vegetation of Yosemite Valley. University of California, Division of Agricultural Sciences.

Gibbens, R. P. and A. M. Schultz 1963. Brush manipulation on a deer winter range. *California Fish and Game* 49(2):95–118.

Gibbs, G. 1860. Journal of the expedition of Colonel Redick M'Kee, United States Indian Agent, through north-western California. Performed in the summer and fall of 1851. In Schoolcraft, H. R., *Archives of aboriginal knowledge, Vol. 3,* pp. 99–177. Philadelphia: J. B. Lippincott.

Gibson, J. L. 1993. *Poverty Point: A terminal archaic culture of the Lower Mississippi Valley,* 2nd ed. Anthropological Study Series. Louisiana Archaeological Survey, Vicksburg.

Gila Reconnaissance Party. 1912. South Diamond Watershed, Gila National Forest, Silver City, NM.

Gilbert, B. 1973. *The trailblazers.* New York: Time-Life Books.

Gilmore, M. R. 1919. Uses of plants by the Indians of the Missouri River Region. Annual Report No. 33. U.S. Bureau of American Ethnology. Washington, DC: U.S. Government Printing Office.

Glassow, M. A., L. Wilcoxon, and J. Erlandson. 1987. Cultural and environmental change during the early period of Santa Barbara Channel prehistory. In Bailey, G. and J. Parkington (eds.), *The archaeology of prehistoric coastlines*, pp. 64–77. New York: Cambridge University Press.

Godman, R. M. and K. Lancaster. 1990. *Tsuga canadensis* (L.) Carr., eastern hemlock. In Burns, R. M. and B. H. Honkala (tech. coord.), *Silvics of North America, Vol. 1, Conifers*, pp. 604–612. Agricultural Handbook No. 654. Washington, DC: USDA Forest Service.

Goetzmann, W. H. 1986. *New lands, new men: America and the second great age of discovery*. New York: Viking Penquin.

Goetzmann, W. H. 1991. *Army exploration in the American West, 1803–1863*. Austin, TX: Texas State Historical Association.

Goetzmann, W. H. 1993. *Exploration and empire*. Austin, TX: Texas State Historical Association.

Gordon, D. T. 1970. Natural regeneration of white and red fir: Influence of several factors. USDA Forest Service Research Paper PSW-58. Pacific Southwest Forest and Range Experiment Station, Berkeley, CA.

Gottfried, G. J. 1992. Ecology and management of the Southwestern pinyon-juniper woodlands. In Ffolliott, P. F., G. J. Gottfried, D. A. Bennett, V. M. Hernandez, A. Ortega-Rubio, and R. H. Hamre (tech. coord.), *Ecology and management of oak and associated woodlands: Perspectives in the southwestern United States and northern Mexico*, pp. 78–86. USDA Forest Service General Technical Report RM-218. Rocky Mountain Forest and Range Experiment Station, Fort Collins, CO.

Goudie, A. 1992. *Environmental change: Contemporary problems in geography*. Oxford: Clarendon Press.

Goudsblom, J. 1992. *Fire and civilization*. London: Penquin Books.

Graham, R. W. 1990. Evolution of new ecosystems at the end of the Pleistocene. In Agenbroad, L. D., J. I Mead, and L. W. Nelson (eds.), *Megafauna & man: Discovery of America's heartland*, Vol. 1, pp. 54–60. Scientific Papers. Hot Springs, SD: The Mammoth Site of Hot Springs.

Graham, R. W. and E. L. Lundelius, Jr. 1996. *FAUNMAP: An electronic database documenting late Quaternary distributions of mammal species*. Springfield: Illinois State Museum.

Graham, R. W., C. V. Haynes, D. L. Johnson, and M Kay. 1981. Kimmswick: A Clovis-mastodon association in eastern Missouri. *Science* 213:1115–1117.

Graham, R. W., E. L. Lundelius, Jr., M. A. Graham, E. K. Schroeder, R. S. Toomey III, E. Anderson, A. D. Barnosky, J. A. Burns, C. S. Churcher, D. K. Grayson, R. D. Guthrie, C. R. Harington, G.T. Jefferson, L. D. Martin,

H. G. McDonald, R. E. Morlan, H. A. Semken, Jr., S. D. Webb, L. Werdelin, and M. C. Wilson. 1996. Spatial response of mammals to late Quaternary environmental fluctuations. *Science* 272:1601–1606.

Graham, S. A. 1941. Climax forests of the upper peninsula of Michigan. *Ecology* 22(4):355–362.

Graham, S. A. and F. B. Knight. 1965. *Principles of forest entomology.* New York: McGraw-Hill.

Grant, M. C. 1993. The trembling giant. *Discover* 14(10):83–88.

Grant, W. L. (ed.). 1952. *Voyages of Samuel de Champlain, 1604–1618.* New York: Barnes & Noble.

Greenlee, J. M. and J. H. Langenheim. 1990. Historic fire regimes and their relation to vegetation patterns in the Monterey Bay area of California. *American Midland Naturalist* 124(2):239–253.

Gregg, J. 1962. *Commerce of the prairies.* (The 1844 Edition Unabridged.) Philadelphia and New York: J. B. Lippincott Co.

Greller, A. M. 1995. Deciduous forest. In Barbour, M. G. and W. D. Billings (eds.), *North American terrestrial vegetation*, pp. 287–316. New York: Cambridge University Press.

Gremillion, K. J. 1996. The paleoethnobotanical record for the Mid-Holocene Southeast. In Sassaman, K. E. and D. G. Anderson (eds.), *Archaeology of the mid-Holocene southeast*, pp. 99–114. Gainesville: Universtiy Press of Florida.

Griffin, J. R. 1977. Oak woodland. In Barbour, M. G. and J. Major (eds.), *Terrestrial vegetation of California*, pp. 383–413. New York: Wiley.

Griffin, J. R. and W. B. Critchfield. 1972. The distribution of forest trees in California. USDA Forest Service Research Paper PSW-82. Pacific Southwest Forest and Range Experiment Station, Berkeley, CA.

Griffin, L. 1993. American chestnut revival. Virginia Cooperative Extension. VPI&SU Extension Forestry Notes.

Griffith, R. S. 1992. *Sequoia sempervirens* [redwood]. In Fischer, W. C. (compiler), *The Fire Effects Information System* [data base]. USDA Forest Service, Intermountain Research Station, Intermountain Fire Sciences Laboratory, Missoula, MT.

Grinnell, G.B. 1972. *The Cheyenne Indians: Their history and ways of life.* Lincoln, NE: University of Nebraska Press.

Grinnell, G. B. 1995. *The fighting Cheyennes.* North Dighton, MA: JG Press.

Gruell, G. E. 1983. Fire and vegetative trends in the northern Rockies: Interpretations from 1871–1982 photographs. USDA Forest Service General Technical Report INT-158. Intermountain Forest and Range Experiment Station, Odgen, UT.

Gruell, G. E. 1985. Indian fires in the interior West: A widespread influence. In

Lotan, J. E., B. M Kilgore, W. C. Fischer, and R. W. Mutch (tech. coord.), *Proceedings of the Symposium and Workshop on Wilderness Fire*, pp. 68–74. USDA Forest Service General Technical Repport INT-182. Intermountain Forest and Range Experiment Station, Odgen, UT.

Gullion, G. W. 1977. Forest manipulation for ruffed grouse. Paper presented at the 42nd North American Wildlife and Natural Resources Conference. Paper No. 1651, Miscellaneous Journal Series. Minnesota Agricultural Experiment Station, St. Paul, MN.

Gunther, E. 1973. *Ethnobotany of western Washington*, rev. ed. Seattle: University of Washington Press.

Gussow, A. 1971. *A sense of place: The artist and the American land*. New York: Seabury Press.

Gustafson, A. M. (ed.). 1966. *John Spring's Arizona*. Tucson: University of Arizona Press.

Guthrie, R. D. 1984. Mosaics, allelochemics and nutrients: An ecological theory of late Pleistocene megafaunal extinctions. In Martin, P. S. and R. G. Klein (eds.), *Quaternary Extinctions: A prehistoric revolution*, pp. 259–298. Tucson: University of Arizona Press.

Guthrie, R. D. 1990. Late Pleistocene faunal revolution—a new perspective on the extinction debate. In Agenbroad, L. D., J. I Mead, and L. W. Nelson (eds.), *Megafauna & man: Discovery of America's heartland*, Vol. 1, Scientific Papers, pp. 42–53. Hot Springs, SD: The Mammoth Site of Hot Springs.

Habeck, J. R. 1976. Forests, fuels and fire in the Selway-Bitterroot Wilderness, Idaho. Proceedings, Tall Timbers Fire Ecology Conference and Fire and Land Management Symposium. Number 14, pp. 305–353. Tall Timbers Research Station. Tallahassee, FL.

Habeck, J. R. and R. W. Mutch. 1973. Fire-dependent forests in the northern Rocky Mountains. *Quaternary Research* 3:408–424.

Habeck, R. J. 1992a. *Pinus ponderosa var. ponderosa* [ponderosa pine]. In Fischer, W. C. (compiler), *The Fire Effects Information System* [data base]. USDA Forest Service, Intermountain Research Station, Intermountain Fire Sciences Laboratory, Missoula, MT.

Habeck, R. J. 1992b. *Pinus jeffreyi* [Jeffrey pine]. In Fischer, W. C. (compiler), *The Fire Effects Information System* [data base]. USDA Forest Service, Intermountain Research Station, Intermountain Fire Sciences Laboratory, Missoula, MT.

Habeck, R. J. 1992c. *Pinus lambertiana* [sugar pine]. In Fischer, W. C. (compiler), *The Fire Effects Information System* [data base]. USDA Forest Service, Intermountain Research Station, Intermountain Fire Sciences Laboratory, Missoula, MT.

Hadley, D. and T. E. Sheridan. 1995. Land use history of the San Rafael Valley, Arizona (1540–1960). USDA Forest Service General Technical Report

RM-GTR-269. Rocky Mountain Forest and Range Experiment Station, Fort Collins, CO.

Hafsten, U. 1961. Pleistocene development of vegetation and climate in the southern High Plains as evidenced by pollen analysis. In Wendorf, F. (ed.), *Paleoecology of the Llano Estacado*, pp. 59–91. Santa Fe: Museum of New Mexico Press.

Haines, A. L. (ed.). 1965. *Journal of a trapper: Russell Osborne*. Lincoln, NE: University of Nebraska Press.

Haines, A. L. (ed.). 1971. *The Snake Country expedition of 1830–1831. John Work's field journal*. Norman, OK: University of Okalahoma Press.

Haines, D. A., V. J. Johnson, and W. A. Main. 1975. Wildfire atlas of the northeastern and north central states. USDA Forest Service General Technical Report NC-16. North Central Forest Experiment Station. St. Paul, MN.

Hall, D. A. 1997. Bering land bridge was open until after 11,000 years ago. *Mammoth Trumpet* 12(4). Center for the Study of the First Americans. Oregon State University, Corvallis.

Hall, D. A. 1998. Great Lakes people lived 2,000 years with glacier. *Mammoth Trumpet* 13(2). Center for the Study of the First Americans. Oregon State University, Corvallis.

Hall, S. A. 1995. Grassland vegetation in the southern Great Plains during the last glacial maximum. *Quaternary Research* 44:237–245.

Hallin, W. E. 1959. The application of unit area control in the management of ponderosa-Jeffrey pine at Blacks Mountain Experimental Forest. USDA Forest Service Technical Bulletin Number 1191. Washington, DC.

Halpern, C. B. and J. F. Franklin. 1990. Physiognomic development of *Pseudotsuga* forests in relation to initial structure and disturbance intensity. *Journal of Vegetation Science* 1:475–482.

Hammett, J. E. 1992. The shapes of adaptation: Historical ecology of anthropogenic landscapes in the Southeastern United States. *Landscape Ecology* 7(2): 121–135.

Hammond, G. P. and A. Rey (trans. & eds.). 1928. *Obregón's history of sixteenth century explorations in western America*. Los Angeles: Wetzel Publishing.

Hammond, G. P. and A. Rey (trans. & eds.). 1966. *The discovery of New Mexico 1580–1594: The explorations of Chamuscado, Espejo, Castaño de Sosa, Morlete, and Leyva de Bonilla and Humaña*. Albuquerque: University of New Mexico Press.

Hanes, T. L. 1971. Succession after fire in the chaparral of southern California. *Ecological Monographs* 41(1):27–52.

Hanes, T. L. and H. W. Jones. 1967. Postfire chaparral succession in southern California. *Ecology* 48(2):259–264.

Hansen, A. J., T. A. Spies, F. J. Swanson, and J. L Ohmann. 1991. Conserving biodiversity in managed forests. *BioScience* 41(6):382–392.

Hansen, H. L., V. Kurmis, and D. D. Ness. 1974. The ecology of upland forest communities and implications for management in Itasca State Park, Minnesota. Technical Bulletin 298, Forestry Series 16. Agricultural Experiment Station, Minneapolis: University of Minnesota, St. Paul.

Harbinger, L. J. 1964. The importance of food plants in the maintenance of Nez Perce cultural identity. Master's Thesis, Anthropology Department, Washington State University, Pullman.

Harcombe, P. A., J. S. Glitzenstein, R. G. Knox, S. L. Orzell, and E. L. Bridges. 1993. Vegetation of the longleaf pine region of the West Gulf Coastal Plain. In Hermann, S. M. (ed.), *The longleaf pine ecosystem: Ecology, restortion and management*, pp. 83–104. Proceedings of the Tall Timbers Fire Ecology Conference No. 18. Tall Timbers Research Station. Tallahassee, FL.

Harlow, W. M., E. S. Harrar, J. W. Hardin, and F. M. White. 1996. *Textbook of dendrology*. New York: McGraw-Hill.

Harington, C. R. 1970. Ice Age mammal research in the Yukon Territory and Alaska. In Smith, R. A. and J. W. Smith (eds.), *Early man and environments in northwest North America*, pp. 35–51. Proceedings of the 2nd Annual Paleo-Environmental Workshop of the University of Calgary Archaeological Association.

Harrington, C. R. 1996. North American short-faced bears. Yukon Beringia Interpretive Centre, Whitehorse, Yokon Territory.

Harrington, H. D. 1967. *Edible native plants of the Rocky Mountains*. Albuquerque: University of New Mexico Press.

Harrington, J. P. 1932. Tobacco among the Karuk Indians of California. *Bureau of American Ethnology Bulletins* 94:1–284.

Harrington, M. G. and F. G. Hawksworth. 1988. Interactions of fire and dwarf mistletoe on mortality of southwestern ponderosa pine. In Krammes, J. S. (tech. coord.), *Effects of fire management of southwestern natural resources*, pp. 234–240. USDA Forest Service General Technical Report RM-191. Rocky Mountain Forest and Range Experiment Station, Fort Collins, CO.

Harrington, M. G. and R. G. Kelsey. 1979. Influence of some environmental factors on initial establishment and growth of ponderosa pine seedlings. USDA Forest Service General Research Paper INT-230. Intermountain Research Station, Odgen, UT.

Harrington, M. G. and S. S. Sackett. 1988. Using fire as a management tool in southwestern ponderosa pine. In Krammes, J. S. (tech. coord.), *Effects of fire management of southwestern natural resources*, pp. 122–133. USDA Forest Service General Technical Report RM-191. Rocky Mountain Forest and Range Experiment Station, Fort Collins, CO.

Harrington, M. G. and S. S. Sackett. 1992. Past and present fire influences on southwestern ponderosa pine old growth. In Kaufmann, M. R., W. H. Moir, and R. L. Bassett (tech. coord.), *Old-growth forests in the Southwest and Rocky*

Mountain regions, proceedings of a workshop, pp. 44–50. USDA Forest Service General Technical Report RM-213. Rocky Mountain Forest and Range Experiment Station, Fort Collins, CO.

Hart, J. 1992. *Montana native plants and early peoples.* Helena: Montana Historical Society Press.

Hart, J. A. 1981. The ethnobotany of the Northern Cheyenne Indians of Montana. *Journal of Ethnopharmacology* 4:1–55.

Hartesveldt, R. J. and H. T. Harvey. 1967. The fire ecology of sequoia regeneration. Proceedings, Tall Timbers Fire Ecology Conference Number 7, pp. 65–77. Tall Timbers Research Station. Tallahassee, FL.

Hartesveldt, R. J., H. T. Harvey, H. S. Shellhammer, and R. E. Stecker. 1975. *The giant sequoia of the Sierra Nevada.* Washington, DC: USDI National Park Service.

Harvey, H. T. 1980a. Vegetational changes. In Harvey, H. T., H. S. Shellhammer, and R. E. Stecker, *Giant sequoia ecology: Fire and reproduction,* pp. 25–40. Scientific Monograph Series No. 12. Washington, DC: USDI National Park Service.

Harvey, H. T. 1980b. Giant sequoia reproduction, survival, and growth. In Harvey, H. T., H. S. Shellhammer, and R. E. Stecker, *Giant sequoia ecology: Fire and reproduction,* pp. 41–68. Scientific Monograph Series No. 12. Washington, DC: USDI National Park Service.

Harvey, H. T. 1986. Evolution and history of giant sequoia. In Weatherspoon, C. P., Y. R. Iwamoto, and D. D. Piirto (tech. coords.), *Proceedings of the Workshop on Management of Giant Sequoia,* pp. 1–3. USDA Forest Service General Technical Report PSW-95. Pacific Southwest Research Station, Albany, CA.

Harvey, H. T. and H. S. Shellhammer. 1991. Survivorship and growth of giant sequoia (*Sequoiadendron giganteum* [Lindl.] Buchh.) seedlings after fire. *Madroño* 38(1):14–20.

Hastings, M. S., S. Barnhart, and J. R. McBride. 1997. Restoration management of northern oak woodlands. In Pillsbury, N. H., J. Verner, and W. D. Tietje (tech. coord.), *Proceedings of a Symposium on Oak Woodlands: Ecology, Management, and Urban Interface Issues,* pp. 275–279. USDA Forest Service General Technical Report PSW-GTR-160. Pacific Southwest Research Station, Albany, CA.

Hawes, A. F. 1923. New England forests in retrospect. *Journal of Forestry.* 21:209–224.

Hayden, F. V. 1872. *Preliminary report of the United States geological survey of Montana and portions of adjacent territories. Fifth Annual Report of Progress.* Washington, DC: U. S. Government Printing Office.

Haynes, G. and D. J. Stanford. 1984. On the possible utilization of Camelops by early man in North America. *Quaternary Research* 22(2):216–230.

Hebda, R. J. and R. W. Mathewes. 1984. Holocene history of cedar and native Indian cultures of the North American Pacific Coast. *Science* 225:711–712.

Heidenreich, C. E. 1971. *Huronia: A history and geography of the Huron Indians, 1600–1650.* Toronto: McClelland and Stewart.

Heidmann, L. J., T. N. Johnsen, Jr., Q. W. Cole, and G. Cullum. 1982. Establishing natural regeneration of ponderosa pine in central Arizona. *Journal of Forestry* 80(2):77–79.

Heinselman, M. L. 1973. Fire in the virgin forests of the Boundary Waters Canoe Area, Minnesota. *Quaternary Research* 3:329–382.

Heinselman, M. L. 1981. Fire and succession in the conifer forests of northern North America. In West, D. C., H. H. Shugart, and D. B. Botkin (eds.), *Forest succession: Concepts and application*, pp. 374–405. New York: Springer.

Heinselman, M. L. 1985. Fire regimes and management options in ecosystems with large high-intensity fires. In Lotan, J. E., B. M Kilgore, W. C. Fischer, and R. W. Mutch (tech. coord.), *Proceedings of the Symposium and Workshop on Wilderness Fire*, pp. 101–109. USDA Forest Service General Technical Report INT-182. Intermountain Forest and Range Experiment Station, Odgen, UT.

Heizer, R. F. (ed.). 1975. *Narrative of the adventures and sufferings of John R. Jewitt while held as a captive of the Nootka Indians of Vancouver Island, 1803 to 1805.* Menlo Park, CA: Ballena Press Publications in Archaeology, Ethnology and History No. 5.

Heizer, R. F. and A. B. Elsasser. 1980. *The natural world of the California Indians.* California Natural History Guides: 46. Berkeley: University of California Press.

Hellson, J. C. 1974. *Ethnobotany of the Blackfoot Indians.* Ottawa: National Museums of Canada. Mercury Series.

Helms, J. A. 1995. The California region. In Barrett, J. W. (ed.), *Regional silviculture of the United States*, 3rd ed., pp. 441–497. New York: Wiley.

Henderson, J. A. 1982. Succession on two habitat types in western Washington. In Means, J. E. (ed.), *Forest succession and stand development research in the Northwest*, pp. 80–86. Forest Research Laboratory. Corvallis: Oregon State University.

Hendricks, J. H. 1968. Control burning for deer management in chaparral in California. Proceedings, Annual Tall Timbers Fire Ecology Conference, Number 8, pp. 219–233. Tall Timbers Research Station. Tallahassee, FL.

Hennepin, Father Louis. 1880. *Description of Louisiana*, pp. 69–258. New York: John G. Shea.

Hennon, P. E. 1995. Are heart rot fungi major factors of disturbance in gap-dynamic forests? *Northwest Science* 69(4):284–293.

Henry, J. D. and J. M. A. Swan. 1974. Reconstructing forest history from live and dead plant material—an approach to the study of forest succession in southwest New Hampshire. *Ecology* 55:772–783.

Henry, J. F. 1984. *Early maritime artists of the Pacific Northwest Coast, 1741–1841.* Seattle: University of Washington Press.

Hermann, R. K and D. P. Lavender. 1990. *Pseudotsuga menziesii* (Mirb.) Franco. Douglas-fir. In Burns, R. M. and B. H. Honkala (tech. coord.), *Silvics of North America, Vol. 1, Conifers,* pp. 527–540. Agricultural Handbook No. 654. Washington, DC: USDA Forest Service.

Herrick, J. W. 1977. Iroquois medical botany. Ph.D. dissertation, State Department of Anthropology, University of New York, Albany.

Hessburg, P. F., R. G. Mitchell, and G. M. Filip. 1994. Historical and current roles of insects and pathogens in eastern Oregon and Washington forested landscapes. USDA Forest Service General Technical Report PNW-GTR-327. Pacific Northwest Research Station. Portland, OR.

Heusser, C. J. 1983. Vegetation history of the Northwestern United States including Alaska. In Porter, S. C. (ed.), *Late-Quaternary environments of the United States, Vol. 1, The Pleistocene,* pp. 239–258. Minneapolis: University of Minnesota Press.

Hibbs, D. E. 1982. Gap dynamics in a hemlock-hardwood forest. *Canadian Journal of Forest Research* 12(3):522–527.

Higgins, K. F. 1986. Interpretation and compendium of historical fire accounts in the Northern Great Plains. Resource Publication #161. U.S. Department of the Interior. Washington, DC.

Hinds, T. E. 1985. Diseases. In DeByle, N. V. and R. P. Winokur (eds.), *Aspen: Ecology and management in the western United States,* pp. 87–106. USDA Forest Service General Technical Report RM-119. Rocky Mountain Forest and Range Experiment Station, Fort Collins, CO.

Hines, Gustavus (Rev.). 1850. *A voyage round the world: With a history of the Oregon Mission.* Buffalo: George H. Derby.

Hoblitt, R. P., C. D. Miller, and W. E. Scott. 1996. Glacier Peak, Washington—Eruptive history. In *Volcanic hazards with regard to siting nuclear-power plants in the Pacific Northwest.* USGS/Cascades Volcano Observatory, Vancouver, WA. Open-File Report 87-297.

Hodge, F. W. and T. H. Lewis (eds.). 1990a. *Spanish explorers in the Southern United States. Complete narrative of Alvar Nunez Cabeza de Vaca (1527–1537),* pp. 3–126. Austin, TX: Texas State Historical Association.

Hodge, F. W. and T. H. Lewis (eds.). 1990b. Spanish explorers in the Southern United States. *The narrative of the expedition of Hernando de Soto, by the gentleman of Elvas (1539–1542),* pp. 133–272. Austin, TX: Texas State Historical Association.

Hodge, F. W. and T. H. Lewis (eds.). 1990c. Spanish explorers in the Southern United States. *The narrative of the expedition of Coronado, by Pedro de Castaneda. (1540–1542),* pp. 2811–387. Austin, TX: Texas State Historical Association.

Hoffmann, R. S. and J. K. Jones. 1970. Influence of late-glacial and post-glacial events on the distribution of recent mammals on the northern Great Plains. In Dort, W., Jr. And J. K. Jones, Jr. (eds.), *Pleistocene and recent environments of the central great plains*, pp. 341–354. Lawrence: University Press of Kansas.

Holsinger, S. J. 1902. The boundary line between the desert and the forest. *Forestry & Irrigation* 8:21–27.

Hooper, R. G. and C. J. McAdie. 1995. Hurricanes and the long-term management of the red-cockaded woodpecker. In Kulhavy, D. L., R. G. Hooper, and R. Costa (eds.), *Red-cockaded woodpecker: Recovery, ecology and management*, pp. 148–166. Austin, TX: Center for Applied Studies in Forestry, College of Forestry, Stephen F. Austin State University.

Hooper, R. G., A. F. Robinson, and J. A. Jackson. 1996. The red-cockaded woodpecker: notes on life history and management. USDA Forest Service General Report SA-GR 9. State and Private Forestry, Southeastern Area. Atlanta, GA.

Hopkins, D. M. 1967. The Cenozoic history of Beringia—a synthesis. In Hopkins, D. M. (ed.), *The Bering land bridge*, pp. 451–484. Stanford, CA: Stanford University Press.

Hostetler, S. W., F. Giorgi, G. T. Bates, and P. J. Bartlein. 1994. Lake-atmosphere feedbacks associated with paleolakes Bonneville and Lahontan. *Science* 263(5147)665–668.

Houston, D. B. 1973. Wildfires in northern Yellowstone National Park. *Ecology* 54(5):1111–1117.

Howard, J. L. n.d.(a). *Quercus douglasii* [blue oak]. In Fischer, W. C. (compiler), *The Fire Effects Information System* [data base]. USDA Forest Service, Intermountain Research Station, Intermountain Fire Sciences Laboratory, Missoula, MT.

Howard, J. L. n.d.(b). *Quercus garryana* [Oregon white oak]. In Fischer, W. C. (compiler), *The Fire Effects Information System* [data base]. USDA Forest Service, Intermountain Research Station, Intermountain Fire Sciences Laboratory, Missoula, MT.

Howard, J. L. 1992a. *Umbellularia californica* [California laurel or bay]. In Fischer, W. C. (compiler), *The Fire Effects Information System* [data base]. USDA Forest Service, Intermountain Research Station, Intermountain Fire Sciences Laboratory, Missoula, MT.

Howard, J. L. 1992b. *Quercus lobata* [valley oak]. In Fischer, W. C. (compiler), *The Fire Effects Information System* [data base]. USDA Forest Service, Intermountain Research Station, Intermountain Fire Sciences Laboratory, Missoula, MT.

Howard, J. L. and D. Tirmenstein. 1996. *Populus tremuloides* [quaking aspen]. In Simmerman, D. G (compiler), *The Fire Effects Information System* [data base]. USDA Forest Service, Intermountain Research Station, Intermountain Fire Sciences Laboratory, Missoula, MT.

Hughes, J. W. and D. A. Bechtel. 1997. Effect of distance from forest edge on regeneration of red spruce and balsam fir in clearcuts. *Canadian Journal of Forest Research* 27(12):2088–2096.

Hunn, E. S. and D. H. French. 1981 *Lomatium*: A key resource for Columbia Plateau native subsistence. *Northwest Science* 55(2):87–94.

Hunter, J. C. and V. T. Parker. 1993. The disturbance regime of an old-growth forest in coastal California. *Journal of Vegetation Science* 4(1):19–24.

Huntley, B. and T. Webb III. 1989. Migration: Species' response to climatic variations caused by changes in the earth's orbit. *Journal of Biogeography* 16:5–19.

Hurt, R. D. 1987. *Indian agriculture in America: Prehistory to the present.* Lawrence, KS: University Press of Kansas.

Hutchings, J. M. 1859a. Scenes in the valleys and mountains of California. Hutchings' California Magazine. Vol. III. N. 11. In Olmsted, R. R. (ed.). 1962, *Scenes of wonder & curiosity from Hutchings' California Magazine, 1856–1861*, pp. 152–167. Berkeley, CA: Howell-North.

Hutchings, J. M. 1859b. The mammoth trees of California. Hutchings' California Magazine. Vol. III. N. 9. In Olmsted, R. R. (ed.). 1962, *Scenes of wonder & curiosity from Hutchings' California Magazine, 1856–1861*, pp. 205–217. Berkeley, CA: Howell-North.

Illg, C. and G. Illg. 1994. The ponderosa and the flammulated. *American Forests* 100(3&4):36–37, 58–59.

Illinois State Museum. 1995. Midwestern U.S. 16,000 years ago [exhibit]. Springfield, IL.

Imbrie, J. and K. P. Imbrie, 1979. *Ice ages: Solving the mystery.* Short Hills, NJ: Enslow Publishers.

Irving, W. 1886. *A tour of the prairies.* New York: John B. Alden.

Irwin, L. L., J. G. Cook, R. A. Riggs, and J. M. Skovlin. 1994. Effects of long-term grazing by big game and livestock in the Blue Mountains forest ecosystems. USDA Forest Serive General Technical Report PNW-GTR-325. Pacific Northwest Research Station. Portland, OR.

Ives, J. C. 1861. Report upon the Colorado River of the West, explored in 1857 and 1858. United States Army Corps of Topographical Engineers. 36th Congress, 1st Session. U.S. Senate. Executive Document. U.S. Government Printing Office. Washington, DC.

Jackson, D. 1966. *Custer's gold, the United States cavalry expedition of 1874.* Lincoln, NE: University of Nebraska Press.

Jackson, S. T., D. R. Whitehead, and G. D. Ellis. 1986. Late-glacial and early Holocene vegetational history at the Kolarik Mastodon Site, northwestern Indiana. *American Midland Naturalist* 115:361–373.

Jacobs, D. F., D. W. Cole, and J. R. McBride. 1985. Fire history and perpetuation of natural coast redwood ecosystems. *Journal of Forestry* 83(8):494–497.

Jacobson, G. L., Jr., and A. Dieffenbacher-Krall. 1995. White pine and climate change: Insights from the past. *Journal of Forestry* 93(2):39–42.

Jaeger, E. 1966. *Wildwood wisdom.* New York: MacMillan.

Jaffe, A. J. 1992. *The first immigrants from Asia*, pp. 47–49. New York: Plenum Press.

James, E. 1844. Surveyors field notes (for section lines, Townships 38N, Range 22W). *United States Federal Land Surveys.* Missouri State Archives. Jefferson City, MO.

James, E. 1822. Account of an expedition from Pittsburgh to the Rocky Mountains. University Microfilm, Ann Arbor, MI, 1966.

James, T. 1846. Three years among the Indians and Mexicans. Printed at the office of the "War Eagle," Waterloo, IL (book reproduced by The American Mountain Men).

Janetski, J. C. 1987. *Indians of Yellowstone Park.* Salt Lake City: Bonneville Books, University of Utah Press.

Jennings, F. 1975. *The invasion of America.* New York: W. W. Norton.

Jepson, W. L. 1967. Trees, shrubs and flowers of the redwood region. Save-the-Redwoods League. San Francisco.

Johnson, A. S., P. E. Hale, W. M. Ford, J. M. Wentworth, J. R. French, O. F. Anderson, and G. B. Pullen. 1995. White-tailed deer foraging in relation to successional stage, overstory type and management of southern Appalachian forests. *American Midland Naturalist* 133:18–35.

Johnson, A. W. and J. G. Packer. 1967. Distribution, ecology, and cytology of the Ogotoruk Creek Flora and the history of Beringia. In Hopkins, D. M. (ed.), *The Bering land bridge*, pp. 245–265. Stanford, CA: Stanford University Press.

Johnson, E. 1654. Johnson's Wonder-Working Providence, 1628–1651. *Original Narratives of Early American History* 9 (1910):1–285.

Johnson, J. K. 1994. Prehistoric exchange in the Southeast. In Baugh, T. G. and J. E. Ericson (eds.), *Prehistoric exchange systems in North America*, pp. 99–125. New York: Plenum Press.

Johnson, O. 1969. *Flathead and Kootenai*, pp. 199–259. New York: Arthur H. Clark.

Johnson, O. and W. H. Winter. 1846. *Route across the Rocky Mountains, with a description of Oregon and California ... of the emigration of 1843.* Lafayette, IN: John B. Semans. University Microfilms, Ann Arbor, MI.

Johnson, P. S. 1993a. Perspectives on the ecology and silviculture of oak-dominated forests in the central and eastern states. USDA Forest Service General Technical Report NC-153. North Central Forest Experiment Station. St. Paul, MN.

Johnson, P. S. 1993b. Sources of oak reproduction. In Loftis, D. and C. E. McGee (eds.), *Oak regeneration: Serious problems, practical recommendations,* pp. 112–131. Symposium Proceedings, September 8–10, Knoxville, Tennessee. USDA Forest Servie General Technical Report SE–84. Southeastern Forest Experiment Station. Asheville, NC.

Johnson, W. C. and C. S. Adkisson. 1985. Dispersal of beech nuts by blue jays in fragmented landscapes. *American Midland Naturalist* 113:319–324.

Johnson, W. C. and T. Webb III. 1989. The role of blue jays (*Cyanocitta cristata* L.) in the postglacial dispersal of fagacous trees in eastern North America. *Journal of Biogeography* 16:561–571.

Johnston, A. 1970. Blackfoot Indian utilization of the flora of the Northwestern Great Plains. *Economic Botany* 24:301–324.

Johnston, W. F. 1990. *Larix laricina* (Du Roi) K. Koch, Tamarack. In Burns, R. M. and B. H. Honkala (tech. coord.), *Silvics of North America, Vol. 1, Conifers*, pp. 141–151. Agricultural Handbook No. 654. Washington, DC: USDA Forest Service.

Jones, E. W. 1945. The structure and reproduction of the virgin forest of the north temperate zone. *New Phytololgist* 44(2):130–148.

Jones, J. R. 1985. Distribution. In DeByle, N. V. and R. P. Winokur (eds.), *Aspen: Ecology and management in the western United States*, pp. 9–10. USDA Forest Service General Technical Report RM-119. Rocky Mountain Forest and Range Experiment Station, Fort Collins, CO.

Jones, J. R. and N. V. DeByle. 1985a. Fire. In DeByle, N. V. and R. P. Winokur (eds.), *Aspen: Ecology and management in the western United States*, pp. 77–81. USDA Forest Service General Technical Report RM-119. Rocky Mountain Forest and Range Experiment Station, Fort Collins, CO.

Jones, J. R. and N. V. DeByle. 1985b. Morphology. In DeByle, N. V. and R. P. Winokur (eds.), *Aspen: Ecology and management in the western United States*, pp. 11–18. USDA Forest Service General Technical Report RM-119. Rocky Mountain Forest and Range Experiment Station, Fort Collins, CO.

Jones, J. R., N. V. DeByle, and D. M. Bowers. 1985. Insects and other invertebrates. In DeByle, N. V. and R. P. Winokur (eds.), *Aspen: Ecology and management in the western United States*, pp. 107–114. USDA Forest Service General Technical Report RM-119. Rocky Mountain Forest and Range Experiment Station, Fort Collins, CO.

Joutel, H. 1962. *A journal of La Salle's last voyage*. New York: Corinth Books.

Kaib, M., C. H. Baisan, H. D. Grissino-Mayer, and T. W. Swetnam. 1996. Fire history of the gallery pine-oak forests and adjacent grasslands of the Chiricahua Mountains of Arizona. In Ffolliott, P. F., L. F. DeBano, M. B. Baker, Jr., G. J. Gottfried, G. Solis-Garza, C. B. Edminster, D. G. Neary, L. S. Allen, and R. H. Hamre (tech. coord.), *Effects of fire on Madrean Province ecosystems*, pp. 253–264. USDA Forest Service General Technical Report RM-GTR-289. Rocky Mountain Forest and Range Experiment Station, Fort Collins, CO.

Kauffman, B. E. 1977. Preliminary analysis of seeds from the Shawnee-Minisink site. Master's Thesis, American University, Washington, DC.

Kay, C. E., B. Patton, and C. A. White. 1994. Assessment of long-term terrestrial ecosystem states and processes in Banff National Park and the central Canadian Rockies. Banff National Park, Resource Conservation, Alberta, Canada.

Keeley, J. E. and S. C. Keeley. 1995. Chaparral. In Barbour, M. G. and W. D. Billings (eds.), *North American terrestrial vegetation*, pp. 165–207. New York: Cambridge University Press.

Keen, F. P. 1952. *Insect enemies of western forests.* Washington, DC: USDA Forest Service Publication No. 273.

Keener, C. and E. Kuhns. 1997. The impact of Iroquoian populations on the northern distribution of pawpaws in the Northeast. *North American Archaeologist* 18(4):327–342.

Keever, C. 1953. Present composition of some stands of the former oak chestnut forests in the southern Blue Ridge Mountains. *Ecology* 34:44A5.

Kelly, R. L. and L. C. Todd. 1988. Coming into the country: Early Paleoindian hunting and mobility. *American Antiquity* 53(2):231–244.

Kennedy, M. S. 1961. *The Assiniboines.* Norman, OK: Univeristy of Oklahoma Press.

Kenyon, W. A. 1986. Mounds of scared earth, burial mounds in Ontario. Archaeology Monograph 9. Toronto, Canada: Royal Ontario Museum.

Kephart, H. (ed.). 1916. *Wild life in the Rocky Mountains, by George Frederick Ruxton; a true tale of rough adventure in the days of the Mexican war.* New York: Outing Publishing (book reproduced by The American Mountain Men).

Kerr, R. A. 1993a. Ancient climates. *Science* 262:1972–1973.

Kerr, R. A. 1993b. Earth science. *Science* 261:292.

Kerr, R. A. 1994. Ancient tropical climates warm San Francisco gathering. *Science* 263:173–175.

Kerr, R. A. 1995. Meeting briefs. *Science* 267:27–28.

Keter, T. S. 1987. Indian burning: Managing the environment before 1865 along the North Fork. Paper presented to The Society for California Archaeology, April 16. Fresno, California. USDA Forest Service Reprint, Eureaka, CA.

Kidwell, B. 1989. Redwoods of the East. *Progressive Farmer* February, pp. 26–27.

Kilgore, B. M. and H. H. Biswell. 1971. Seedling germination following fire in a giant sequoia forest. *California Agriculture* 25(2):8–10.

Kilgore, B. M. and R. W. Sando. 1975. Crown-fire potential in a sequoia forest after prescribed burning. *Forest Science* 21(1):83–87.

Kilgore, B. M. and D. Taylor. 1979. Fire history of a sequoia-mixed conifer forest. *Ecology* 60:129–142.

Kimmey, J. W. 1950. Cull factors for forest tree species in northwestern California.

USDA Forest Service. Pacific Southwest Forest and Range Experiment Station Survey Release No. 7.

Kindscher, K. 1987. *Edible wild plants of the prairie: an ethnobotanical guide.* Lawrence, KS: University of Kansas Press.

King, C. 1874. *Mountaineering in the Sierra Nevada.* Boston, MA: James R. Osgood.

Koch, P. 1996. Lodgepole pine commercial forests: An essay comparing the natural cycle of insect kill and subsequent wildfire with management for utilization and wildlife. USDA Forest Service General Technical Report INT-GTR-342. Intermountain Research Station, Odgen, UT.

Koehler, G. M. and J. D. Brittell. 1990. Managing spruce-fir habitat for lynx and snowshoe hares. *Journal of Forestry* 88(10):10–14.

Koenig, W. D. and J. M. H. Knops. 1997. Patterns of geographic synchrony in growth and reproduction of oaks within California and beyond. In Pillsbury, N. H., J. Verner, and W. D. Tietje (tech. coord.), *Proceedings of a Symposium on Oak Woodlands: Ecology, Management, and Urban Interface Issues,* pp. 101–108. USDA Forest Service General Technical Report PSW-GTR-160. Pacific Southwest Research Station, Albany, CA.

Kohler, T. A. 1992. Prehistoric human impact on the environment in the upland North American Southwest. *Population and Environment: Journal of Interdisciplinary Studies* 13(4):255–268.

Kohler, T. A. and M. H. Matthews. 1988. Long-term Anasazi land use and forest reduction: A case study from southwest Colorado. *American Antiquity* 53(3):537–564.

Komarek, E. V., Sr. 1967. The nature of lightning fires. Proceedings, Tall Timbers Fire Ecology Conference Number 7, pp. 5–41. Tall Timbers Research Station, Tallahassee, FL.

Komarek, E. V., Sr. 1968. Lightning and lightning fires as ecological forces. Proceedings, Annual Tall Timbers Fire Ecology Conference, Vol. 8, pp. 169–197. Tall Timbers Research Station. Tallahassee, FL.

Komarek, E. V., Sr. 1969. Fire and animal behavior. Proceedings, Annual Tall Timbers Fire Ecology Conference, Vol. 9, pp. 160–207. Tall Timbers Research Station. Tallahassee, FL.

Kudish, M. 1992. *Adirondack upland flora: An ecological perspective.* Saranac, NY: Chauncy Press.

Kunz, M. L. and R. E. Reanier. 1994. Paleoindians in Beringia: Evidence from arctic Alaska. *Science* 263(5147):660–662.

Laacke, R. J. 1990. *Abies concolor* (Gord. & Glend.) Lindl. Ex Hildebr. White Fir. In Burns, R. M. and B. H. Honkala (tech. coord.), *Silvics of North America, Vol. 1, Conifers,* pp. 36–46. Agricultural Handbook No. 654. Washington, DC: USDA Forest Service.

Lafferty, R. H. III. 1994. Prehistoric exchange in the lower Mississippi Valley. In

Baugh, T. G. and J. E. Ericson, *Prehistoric exchange systems in North America*, pp. 177–213. New York: Plenum Press.

Lahren, L. and R. Bonnichsen. 1974. Bone foreshafts from a Clovis burial in southwestern Montana. *Science* 186:147–150.

Landers, J. L., D. H. Van Lear, and W. D. Boyer. 1995. The longleaf pine forests of the Southeast: Requiem or renaissance? *Journal of Forestry* 93(11):39–44.

Lang, D. M. and S. S. Stewart. 1910. Reconnaissance of the Kaibab National Forest. USDA Forest Service Timber Survey, North Kaibab Ranger District. Administrative Report.

Langford, N. P. 1873. Report of the superintendent of the Yellowstone National Park for the year 1872. 42nd Congress, 3rd Session, U. S. Senate Executive Document No. 35. Washington, DC.

Lanman, Charles. 1856. *Adventures in the wilds of the United States and British American provinces*. Philadelphia: J. W. Moore.

Larsen, L. T. and T. D. Woodbury. 1916. Sugar pine. U. S. Department of Agriculture Bulletin Number 426. Washington, DC.

Laughlin, W. S. 1967. Human migration and permanent occupation in the Bering Sea area. In Hopkins, D. M. (ed.), *The Bering land bridge*, pp. 409–450. Stanford, CA: Stanford University Press.

Laven, R. D., P. N. Omi, J. G. Wyant, and A. S. Pinkerton. 1980. Interpretation of fire scar data from a ponderosa pine ecosystem in the central Rocky Mountains, Colorado. Proceedings, Fire History Workshop, pp. 46–49. USDA Forest Service General Technical Report RM-81. Rocky Mountain Forest and Range Experiment Station, Fort Collins, CO.

Lawson, E. R. 1990. *Pinus echinata* Mill. Shortleaf Pine. In Burns, R. M. and B. H. Honkala (tech. coord.), *Silvics of North America, Vol. 1, Conifers*, pp. 316–326. Agricultural Handbook No. 654. Washington, DC: USDA Forest Service.

Lawton, H. W. , P. J. Wilke, M De Decker, and W. M. Mason. 1993. Agriculture among the Paiute of Owens Valley. In Blackburn, T. C. and M. K. Anderson (eds.), *Before wilderness: Environmental management by native Californians*, pp. 329–377. Menlo Park, CA: Ballena Press.

Leak, W. B. 1991. Secondary forest succession in New Hampshire, USA. *Forest Ecology and Management* 43:69–86.

Leak, W. B. and S. M. Filip. 1977. Thirty-eight years of group selection in new England northern hardwoods. *Journal of Forestry* 75(10):641–643.

Leak, W. B. and J. H. Gottsacker. 1985. New approaches to uneven-age management in New England. *Northern Journal of Applied Forestry* 2:28–31.

Leaphart, C. D. 1963. Dwarf mistletoes: A silvicultural challenge. *Journal of Forestry* 61(1):40–46.

Le Conte, J. 1971. *A journal of ramblings through the high Sierra of California*. New York: Ballantine Books.

Lee, R. G. and T. M. Bonnicksen. 1978. Brushland watershed fire management policy in southern California: Biosocial considerations. California Water Resources Center Contribution No. 172.

Lefler, H. T. 1967. *A new voyage to Carolina, by John Lawson* (reprint of 1709 edition). Chapel Hill: University of North Carolina Press.

Lehmann, V. W. 1965. Fire in the range of Attwater's prairie chicken. Proceedings, Fourth Annual Tall Timbers Fire Ecology Conference, pp. 127–143. Tall Timbers Research Station, Tallahassee, FL.

Leiberg, J. B. 1900. San Gabriel, San Bernardino, San Jacinto foret reserves. In Gannett, H. (ed.)., *Twentieth Annual Report of the U.S. Geological Survey to the Secretary of Agriculture. Part 5, Forest Reserves*, pp. 411–479. Washington, DC: U.S. Government Printing Office.

Leiberg, J. B. 1904. *Forest conditions in the Absaroka Division of the Yellowstone Forest Reserve, Montana, and the Livingston and Big Timber Quadrangles.* Document No. 717. Washington, DC: Government Printing Office.

Leiberg, J. B., T. F. Rixon, and A. Dodwell. 1904. Forest conditions in the San Francisco Mountains Forest Reserve, Arizona. US Geological Survey Professional Paper No. 22. Washington, DC: US Government Printing Office.

Leighton, A. L. 1985. *Wild plant use by the woods Cree (Nihithawak) of East-Central Saskatchewan.* Ottawa: National Museums of Canada. Mercury Series.

Leitner, L. A., C. P. Dunn, G. R. Guntenspergen, F. Stearns, and D. M. Sharpe. 1991. Effects of site, landscape features, and fire regime on vegetation patterns in presettlement southern Wisconsin. *Landscape Ecology* 5(4):203–217.

Lewis, H. T. 1977. Maskuta: The ecology of Indian fires in northern Alberta, *Western Canadian Journal of Anthropology* 7:15–52.

Lewis, H. T. 1982a. A time for burning. Boreal Institute for Northern Studies, Occasional Publication Number 17. The University of Alberta, Edmonton, Canada.

Lewis, H. T. 1982b. Fire technology and resource management in aboriginal North America and Australia. In Hunn, E. and N. Williams (eds.), *The regulation of environmental resources in food collecting societies*, pp. 45–67. American Association for the Advancement of Science Selected Symposium Series No. 67. Boulder, CO: Westview Press.

Lewis, H. T. 1993. Patterns of Indian burning in California: Ecology and ethnohistory. In Blackburn, T. C. and M. K. Anderson (ed.), *Before wilderness: Environmental management by native Californians*, pp. 55–116. Menlo Park, CA: Ballena Press.

Lewis, H. T. 1994. Management fires vs. corrective fires in northern Australia: An analogue for environmental change. *Chemosphere* 29(5):949–963.

Lewis, H. T. and T. A. Ferguson. 1988. Yards, corridors, and mosaics: How to burn a boreal forest. *Human Ecology* 16(1):57–77.

Lewis, H. T. and C. Schweger. 1973. Paleoindian uses of fire during the late Pleistocene: The human factor in environmental change. Paper presented at the IXth International Union for Quarternary Research. University of Canterbury, Christchurch, New Zealand. December 2–10.

Little, E. L. 1980. *The Audubon Society field guide to North American trees: Eastern region.* New York: Alfred A. Knopf.

Loope, L. L. and G. E. Gruell. 1973. The ecological role of fire in the Jackson Hole area, northwestern Wyoming. *Quaternary Research* 3:425–443.

Lorimer, C. G. 1977. The presettlement forest and natural disturbance cycle of Northeastern Maine. *Ecology* 58:139–149.

Lorimer, C. G. 1980a. Age structure and disturbance history of a southern Appalachian virgin forest. *Ecology* 61(5):1169–1184.

Lorimer, C. G. 1980b. The use of land survey records in estimating presettlement fire frequency. Proceedings, Fire History Workshop. USDA Forest Service General Technical Report RM-81, pp. 57–62. Rocky Mountain Forest and Range Experiment Station, Fort Collins, CO.

Lorimer, C. G. 1983. Eighty-year development of northern red oak after partial cutting in a mixed-species Wisconsin forest. *Forest Science* 29(2):371–383.

Lorimer, C. G. 1985. The role of fire in the perpetuation of oak forests. In Johnson, J. E. (ed.), *Proceedings, Challenges in Oak Management and Utilization*, pp. 8–25. Cooperative Extension Service. Madison: University of Wisconsin.

Lorimer, C. G. 1993. Causes of the oak regeneration problem. In Loftis, D. and C. E. McGee (eds.), *Oak regeneration: Serious problems, practical recommendations*, pp. 14–39. Symposium Proceedings, September 8–10, Knoxville, Tennessee. USDA Forest Servie General Technical Report SE-84. Southeastern Forest Experiment Station, Asheville, NC.

Lorimer, C. G. 1994. Timber harvest effects on nongame birds.. what does the evidence show? Forestry Facts No. 77. Department of Forestry, University of Wisconsin-Madison.

Los Angeles Star. 1859. Expedition to Owen's Lake. August 27, p. 2.

Lotan, J. E. 1976. Cone serotiny—fire relationships in lodgepole pine. Proceedings, Tall Timbers Fire Ecology Conference and Fire and Land Management Symposium, Number 14, pp. 267–278. Tall Timbers Research Station. Tallahassee, Florida.

Lotan, J. E. and W. B. Critchfield. 1990. *Pinus contorta* Dougl. ex. Loud. Lodgepole Pine. In Burns, R. M. and B. H. Honkala (tech. coord.), *Silvics of North America, Vol. 1, Conifers*, pp. 302–315. Agricultural Handbook No. 654. Washington, DC: USDA Forest Service.

Loucks, O. L. 1983. New light on the changing forest. In Flader, S. L. (ed.), *The Great Lakes Forest: An environmental and social history*, pp. 17–32. Minneapolis: University of Minnesota Press.

Loud, L. L. 1918. Ethnogeography and archaeology of the Wiyot territory.

University of California Publications in American Archaeology and Ethnology 14(3):221–423.

Lowe, P. O., P. F. Ffolliott, J. H. Dieterich, and D. R. Patton. 1978. Determining potential wildlife benefits from wildfire in Arizona ponderosa pine forests. USDA Forest Service General Technical Report RM-52. Rocky Mountain Forest and Range Experiment Station, Fort Collins, CO.

Ludlow, W. 1875. *Report of a reconnaissance of the Black Hills of Dakota, made in the summer of 1874.* Washington, DC: U.S. Government Printing Office.

Ludlow, W. 1876. *Report of a reconnaissance from Carroll, Montana territory, on the upper Missouri, to the Yellowstone national park, and return, made in the summer of 1875.* Washington, DC: U.S. Government Printing Office.

Lundelius, E. L., Jr. 1988. What happened to the mammoth? The climatic model. In R. C. Carlisle (ed.), *Americans before Columbus: Ice-age origins*, pp. 75–82. Ethnology Monographs 12. Department of Anthropology, Pittsburgh: University of Pittsburgh.

Lutz, H. J. 1930a. The vegetation of Heart's Content, a virgin forest in northwestern Pennsylvania. *Ecology* 11(1):2–29.

Lutz, H. J. 1930b. Original forest composition in northwestern Pennsylvania as indicated by early land survey notes. *Journal of Forestry* 28:1098–1103.

Lutz, H. J. 1959. Aboriginal man and white man as historical causes of fires in the boreal forest, with particular reference to Alaska. Yale School of Forestry Bulletin No. 65. Yale University, New Haven, CT.

Lyman, H. 1900. Indian names. *Oregon Historical Quarterly* 1(3):316–326.

Lynham, T. J. and B. J. Stocks. 1989. The natural fire regime of an unprotected section of the boreal forest in Canada. Proceedings, 17th Tall Timbers Fire Ecology Conference, High Intensity Fire in Wildlands: Management Challenges and Options, pp. 99–109. Tall Timbers Research Station. Tallahassee, FL.

Lynn, A. (ed.). 1996. Ancient Indians in Iowa may have grown weeds as crops. ACES *News, University of Illinois at Urbana-Champaign.* November.

Lysek, C. A. 1997. Ancient Alaskan bones may help to prove coast migration theory. *Mammoth Trumpet* 12(4). Center for the Study of the First Americans. Oregon State University, Corvallis.

Mackenzie, A. 1904. *Voyages from Montreal through the continent of North America to the frozen and Pacific Oceans in 1789 and 1793.* New York: Allerton Book.

Macoun, J. 1882. *Manitoba and the great North-West.* Guelph, Ont: World Publishing.

Mahar, J. M. 1953. Ethnobotany of the Oregon Paiutes of the Warm Springs Indian Reservation. B.A. Thesis, Reed College, p. 41.

Maissurow, D. K. 1935. Fire as a necessary factor in the perpetuation of white pine. *Journal of Forestry* 33:373–378.

Maissurow, D. K. 1941. The role of fire in the perpetuation of virgin forests of Northern Wisconsin. *Journal of Forestry* 39(2):201–207.

Mandryk, C. A. 1990. Could humans survive the ice-free corridor? Late-glacial vegetation and climate in West Central Alberta. In Agenbroad, L. D., J. I Mead, and L. W. Nelson (eds.), *Megafauna & man: Discovery of America's heartland, Vol. 1, Scientific Papers,* pp. 67–79. Hot Springs, SD: The Mammoth Site of Hot Springs.

Marryat, F. 1855. *Mountains, molehills and recolletions of a burnt journal.* London: Longman, Brown, Green, and Longmans. London (reprinted 1982 by Time-Life Books).

Martin, G. 1996. Keepers of the oaks. *Discover* 17(8):45–50.

Martin, P. S. 1967. Pleistocene overkill. *Natural History* 76(9):32–38.

Martin, P. S. 1973. The discovery of America. *Science* 179:969–974.

Martin, P. S. 1975. Vanishings, and the future of the prairie. In Perkins, B. F. (ed.), *Grasslands Ecology: A Symposium,* pp. 39–49. *Geoscience and Man,* Vol. X. School of Geoscience, Louisiana State University, Baton Rouge.

Martin, P. S. 1990. Who or what destroyed out mammoths? In Agenbroad, L. D., J. I Mead, and L. W. Nelson (eds.), *Megafauna & man: Discovery of America's heartland, Vol. 1, Scientific Papers,* pp. 109–117. Hot Springs, SD: The Mammoth Site of Hot Springs.

Martin, R. E. 1982. Fire history and its role in succession. In Means, J. E. (ed.), *Forest succession and stand development research in the Northwest,* pp. 92–99. Forest Research Laboratory. Corvallis, OR: Oregon State University.

Martin, R. E., D. D. Robinson, and W. H. Schaeffer. 1974. Fire in the Pacific Northwest—perspectives and problems. Proceedings, Tall Timbers Fire Ecology Conference Number 15, pp. 1–23. Tall Timbers Research Station. Tallahassee, FL.

Maser, C., R. G. Anderson, K. Cromack, Jr., J. T. Williams, and R. E. Martin. 1979. Dead and down woody material. In Thomas, J. W. (tech. ed.), *Wildlife habitats in managed forests: The Blue Mountains of Oregon and Washington,* pp. 78–95. Agriculture Handbook No. 553. Washington, DC: USDA Forest Service.

Maxwell, H. 1910. The use and abuse of forests by the Virginia Indians. *William and Mary College Quarterly Historical Magazine* 19(2):73–105.

May, G. M. 1953. The Indian as an ecological factor in the northeastern forest. *Ecology* 34(2):329–346.

Mayle, F. E. and L. C. Cwynar. 1995. Impact of the Younger Dryas cooling event upon lowland vegetation of maritime Canada. *Ecological Monographs* 65(2):129–154.

McAndrews, J. H. 1966. Postglacial history of prairie, savanna, and forest in northwestern Minnesota. *Memoirs of The Torrey Botanical Club,* 22(2).

McBride, J. R. and D. F. Jacobs. 1980. Land use and fire history in the mountains of southern California. Proceedings, Fire History Workshop. USDA Forest Service General Technical Report RM-81, pp. 85–88. Rocky Mountain Forest and Range Experiment Station, Fort Collins, CO.

McCarthy, H. 1993. Managing oaks and the acorn crop. In Blackburn, T. C. and M. K. Anderson (eds.), *Before the wilderness: Environmental management by native Californians*, pp. 213–228. Menlo Park, CA: Ballena Press.

McComb, W. C., T. A. Spies, and W. H. Emmingham. 1993. Douglas-fir forests: Managing for timber and mature-forest habitat. *Journal of Forestry* 91(12):31–42.

McCracken, H. 1959. *George Catlin and the old frontier*. New York: Bonanza Books.

McCune, B. and G. Cottam. 1985. The successional status of a southern Wisconsin oak woods. *Ecology* 66(4):1270–1278.

McDonald, K. A. 1995. Early man's refrigerator. *Chronicle of Higher Education* May 12, pp. A12, A18.

McDonald, P. M., P. J. Anderson, and G. O. Fiddler. 1997. Vegetation in group-selection openings: early trends. USDA Forest Service Research Note PSW-RN–421. Pacific Southwest Research Station, Albany, CA.

McGarigal, K. 1994. Relationship between landscape structure and avian abundance patterns in the Oregon Coast Range. Ph.D. Dissertation, Oregon State University, Corvalis, OR.

McGarigal, K. and W. C. McComb. 1995. Relationships between landscape structure and breeding birds in the Oregon Coast Range. *Ecological Monographs* 65(3):235–260.

McIntosh, R. P. 1962. The forest cover of the Catskill Mountain region, New York, as indicated by land survey records. *American Midland Naturalist* 68(2):409–423.

McKelvey, K. S. and J. D. Johnston. 1992. Historical perspectives on forests of the Sierra Nevada and the Transverse Ranges of southern California: Forest conditions at the turn of the century. In Verner, J. K., B. Noon, K. McKelvey, R. Gutierrez, G. Gould, and T. Beck (tech. coord.), *The California spotted owl: A technical assessment of its current status*, pp. 225–246. USDA Forest Service General Technical Report PSW-GTR-133. Pacific Southwest Research Station, Albany, CA.

McMartin, D., Jr. 1823. Survey minutes of Township IV in the Moose River principally made in the year 1821 and finished in 1823. New York Department of Public Works Field Book, Albany.

McMillan, R. B. 1976. The Pomme de Terre study locality: Its setting. In Wood, W. R. and R. B. McMillan (eds.), *Prehistoric man and his environments: A case study in the Ozark Highland*, pp. 13–44. New York: Academic Press.

McMillan, R. B. and W. R. Wood. 1976. A summary of environmental and cultural change in the western Missouri Ozarks. In Wood, W. R. and R. B. McMillan (eds.), *Prehistoric man and his environments: A case study in the Ozark Highland*, pp. 235–240. New York: Academic Press.

McMurray, N. E. 1986a. *Pinus edulis* [piñion pine]. In Fischer, W. C. (compiler), *The Fire Effects Information System* [data base]. USDA Forest Service, Intermountain Research Station, Intermountain Fire Sciences Laboratory, Missoula, MT.

McMurray, N. E. 1986b. *Pinus monophylla* [singleleaf piñion pine]. In Fischer, W. C. (compiler), *The Fire Effects Information System* [data base]. USDA Forest Service, Intermountain Research Station, Intermountain Fire Sciences Laboratory, Missoula, MT.

McMurray, N. E. 1989a. *Lithocarpus densiflora* [tanoak]. In Fischer, W. C. (compiler), *The Fire Effects Information System* [data base]. USDA Forest Service, Intermountain Research Station, Intermountain Fire Sciences Laboratory, Missoula, MT.

McMurray, N. E. 1989b. *Arbutus menziesii* [Pacific madrone]. In Fischer, W. C. (compiler), *The Fire Effects Information System* [data base]. USDA Forest Service, Intermountain Research Station, Intermountain Fire Sciences Laboratory, Missoula, MT.

McMurray, N. E. 1990. *Rhamnus californica* [California coffeeberry]. In Fischer, W. C. (compiler), *The Fire Effects Information System* [data base]. USDA Forest Service, Intermountain Research Station, Intermountain Fire Sciences Laboratory, Missoula, MT.

Mead, J. I. and D. J. Meltzer. 1984. North American Late Quaternary extinctions and the radiocarbon record. In Martin, P. S. and R. G. Klein (eds.), *Quaternary extinction: A prehistoric revolution*, pp. 440–450. Tucson: University of Arizona Press.

Means, J. E. 1982. Developmental history of dry coniferous forests in the central western Cascade Range of Oregon. In Means, J. E. (ed.), *Forest succession and stand development research in the Northwest*, pp. 142–158. Forest Research Laboratory, Corvallis, OR: Oregon State University.

Means, J. E., J. H. Cissel, and F. J. Swanson. 1996. Fire history and landscape restoration in Douglas-fir ecosystems of western Oregon. In Hardy, C. C. and S. F. Arno (eds.), *The use of fire in forest restoration*, pp. 61–67. USDA Forest Service General Technical Report INT-GTR-341. Intermountain Research Station, Odgen, UT.

Meares, J. 1790. *Voyages made in the years 1788 and 1789, from China to the Northwest Coast of America*. London: Logographic Press.

Meggers, B. J. 1996. Letter to the editor. *Science* 274(5294):1820–1825.

Meighan, C. and C. V. Haynes,. 1970. The Borax Site revisited. *Science* 167:1213–1221.

Meltzer, D. J. 1997. Monte Verde and the Pleistocene peopling of the Americas. *Science* 276(5313):754–755.

Meltzer, D. J. and J. I. Mead. 1983. The timing of Late Pleistocene mammalian extinctions in North America. *Quaternary Research* 19(1):130–135.

Menon, S. 1998. Clovis R.I.P. *Discover* 18(1):100–101.

Merriam, C. H. 1966. Ethnographic notes on California Indian tribes. University of California Archaeological Research Facility, Berkeley.

Miller, G. W. and J. N. Kochenderfer. 1998. Maintaining species diversity in the central Appalachians. *Journal of Forestry* 96(7):28–33.

Miller, R. F. 1995. Pushing back juniper. *Restoration & Management Notes* 13(1):51–52.

Miller, R. F. and J. A. Rose. 1995. Historic expansion of *Juniperus occidentalis* (western juniper) in southeastern Oregon. *Great Basin Naturalist* 55(1):37–45.

Miller, R. F. and P. E. Wigand. 1994. Holocene changes in semiarid pinyon-juniper woodlands. *BioScience* 44(7):465–474.

Miller, R. K. and S. K. Albert. 1993. Zuni cultural relationships to piñon-juniper woodlands. In Aldon, E. F. and D. W. Shaw (tech. coord.), *Managing piñon-juniper ecosystems for sustainability and social needs*, pp. 74–78. USDA Forest Service General Technical Report RM-236. Rocky Mountain Forest and Range Experiment Station, Fort Collins, CO.

Ministry of Forests. 1995. *Defoliator management guidebook*. Victoria, BC.

Minnich, R. A. 1983. Fire mosaics in Southern California and Northern Baja California. *Science* 219:1287–1294.

Minnich, R. A. 1987. Fire behavior in southern California chaparral before fire control: The Mount Wilson burns at the turn of the century. *Annals of the Association of American Geographers* 77(4):599–618.

Minshall, G. W., J. T. Brock, and J. D. Varley. 1989. Wildfires and Yellowstone's stream ecosystems. *BioScience* 39(10):707–715.

Minshall, G. W., C. T. Robinson, T. V. Royer, and S. R. Rushforth. 1995. Benthic community structure in two adjacent streams in Yellowstone National Park five years after the 1988 wildfires. *Great Basin Naturalist* 55(3):193–200.

Monleon, V. J., K. Cromack, Jr., and J. D. Landsberg. 1997. Short- and long-term effects of prescribed underburning on nitrogen availability in ponderosa pine stands in central Oregon. *Canadian Journal of Forest Research* 27(3):369–378.

Morgan, D. L. 1969. *Jedediah Smith*. Lincoln, NE: University of Nebraska Press.

Morgan, T. 1993. *Wilderness at dawn: The settling of the North American continent*. New York: Simon & Schuster.

Morison, S. E. 1971. *The European discovery of America: The northern voyages A.D. 500–1600*. New York: Oxford University Press.

Morison, S. E. 1974. *The European discovery of America: The southern voyages A.D. 1492–1616*. New York: Oxford University Press.

Morris, E. 1979. *The rise of Theodore Roosevelt*. New York: Ballantine.

Morrow, T. 1996. *The late Paleoindian/early archaic period.* University of Iowa, Iowa City, Office of the Iowa State Archaeologist.

Morton, T. 1632. New English Canaan: or New Canaan. In Force, P. (ed.), *Tracts and other papers . . . of the colonies in North America,* Vol. II. Collected in 1838. Washington, DC.

Mueggler, W. F. 1985. Forage. In DeByle, N. V. and R. P. Winokur (eds.), *Aspen: Ecology and management in the western United States,* pp. 129–152. USDA Forest Service General Technical Report RM-119. Rocky Mountain Forest and Range Experiment Station, Fort Collins, CO.

Muhlstein, A. 1994. *La Salle, explorer of the North American frontier.* New York: Arcade Publishing.

Muir, J. 1901. *Our national parks.* Boston: Houghton Mifflin. (John Muir Exhibit, Sierra Club.)

Muir, J. 1918. Steep trails. In Bade, W. F. (ed.). Editied from field notes in 1888. Etext 326, September 1995. Project Gutenberg, University of Illinois.

Muir, J. 1961. *The mountains of California.* Garden City, NY: Anchor Books.

Mullan, J. 1861. Report of Lieutenant Mullan, in charge of the construction of the military road from Fort Benton to Fort Walla Walla. 36th Congress, 2nd Session, House Executive Document No. 44. Washington, DC.

Muller-Beck, H. J. 1967. On migrations of hunters across the Bering land bridge in the upper Pleistocene. In Hopkins, D. M. (ed.), *The Bering land bridge,* pp. 373–408. Stanford, CA: Stanford University Press.

Murphy, E. V. 1959. *Indian uses of native plants.* Fort Bragg, CA: Mendocino County Historical Society.

Nabhan, G. P. 1985. *Gathering the desert.* Tucson: University of Arizona Press.

Nash, R. F. 1967. *Wilderness and the American mind.* New Haven, CT: Yale University Press.

Natural Resources Canada. 1998. Compendium of Canadian Forestry Statistics, 1996. Canada's Forest Inventory (1994 version). Canadian Council of Forest Ministers, National Forestry Database Program.

Neiland, B. J. and L. A. Viereck. 1977. Forest types and ecosystems. Proceedings, North American Forest Lands at Latitudes North of 60 Degrees, pp. 109–136. University of Alaska, Fairbanks.

Nelson, J. C., L. Arndt, J. Ruhser, and L. Robinson. 1996. LUHNA pilot project—presettlement and contemporary vegetation patterns along upper Mississippi River Reaches 25 and 26. Land Use History of North America (LUHNA). Biological Resources Division, US Geological Survey.

Neumann, T. W. 1985. Human-wildlife competition and the passenger pigeon: Population growth from system destabilization. *Human Ecology* 4:389–410.

Neumann, T. W. 1995. The structure and dynamics of the prehistoric ecological system in the eastern woodlands: Ecological reality versus cultural myths. *Journal of Middle Atlantic Archaeology* 11:125–138.

Nevin, D. 1974. *The soldiers*. New York: Time-Life Books.

Newcomb, W. W., Jr. 1993. *The Indians of Texas*. Austin: University of Texas Press.

Nienstaedt, H. and J. C. Zasada. 1990. *Picea glauca* (Moench) Voss., White Spruce. In Burns, R. M. and B. H. Honkala (tech. coord.), *Silvics of North America, Vol. 1, Conifers*, pp. 204–226. Agricultural Handbook No. 654. Washington, DC: USDA Forest Service.

Nilsson, T. 1983. *The Pleistocene: Geology and life in the Quaternary Ice Age*. Boston, MA: D. Reidel.

Nordhoff, C. 1875. *Northern California, Oregon, and the Sandwich Islands*. New York: Harper & Brothers.

Nowacki, G. J. and M. D. Abrams. 1992. Community, edaphic, and historical analysis of mixed oak forests of the Ridge and Valley Province in central Pennsylvania. *Canadian Journal of Forest Research* 22:790–800.

Nowacki, G. J. and M. D. Abrams. 1994. Forest composition, structure, and disturbance history of the Alan Seeger Natural Area, Huntington County, Pennsylvania. *Bulletin of the Torrey Botanical Club* 12(3):277–291.

Nuzzo, V. A. 1986. Extent and status of midwest oak savanna: Presettlement and 1985. *Natural Areas Journal* 6(2):6–36.

O'Hagan, T. 1896. In the land of the Jesuit Martyrs. *The Catholic World* LXIII:71–82.

Oliver, C. D. and B. C. Larson. 1990. *Forest stand dynamics*. New York: McGraw-Hill.

Oliver, C. D. and E. P. Stephens. 1977. Reconstruction of a mixed-species forest in central New England. *Ecology* 58(3):562–572.

Oliver, W. W. 1995. Is self-thinning in ponderosa pine ruled by *Dendroctonus* bark beetles? In Eskew, L. G. (compiler), *Forest health through silviculture*, pp. 213–218. USDA Forest Service General Technical Report RM-GTR-267. Rocky Mountain Forest and Range Experiment Station, Fort Collins, CO.

Oliver, W. W. and R. A. Ryker. 1990. *Pinus ponderosa* Dougl. ex Laws. Ponderosa Pine. In Burns, R. M. and B. H. Honkala (tech. coord.), *Silvics of North America, Vol. 1, Conifers*, pp. 413–424. Agricultural Handbook No. 654. Washington, DC: USDA Forest Service.

Olmsted, F. L. 1857. *A journey through Texas, or a saddle-trip on the southwestern frontier: With a statistical appendix*. New York: Dix, Edwards.

Olson, D. F., Jr., D. F. Roy, and G. A. Walters. 1990. *Sequoia sempervirens* (D. Don) Endl. Redwood. In Burns, R. M. and B. H. Honkala (tech. coord.), *Silvics of North America, Vol. 1, Conifers*, pp. 541–551. Agricultural Handbook No. 654. Washington, DC: USDA Forest Service.

Omi, P. N. and K. D. Kalabokidis. 1990. Fire damage on extensively vs. intensively managed forest stands within the North Fork Fire, 1988. Proceedings, George Fahnestock Memorial Fire Science Symposium. March 23, 1990, pp. 149–157, Corvallis, OR.

O'Neale, L. M. 1932. Yurok-Karok basket weavers. *University of California Publications in American Archaeology and Ethnology* 32(1):1–184.

Ornduff, R. 1994. A botanist's view of the big tree. In Aune, P. S. (tech. coord.), *Proceedings of the Symposium on Giant Sequoias: Their Place in the Ecosystem and Society*, pp. 11–14. USDA Forest Service General Technical Report PSW-151. Pacific Southwest Research Station, Albany, CA.

Ortiz, B. 1993. Contemporary California Indian basketweavers and the environment. In Blackburn, T. C. and M. K. Anderson (eds.), *Before the wilderness: Environmental management by native Californians*, pp. 195–211. Menlo Park, CA: Ballena Press.

Overpeck, J. T. 1996. Warm climate surprises. *Science* 271:1820.

Packard, S. 1988. Just a few oddball species: Restoration and the rediscovery of the tallgrass savanna. *Restoration & Management Notes* 6(1):13–20.

Packee, E. C. 1990. *Tsuga heterophylla* (Raf.) Sarg. Western Hemlock. In Burns, R. M. and B. H. Honkala (tech. coord.), *Silvics of North America, Vol. 1, Conifers*, pp. 613–622. Agricultural Handbook No. 654. Washington, DC: USDA Forest Service.

Paillet, F. L. 1988. Character and distribution of American chestnut sprouts in southern New England woodlots. *Bulletin of the Torrey Botanical Club* 115:32–44.

Paillet, F. L. 1994. Ecology and paleoecology of American chestnut in eastern North American forests. *Journal of the American Chestnut Foundation.* VIII(2).

Paillet, F. L. 1996a. The archaeology of the chestnut: Investigating the history of American chestnut on a New England woodlot. *Journal of the American Chestnut Foundation.* X(1).

Paillet, F. L. 1996b. One of the world's last truly wild chestnut forests—the Achipsee River Valley of Russia's Caucasus State Biosphere Preserve. *Journal of the American Chestnut Foundation.* IX(2).

Palik, B. J. and K. S. Pregitzer. 1992. A comparison of presettlement and present-day forests on two bigtooth aspen-dominated landscapes in northern Lower Michigan. *American Midland Naturalist* 127:327–338.

Palik, B. J., R. J. Mitchell, G. Houseal, and N. Pederson. 1997. Effects of canopy structure on resource availability and seedling responses in a longleaf pine ecosystem. *Canadian Journal of Forest Research* 27:1458–1464.

Palmer, J. 1847. *Journal of travels over the Rocky Mountains.* Cincinnati: J. A. and U. P. James.

Parker, A. A. 1836. *Trip to the West and Texas.* 2nd edition. Concord, NH: William White.

Parker, A. C. 1910. *Iroquois uses of maize and other food plants.* Albany, NY: University of the State of New York.

Parker, A. C. 1913. The Code of Handsome Lake. New York State Museum Bulletin 163.

Parker, A. J. 1988. Stand structure in subalpine forests of Yosemite National Park, California. *Forest Science* 34(4):1047–1058.

Patterson, R. 1992. Fire in the oaks. *American Forests* 98(11):32–34, 58–59.

Patton, D. R. 1974. Patch cutting increases deer and elk use of a pine forest in Arizona. *Journal of Forestry* 72(12):764–766.

Patton, D. R. 1977. Managing southwestern ponderosa pine for the Albert squirrel. *Journal of Forestry* 75(5):264–267.

Pavek, D. S. 1992. *Epilobium angustifolium* [fireweed]. In Fischer, W. C. (compiler), *The Fire Effects Information System* [data base]. USDA Forest Service, Intermountain Research Station, Intermountain Fire Sciences Laboratory, Missoula, MT.

Payandeh, B. 1974. Spatial pattern of trees in the major forest types of northern Ontario. *Canadian Journal of Forest Research* 4:8–14.

Peet, R. K. 1989. Forests of the Rocky Mountains. In Barbour, M. G. and W. D. Billings (eds.), *North American terrestrial vegetation*, pp. 63–101. New York: Cambridge University Press.

Peet, R. K. and D. J. Allard. 1993. Longleaf pine vegetation of the southern Atlantic Gulf Coast regions: a preliminary classification. In Hermann, S. M. (ed.), *The longleaf pine ecosystem: Ecology, restoration and management*, pp. 45–81. Proceedings of the Tall Timbers Fire Ecology Conference No. 18. Tall Timbers Research Station, Tallahassee, FL.

Percy, G. 1625. Observations gathered out of a discourse of the plantation of the southerne colonie in Virginia by the English, 1606. In. S. Purchas, *Purchas his pilgrimes*, pp. 1685–1690. London: Wm. Strachey.

Peri, D. W. and S. M. Patterson. 1978. An ethnographic and ethnohistoric survey and an assessment of Native American interests in the SUNEDCO geothermal leasehold, Mendocino County, California. A report prepared for Ecoview Environmental Consultants, Napa, California. (Ms. on file at the Ethnographic Laboratory, Sonoma State University.)

Peri, D. W. and S. M. Patterson. 1979. Ethnobotanical resources of the Warm Springs Dam–Lake Sonoma Project Area, Sonoma County, California. Final Report for U.S. Army Corps of Engineers, San Francisco District.

Peri, D. W., S. M. Patterson and J. L. Goodrich. 1982. Ethnobotanical mitigation of Warm Springs Dam–Lake Sonoma, California. Unpublished report prepared for the U.S. Army Corps of Engineers, San Francisco District.

Peters, E. F. and S. C. Buntin. 1994. Proceedings—Ecology and Management of

Annual Rangelands; 1992 May 18–21, Boise, ID, pp. 31–36. General Technical Report INT-GTR-313. USDA Forest Service, Intermountin Research Station, Odgen, UT.

Peters, R. L. II. 1988. The effect of global climatic change on natural communities. In Wilson, E. O. (ed.), *Biodiversity*, pp. 450–461. Washington, DC: National Academy Press.

Peterson, C. J. and S. T. A. Pickett. 1991. Treefall and resprouting following catastrophic windthrow in an old-growth hemlock-hardwoods forest. *Forest Ecology and Management* 42:205–217.

Peterson, C. J. and E. R. Squiers. 1995. An unexpected change in spatial pattern across 10 years in an aspen-white pine forest. *Journal of Ecology* 83:847–855.

Peterson, D. L. 1984. Predicting fire-caused mortality in four northern Rocky Mountain conifers. Proceedings, 1983 Society of American Foresters National Conference. October 16–20, pp. 276–280. Portland, OR.

Peterson, J. 1994. Voices in the forest. Evergreen Magazine. Evergreen Foundation, Medford, Oregon. March, April, pp. 14–15.

Peterson, J. 1994–1995. The 1910 fire. Evergreen Magazine. Evergreen Foundation, Medford, Oregon. pp. 8–18.

Peterson, R. T. 1947. *A field guide to the birds, eastern land and water birds.* Boston: Houghton Mifflin Company.

Peterson, R. T. 1961. *A field guide to western birds.* Cambridge, MA: Riverside Press.

Phillips, P. C. (ed.). 1940. *Life in the Rocky Mountains: A diary of wanderings on the sources of the rivers Missouri, Columbia, and Colorado from February, 1830, to November, 1835 by W A. Ferris, then in the employ of the American Fur Company.* Denver: Old West Publishing.

Phillips, P. C. (ed.). 1957. *Forty years on the frontier as seen in the journals and reminiscences of Granville Stuart.* Glendale, CA: Arthur Clark Company.

Philpot, C. A. 1973. The changing role of fire on chaparral lands. In *Symposium on Living with the Chaparral*, pp. 131–150. March 30–31, University of California, Riverside.

Pielou, E. C. 1991. *After the ice age.* Chicago: University of Chicago Press.

Pieper, R. D. 1977. The Southwestern pinyon-juniper ecosystem. In Aldon, E. F. and T. J. Loring (tech. coord.), *Ecology, uses, and management of pinyon-juniper woodlands*, pp. 1–6. USDA Forest Service General Technical Report RM-39. Rocky Mountain Forest and Range Experiment Station, Fort Collins, CO.

Pieper, R. D. 1993. Spatial variation of piñion-juniper woodlands in New Mexico. In Aldon, E. F. and D. W. Shaw (tech. coord.), *Managing piñion-juniper ecosystems for sustainability and social needs*, pp. 89–96. USDA Forest Service General Technical Report RM-236. Rocky Mountain Forest and Range Experiment Station, Fort Collins, CO.

Pieper, R. D. and R. D. Wittie. 1990. Fire effects in southwestern chaparral and pinyon-juniper vegetation. In Drammes, J. S. (tech. coord.), *Effects of fire management of southwestern natural resources*, pp. 87–93. USDA Forest Service General Technical Report RM-191. Rocky Mountain Forest and Range Experiment Station, Fort Collins, CO.

Piirto, D. D. 1996. Reference variability for giant sequoia: an annotated review of literature. USDA Forest Service. Sequoia National Forest, CA.

Pilarski, M. (ed.). 1994. *Restoration forestry*, pp. 301–304. Durango, CO: Kivaki Press.

Platt, W. J. and S. L. Rathbun. 1993. Dynamics of an old-growth longleaf pine population. In Hermann, S. M. (ed.), *The longleaf pine ecosystem: Ecology, restoration and management*, pp. 275–297. Proceedings of the Tall Timbers Fire Ecology Conference No. 18. Tall Timbers Research Station, Tallahassee, FL.

Platt, W. J., G. W. Evans, and S. L. Rathbun. 1988. The population dynamics of a long-lived conifer (Pinus palustris). *American Naturalist* 131:491–525.

Platt, W. J., J. S. Glitzenstein, and D. R. Streng. 1991. Evaluating pyrogenicity and its effects on vegetation in longleaf pine savannas. Proceedings, 17th Tall Timbers Fire Ecology Conference, High Intensity Fire in Wildlands: Management Challenges and Options, pp. 143–161. Tall Timbers Research Station, Tallahassee, FL.

Porter, S. C. 1988. Landscapes of the last ice age in North America. In R. C. Carlisle (ed.), *Americans before Columbus: Ice-age origins*, pp. 1–24. Ethnology Monographs 12, Department of Anthropology. Pittsburgh: University of Pittsburg.

Pringle, H. 1997. New respect for metal's role in ancient Arctic cultures. *Science* 277(5327):766–767.

Progulske, D. R. 1974. Yellow ore, yellow hair, and yellow pine: A photographic study of a century of forest ecology. South Dakota State University, Brookings. Agricultural Experiment Station Bulletin 616.

Purchas, S. 1906. *Hakluytus Posthusmus*, or Purchas His Pilgrimes, 20 vols. Glasgow: James Maclehose and Sons.

Pyne, S. J. 1982. *Fire in America: A cultural history of wildland and rural fire.* Princeton, NJ: Princeton University Press.

Quaife, M. M. 1921. *Alexander Henry's travels and adventures in the years 1760–1776.* Chicago, IL: The Lakeside Press.

Quaife, M. M. (ed.). 1978. *Adventures of a mountain man: The narrative of Zenas Leonard.* Lincoln: University of Nebraska Press.

Quick, C. R. 1956. Viable seed from the duff and soil of sugar pine forests. *Forest Science* 2:36–42.

Quick, C. R. and A. S. Quick. 1961. Germination of *Ceanothus* seeds. *Madroño* 16:23–30.

Quinby, P. A. 1991. Self-replacement in old-growth white pine forests of Temagami, Ontario. *Forest Ecology and Management* 41:95–109.

Quirk, W. A. and D. J. Sykes. 1971. White spruce stringers in a fire-patterned landscape in interior Alaska. In Slaughter, C. W., R. J. Barney, and G. M. Hansen (eds.), *Fire in the northern environment—A Symposium*, pp. 179–197. USDA Forest Service, Pacific Northwest Forest and Range Experiment Station, Portland, OR.

Ralegh, Sir Walter (1751). The Cabinet Council. Chapter 25. In *The Columbia Dictionary of Quotations*. Columbia University Press. 1993.

Ramenofsky, A.F. 1987. *Vectors of death: The archaeology of European contact.* Albuquerque, NM: University of New Mexico Press.

Rebertus, A. J., G. B. Williamson, and W. J. Platt. 1993. Impact of temporal variation in fire regime on savanna oaks and pines. In Hermann, S. M. (ed.), *The longleaf pine ecosystem: Ecology, restortion and management*, pp. 215–225. Proceedings of the Tall Timbers Fire Ecology Conference No. 18. Tall Timbers Research Station, Tallahassee, FL.

Regelbrugge, J. C. and D. W. Smith. 1994. Postfire tree mortality in relation to wildfire severity in mixed oak forests in the Blue Ridge of Virginia. *Northern Journal of Applied Forestry* 11(3):90–97.

Reiners, W. A. and G. E. Lang. 1979. Vegetational patterns and processes in the balsam fir zone, White Mountains, New Hampshire. *Ecology* 60(2):403–417.

Reynolds, R. D. 1959. The effect upon the forest of natural fire and aboriginal burning in the Sierra Nevada. M.A. Thesis, University of California, Berkeley.

Reynolds, R. T. and B. D. Linkhart. 1992. Flammulated owls in ponderosa pine. In Kaufmann, M. R., W. H. Moir, and R. L. Bassett (tech. coord.), *Old-growth forests in the Southwest and Rocky Mountain regions, Proceedings of a Workshop*, pp. 166–169. USDA Forest Service General Technical Report RM-213. Rocky Mountain Forest and Range Experiment Station, Fort Collins, CO.

Reynolds, R. T., R. T. Graham, M. H. Reiser, R. L. Bassett, P. L. Kennedy, D. A. Boyce, Jr., G. Goodwin, R. Smith, and E. Leon Fisher. 1992. Management recommendations for the northern goshawk in the Southwestern United States. USDA Forest Service General Technical Report RM-217. Rocky Mountain Forest and Range Experiment Station, Fort Collins, CO.

Rhode, D. and D. B. Madsen. 1995. Late Wisconsin/Early Holocene vegetation in the Bonneville Basin. *Quaternary Research* 44:246–256.

Rice, E. L. and W. T. Penfound. 1959. The upland forests of Oklahoma. *Ecology* 40:593–608.

Richards, R. T. 1997. What the natives know: Wild mushrooms and forest health. *Journal of Forestry* 95(9):4–10.

Riddle, G. 1920. *History of early days in Oregon.* Seattle: Shorey Book reprint, 1968.

Ringland, A. C. 1905. Report on the resources and needs of the Lincoln Forest Reserve. Lincoln National Forest. Alamogordo, NM.

Ripple, W. J. 1994. Historic spatial patterns of old forests in western Oregon. *Journal of Forestry* 92(11):45–49.

Ritchie, J. C. and G. M. MacDonald. 1986. The patterns of post-glacial spread of white spruce. *Journal of Biogeography* 13:527–540.

Rodesch, J. C. 1984. Jean Nicolet. *Voyageur. Historical Review of Brown County and Northeast Wisconsin* Spring, pp. 4–8.

Rogers, R. S. 1978. Forests dominated by hemlock (*Tusga canadensis*): Distribution as related to site and potsettlement history. *Canadian Journal of Botany* 56(7):843–854.

Rollins, P. A. 1995. *The discovery of the Oregon Trail: Robert Stuart's narratives of his overland trip eastward from Astoria 1812–13*. Lincoln: University of Nebraska Press.

Romme, W. H. 1980. Fire frequency in subalpine forests of Yellowstone National Park. Proceedings, Fire History Workshop, pp. 27–30. USDA Forest Service General Technical Report RM-81. Rocky Mountain Forest and Range Experiment Station, Fort Collins, CO.

Romme, W. H. 1982. Fire and landscape diversity in subalpine forests of Yellowstone National Park. *Ecological Monographs* 52(2):199–221.

Romme, W. H. and D. G. Despain. 1989. The long history of fire in the Greater Yellowstone ecosystem. *Western Wildlands* 15:10–17.

Roosevelt, A. C., M. Lima da Costa, C. Lopes Machado, M. Michab, N. Mercler, H. Valladas, J. Feathers, W. Barnett, M. Imazio da Silveira, A. Henderson, J. Silva, B. Chernoff, D. S. Reese, J. A. Holman, N. Toth, and K. Schick. 1996. Paleoindian cave dwellers in the Amazon: The peopling of the Americas. *Science* 272:373–384.

Rosier, J. 1906. A true relation of the voyage of Captain George Waymouth, 1605. In Burrage, H. S. (ed.), *Early English and French voyages*. New York: Barnes & Noble.

Rostlund, E. 1957. The myth of a natural prairie belt in Alabama: An interpretation of historical records. *Annals of the Association of American Geographers* 47(4):392–411.

Rostlund, E. 1960. The geographic range of the historic bison in the Southeast. *Annals of the Association of American Geographers* 50(4):395–407.

Rothermel, R. C. 1991. Predicting behavior and size of crown fires in the northern Rocky Mountains. USDA Forest Service Research Paper INT-438. Intermountain Research Station, Odgen, UT.

Rothwell, R. L. 1991. *Henry David Thoreau, an American landscape*. New York: Paragon House.

Rowe, J. S. 1970. Spruce and fire in northwest Canada and Alaska. Proceedings, Annual Tall Timbers Fire Ecology Conference, Number 10, pp. 245–254. Tall Timbers Research Station, Tallahassee, FL.

Rowe, J. S. and G. W. Scotter. 1973. Fire in the boreal forest. *Quaternary Research* 3: 444–464.

Rube, R. V. 1983. Depositional environment of late Wisconsin loess in the mid-continental United States. In Porter, S. C. (ed.), *Late-Quaternary environments of the United States, Vol. 1, The Pleistocene*, pp. 130–137. Minneapolis: University of Minnesota Press.

Rudolph, T. D. and P. R. Laidly. 1990. *Pinus banksiana* Lamb., Jack Pine. In Burns, R. M. and B. H. Honkala (tech. coord.), *Silvics of North America, Vol. 1, Conifers*, pp. 280–293. Agricultural Handbook No. 654. Washington, DC: USDA Forest Service.

Rudolph, V. J. and W. A. Lemmien. 1976. Silvicultural cuttings in an oak-hickory stand in Michigan: 21-year results. In Fralish, J. S., G. T. Weaver, and R. C. Schlesinger (eds.), *Proceedings of the Central Hardwood Forest Conference*, pp. 431–453. Carbondale, IL: Southern Illinois University.

Rumble, M. A. and J. E. Gobeille. 1995. Wildlife associations in Rocky Mountain juniper in the northern Great Plains, South Dakota. In Shaw, D. W., E. F. Aldon, and C. LoSapio (tech. coord.), *Desired future conditions for piñon-juniper ecosystems*, pp. 80–90. USDA Forest Service General Technical Report RM-258. Rocky Mountain Forest and Range Experiment Station, Fort Collins, CO.

Rundel, P. W. 1971. Community structure and stability in the giant sequoia groves of the Sierra nevada, California. *American Midland Naturalist* 85(2):478–492.

Rundel, P. W. 1972. Habitat restriction in giant sequoia: The environmental control of grove boundaries. *American Midland Naturalist* 87(1):81–99.

Rundel, P. W., D. J. Parsons, and D. T. Gordon. 1977. Montane and subalpine vegetation of the Sierra Nevada and Cascade ranges. In Barbour, M. G. and J. Major (eds.), *Terrestrial vegetation of California*, pp. 559–599. New York: Wiley.

Runkle, J. R. 1981. Gap regeneration in some old-growth forests of the eastern United States. *Ecology* 62(4):1041–1051.

Runkle, J. R. 1991. Natural disturbance regimes and the maintenance of stable regional floras. In Henderson D. and L. D. Hedrick (eds.), *Restoration of old growth forests in the interior highlands of Arkansas and Oklahoma—Proceedings of the Conference.* pp. 31–47. Ouachita National Forest and Winrock International Institute for Agricultural Development, Hot Springs, AK.

Russell, E. W. B. 1983. Indian-set fires in the forests of the Northeastern United States. *Ecology* 64(1):78–88.

Russell, H. S. 1980. *Indian New England before the Mayflower*. Hanover, NH: University Press of New England.

Sackett, S. S. 1984. Observations on natural regeneration in ponderosa pine following a prescribed fire in Arizona. USDA Forest Service Research Note RM-435. Rocky Mountain Forest and Range Experiment Station, Fort Collins, CO.

Sackett, S. S. and S. M. Hasse. 1996. Fuel loadings in southwestern ecosystems of the United States. In Ffolliott, P. F., L. F. DeBano, M. B. Baker, Jr., G. J. Gottfried, G. Solis-Garza, C. B. Edminster, D. G. Neary, L. S. Allen, and R. H. Hamre (tech. coord.), *Effects of fire on Madrean Province ecosystems*, pp. 187–192. USDA Forest Service General Technical Report RM-GTR-289. Rocky Mountain Forest and Range Experiment Station, Fort Collins, CO.

Saleeby, B. M. 1983. Prehistoric settlement patterns in the Portland Basin of the Lower Columbia River. Ph.D. Dissertation, University of Oregon, Eugene.

Samson, F. and F. Knopf. 1994. Prairie conservation in North America. *BioScience* 44(6):418–421.

Sander, I. L. and D. L. Graney. 1993. Regenerating oaks in the Central States. In Loftis, D. and C. E. McGee (eds.), *Oak regeneration: Serious problems, practical recommendations*, pp. 174–183. USDA Forest Service General Technical Report SE-84. Southeastern Forest Experiment Station. Asheville, NC.

Sapir, E. 1907. Notes on the Takelma Indians. *American Anthropologist* 9:251–275.

Sauer, C. O. 1971. *Sixteenth century North America*. Berkeley: University of California Press.

Saunders, J. W., R. D. Mandel, R. T. Saucier, E. T. Allen, C. T. Hallmark, J. K. Johnson, E. H. Jackson, C. M. Allen, G. L. Stringer, D. S. Frink, J. K. Feathers, S. Williams, K. J. Gremillion, M. F. Vidrine, and R. Jones. 1997. A mound complex in Louisiana at 5400–5000 years before the present. *Science* 277(5333):1796–1799.

Sawyer, J. O and D. A. Thornburgh. 1977. Montane and subalpine vegetation of the Klamath Mountains. In Barbour, M. G. and J. Major (eds.), *Terrestrial vegetation of California*, pp. 699–732. New York: Wiley.

Sawyer, J. O, D. A. Thornburgh, and J. R. Griffin. 1977. Mixed evergreen forest. In Barbour, M. G. and J. Major (eds.), *Terrestrial vegetation of California*, pp. 359–381. New York: Wiley.

Schenk, S. M. and E. W. Gifford. 1952. Karok ethnobotany. *Anthropological Records* 13(6):377–392.

Schier, G. A., J. R. Jones, and R. P. Winokur. 1985. Vegetative regeneration. In DeByle, N. V. and R. P. Winokur (eds.), *Aspen: Ecology and management in the western United States*, pp. 29–33. USDA Forest Service General Technical Report RM-119. Rocky Mountain Forest and Range Experiment Station, Fort Collins, CO.

Schlick, M. D. 1994. *Columbia River basketry, gift of the ancestors, gift of the earth.* Seattle: University of Washington Press.

Schoolcraft, H. R. 1966. *Travels through the northwestern regions of the United States* (originally published in 1821). Ann Arbor, MI: University Microfilms.

Schoonmaker, P. and A. McKee. 1988. Species composition and diversity during secondary succession of coniferous forests in the western Cascade Mountains of Oregon. *Forest Science* 34(4):960–979.

Schubert, G. H. 1956. Early survival and growth of sugar pine and white fir in clear-cut openings. USDA Forest Service Research Note Number 117. Pacific Southwest Forest and Range Experiment Station. Berkeley, CA.

Schubert, G. H. 1974. Silviculture of southwestern ponderosa pine: A status of our knowledge. USDA Forest Service Research Paper RM-123. Rocky Mountain Forest and Range Experiment Station, Fort Collins, CO.

Scott, V. E. 1978. Characteristics of ponderosa pine snags used by cavity-nesting birds in Arizona. *Journal of Forestry* 76(1):26–28.

Scott, V. E., K. E. Evans, D. R. Patton, and C. P. Stone. 1977. *Cavity-nesting birds of North American forests. Agriculture Handbook 511.* Washington, DC: USDA Forest Service.

Scotter, G. W. 1970. Wildfires in relation to the habitat of barren-ground caribou in the taiga of northern Canada. Proceedings, Annual Tall Timbers Fire Ecology Conference, Number 10, pp. 85–105. Tall Timbers Research Station, Tallahassee, FL.

Scurlock, D. and D. M. Finch. 1997. A historical review. In Block, W. M. and D. M. Finch (tech. eds.), *Songbird ecology in southwestern ponderosa pine forests*, pp. 43–68. USDA Forest Service General Technical Report RM-GTR-292. Rocky Mountain Forest and Range Experiment Station, Fort Collins, CO.

Seip, D. and J. P. Savard. 1995. Wildlife diversity in old growth forests and management stands. Ministry of Forests, Prince George Forest Region, British Columbia.

Seklecki, M. T., H. D. Grissino-Mayer, and T. W. Swetnam. 1996. Fire history and the possible role of Apache-set fires in the Chiricahua Mountains of southeastern Arizona. In Ffolliott, P. F., L. F. DeBano, M. B. Baker, Jr., G. J. Gottfried, G. Solis-Garza, C. B. Edminster, D. G. Neary, L. S. Allen, and R. H. Hamre (tech. coord.), *Effects of fire on Madrean Province ecosystems*, pp. 238–246. USDA Forest Service General Technical Report RM-GTR-289. Rocky Mountain Forest and Range Experiment Station, Fort Collins, CO.

Sellers, J. A. 1970. Mixed conifer forest ecology: A quantitative study in Kings Canyon National Park, Fresno County, California. M.A. Thesis, Department of Biology, Fresno State College, Fresno, CA.

Semken, H. A., Jr. 1983. Holocene mammalian biogeography and climatic change in the eastern and central United States. In Wright, H. E., Jr. (ed.), *Late-Quaternary environments of the United States, Vol. 2, The Holocene*, pp. 182–207. Minneapolis: University of Minnesota Press.

Seno, W. J. 1985. *Up country: Voices from the midwestern wilderness.* Madison, WI: Round River Publishing.

Severson, K. E. and A. L. Medina. 1983. Deer and elk habitat management in the Southwest. *Journal of Range Management Monograph No. 2.*

Seymour, R. S. 1995. The northeastern region. In Barrett, J. W. (ed.), *Regional silviculture of the United States*, 3rd ed., pp. 31–79. New York: Wiley.

Shantz, H. L. and R. Zon. 1924. Natural vegetation. In *Atlas of American Agriculture*. Washington, DC: US Government Printing Office.

Shaw, B. K. 1997. The huckleberry story: A bridge between culture and science. Oregon State University Extension Service. Warm Springs Indian Reservation, OR.

Shellhammer, H. S. 1980a. Birds and mammals, fire, and giant sequoia reproduction. In Harvey, H. T., H. S. Shellhammer, and R. E. Stecker, *Giant sequoia ecology: Fire and reproduction*, pp. 101–118. Scientific Monograph Series No. 12. Washington, DC: USDI National Park Service.

Shellhammer, H. S. 1980b. Douglas squirrels and sequoia regeneration. In Harvey, H. T., H. S. Shellhammer, and R. E. Stecker (eds.), *Giant sequoia ecology: Fire and reproduction*, pp. 119–142. Scientific Monograph Series No. 12. Washington, DC: USDI National Park Service.

Shirreff, Patrick. 1835. *A tour through North America; together with a comprehensive view of the Canadas and the United States as adopted for agricultural emigration.* Edinburgh.

Short, H. L. and C. Y. McCulloch. 1977. Managing pinyon-juniper ranges for wildlife. USDA Forest Service General Technical Report RM-47. Rocky Mountain forest and Range Experiment Station, Fort Collins, CO.

Short, H. L., W. Evans, and E. L. Boeker. 1977. The use of natural and modified pinyon pine-juniper woodlands by deer and elk. *Journal of Wildlife Management* 41(3):543–559.

Shotola, S. J., G. T. Weaver, P. A. Robertson, and W. C. Ashby. 1992. Sugar maple invasion of an old-growth oak-hickory forest in southwestern Illinois. *American Midland Naturalist* 127:125–138.

Show, S. B. and E. I. Kotok. 1924. The role of fire in the California pine forests. USDA Department Bulletin 1294, Washington, DC.

Show, S. B. and E. I. Kotok. 1929. Cover type and fire control in the national forests of northern California. USDA Department Bulletin 1495, Washington, DC.

Shugart, H. H., D. C. West, and W. R. Emanuel. 1981. Patterns and dynamics of forests: An application of simulation models. In West, D. C., H. H. Shugart, and D. B. Botkin (eds.), *Forest succession: Concepts and application*, pp. 74–94. New York: Springer.

Siccama, T. G. 1971. Presettlement and present forest vegetation in northern Vermont with special reference to Chittenden County. *American Midland Naturalist* 85:153–172.

Silverberg, R. 1970. *Mammoths, mastodons and man.* New York: McGraw-Hill.

Simpson, L. B. 1938. California in 1792: *The expedition of Jose Longinos Martinez.* San Marino, CA: Huntington Library.

Sitgreaves, L. 1853. *Report of an expedition down the Zuni and Colorado Rivers.* United States Army Corps of Topographcial Engineers, 32nd Congress, 2nd Session, U.S. Senate, Executive Document No. 59. Washington, DC: Robert Armstrong, Public Printer.

Smith, B. D. 1997. The initial domestication of *Cucurbita pepo* in the Americas 10,000 years ago. *Science* 276(5314):932–934.

Smith, G. I. and F. A. Street-Perrott. 1983. Pluvial lakes of the western United States. In Porter, S. C. (ed.), *Late-Quaternary environments of the United States, Vol. 1, The Pleistocene,* pp. 190–212. Minneapolis: University of Minnesota Press.

Smith, H. C. 1980. An evlauation of four uneven-age cutting practices in central Appalachian hardwoods. *Southern Journal of Applied Forestry* 4(4):193–200.

Smith, H. H. 1933. Ethnobotany of the Forest Potawatomi Indians. *Bulletin of the Public Museum of the City of Milwaukee* 7:1–230.

Smith, J. K. and W. C. Fischer. 1997. Fire ecology of the forest habitat types of northern Idaho. USDA Forest Service Research Paper INT-GTR-363. Intermountain Research Station, Odgen, UT.

Smith, John. 1624a. The generall historie of Virginia, New England, and the Summer Isles. Book II. In Lankford, J. (ed.). 1967. *Captain John Smith's America: Selections from his writings,* pp. 3–34. New York: Harper & Row.

Smith, John. 1624b. The generall historie of Virginia, New England, and the Summer Isles. Book III. In Lankford, J. (ed.). 1967. *Captain John Smith's America: Selections from his writings,* pp. 35–124. New York: Harper & Row.

Smith, John. 1624c. The generall historie of Virginia, New England, and the Summer Isles. Book VI. In Lankford, J. (ed.). 1967. *Captain John Smith's America: Selections from his writings,* pp. 125–148. New York: Harper & Row.

Smith, John. 1631. Advertisements for the unexperienced planters of New-England, or any where. OR The path-way to experience to errect a Plantation. In Lankford, J. (ed.). 1967. *Captain John Smith's America: Selections from his writings,* pp. 164–195. New York: Harper & Row.

Smith, K. G. and J. C. Neal. 1991. Pre-settlement birds and mammals of the interior highlands. In Henderson D. and L. D. Hedrick (eds.), *Restoration of old growth forests in the interior highlands of Arkansas and Oklahoma—Proceedings of the Conference.* pp. 77–103. Ouachita National Forest and Winrock International Institute for Agricultural Development. Hot Springs, AK.

Smith, W. R. 1854a. *The history of Wisconsin. In three parts, historical, documentary, and descriptive. Part I, Historcial, Vol. I.* Madison, WI: B. Brown.

Smith, W. R. 1854b. *The history of Wisconsin. In three parts, historical, documentary, and descriptive. Part II, Documentary, Vol. III.* Madison, WI: B. Brown.

Snyder, S. A. 1991a. *Alces alces* [moose]. In Fischer, W C. (compiler), *The Fire Effects Information System* [data base]. USDA Forest Service, Intermountain Research Station, Intermountain Fire Sciences Laboratory, Missoula, MT.

Snyder, S. A. 1991b. *Odocoileus hemionus* [mule deer]. In Fischer, W C. (compiler), *The Fire Effects Information System* [data base]. USDA Forest Service, Intermountain Research Station, Intermountain Fire Sciences Laboratory, Missoula, MT.

Snyder, S. A. 1991c. *Cervus elaphus* [elk]. In Fischer, W C. (compiler), *The Fire Effects Information System* [data base]. USDA Forest Service, Intermountain Research Station, Intermountain Fire Sciences Laboratory, Missoula, MT.

Snyder, S. A. 1991d. *Rangifer tarandus* [caribou]. In Fischer, W. C. (compiler), *The Fire Effects Information System* [data base]. USDA Forest Service, Intermountain Research Station, Intermountain Fire Sciences Laboratory, Missoula, MT.

Society of American Foresters. 1967. *Forest cover types of North America*. Washington, DC.

Solomon, A. M. and T. Webb, III. 1985. Computer-aided reconstruction of late-Quaternary landscape dynamics. *Annual Review of Ecology and Systematics.* 16:63–84.

Southern Appalachian Man and the Biosphere (SAMAB). 1996. *The southern Appalachian assessment terrestrial technical report*, Report 5 of 5. Atlanta: USDA Forest Service, Southern Region.

Spaulding, W. G., E. B. Leopold, and T. R. Van Devender. 1983. Late Wisconsin paleoecology of the American Southwest. In Porter, S. C. (ed.), *Late-Quaternary environments of the United States, Vol. 1, The Pleistocene*, pp. 259–293. Minneapolis: University of Minnesota Press.

Spear, R. W., M. B. Davis, and L. C. K. Shane. 1994. Late Quaternary history of low- and mid-elevation vegetation in the White Mountains of New Hampshire. *Ecological Monographs* 64(1):85–109.

Speck, F. G. 1941. A list of plant curatives obtained from the Houma Indians of Louisiana. *Primitive Man* 14:49–75.

Spoerl, P. M. and J. C. Ravesloot. 1995. From Casas Grandes to Casa Grande: Prehistoric human impacts in the Sky Islands of southern Arizona and northwestern Mexico. In *Biodiversity and management of the Madrean Archipelago: The Sky Islands of southwestern United States and Northwestern Mexico*, pp. 492–501. USDA Forest Service General Technical Report RM-GTR-264. Rocky Mountain Forest and Range Experiment Station, Fort Collins, CO.

Sprugel, D. G. 1976. Dynamic structure of wave-regenerated *Abies Balsamea* forests in the North-Eastern United States. *Journal of Ecology* 64:889–911.

Spurr, S. H. and B. V. Barnes. 1980. *Forest ecology*. New York: Wiley.

Standiford, R. B., J. Klein, and B. Garrison. 1996. Sustainability of Sierra Nevada hardwood rangelands. In *Status of the Sierra Nevada*, Vol. III, pp. 637–680. Sierra Nevada Ecosystem Project, Final Report to Congress, Wildland Resources Center Report No. 38, Centers for Water and Wildland Resources. Davis, CA: University of California.

Stark, F. 1948. Perseus in the Wind, Chapter 14. In *The Columbia Dictionary of Quotations*, Columbia University Press. 1993.

Stark, N. 1965. Natural regeneration of Sierra Nevada mixed conifers after logging. *Journal of Forestry* 63:456–461.

Startwell, C., R. F. Schmitz, and W. J. Buckhorn. 1971. Pine engraver, *Ips pini*, in the western states. USDA Forest Service Pest Leaflet No. 122.

Stearns, F. W. 1949. Ninety years change in a northern harwood forest in Wisconsin. *Ecology* 30(3):350–358.

Stearns, F. W. 1950. The composition of a remnant of white pine forest in the Lake States. *Ecology* 31(2):290–292.

Stearns, F. W. 1951. The composition of the sugar maple-hemlock-yellow birch association in northern Wisconsin. *Ecology* 32(2):245–265.

Stecker, R. E. 1980. The role of insects in giant sequoia reproduction. In Harvey, H. T., H. S. Shellhammer, and R. E. Stecker, *Giant sequoia ecology: Fire and reproduction*, pp. 83–100. Scientific Monograph Series No. 12. Washington, DC: USDI National Park Service.

Steele, C. 1985. A journey through Texas in 1767. *El Campanario* 16(1):1–28.

Stephenson, N. L. 1994. Long-term dynamics of giant sequoia populations: Implications for managing a pioneer species. In Aune, P. S. (tech. coord.), *Proceedings of the Symposium on Giant Sequoias: Their Place in the Ecosystem and Society*, pp. 56–63. USDA Forest Service General Technical Report PSW–151. Pacific Southwest Research Station, Albany, CA.

Stephenson, N. L., D. J. Parsons, and T. W. Swetnam. 1991. Restoring natural fire to the sequoia-mixed conifer forest: Should intense fire play a role? Proceedings, 17th Tall Timbers Fire Ecology Conference, *High Intensity Fire in Wildlands: Management Challenges and Options*, pp. 321–337. Tall Timbers Research Station, Tallahassee, FL.

Stephenson, S. L. 1986. Changes in a former chestnut-dominated forest after a half century of succession. *American Midland Naturalist* 116(1):173–179.

Stepp, D. 1997. California lake site rich in fluted projectile points. Mammoth Trumpet 12(4). Center for the Study of the First Americans. Oregon State University, Corvallis.

Stevens, I. I. 1855. *Explorations for a route for the Pacific Railroad near the fortyseventh and forty-ninth parallels of north latitude from St. Paul to Puget Sound.* United States War Department. Washington, DC: A. O. P. Nicholson, Printer.

Stevens, I. I. 1860. *Reports of explorations and surveys, to ascertain the most practicable and economical route for a railroad from the Mississippi River to the Pacific Ocean, 1853–54.* Volume XII, Book I. 36th Congress, 1st Session. Executive Document. Washington, DC: Thomas H. Ford, Printer.

Stewart, O. C. 1955. Fire as the first great force employed by man. In Thomas, W. L. (ed.), *Man's role in changing the face of the earth*, pp. 115–133. Chicago: University of Chicago Press.

Stewart, O. C. 1963. Barriers to understanding the influence of use of fire by aborigines on vegetation. *Annual Proceedings, Tall Timbers Fire Ecology Conference* 2:117–126.

Stocking, S. K. and J. A. Rockwell. 1969. Widlflowers of Sequoia and Kings Canyon National Parks. Sequoia Natural History Association, Three Rivers, CA.

Stohlgren, T. J. 1991. Size distributions and spatial patterns of giant sequoia (*Sequoiadendron giganteum*) in Sequoia and Kings Canyon National Parks. Cooperative National Park Resources Studies Unit Technical Report No. 43. USDI National Park Service, Davis, CA.

Stoltman, J. B. and D. A. Baerreis. 1983. The evolution of human ecosystems in the eastern United States. In Wright, H. E., Jr. (ed.), *Late-Quaternary environments of the United States. Vol. 2, The Holocene*, pp. 252–268. Minneapolis: University of Minnesota Press.

Stone, E. C. 1968. Preservation of coast redwoods on alluvial flats. *Science* 159(3811): 157–161.

Stone, E. C., R. F. Grah, and P. J. Zinke. 1972. Preservation of primeval redwoods in the Redwood National Park, Part I. *American Forests* 78(4):50–55.

Storer, T. I. and R. L. Usinger. 1968. *Sierra Nevada natural history.* Berkeley: University of California Press.

Stritch, L. R. 1990. Landscape-scale restoration of barrens-woodland within the oak-hickory forest mosaic. *Restoration and Management Notes* 8(2):73–77.

Sudworth, G. B. 1900. Stanislaus and Lake Tahoe Forest Reserves, California, and adjacent territory. In Walcott, C. D., *Twenty-First Annual Report of the United States Geological Survey, Part V.* 56th Congress, 2nd Session. House of Representatives. Document No. 5. Washington, DC: Government Printing Office.

Suffling, R. 1983. Stability and diversity in boreal and mixed temperate forests: A demographic approach. *Journal of Environmental Management* 17(4):359–371.

Suffling, R., B. Smith, and J. D. Molin. 1982. Estimating past forest age distributions and disturbance rates in north-western Ontario: A demographic approach. *Journal of Environmental Management* 14(1):45–56.

Sugihara, N. G. and L. J. Reed 1987. Prescribed fire for restoration and maintenance of Bald Hills oak woodlands. In Plumb, T. R. and N. H. Pillsbury (tech. coords.), *Proceedings of the Symposium on Multiple-Use Management of California's Hardwood Resources*, pp. 446–451. USDA Forest Service General Technical Report PSW–100. Pacific Southwest Forest and Range Experiment Station, Berkeley, CA.

Sugihara, N. G., L. J. Reed, and J. M. Lenihan. 1987. Vegetation of the Bald Hills oak woodlands, Redwood National Park, California. *Madrono* 34(3):193–208.

Sullivan, J. 1994. *Betula alleghaniensis* [yellow birch]. In Fischer, W. C. (compiler), *The Fire Effects Information System* [data base]. USDA Forest Service, Intermountain Research Station, Intermountain Fire Sciences Laboratory, Missoula, MT.

Sullivan, J. 1995a. Oak-hickory forest. In Simmerman, D. G. (compiler), *The Fire Effects Information System* [data base]. USDA Forest Service, Intermountain Research Station, Intermountain Fire Sciences Laboratory, Missoula, MT.

Sullivan, J. 1995b. Mosaic of bluestem prairie and oak-hickory forest. In Fischer, W. C. (compiler), *The Fire Effects Information System* [data base]. USDA Forest Service, Intermountain Research Station, Intermountain Fire Sciences Laboratory, Missoula, MT.

Suphan, R. J. 1974. An ethnological report on the identity and localization of certain native peoples of northwestern Oregon. In Indian Claims Commission, *Oregon Indians I*, pp. 167–256. New York: Garland Publishing.

Swain, A. M. 1973. A history of fire and vegetation in northeastern Minnesota as recorded in lake sediments. *Quaternary Research* 3:383–396.

Swan, J. G. 1857. The Northwest Coast; or three years' residence in Washington Territory, New York. pp. 79–82. In Gates, C. M. (ed.). 1941. *Readings in Pacific Northwest history: Washington, 1790–1895*, pp. 34–36. Seattle: University Bookstore.

Swetnam, T. W. 1988. Fire history and climate in the southwestern United States. In Krammes, J. S. (tech. coord.), *Effects of fire management of southwestern natural resources*, pp. 6–17. USDA Forest Service General Technical Report RM-191. Rocky Mountain Forest and Range Experiment Station, Fort Collins, CO.

Swetnam, T. W. 1993. Fire history and climate change in giant sequoia groves. *Science* 262:885–889.

Swetnam, T. W. and C. H. Baisan. 1996. Fire histories of montane forests in the Madrean Borderlands. In Ffolliott, P. F., L. F. DeBano, M. B. Baker, Jr., G. J. Gottfried, G. Solis-Garza, C. B. Edminster, D. G. Neary, L. S. Allen, and R. H. Hamre (tech. coord.), *Effects of fire on Madrean Province ecosystems*, pp. 15–36. USDA Forest Service General Technical Report RM-GTR-289. Rocky Mountain Forest and Range Experiment Station, Fort Collins, CO.

Synder, N. F. and H. A. Snyder. 1989. Biology and conservation of the California condor. *Current Ornithology* 6:175–267.

Taber, R. D. and R. F. Dasmann. 1958. The black-tailed deer of the chaparral. California Department of Fish and Game, Game Bulletin No. 8.

Tackle, D. and D. F. Roy. 1953. Site preparation as related to ground cover density in natural regeneration of ponderosa pine. USDA Forest Service. California Forest and Range Experiment Station Technical Paper No. 4.

Tanner, O. 1977. *The Canadians*. New York: Time-Life Books.

Tantaquidgeon, G. 1942. A study of Delaware Indian medicine practice and folk beliefs. New York: AMS Press.

Taylor, A. R. 1971. Lightning—agent of change in forest ecosystems. *Journal of Forestry* 69(8):477–480.

Taylor, D. L. 1979. Forest fires and the tree-hole nesting cycle in Grand Teton and Yellowstone National Parks. In Linn, R. M. (ed.), *Proceedings, First Conference on Scientific Research in the National Parks*, Vol. I, pp. 509–511. Transactions and Proceedings Series Number Five. Washington, DC: USDI National Park Service.

Taylor, H. C., Jr. 1974. Anthropological investigation of the Tillamook Indians relative to tribal identity and aboriginal possession of lands. In Indian Claims Commission, *Oregon Indians I*, pp. 25–102. New York: Garland Publishing.

Teensma, D. A., J. T. Rienstra, and M. A. Yeiter. 1991. Preliminary reconstruction and analysis of change in forest stand age classes of the Oregon Coast Range from 1850 to 1940. USDI Bureau of Land Management Technical Note T/N OR–9.

Temple, S. A., M. J. Mossman, and B. Ambuel. 1979. The ecology and management of avian communities in mixed hardwood-coniferous forests. In *Management of North Central and Northeastern forests for nongame birds*, pp. 132–151. USDA Forest Service General Technical Report NC–51. North Central Forest Experiment Station, St. Paul, MN.

Tesch, S. D. 1995. The Pacific Northwest region. In Barrett, J. W. (ed.), *Regional silviculture of the United States*, 3rd ed., pp. 499–558. New York: Wiley.

Tesky, J. L. 1992a. *Cladonia spp.* [reindeer lichen]. In Fischer, W. C. (compiler), *The Fire Effects Information System* [data base]. USDA Forest Service, Intermountain Research Station, Intermountain Fire Sciences Laboratory, Missoula, MT.

Tesky, J. L. 1992b. *Thuja plicata* [western redcedar]. In Fischer, W. C. (compiler), *The Fire Effects Information System* [data base]. USDA Forest Service, Intermountain Research Station, Intermountain Fire Sciences Laboratory, Missoula, MT.

Tesky, J. L. 1992c. *Tsuga heterophylla* [western hemlock]. In Fischer, W. C. (compiler), *The Fire Effects Information System* [data base]. USDA Forest Service, Intermountain Research Station, Intermountain Fire Sciences Laboratory, Missoula, MT.

Tesky, J. L. 1993. *Castor canadensis* [beaver]. In Fischer, W. C. (compiler), *The Fire Effects Information System* [data base]. USDA Forest Service, Intermountain Research Station, Intermountain Fire Sciences Laboratory, Missoula, MT.

Thomas, D. H. 1994. *Exploring ancient native America.* New York: MacMillan.

Thomas, J. W., C. Maser, and J. E. Rodiek. 1978. Edges—their interspersion, resulting diversity and its measurement. In DeGraaf, R. M. (tech. coord.), *Proceedings of the Workshop on Nongame Bird Habitat Management in the Conifer-*

ous Forests of the Western United States, pp. 91–100. USDA Forest Service General Technical Report PNW–64. Pacific Northwest Forest and Range Experiment Station, Portland, OR.

Thomas, J. W., R. J. Miller, C. Maser, R. G. Anderson, and B. E. Carter. 1979a. Plant communities and successional stages. In Thomas, J. W. (tech. ed.), *Wildlife habitats in managed forests: The Blue Mountains of Oregon and Washington*, pp. 22–39. Agriculture Handbook No. 553. Washington, DC: USDA Forest Service.

Thomas, J. W., R. G. Anderson, C. Maser, and E. L. Bull. 1979b. Snags. In Thomas, J. W. (tech. ed.), *Wildlife habitats in managed forests: The Blue Mountains of Oregon and Washington*, pp. 60–77. Agriculture Handbook No. 553. Washington, DC: USDA Forest Service.

Thompson, D. Q. and R. H. Smith. 1970. The forest primeval in the Northeast—a great myth? Proceedings, Annual Tall Timbers Fire Ecology Conference. Vol. 10, pp. 255–265. Tall Timbers Research Station, Tallahassee, FL.

Thompson, K. 1961. Riparian forests of the Sacramento Valley, California. *Annals of the Association of American Geographers* 51(3):294–315.

Thornburgh, D. H. 1982. Succession in the mixed evergreeen forests of northwestern California. In Means, J. E. (ed.), *Forest succession and stand development research in the Northwest*, pp. 87–91. Forest Research Laboratory. Corvallis: Oregon State University.

Thwaites, R. G. (ed.). 1898a. *Jesuit relations and allied documents*, Vol. 55, pp. 190–199, Vol. 56, pp. 121–133. Cleveland: Burrows Brothers.

Thwaites, R. G. 1898b. *Jesuit relations and allied documents*. Vol. II, Acadia 1612–1614. Cleveland: Burrows Brothers.

Thwaites, R. G. (ed.). 1899. *Jesuit relations and allied documents*, Vol. 48, pp. 115–143. Cleveland: Burrows Brothers.

Thwaites, R. G. (ed.). 1959. *Original journals of the Lewis and Clark expedition*. New York: Antiquarian Press.

Thwaites, R. G. (ed.). 1966. *Early western travels 1748–1846*, Vol. 23. New York: AMS Press.

Timbrook, J., J. R. Johnson, and D. D. Earle. 1993. Vegetation burning by the Chumash. In Blackburn, T. C. and M. K. Anderson (ed.), *Before wilderness: Environmental management by native Californians*, pp. 117–149. Menlo Park, CA: Ballena Press.

Tinker, D. B., W. H. Romme, W. W. Hargrove, R. H. Gardner, and M. G. Turner. 1994. Landscape-scale heterogeneity in lodgepole pine serotiny. *Canadian Journal of Forest Research* 24(5):897–903.

Tirmenstein, D. A. 1986. *Juniperus occidentalis* [western juniper]. In Fischer, W. C. (compiler), *The Fire Effects Information System* [data base]. USDA Forest Service, Intermountain Research Station, Intermountain Fire Sciences Laboratory, Missoula, MT.

Tirmenstein, D. A. 1988. *Quercus macrocarpa* [bur oak]. In Fischer, W. C. (compiler), *The Fire Effects Information System* [data base]. USDA Forest Service, Intermountain Research Station, Intermountain Fire Sciences Laboratory, Missoula, MT.

Tirmenstein, D. A. 1989a. *Ceanothus integerrimus* [deer brush]. In Fischer, W C. (compiler), *The Fire Effects Information System* [data base]. USDA Forest Service, Intermountain Research Station, Intermountain Fire Sciences Laboratory, Missoula, MT.

Tirmenstein, D. A. 1989b. *Rubus spectabilis* [salmonberry]. In Fischer, W C. (compiler), *The Fire Effects Information System* [data base]. USDA Forest Service, Intermountain Research Station, Intermountain Fire Sciences Laboratory, Missoula, MT.

Tirmenstein, D. A. 1989c. *Ceanothus cuneatus* [wedgeleaf ceanothus]. In Fischer, W. C. (compiler), *The Fire Effects Information System* [data base]. USDA Forest Service, Intermountain Research Station, Intermountain Fire Sciences Laboratory, Missoula, MT.

Tirmenstein, D. A. 1990a. *Vaccinium parvifolium* [red huckleberry]. In Fischer, W C. (compiler), *The Fire Effects Information System* [data base]. USDA Forest Service, Intermountain Research Station, Intermountain Fire Sciences Laboratory, Missoula, MT.

Tirmenstein, D. A. 1990b. *Gaultheria shallon* [salal]. In Fischer, W. C. (compiler), *The Fire Effects Information System* [data base]. USDA Forest Service, Intermountain Research Station, Intermountain Fire Sciences Laboratory, Missoula, MT.

Tirmenstein, D. A. 1991a. *Quercus rubra* [northern red oak]. In Fischer, W. C. (compiler), *The Fire Effects Information System* [data base]. USDA Forest Service, Intermountain Research Station, Intermountain Fire Sciences Laboratory, Missoula, MT.

Tirmenstein, D. A. 1991b. *Quercus alba* [white oak]. In Fischer, W. C. (compiler), *The Fire Effects Information System* [data base]. USDA Forest Service, Intermountain Research Station, Intermountain Fire Sciences Laboratory, Missoula, MT.

Tirmenstein, D. A. 1991c. *Acer saccharum* [sugar maple]. In Fischer, W. C. (compiler), *The Fire Effects Information System* [data base]. USDA Forest Service, Intermountain Research Station, Intermountain Fire Sciences Laboratory, Missoula, MT.

Tirmenstein, D. A. 1991d. *Vaccinium vitis-idaea* [mountain cranberry]. In Fischer, W. C. (compiler), *The Fire Effects Information System* [data base]. USDA Forest Service, Intermountain Research Station, Intermountain Fire Sciences Laboratory, Missoula, MT.

Tirmenstein, D. A. 1991e. *Carya ovata* [shagbark hickory]. In Fischer, W. C. (compiler), *The Fire Effects Information System* [data base]. USDA Forest Service, Intermountain Research Station, Intermountain Fire Sciences Laboratory, Missoula, MT.

Touchan, R., C. D. Allen, and T. W. Swetnam. 1996. Fire history and climatic patterns in ponderosa pine and mixed-conifer forests of the Jemez Mountains, northern New Mexico. In Allen, C. D. (tech. ed.), *Fire effects in southwestern forests*, pp. 33–46. USDA Forest Service General Technical Report RM-GTR-286. Rocky Mountain Forest and Range Experiment Station, Fort Collins, CO.

Tralau, H. 1973. Some Quaternary plants. In Hallam, A. (ed.), *Atlas of paleobiogeography*, pp. 499–503. Amsterdam: Elsevier Scientific.

Truett, J. 1996. Bison and elk in the American Southwest: In search of the pristine. *Environmental Management* 20(2):195–206.

Turner, N. C. and M. A. M. Bell. 1973. Ethnobotany of southern Kwakiutl Indians of British Columbia. *Economic Botany* 27:257–310.

Turner, N. J. and H. V. Kuhnlein. 1983. Camas (*Camassia* spp.) and riceroot (*Fritillaria* spp.): Two liliaceous "root" foods of the Northwest Coast Indians. *Ecology of Food and Nutrition* 13:199–219.

Twining, C. E. 1983. The lumbering frontier. In Flader, S. L. (ed.), *The Great Lakes forest: An environmental and social history*, pp. 121–136. Minneapolis: University of Minnesota Press.

Uchytil, R. J. 1989. *Acer circinatum* [vine maple]. In Fischer, W. C. (compiler), *The Fire Effects Information System* [data base]. USDA Forest Service, Intermountain Research Station, Intermountain Fire Sciences Laboratory, Missoula, MT.

Uchytil, R. J. 1991a. *Picea glauca* [white spruce]. In Fischer, W. C. (compiler), *The Fire Effects Information System* [data base]. USDA Forest Service, Intermountain Research Station, Intermountain Fire Sciences Laboratory, Missoula, MT.

Uchytil, R. J. 1991b. *Picea mariana* [black spruce]. In Fischer, W. C. (compiler), *The Fire Effects Information System* [data base]. USDA Forest Service, Intermountain Research Station, Intermountain Fire Sciences Laboratory, Missoula, MT.

Uchytil, R. J. 1991c. *Abies balsamea* [balsam fir]. In Fischer, W. C. (compiler), *The Fire Effects Information System* [data base]. USDA Forest Service, Intermountain Research Station, Intermountain Fire Sciences Laboratory, Missoula, MT.

Uchytil, R. J. 1991d. *Betula papyrifera* [paper birch]. In Fischer, W. C. (compiler), *The Fire Effects Information System* [data base]. USDA Forest Service, Intermountain Research Station, Intermountain Fire Sciences Laboratory, Missoula, MT.

Uchytil, R. J. 1991e. *Pseudotsuga menziesii var. menziesii* [Pacific Douglas-fir]. In Fischer, W. C. (compiler), *The Fire Effects Information System* [data base].

USDA Forest Service, Intermountain Research Station, Intermountain Fire Sciences Laboratory, Missoula, MT.

Uchytil, R. J. 1992. *Pinus contorta var. latifolia* [lodgepole pine]. In Fischer, W. C. (compiler), *The Fire Effects Information System* [data base]. USDA Forest Service, Intermountain Research Station, Intermountain Fire Sciences Laboratory, Missoula, MT.

USDA Forest Service. 1989. An analysis of the land base situation in the United States 1989–2040. General Technical Report RM-181. Rocky Mountain Forest and Range Experiment Station, Fort Collins, CO.

USDA Forest Service. 1993. *Forest type groups of the United States.* Washington, DC.

U.S. Department of Agriculture. 1973. Silvicultural systems for the major forest types of the United States. USDA Handbook 445. Washington, DC.

U.S. National Park Service. 1996a. The Archaic Period. In *Outline of prehistory and history in the southeastern U.S. and Caribbean culture area with reference to relevant national park units in the southeast field area.* Adapted from Regionwide Archeological Survey Plan, Southeast Archeological Center, Tallahassee, FL.

U.S. National Park Service. 1996b. The Mississippian and late Prehistoric period. In *Outline of prehistory and history in the southeastern U.S. and Caribbean culture area with reference to relevant national park units in the southeast field area.* Adapted from Regionwide Archeological Survey Plan, Southeast Archeological Center, Tallahassee, FL.

U.S. War Department. 1854. *Exploration of the Red River of Louisiana in the year 1852.* Washington, DC: B. Tucker, Senate Printer.

Utley, R. M. 1997. *A life wild and perilous.* New York: Henry Holt.

Van Cleve, K., F. S. Chapin III, C. T. Dyrness, and L. A. Viereck. 1991. Element cycling in taiga forests: State-factor control. *BioScience* 41(2):78–88.

Vancouver, G. 1798. *A voyage of discovery to the north Pacific Ocean and round the world* (3 vols.). London: Robinson and Edwards.

Van der Donck, A. 1841. A description of the New Netherlands. *Collections of the New York Historical Society* (second series) 1:125–242.

Van Doren, M. (ed.). 1928. *Travels of William Bartram.* New York: Dover.

Vankat, J. L. 1977. Fire and man in Sequoia National Park. *Annals of the Association of American Geographers* 67(1):17–27.

Vankat, J. L. 1979. *The natural vegetation of North America.* New York: Wiley.

Vankat, J. L. 1985. General patterns of lightning in Sequoia National Park, California. In Lotan, J. E., B. M Kilgore, W. C. Fischer, and R. W. Mutch (tech. coord.), *Proceedings of the Symposium and Workshop on Wilderness Fire*, pp. 408–411. USDA Forest Service General Technical Repport INT-182. Intermountain Forest and Range Experiment Station, Odgen, UT.

Vankat, J. L. and J. Major. 1978. Vegetation changes in Sequoia National Park, California. *Journal of Biogeography* 5:377–402.

Vankat, J. L., W. H. Blackwell, Jr., and W. E. Hopkins. 1975. The dynamics of Hueston Woods and a review of the question of the successional status of the southern beech-maple forest. *Castanea* 40:290–308.

Van Lear, D. H. 1993. The role of fire in oak regeneration. In Loftis, D. and C. E. McGee (eds.), *Oak regeneration: Serious problems, practical recommendations,* pp. 66–78. USDA Forest Servie General Technical Report SE-84. Southeastern Forest Experiment Station. Asheville, NC.

Van Tramp, J. C. 1870. *Praire and Rocky Mountain adventures, or, life in the West.* Columbus, OH: Segner & Condit.

Van Wagner, C. E. 1970. Fire and red pine. Proceedings, Annual Tall Timbers Fire Ecology Conference, Number 10, pp. 211–219. Tall Timbers Research Station, Tallahassee, FL.

Van Wagner, C. E. 1977. Conditions for the start and spread of crown fire. *Canadian Journal of Forest Research* 7(1):23–34.

Van Wagner, C. E. 1978. Age-class distribution and the forest fire cycle. *Canadian Journal of Forest Research* 8(2):220–227.

Vastoka, J. M. 1969. Architecture and environment, the importance of the forest to the Northwest Coast Indian. *Forest History* 13(3):13–21.

Veirs, S. D., Jr. 1980a. The role of fire in northern coast redwood forest dynamics. Proceedings, First Conference on Scientific Research in the National Parks, Vol. 10, pp. 190–209. Washington, DC: U.S. National Park Service.

Veirs, S. D., Jr. 1980b. The influence of fire in coast redwood forests. In Stokes, M. A. and J. H. Dieterich (tech. coord.), *Proceedings of the Fire History Workshop,* pp. 93–95. USDA Forest Service General Technical Report RM-81. Rocky Mountain Forest and Range Experiment Station, Fort Collins, CO.

Veirs, S. D., Jr. 1982. Coast redwood forest: Stand dynamics, successional status, and the role of fire. In Means, J. E. (ed.), *Forest succession and stand development research in the Northwest,* pp. 119–141. Forest Research Laboratory. Corvallis: Oregon State University.

Verner, J. and A. S. Boss (tech. coord.). 1980. California wildlife and their habitats: western Sierra Nevada. USDA Forest Service, General Technical Report PSW–37. Pacific Southwest Forest and Range Experiment Station, Berkeley, CA.

Viereck, L. A. 1975. Forest ecology of the Alaska taiga. Proceedings, Circumpolar Conference on Northern Ecology. September 15–18. Ottawa, Canada. USDA Forest Service reprint.

Viereck, L. A. and W. F. Johnston. 1990. *Picea mariana* (Mill.) B.S.P., Black Spruce. In Burns, R. M. and B. H. Honkala (tech. coord.), *Silvics of North America, Vol. 1, Conifers,* pp. 227–237. Agricultural Handbook No. 654. Washington, DC: USDA Forest Service.

Vogl, R. J. 1967. Fire adaptations of some southern California plants. Proceedings, Annual Tall Timbers Fire Ecology Conference, Number 7, pp. 79–109. Tall Timbers Research Station. Tallahassee, FL.

Vogl, R. J. 1970. Fire and the northern Wisconsin pine barrens. Proceedings, Annual Tall Timbers Fire Ecology Conference, Number 10, pp. 175–209. Tall Timbers Research Station. Tallahassee, FL.

Vogl, R. J. 1972. Fire in the southeastern grasslands. Proceedings, Annual Tall Timbers Fire Ecology Conference, Vol. 12, pp. 175–198. Tall Timbers Research Station, Tallahassee, FL.

Wadleigh, L. and M. J. Jenkins. 1996. Fire frequency and the vegetative mosaic of a spruce-fir forest in northern Utah. *Great Basin Naturalist* 56(1):28–37.

Wagener, W. W. and M. S. Cave. 1946. Pine killing by the root fungus, *Fomes annosus*, in California. *Journal of Forestry* 44:47–54.

Wahlenberg, W. G. 1960. Loblolly pine, its use, ecology, regeneration, protection, growth and management. Duke University, School of Forestry. Durham, NC.

Waitt, R. B., Jr., and R. M. Thorson. 1983. The Cordilleran ice sheet in Washington, Idaho, and Montana. In Porter, S. C. (ed.), *Late-Quaternary environments of the United States, Vol. 1, The Pleistocene*, pp. 53–70. Minneapolis: University of Minnesota Press.

Waldman, C. 1985. *Atlas of the North American Indian*. New York: Facts on File.

Walker, J. 1985. Species diversity and production in pine-wiregrass savannas of the Green Swamp, North Carolina. Ph.D. Dissertation. Department of Botany, University of North Carolina at Chapel Hill.

Walker, J. S. and R. E. F. Escano. 1992. Longleaf pine ecosystem management for red-cockaded woodpeckers and other resources. In Murphy, D. (Comp.), *Getting to the future through silviculture—Workshop Proceedings*, pp. 77–83. USDA Forest Service General Technical Report INT-291. Intermountain Research Station, Odgen, UT.

Wallace, W. S. (ed.). 1932. *John McLeans's notes of a twenty-five year's service in the Hudson's Bay Territory*. Toronto: Champlain Society.

Walstad, J. D., S. R. Radosevich, and D. V. Sandberg. 1990. Introduction to natural and prescribed fire in Pacific Northwest forests. In Walstad, J. D., S. R. Radosevich, and D. V. Sandberg (eds.), *Natural and prescribed fire in Pacific Northwest forests*, pp. 3–5. Corvallis: Oregon State University Press.

Ward, J. and G. Parker. 1989. Spatial dispersion of woody regeneration in an old-growth forest. *Ecology* 70(5):1279–1285.

Warner, T. E. 1980. Fire history in the yellow pine forest of Kings Canyon National Park. Proceedings, Fire History Workshop. USDA Forest Service General Technical Report RM-81, pp. 89–95. Rocky Mountain Forest and Range Experiment Station, Fort Collins, CO.

Watt, A. S. 1947. Pattern and process in the plant community. *Journal of Ecology* 35:1–22.

Watts, W. A. 1979. Late Quaternary vegetation of central Appalachia and the New Jersey Coastal Plain. *Ecological Monographs* 49:427–469.

Watts, W. A. 1983. Vegetational history of the Eastern United States 25,000 to 10,000 years ago. In Porter, S. C. (ed.), *Late-Quaternary environments of the United States, Vol. 1, The Pleistocene*, pp. 115–129. Minneapolis: University of Minnesota Press.

Watts, W. A., E. C. Grimm, and T. C. Hussey. 1996. Mid-Holocene forest history of Florida and the coastal plain of Georgia and South Carolina. In Sassaman, K. E. and D. G. Anderson (eds.), *Archaeology of the Mid-Holocene Southeast*, pp. 28–38. Gainesville: Universtiy Press of Florida.

Weatherspoon, C. P. 1990. *Sequoiadendron giganteum* (Lindl.) Buchholz Giant Sequoia. In Burns, R. M. and B. H. Honkala (tech. coord.), *Silvics of North America, Vol. 1, Conifers*, pp. 552–562. Agricultural Handbook No. 654. Washington, DC: USDA Forest Service.

Weaver, H. 1951. Fire as an ecological factor in the southwestern ponderosa pine forests. *Journal of Forestry* 49(2):93–98.

Weaver, J. E. and F. E. Clements. 1938. *Plant ecology*. New York: McGraw-Hill.

Webb, S. L. 1986. Potential role of passenger pigeons and other vertebrates in the rapid Holocene migrations of nut trees. *Quaternary Research* 26:367–375.

Webb, T., III. 1987. The appearance and disappearance of major vegetaional assemblages: Long-term vegetational dynamics in eastern North America. *Vegetatio* 69:177–187.

Webb, T., III, E. J. Cushing, and H. E. Wright, Jr. 1983. Holocene changes in the vegetation of the Midwest. In Wright, H. E., Jr. (ed.), *Late-Quaternary environments of the United States. Vol. 2, The Holocene*, pp. 142–165. Minneapolis: University of Minnesota Press.

Weedon, R. R. and P. M. Wolken. 1990. The Black Hills environment. In Agenbroad, L. D., J. I Mead and L. W. Nelson (eds.), *Megafauna & man: Discovery of America's heartland*, pp. 123–135. *Vol. 1, Scientific Papers*. Hot Springs, SD: The Mammoth Site of Hot Springs.

Wein, R. W. and J. M. Moore. 1977. Fire history and rotations in the New Brunswick Acadian forest. *Canadian Journal of Forest Research* 7(2):285–294.

Wells, P. V. 1965. Scarp woodlands, transported grassland soils, and concept of grassland climate in the Great Plains Region. *Science* 148(3667):246–249.

Wells, P. V. 1970. Postglacial vegetation history of the Great Plains. *Science* 167:1574–1582.

Wells, P. V. and J. D. Stewart. 1987. Cordilleran-boreal taiga and fauna on the central Great Plains of North America, 14,000–18,000 years ago. *American Midland Naturalist* 118(1):94–106.

Welsh, C. J. E. and W. M. Healy. 1993. Effect of even-aged timber management on bird species diversity and composition in northern hardwoods of New Hampshire. *Wildlife Society Bulletin* 21:143–154.

Wendel, G. W. and H. C. Smith. 1990. *Pinus strobus* L. Eastern White Pine. In Burns, R. M. and B. H. Honkala (tech. coord.), *Silvics of North America, Vol. 1, Conifers,* pp. 476–488. Agricultural Handbook No. 654. Washington, DC: USDA Forest Service.

Wendorf, F. 1961. An interpretation of late Pleistocene environments of the Llano Estacado. In Wendorf, F. (ed.), *Paleoecology of the Llano Estacado,* pp. 115–133. Santa Fe: Museum of New Mexico Press.

West, F. H. 1983. The antiquity of man in America. In Porter, S. C. (ed.), *Late-Quaternary environments of the United States, Vol. 1, The Pleistocene,* pp. 364–382. Minneapolis: University of Minnesota Press.

West, G. J. 1989. Holocene fossil pollen records of Douglas-fir in northwestern California: Reconstruction of past climate. Paper presented at the Sixth Annual Pacific Climate (PACLIM) Workshop. Asilomar, California, March 5–8.

West, N. E. 1995. Intermountain deserts, shrub steppes, and woodlands. In Barbour, M. G. and W. D. Billings (eds.), *North American terrestrial vegetation,* pp. 209–230. New York: Cambridge University Press.

Westveld, M. W. (coord.). 1956. Natural forest vegetation zones of New England. *Journal of Forestry* 54:332–338.

Wheeler, G. M. 1878. *Report upon United States Geographical Surveys west of the 100th meridian, Vol. 1, Topography.* Washington, DC: U.S. Government Printing Office.

Wheeler, H. W. 1990. *Buffalo days, the personal narrative of a cattleman, Indian fighter and army officer.* Lincoln: University of Nebraska Press.

White, A. S. 1985. Presettlement regeneration patterns in a southwestern ponderosa pine stand. *Ecology* 66:589–594.

White, C. A. 1985a. Fire and biomass in Banff National Park closed forests. Masters Thesis, Department of Forest and Wood Sciences, Colorado State University. Fort Collins, CO.

White, C. A. 1985b. Wildland fires in Banff National Park, 1880–1980. Occasional Paper No. 3. National Parks Branch, Parks Canada, Ottawa.

White, J. 1995. Woodlands in early Illinois. In Stearns, F. and K. Holland (eds.), *Proceedings, Midwest Oak Savanna Conference.* U.S. Environmental Protection Agency, Washington, DC.

White, M. A. and J. L. Vankat. 1993. Middle and high elevation coniferous forest communities of the North Rim region of Grand Canyon National Park, Arizona, USA. *Vegetatio* 109:161–174.

White, P. S. and R. D. White. 1996. Old-growth oak and oak-hickory forests. In Davis, M. B. (ed.), *Eastern old-growth forests,* pp. 178–198. Washington, DC: Island Press.

White, P. S., M. D. MacKenzie, and R. T. Busing. 1985. Natural disturbance and gap phase dynamics in southern Appalachian spruce-fir forests. *Canadian Journal of Forest Research* 15:233–240.

White, Z. W. 1984. Loblolly pine, with emphasis on its history. In Karr, B. L. and J. B. Baker (eds.), *Proceedings of the Symposium on the Loblolly Pine Ecosystem (West Region)*, pp. 3–16. Extension Service of Mississippi State University, Publication 1454. Starkville, MI.

Whitehead, D. R. 1967. Studies of full-glacial vegetation and climate in southeastern United States. In Cushing, E. J. and H. E. Wright, Jr. (eds.), *Quaternary paleoecology*, pp. 237–248. Volume 7 of the *Proceedings of the VII Congress of the International Association for Quaternary Research.* New Haven, CT: Yale University Press.

Whitehead, D. R. 1973. Late-Wisconsin vegetational changes in unglaciated eastern North America. *Quaternary Research* 3:621–631.

Whitlock, C. 1993. Postglacial vegetation and climate of Grand Teton and Southern Yellowstone National Parks. *Ecological Monographs* 63(2):173–198.

Whitlock, C. 1995. The history of *Larix occidentalis* during the last 20,000 years of environmental change. In Schmidt, W. C. and K. J. McDonald (comps.), *Ecology and management of larix forests: A look ahead*, pp. 83–90. USDA Forest Service General Technical Report INT-GTR-319. Intermountain Research Station, Odgen, UT.

Whitmore, F. C., Jr., H. B. S. Cooke, and D. J. P. Swift. 1967. Elephant teeth from the Atlantic continental shelf. *Science* 156:1477–1481.

Whitney, G. G. 1990. The history and status of the hemlock-hardwood forests of the Allegheny Plateau. *Journal of Ecology* 78:443–458.

Whitney, G. G. and W. C. Davis. 1986. Thoreau and the forest history of Concord, Massachusetts. *Journal of Forest History* 30:70–81.

Whitney, S. 1986. *Western forests.* New York: Alfred A. Knopf.

Wicander, R. and J. S. Monroe. 1993. *Historical geology.* St. Paul, MN: West Publishing.

Wilcox, F. N. 1970. *Ohio Indian trails.* Kent, OH: Kent State University Press.

Wilke, P. J. 1993. Bow staves harvested from juniper trees by Indians of Nevada. In Blackburn, T. C. and M. K. Anderson (eds.), *Before the wilderness: Environmental management by native Californians*, pp. 241–277. Menlo Park, CA: Ballena Press.

Wilkes, C. 1849. *Narrative of the U.S. Exploring Expedition of 1838–1842, Vols. 4 and 5.* Philadelphia: C. Sherman.

Wilkinson, W. H., P. M. McDonald, and P. Morgan. 1997. Tanoak sprout development after cutting and burning in a shade environment. *Western Journal of Applied Forestry* 12(1):21–26.

Will, G. F. and G. E. Hyde. 1964. *Corn among the Indians of the Upper Missouri.* Lincoln: University of Nebraska Press.

Williams, M. 1989. *Americans and their forests.* New York: Cambridge University Press.

Williams, R. 1643. *A key into the language of America.* (Reprinted 1936.) Rhode Island Tercentenary Committee. Providence, RI.

Williams, R. L. 1976. *The loggers.* New York: Time-Life Books.

Wilson, M. R. 1978. Notes on ethnobotany in Inuktitut. *Western Canadian Journal of Anthropology* 8:180–196.

Winchell, L. A. 1933. *History of Fresno County and the San Joaquin Valley.* Fresno, CA: A. H. Cawston.

Wislizenus, F. A. 1912. *A journey into the Rocky Mountains in the year 1839.* Spokane, WA: Arthur H. Clark.

Wisner, G. 1998a. Living on the rim: California island cave offers tantalizing clues to Paleoindian life, oldest site yet on Pacific Coast yields cordage. *Mammoth Trumpet* 13(2). Center for the Study of the First Americans. Oregon State University, Corvallis.

Wisner, G. 1998b. Ohio cave, sealed since Ice Age, yields data on Paleo-Americans. *Mammoth Trumpet* 13(1). Center for the Study of the First Americans. Oregon State University, Corvallis.

Wissler, C. 1989. *Indians of the United States.* New York: Anchor Books, Doubleday.

Wolfe, C. W. 1972. Effects of fire on a Sand Hills grassland environment. Proceedings, Annual Tall Timbers Fire Ecology Conference, Number 12, pp. 241–255. Tall Timbers Research Station, Tallahassee, FL.

Wolff, J. O. 1980. The role of habitat patchiness in the population dynamics of snowshoe hares. *Ecological Monographs* 50(1):111–130.

Wood, W. 1633. *New England Propsect.* London: reprint Boston, MA: Prince Society, 1865.

Woodford, A. O. 1965. *Historical geology.* New York: W. H. Freeman.

Woods, K. D. and R. H. Whittaker. 1981. Canopy-understory interaction and the internal dynamics of mature hardwood and hemlock-hardwood forests. In West, D. C., H. H. Shugart, and D. B. Botkin (eds.), *Forest succession: Concepts and application*, pp. 305–323. New York: Springer.

Woods, W. I. 1996. The fall of Cahokia. Paper presented at the Illinois History Symposium, Springfield. December 7, 1996.

Woolfenden, W. B. 1996. Quaternary vegetation history. In *Status of the Sierra Nevada*, Vol. II, pp. 47–69. Sierra Nevada Ecosystem Project, Final Report to Congress. Wildland Resources Center Report No. 37. Centers for Water and Wildland Resources. Davis, CA: University of California.

Worona, M. A. and C. Whitlock. 1995. Late Quaternary vegetation and climate history near Little Lake, central Coast Range, Oregon. *Geological Society of America Bulletin* 107(7):867–876.

Wright, H. A. and A. W. Bailey. 1980. Fire ecology and prescribed burning in the Great Plains—a research review. USDA Forest Service Gen. Tech.

Rep. INT-77. Intermountain Forest and Range Experiment Station, Ogden, UT.

Wright, H. A., L. F. Neuenschwander, and C. M Britton. 1979. The role and use of fire in sagebrush-grass and pinyon-juniper plant communities. USDA Forest Service General Technical Report INT-58. Intermountain Forest and Range Experiment Station, Odgen, UT.

Wright, H. E., Jr. 1976. The dynamic nature of Holocene vegetation, a problem in paleoclimatology, biogeography, and stratigraphic nomenclature. *Quaternary Research* 6:581–596.

Wright, H. E., Jr. 1981. Vegetation east of the Rocky Mountains 18,000 years ago. *Quaternary Research* 15:113–125.

Wright, H. E., Jr. 1984. Sensitivity and response time of natural systems to climatic change in the late Quaternary. *Quaternary Science Reviews* 3:91–131.

Wright, J. V. 1994. Before European contact. In Rogers, E. S. and D. B. Smith (eds.), *Aboriginal Ontario, historical perspectives on the First Nations*, pp. 21–38. Ontario Historical Studies Series for the Government of Ontario. Toronto: Dundurn Press.

Wright, K. 1993. Revelations of rat scat. *Discover* 14(9):64–71.

Wyman, L. C. and S. K. Harris. 1951. *Ethnobotany of the Kayenta Navaho*. Albuquerque: University of New Mexico Press.

Young, F. G. (ed.). 1899. *The correspondence and journals of Captain Nathaniel J. Wyeth, 1831–6*. Eugene, OR: Oregon University Press.

Zigmond, M.L. 1981. *Kawaiisu Ethnobotany*. Salt Lake City: University of Utah Press.

Zimov, S. A., V. I. Chuprynin, A. P. Oreshko, F. S. Chapin, III, J. F. Reynolds, and M. C. Chapin. 1995. Steppe-tundra transition: A herbivore-driven biome shift at the end of the Pleistocene. *American Naturalist* 146(5):765–794.

Zinke, P. J. 1977. The redwood forest and associated north coast forests. In Barbour, M. G. and J. Major (eds.), *Terrestrial vegetation of California*, pp. 679–698. New York: Wiley.

Zybach, R. 1993. Native forests of the Northwest: 1788–1856 American Indians, cultural fire, and wildlife habitat. *Northwest Woodlands* 9(2):14–15, 30–31.

Index

565